Production, U

Sustainability of Groundwater

Production, Use, and Sustainability of Groundwater

Charles A. Job

CRC Press
Taylor & Francis Group
Boca Raton London New York

CRC Press is an imprint of the
Taylor & Francis Group, an **informa** business

Second edition published 2022
by CRC Press
6000 Broken Sound Parkway NW, Suite 300, Boca Raton, FL 33487-2742

and by CRC Press
2 Park Square, Milton Park, Abingdon, Oxon, OX14 4RN

First edition published by CRC Press 2009

CRC Press is an imprint of Taylor & Francis Group, LLC

Library of Congress Cataloging-in-Publication Data
Names: Job, Charles A., author.
Title: Groundwater economics, two-volume set / Charles A. Job.
Description: Second edition. | Boca Raton : CRC Press, 2021. | Includes bibliographical references and index. |
Contents: volume 1. Production, use, and sustainability of groundwater -- volume 2. Cost-benefit analysis of groundwater policy and projects, with case studies. | Summary: "The competition for groundwater sources as a water supply reinforces the need for a strong economic rationale in decision-making. Evaluating economic decisions in the context of total water management and life-cycle water use is essential to making critical development and remediation choices. This revised volume provides fundamental economic and policy concepts related to groundwater, discusses important factors in cost-benefits and life-cycle evaluation, and explains triple-bottom-line analysis for different groundwater projects. It includes new and updated case studies on groundwater issues with solutions for a range of situations based on economic data"– Provided by publisher.
Identifiers: LCCN 2021021261 (print) | LCCN 2021021262 (ebook) | ISBN 9780367205478 (hardback) | ISBN 9780367206314 (hardback) | ISBN 9781003164081 (ebook) | ISBN 9780429262197 (ebook) |
Subjects: LCSH: Groundwater–Economic aspects. | Water resources development–Economic aspects.
Classification: LCC HD1691 .J63 2021 (print) | LCC HD1691 (ebook) | DDC 333.91/04–dc23
LC record available at https://lccn.loc.gov/2021021261
LC ebook record available at https://lccn.loc.gov/2021021262

ISBN: 978-0-367-20547-8 (hbk)
ISBN: 978-0-367-75799-1 (pbk)
ISBN: 978-0-429-26219-7 (ebk)

Typeset in Times
by SPi Global, India

This book is dedicated to
Avery, Gavin, Sadie, Kylie, Rita and
all the children of the world
with the hope that they will have the water they need!

He opened the rock, and water gushed out;
like a river it flowed in the desert.

Psalm 105:41

. . . *And he who waters will also be watered himself.*

Proverbs 11:25

Contents

Preface

Groundwater increases in importance and value while being ubiquitous under the continents, varying significantly in depth, quality, accessibility and availability. Essential to many forms of life, groundwater has substantial value where alternative water sources do not exist. Groundwater discharge is critical to maintaining the flow of major streams and the quality of many rivers, lakes, wetlands and coastal waters. But while it is present widely, many people do not understand it or its economics – in the face of significant competition for surface and groundwater, for the subsurface environment, from growing populations with tremendous thirst and need for food, we will all experience the value of groundwater at some point in time most likely. Groundwater may therefore be involved directly or indirectly in many decisions associated with private and public economic investment.

I recall my parents being concerned about whether we could pump water from our well when the creek behind our house on a hill would go dry. They certainly had a basic sense of the hydrologic environment on which we depended as well as a cost of its potential changed condition. They were also concerned about whether our septic system was operating properly at different times to ensure safe water – both fixing the system or replacing our water source had costs and benefits.

Two hazardous waste sites were discovered to have contaminated our community's groundwater. One site was a quarter mile from our house. Prior to that time, it was commonplace and accepted practice to pour or dump liquid chemical waste in low areas or into underground drain galleries. These industrial operations were declared Superfund sites and eventually placed in remedial status for many years. Basically, the disposal capacity of the subsurface was treated as having no cost to use but some value in using it. In these cases, people were sickened when their wells were contaminated as well as a local stream impacted by groundwater pollutant discharge. These conditions occurred at that time across the country in communities that were trying to do their best with the understanding and resources they had. However, at the time, if it was out of sight, it was out of mind. Decisions were made – without understanding the impacts on the water cycle with consequent human and economic effects – and the marginal cost was high. Fortunately, now, wellhead protection is in place reflecting the value of groundwater to the community.

Now today we have a more measured use of groundwater but communities and regions experiencing significant drought and continued arid conditions are depleting their aquifers and creating a water resource challenge for the future. The value of groundwater will grow and we need to make sure that it is available to those who need it and not waste it. Demand for groundwater continues to increase.

This text examines the production, use, cost and benefit of groundwater services, the aquifers that contain groundwater and the ecosystem that cycles it for many purposes, some of which we may not currently understand. Importantly, the text provides a perspective and approach to evaluating the supply and use of groundwater for a range of purposes through the tool of economics – focusing on the cost and benefit of producing and using groundwater at the margin. Volume 1 addresses production, use and sustainability for the major purposes to which groundwater is applied. Volume 2 provides an approach through life-cycle cost–benefit analysis to evaluating and informing decisions for groundwater projects.

The text is intended to be an applied economics reference, handbook and educational text. It attempts to draw together perspectives and material regarding groundwater economics comprehensively but not exhaustively. This text assembles in one place existing salient concepts related to the economics of production, use and sustainability of groundwater, including its key role in agriculture. Another purpose of the text is to encourage reporting of the application of economics to inform decisions on costs and benefits of groundwater development. Many projects do not generate widely available reports on the costs and benefits of groundwater and subsurface development for broader understanding of decisions related to groundwater use including the effects of its production and of waste disposal underground.

The text explores and describes the economics of different policy instruments and management approaches to dealing with groundwater supply, use and protection. Legal aspects, natural capital factors, groundwater markets and managed aquifer recharge are covered. The final chapters of the first volume consider the longer term and the need to look for approaches that will sustain groundwater as one resource among many on which we rely as individuals, corporations and nations – and a resource that our children and their children may increasingly turn to in a very different world of the future. Volume 2 focuses on evaluating groundwater projects and programs through the tool of life-cycle cost-benefit analysis. Also included in Volume 2 is special attention to transboundary, international, climate change and macroeconomic considerations as well as groundwater economic research needs. The final chapter of Volume 2 provides a range of case studies of different economic analyses of groundwater production and use and subsurface disposal to promote broad thinking about the application of economics in decisions affecting future groundwater development and in wisely building on lessons learned from past decisions.

Hopefully, others will expand on the content of this text and improve the concepts it addresses as a helpful input to any community's or nation's sustainable natural resource and ecosystem planning for the next generations. As more people, businesses and governments compete for the resources that the ecosystem provides, those resources will become more scarce and valuable – and we must become more innovative and creative in their economic allocation and use to ensure that they will continue to be available sustainably to all who and which need them.

Thank you for your interest in obtaining a perspective regarding groundwater economics.

Charles A. Job
July 9, 2020

Acknowledgments

Many people contributed to this book knowingly and unknowingly – and I thank them all but do not have sufficient space to recognize them completely. Several of them I need to identify specifically for their encouragement, insight and suggestions. Most significantly, I am indebted to my wife, Susan, and my family for allowing me to pursue and revise an idea that I had nurtured for 35 years. Next, my father, Richard, encouraged me to enter the environmental field when it was just beginning to be recognized as a focus of study and professional application. My groundwater geology professor, Dwight Baldwin, provided further encouragement. Fletcher Driscoll gave heartening words at different times that supplied meaning to what I was doing. Mary Jo Kealy, a co-worker in environmental and resource economics, provided positive feedback on the project at the beginning and conclusion of the first edition. John Bergstrom, a groundwater economics professor and fellow co-author of an article on groundwater economics, gave key perspectives on completeness of aspects of the text. George Hallberg provided encouragement and feedback to move forward. Mike Wireman, a hydrogeology peer and frequent collaborator, was a source of continuing insight. Kevin McCray gave me opportunity to gain further perspective of the groundwater industry. Pastor Dale O'Shields' request provided significant God-inspired guidance that gratefully affected the text's content. Irma Shagla and Rebecca Pringle at Taylor and Francis have been most supportive when I needed them to be. To you all, I thank more than written words can express!

I acknowledge that any errors in the text are mine and are unintentional. I appreciate readers sending any observed errors to the publisher for research and possible future correction.

Author

Charles A. Job grew up in southwestern Ohio and drank water from three domestic wells as his family moved locally in the community of Beavercreek, remembering the great attention the wells and water levels received during dry years. He earned a BS in social science at Michigan State University, a master of environmental science at Miami University, further graduate study in water management at the University of Arizona and a master of applied economics at the University of Michigan. Early in his career, Chuck worked on Great Lakes water issues and then briefly for two Fortune 500 companies, consulted on groundwater matters, and joined the Environmental Protection Agency in 1984 in its Chicago Regional Office and then in 1988 in Washington, DC. He helped others launch a field initially known as "groundwater ecology" and completed one of the first published cost–benefit analyses of wellhead protection. His work at EPA took him from groundwater protection to manage information systems and drinking water system financial assistance, as well as to serve as a charter member of an interagency group that established the National Ground Water Monitoring Network with his focus on the cost of state monitoring data collaboration. Chuck has authored or coauthored over 50 articles and publications on groundwater and water resources. He is a member of the International Society for Ecological Economics, the National Ground Water Association and the American Water Works Association, holds the LEED BD&C sustainability professional credential and currently serves as Regulatory Affairs Manager for the National Ground Water Association.

1 An Introduction to the Economics of Groundwater

ECONOMICS INFORMS DECISIONS

Decisions about the production and use of groundwater are fundamental – and increasingly so – to the fabric of society. These decisions are economic decisions, with each decision affecting future availability of groundwater. While groundwater may be some distance under our feet, available in most places around the world from a well and viewed by some people as having mysterious qualities, the resource has become well documented with its existence and functions continuing to be researched. Groundwater provides significantly useful as well as ignored services. It is a vast resource being increasingly tapped for people's vital uses, becoming ever scarce and growing in value. It constitutes both an economic opportunity as well as a challenge in decision-making. Groundwater economics informs these decisions.

The complexity of the ecosystem combined with that of the social and institutional environments are substantial factors influencing decisions about the production and use of groundwater. Groundwater exists in varied subsurface conditions. People use this significant resource in a multitude of ways for daily needs. To ensure fair and safe delivery of groundwater for use, rights to use and equipment standards to pump the resource affect groundwater's supply and demand. Different uses require different qualities of water. Although being out of sight and with many tens and even hundreds of feet of soil and rock between land surface and aquifer below, wastes placed underground and other uses of the subterranean environment can affect groundwater quality. Decisions about groundwater use are then affected. The evaluation of the cost and benefit of responding to these decisions encompasses a major aspect of groundwater economics. These decisions can affect people's health and survival.

ECONOMIC SIGNIFICANCE OF GROUNDWATER

Extreme hydrologic events have brought the economics of groundwater into focus affecting people's daily activities. About half of the population of the United States and over one third of people worldwide rely on groundwater for water supply with many aquifers being depleted. Flooding from hurricanes in 2018 potentially affected over 650,000 private wells in the southeastern United States (Beitsch, 2018). Floodwaters can float debris that can damage wells but can also carry pathogens from animal waste lagoons that can enter wells not properly maintained and result in waterborne disease. These wells can be disinfected, saving people from illness and lost wages. With the long drought in the western United States, particularly in California, farmers turned from reservoir and canal water to aquifers to irrigate farm fields. Agriculture is a $45 billion industry in California with half the fruits, vegetables and nuts for the United States produced there (Alley and Alley, 2017, p. 97). Drilling new wells was a significant factor in feeding the nation (Sabalow et al., 2016).

With attention on greater availability and sustainability of groundwater, the United States has more than 500 aquifer storage and recovery wells (NGWA, 2016) and 260 managed aquifer recharge projects (IGRAC, 2018) with 46 more having been proposed (USEPA, 2018). These projects provide water for municipal supply as well as agriculture, industry and ecological purposes and involve decisions for millions of dollars of investment in each case. At the same time, scientists have raised cautions about use of stormwater for recharge because of the need to provide protection from pathogens and trace metals if soil and vadose zones do not provide adequate filtration and degradation

(Brumley et al., 2018). This concern raises the potential need for greater investment for groundwater quality protection at recharge sites that short-circuit subsurface treatment capability. In addition to being used directly from wells and springs with and without treatment as dependent on quality, groundwater maintains the baseflow of streams and thereby contributes to economic surface water availability and use. Consequently, groundwater quantity and quality, and their management, are significant economic inputs to nearly any decision involving water in most people's daily lives and for agricultural and industrial purposes.

ADVANTAGES OF AND DEMAND FOR GROUNDWATER

Groundwater has numerous advantages for water supply as shown in Box 1.1.1 that translate into economic benefits. Most importantly, groundwater is typically available and accessible where people need it, although highly consumptive use makes it less so as water tables decline with greater pumping. Significantly, as demonstrated in California during extreme drought, groundwater serves as a buffer and backstop water supply source when surface water is not available. From an economic perspective, groundwater is typically a less expensive water source due to its local availability.

While groundwater has many advantages, a competition exists for subterranean resource services for water supply, materials transfer and waste disposal. This demand on the underground environment puts groundwater clearly at the top of lists of concerns related to the maintenance of subsurface quality. Box 1.1.2 identifies a number of competing factors that can influence both the availability and quality of groundwater and influence economic considerations related to its use.

Thus, the subsurface environment's use has become more complex – with groundwater used for and impacted by many purposes. Thorough economic analyses of each case of subsurface use should be intentionally and fully informative of the many effects of its potential use. These analyses must consider both immediate as well as more distant impacts on others using water within watersheds and across adjacent and distant watersheds from which groundwater may flow and wastes may migrate.

BOX 1.1.1 ADVANTAGES OF GROUNDWATER FOR WATER SUPPLY

Groundwater has several advantages for water supply (Foster, 1996; Llamas and Custodio, 2003):

- Easily accessible and capable of being captured and managed by individuals or small groups of people.
- Available over extensive areas, most often being developed where needed, rather than installing long transmission lines, and in a staged approach.
- Less expensive and less capital intensive to develop than surface water in most situations.
- More reliable in dry seasons or droughts because of the large storage capability.
- Generally of adequate chemical and microbiological quality with little treatment needed if unpolluted.
- Less affected by catastrophic events.
- Often uniquely provides people with limited means, the capability to sustain life in locations having less essential resources and constrained opportunities.

BOX 1.1.2 SOME EFFECTS OF DEMAND AND COMPETITION FOR GROUNDWATER AND SUBSURFACE RESOURCES

- Expanding groundwater use has resulted in measured significant declines in water tables in every European Union country (AGSE, 2007), and in principal aquifers in nearly the entire United States (USGS, 2009).
- Pollutant degradation of groundwater has been recorded in every state of the United States and in all the countries of the European Union.
- Continued biofuels production may require more irrigation and increase agricultural chemical use.
- Locations of scarce freshwater may need to treat groundwaters containing naturally occurring arsenic and radioactive substances which, when treated, generate wastes that are often disposed underground.
- Heavy groundwater withdrawals of surficial aquifers may reduce baseflow discharge to streams.
- Fully allocated surface sources and drought may eventuate demands for brackish and deep groundwater.
- Disposal of salts and other minerals from brackish, saline and seawater treatment adds to production costs with disposal locations increasingly being considered underground.
- Hydraulic fracturing of deep geologic formations to produce natural gas require enormous amounts of groundwater which, when produced in the gas drilling process, has toxic contaminants to be treated before disposal.
- Use of the subsurface and groundwater for waste and commercial residuals disposal, such as agricultural nitrate fertilizer and highway road salt application, creates problems for use of groundwater in water supply.
- Sea level rise from climate change may motivate the coastal communities to compete for inland groundwater sources owing to saltwater intrusion in coastal aquifers.
- Groundwater brines are sources of important materials, such as lithium for batteries, the processing of which requires substantial groundwater.
- Flora and fauna relying on groundwater may be threatened and endangered in areas of major groundwater production and lowered water tables.

CHARACTERISTICS OF GROUNDWATER

Table 1.1.1 presents a further perspective on the physical, ecological and socioeconomic characteristics of groundwater that may be compared with alternative water sources. The table reflects the points mentioned earlier and suggests that groundwater may be a resource of more flexibility in use with lower developmental cost, while having potential vulnerability.

As a resource, groundwater is often used without charge or cost to the user. Its presence enhances the value of the property under which it lies – yet it can flow away. By flowing in the subsurface – albeit slowly – it carries issues with its use, both as a water source and a waste sink, beyond the property and national boundaries. Thus, it is one of the less visible components of the hydrologic cycle that may supply as much as half of the world's population with drinking water (Llamas and Custodio, 2003, p. 14; UNESCO, 2003, p. 78). Groundwater has become the fastest growing source of additional irrigation water for food production, a critical issue with our planet's expanding population (UNESCO, 2003; Brown, 2004, p. 100; USGS, 2018). Furthermore, groundwater has

TABLE 1.1.1
Physical, Ecological and Socioeconomic Characteristics of Groundwater Sources

Characteristics	Groundwater Description
Hydrogeological	
Storage volumes	Very large storage capacity in aquifers
Resource areas	Ubiquitous under continents
Flow velocities	Very slow, often measured in millimeters, centimeters, or meters per second
Residence times	Measured in decades/centuries depending on the geologic strata and depth
Drought propensity	Typically low depending on the duration of drought
Evaporative losses	Considered low and localized
Resource evaluation	High cost with significant uncertainty with unknown features underground
Production impacts	Occur over periods of time, often dispersed with cumulative effects
Natural quality	Usually high, depending on the nature of the geologic formation
Pollution vulnerability	Variable natural protection, affected by concentrated pollution releases
Pollutant persistence	Often long-lasting
Ecological	
Habitat	Variable with depth
Biodiversity	Wide range, with predominant microscopic species, except in Karst that has macrospecies
Biomass	Very large and vary with depth
Functions	Not fully understood, especially when considered with its containing subsurface environment
Socioeconomic	
Public perception	Resource is mythical, unpredictable and often forgotten
Development cost	Modest investment is required for most individuals and communities, higher for very large volumes of water needed
Development risk	May be less than often perceived, but depends on the location and impacts on others
Style of development	Typically mixed public and private finance with individual or community operation
Project promotion time	Usually short-to-moderate depending on the project size and availability of water
Irrigation efficiency	Frequently high, depending on the rights to access and availability

Source: Adapted from Foster et al., 2000, p. 2, citing Llamas, M.R., Water in 21st Century: A Looming Crisis, UNESCO Congress, Paris, France, 2, 1–20, 1998.

contributed to alleviating poverty and public health problems (Llamas and Custodio, 2003, p. 17), while serving as the foundation of the "green agricultural revolution" in Asia (Foster et al., 2000, p. 1; UNESCO, 2003, p. 78). However, once polluted, it is difficult and expensive to clean up for other uses and, in some cases, may cost billions of dollars (NAS, 2012).

The supply of groundwater is vast – 99% of the world's available freshwater exists underground (Gleick, 1996). If it exists under nearly all land and is so ubiquitous, how can groundwater be an issue worthy of economic analysis and discourse? Yet, as Benjamin Franklin stated: "When the well's dry, we know the worth of water," part of the answer lies in the fact that groundwater is subjected to the vagaries of nature – precipitation (or lack thereof), differences in soil types and geological strata, both locally as well as regionally, and use by other people and industry for a range of purposes – some using a small amount of groundwater, while others requiring enormous quantities. See Box 1.1.3 for another perspective on the value of water. Water is essential to life – not just human life, but all forms of life, existing on the ground surface or thousands of meters below it, including the microorganisms. Groundwater is an alternative source to surface water, which often maintains the streamflow, especially when it is not raining, and may be the only water source for many people, animals and organisms in various locations. Wetlands maintained by groundwater provide habitat for much wildlife, including endangered species. In some places or times, groundwater – or any water – may be scarce, for example, in Saharan Africa or during a drought. Furthermore, rapidly

BOX 1.1.3 ADAM SMITH'S PARADOX OF WATER AND DIAMONDS

Notably, Adam Smith suggested the Paradox of Water and Diamonds, which seems to be well suited to the initial considerations of the application of economics to groundwater. The paradox as stated by Smith is as follows: "The word value, it is to be observed, has two different meanings, and sometimes expresses the utility of some particular object, and sometimes the power of purchasing other goods which the possession of that object conveys. The one may be called "value in use"; the other, "value in exchange." The things which have the greatest value in use have frequently little or no value in exchange; and on the contrary, those which have the greatest value in exchange have frequently little or no value in use. Nothing is more useful than water; but it will purchase scarce any thing; scarce any thing can be had in exchange for it. A diamond, on the contrary, has scarce any value in use; but a very great quantity of other goods may frequently be had in exchange for it" (Smith, 1976, pp. 32–33).

The current economic answer to Smith's paradox is that demand creates value based on utility (or use). An individual tries to maximize their total utility or satisfaction. The answer also depends on how much of one or the other a person starts out with. If the person is very thirsty, a unit of water is greatly valued. Once the thirst is quenched, another unit of water is of less utility and the person will have less demand for it right now. A diamond is a beautiful stone and may satisfy other utilities of the individual, but not thirst. However, its utility may be greater than the second or third or fourth unit of water, and at that point, the diamond has utility value for which the person is willing to incur its price.

Source: Smith, A., An Inquiry into the Nature and Causes of the Wealth of Nations, The University of Chicago Press, Chicago, IL, 1976 (originally 1776), 568.

increasing use of groundwater may make it even scarcer and, therefore, the economic evaluation of groundwater has been a focus in recent years. Irrigation uses the greatest quantity of groundwater, resulting in tremendous demand on the resource in particular regions of the world, often mining and depleting it – increasing its scarcity.

The availability of groundwater has health effects associated with its use. Beyond the human body's basic need for water, groundwater may contain trace amounts of minerals (e.g., potassium) that could enhance health. On the other hand, other minerals in higher concentrations may be harmful to humans (e.g., arsenic that can cause digestive and neurological problems) (ATSDR, 2003) and animals (e.g., selenium that can produce neurological and reproductive effects and cause death) (Cornell University, Department of Animal Science, 2001). As groundwater is often obtained through a well, there may be associated costs for having access to it. As more people install and pump more wells, they may interfere with each other's use, requiring laws to assist in guiding its withdrawal. In situations where wastes are disposed of or chemicals are applied on or under the ground surface, precipitation and percolating water in the soil can carry these substances deep into the earth reaching the groundwater of an aquifer, the zone where the minute spaces between the rock grains are saturated with water. As a result, a cost to remove the unwanted substances may be incurred before groundwater is safe to drink or use for other purposes in these circumstances.

Thus, many factors contribute to an economic analysis – supply, demand, effects on other people and the ecosystem and increasing scarcity – as well as questions, such as: at what cost? What is the value of the resource? In what quantity? For what quality? At what price? Where? When? What is the balance of supply cost and consumer price, irrigation of intensely farmed cropland versus community water supply, or loss of aquatic and wildlife habitat owing to groundwater mining?

This text attempts to aid both the economic analyst and the noneconomist hydrologist, engineer and planner in addressing these and other questions. The presentation of economic information and concepts provides guidance for those not as knowledgeable in the foundations of economics to consider the broader sociopolitical-economic implications of their activities. At the core of considering the relation of economics to groundwater issues is a basic understanding of hydrogeology. These core concepts are included in this text for the economist to relate his or her detailed knowledge of the economics discipline to the groundwater portion of the ecosystem and its hydrologic cycle.

Economic Analysis Approach to Groundwater

The main purpose of economic analysis, whether applied to groundwater development or other evaluations, is to inform a decision maker (Bergstrom and Randall, 2016) – an individual deciding between drinking groundwater from a tap and drinking spring water from a bottle, a company evaluating whether to drill a deeper well for a more reliable source of water or to institute water treatment and reuse, a farmer debating on continued rain-fed agriculture or investment in irrigation equipment, or a government considering the options of raising water prices and adopting tax incentives to promote efficient water use to extend infrastructure life. The application of economics to this range of decisions assists in providing additional information to people or groups about the allocation of scarce productive resources (such as money, time, water, or land) among alternative and competing ends over place and time (Samuelson, 1964; Daly and Farley, 2004). It is the interaction of the people trying to satisfy their needs with the ecosystem that supplies the materials available in limited quantity to address those needs that create the scarcity: which needs should be satisfied, in what priority, applying which and *how many* resources, over what timeframe – what is the balance?

The field of economics has matured since the time of Adam Smith (author of *The Wealth of Nations* in 1776, which provided the principles of political economics) and has a range of subdivisions useful for evaluating the allocation of groundwater resources.

- The *classical economics* school of thought of 1700s to 1900s focused on the theories of market value and the functioning of market economies, emphasizing economic growth, free competition, pursuit of individual self-interest to maximize benefits to society (Farlex, 2005; ENI, 2006) and economy in equilibrium, balancing supply and demand through prices accepted by both buyer and seller.
- *Neoclassical economics* subsequently established mathematical relationships and a model of the function of an economy. This school of thought is focused on the utility and value of adding one more unit at the margin of production or other activity (ENI, 2006). This is referred to as marginalism or marginal analysis (Young, 2005, p. 18).
- In the 1980s, *ecological economics* emerged, becoming a distinct branch of economics addressing the relationships between ecosystems and economic systems. The concepts of ecological economics focus on examining the effects of alternative resource consumption and transformation strategies on the sustainability of the ecosystem, and, in this case, of groundwater resources (Daly and Farley, 2004).

The subdisciplines of microeconomics and macroeconomics operate within each of these branches of economics. *Microeconomics* focuses analyses at the level of the consumer, household and firm, which communicate their decisions by responding to the prices of goods and factors of production in the market – balancing supply and demand – to allocate scarce resources among the alternative uses (Samuelson, 1964; Bannock et al., 1979; Daly and Farley, 2004). *Macroeconomics* comprises the relationships among "broad economic aggregates, the most important of which are national income, aggregate saving and consumers' expenditures, investment, aggregate employment, the quantity of money (money supply), the average price level [across an economy], and the balance of payments [relative to foreign trade]" (Bannock et al., 1979).

A primary focus of macroeconomics is on the evaluation of government policies relative to expenditures, taxation and money supply, including interest rates. Farm income derived from irrigation and subsidies supporting groundwater production are the subjects of national policies of central governments around the world. However, extensive discussions on these economic subdisciplines are beyond the scope of this text.

Within the subdiscipline of microeconomics, various fields of study and analysis have evolved which are also relevant to the consideration of the economics of groundwater: natural resource and environmental economics.

- *Natural resource economics* examines the economic consequences of decisions and policies affecting the biotic and abiotic "endowment," particularly, as these effects relate to the property rights in resources and efficiency of the resource use in specific settings or broadly at a macrolevel (Howe, 1979). A key question for resource economists is the following: How should resources be allocated and then distributed among members of society? (Bergstrom and Randall, 2016)
- *Environmental economics* encompasses the economic significance, causes and incentives to "slow, halt, and reverse" environmental degradation (Turner et al., 1993), such as pollution of groundwater and effects on adjacent water users and on ecosystem processes. An important focus of environmental economics is on governmental policy affecting the actions and property rights of individuals, firms and jurisdictions engaged, or potentially, in activities that may result in environmental damage or change.

Owing to the fact that natural resource, environmental and ecological economics have aspects that are related and overlap each other, the distinction is, to an extent, a matter of emphasis: production, degradation and ecosystem effects, respectively. While these related economics fields reflect the evolution in the field, the issue posed by current circumstances for the relevant fields of economics is the following: Can the tools of economic analysis inform any resulting decisions and policies and assist in providing a sustainable future for humankind in the ecosystem that we call Earth?

The approach of applying economics within these fields considers empirical measurement of effects, referred to as "positive" economics, and policy evaluation, referred to as "normative" economics. Normative economics relies on the results of positive economic observation and measurement, as well as forecasts derived from those measurements, regarding the exchange among individuals, firms and central governing authorities, such as changes in prices or income and their effects on water demand, in this case. Normative economics seeks to evaluate whether improvements resulting from changes in policy or decisions, such as reductions in pollutant release or the addition of another production well, should be conducted from an efficiency perspective. As normative economics has broad societal implications, it is also referred to as "welfare" economics. Normative economics, as practiced, for the most part, has a base in Utilitarian ethics (Young, 2005, p. 18). Thus, it provides the decision maker or policy analyst with information in monetized form of changes in the welfare to society from the proposed actions: how large are the benefits and costs and who receives or bears them? (Young, 2005, pp. 17–18).

Positive normative economics is the fundamental approach of this text on groundwater economics. Many fine texts exist on microeconomics and natural resource and environmental economics and these fields will not be further developed here. Following the positive normative economic perspective, groundwater economics as a subset of natural and environmental economics seeks to analyze resource allocation (Bergstrom and Randall, 2016):

- Identify problem causes
- Examine alternatives as solutions
- Focus on benefits and costs of alternatives
- Examine the incidence of benefits and costs
    - Geographically
    - Among economic sectors
    - Across socioeconomic classes
    - Intertemporally

This approach relies heavily on information to evaluate the economic outcomes that inform resource decisions. Economists working in concert with hydrogeologists and water resource engineers and other planners draw on each other's information to target the input to decision-making for groundwater supply and demand determinations. The perspective of the remainder of this text is that the reader has a basic understanding of the field of economics.

Significant Groundwater Economics Issues

The principal question posed by ecological economics relevant to groundwater is the following: How can the ecosystem be sustained with sufficient groundwater of adequate quality to provide for the economic needs of human and animal populations now and in the future? The amount of water on the earth has been constant, but the world population has grown by 33 times over the last 2000 years (JHSPH, 1998). As of 1998, "31 countries – mostly in Africa and the Near East – face water stress or water scarcity. Population growth alone will push an estimated 17 more countries, with a projected population of 2.1 billion, into these water-short categories within the next 30 years. By the year 2025, 48 countries, with more than 2.8 billion people – 35% of the projected global population in 2025 – will be affected by water stress or scarcity. Another nine countries, including China and Pakistan, will be approaching water stress" (JHSPH, 1998). About 20–40 L of freshwater per person per day is a necessary minimum amount to satisfy drinking and sanitation requirements; however, when water for bathing and cooking is included, the need increases to between 27 and 200L/person/day (Gleik, 1996). From 1999 to 2003, the average per capita water use around the world ranged from 47 L/person/day in Kenya to 294 L/person/day in Phoenix, Arizona (Worldwatch Institute, 2006). Clearly, the need, expectations and technology vary from location to location, and these considerations will affect the supply and demand for groundwater (Figures 1.1.1 and 1.1.2).

The current groundwater issues needing attention are highlighted by Bill Alley and Rosemary Alley (2017) in "High and Dry; Meeting the Challenges of the World's Growing Dependence on Groundwater" summarized in Box 1.1.4. Clearly, the issues address the range of positive normative considerations: geographically, among economic sectors, across socioeconomic classes and intertemporally. The issues set the stage for evaluation of benefits and costs for groundwater supply and use.

FIGURE 1.1.1 Well house, Bethany Beach, Delaware.

FIGURE 1.1.2 Windmill to pump groundwater, Tanana River Valley, Alaska.

BOX 1.1.4 CURRENT GROUNDWATER ISSUES NEEDING ECONOMIC EVALUATION

- No precipitation – without adequate precipitation, must use groundwater for drinking water and crop irrigation
- Green agricultural revolution – in India, drew down groundwater resource supported by government-subsidized electricity to feed nation; crop irrigation is greatest water demand worldwide
- Groundwater use creates costs – if nonrenewable, then inadequate supply; significant production may induce land subsidence beyond point of use
- Inadequate infrastructure - in poorest countries, no or minimal infrastructure to provide basic water need; in these same countries, water is often not safe resulting in typhus, cholera diarrhea and other waterborne diseases
- Aquifers are different from place to place – renewable, nonrenewable; saline needing treatment and brine disposal; karst can carry contaminants easily
- Groundwater law – who owns, how much can be pumped, where can it be used, can water rights be sold depends on locale, state or country; the role of public ownership and of water markets is guided by law
- Streamflow depletion – groundwater pumping affects downstream users and aquatic life
- Water for Nature – watersheds need protection, groundwater and surface water are connected to maintain streamflow, ecological functions on which people rely need preservation
- Land subsidence – a major damaging cost in overpumped regions
- Managed aquifer recharge – treated wastewaters often used, requires surplus water, potable reuse requires advanced treatment, streambank filtration can provide initial treatment
- Waste disposal – uncontrolled chemical releases highly problematic for local groundwater use; employing the subsurface as a waste sink continues to be a problem.

(Continued)

BOX 1.1.4 (Continued)

- Pathogens – sources are human and animal feces from flooding, leaks in sewer lines and septic tanks; benefits of treatment outweigh costs
- Arsenic and other naturally occurring contaminants – in Earth's crust, mobilized in groundwater with right physical/chemical conditions; other natural contaminants include radon, radium, uranium and fluoride; activities that can mobilize natural contaminants are irrigation, managed aquifer recharge, application of fertilizer and recharge from septic tanks
- Fracking (hydraulic fracturing) – uses large volumes of groundwater, large volumes of wastewater need disposal often in impoundments and are highly saline with barium, strontium and radium; chemicals are used as fracture proppant; chemicals and methane may migrate along fractures; improper seal of abandoned frack wells may contribute to contaminant release; insufficient baseline data and conflicting studies of impact create misunderstanding of effects
- Nitrate – applied as ammonia for fertilizer and has reached groundwater where applied demonstrating vulnerability of groundwater to chemicals applied to land surface; other sources include leaking wastewater systems; occurrence demonstrates need to make safe water a priority
- Transboundary aquifers – often sole source of water; since multiple jurisdictions involved, protection can be problematic; often over-exploited and contaminated as a result

In addition to the groundwater issues highlighted above by the Alley's, several other issues are of key interest to economic evaluation:

- Role of the market – in some places is not considered relative to water or groundwater; significant in locations of water scarcity resulting in water being a commodity and in bidding for water rights
- Open access nature of groundwater – affords nearly anyone to install a well and pump; individual interests may supercede collective use of the resource resulting in overuse
- Asymmetric information – causes less information to be available for groundwater than for more accessible resources
- Financing investment in groundwater resources – may demand considerable funding; should take into account the natural capital of groundwater that may lower need for hard infrastructure
- Intergenerational effects – as representatives of the future may not be present or knowledgeable enough to put their interest forward, economic analysis can extrapolate to the future and be supportive of a measure of equity to next generations

BRIEF OVERVIEW OF THE TEXT, VOLUMES 1 AND 2

This first volume of this series describes groundwater in an economics context and is organized differently than many economics texts, considering the context in which groundwater is provided from the ecosystem to meet the basic purposes and economic activities. It focuses first on how we observe groundwater in nature, in the economy, as well as in the results that we encounter each day – but from the perspective of pointing to costs and benefits of groundwater production and use, including access to the resource, legal framework, health considerations, food production, waste sink, water quality, mineral production and subsurface competition. The first volume closes out considering the continuing use of groundwater and focusing on its sustainable development and management.

The second volume addresses the cost-benefit methodology applied to future groundwater decisions.It includes a framework for evaluating policies and projects with a step-by-step evaluation guide for life-cycle cost-benefit analysis. This volume incorporates broadly applicable transboundary, international and climate change considerations that may affect cost-benefit analysis, a macroeconomic perspective focused on throughput and scale of use, development of an approach for the future to improve balancing the range of economic and qualitative factors in the economic management of the resource, and future research needs. The second volume closes with sixteen case studies applying a range of concepts and methods presented in the text.

The key point throughout this series is that for our benefit and that of future generations, we must broadly consider the economic consequences of decisions to produce and use groundwater *sustainably* for water supply or waste disposal, not just considering the immediate effects for the user at the well or the point of disposal but broadly in the ecosystem and its watersheds and the economies that they support into the future.

FOR FURTHER CONSIDERATION

1. How is groundwater involved in your water supply? What advantages does it provide over other potential water sources where you live?
2. What current groundwater issues are most pressing in your locality?
3. Identify the economic characteristics of the issues listed from question 2 above. Do they involve supplying groundwater or demand for (using) groundwater?
4. How would you define sustainability? How does your definition apply to groundwater? Are there well-defined costs and benefits for having a sustainable supply of groundwater?

REFERENCES

Agency for Toxic Substances and Disease Registry (ATSDR). 2003. ToxFAQs for selenium. URL: http://www.atsdr.cdc.gov/tfacts92.html (accessed September 2, 2006).

Alley, W.M. and Alley, R. 2017. *High and Dry; Meeting the Challenges of the World's Growing Dependence on Groundwater*. Yale University Press, New Haven, Connecticut, USA, 294 p.

Association of Geological Surveys of Europe (AGSE). 2007. The role of groundwater in Europe. URL: http://eurogeosurveys.org/download/Groundwater_Policy_bis.pdf (accessed February 10, 2009).

Bannock, G., Baxter, R.E., and Rees, R. 1979. *The Penguin Dictionary of Economics*. Penguin Books Ltd, Harmondsworth, Middlesex, UK, 467 p.

Beitsch, R. 2018. *Few Wells are Tested for Contamination After Massive Flooding from Hurricanes Florence and Michael*. Huffington Post, December 18. URL: https://www.huffingtonpost.com/entry/hurricane-wells-contamination_us_5c13c2ffe4b0cce3ea29d165

Bergstrom, J.C., and Randall, A. 2016. *Resource Economics; An Econoomic Approach to Natural and Environmental Policy*, 4th Edition. Edward Elgar Publishing, Cheltenham, United Kingdom. 461 p.

Brown, L.R. 2004. *Outgrowing the Earth: The Food Security Challenge in an Age of Falling Water Tables and Rising Temperatures*. Earth Policy Institute, W.W. Norton & Company, New York, 239 p.

Brumley, J., Marks, C., Chau, A., Lowrance, R., Huang, J., Richardson, C., Acree, S., Ross, R., and Beak, D. 2018. *The Influence of Green Infrastructure Practices on Groundwater Quality: The State of the Science*. U.S. Environmental Protection Agency, Washington, DC, EPA/600/R-18/227.

Cornell University, Department of Animal Science. 2001. Selenium and livestock: Metabolism, toxicity, and deficiency. URL: http://www.ansci.cornell.edu/plants/toxicagents/selenium/selenium.html (accessed September 2, 2006).

Daly, H.E. and Farley, J. 2004. *Ecological Economics: Principles and Applications*. Island Press, Washington, DC, 454 p.

Farlex, Inc. 2005. Financial dictionary. URL: http://financial-dictionary.thefreedictionary.com/ (accessed April 19, 2006).

Field, B.C. 1994. *Environmental Economics*. McGraw-Hill, Inc., New York, 482 p.

Foster, S. 1996. Ground for concern. Our Planet. 8.3 (October 1996). URL: http://www.ourplanet.com/imgversn/83/foster.html (accessed June 14, 2006).
Foster, S., Chilton, J., Moench, M., Cardy, F., and Schiffler, M. 2000. Groundwater in rural development facing the challenges of supply and resource sustainability. World Bank Technical Paper No. 463, 98 p. URL: http://www-wds.worldbank.org/external/default/WDSContentServer/IW3P/IB/2000/05/02/000094946_00042005553519/Rendered/PDF/multi_page.pdf (accessed July 4, 2006).
Freeman, A.M. 1993. *The Measurement of Environmental and Resource Values*. Resources for the Future, Washington, DC, 516 p.
Gleick, P. 1996. Basic water requirements for human activities: Meeting basic needs. *International Water*, 21 (2), 83–92.
Hirschleifer, J. 1976. *Price Theory and Applications*. Prentice-Hall, Inc., Englewood Cliffs, NJ, 506 p.
Howe, C.W. 1979. *Natural Resource Economics*. John Wiley & Sons, New York, 350 p.
Institute for Learning and Research Technology (ILRT). 2005. *Economics Glossary*. University of Bristol, Bristol, U.K. URL: http://www.bized.ac.uk/cgi-bin/glossarydb (accessed November 12, 2005).
Johns Hopkins School of Public Health (JHSPH). 1998. *Population Reports; Solutions for a Water-Short World*, Vol. XXVI, No. 1. Baltimore, MD. URL: http://www.infoforhealth.org/pr/m14edsum.shtml (accessed April 20, 2006).
Llamas, M.R. 1998. Groundwater overexploitation. *Water in the 21st Century: A Looming Crisis*. UNESCO Congress, Paris, France, Vol. 2, pp. 1–20.
Llamas, R. and Custodio, E. (eds.). 2003. *Intensive Use of Groundwater*. A.A. Balkema Publishers, Lisse, the Netherlands, 478 p.
Mishan, E.J. 1982. *Cost-Benefit Analysis*. George Allen and Unwin (Publishers) Ltd., London, UK, 447 p.
National Academy of Science (NAS). 2012. *Alternatives for Managing the Nation's Complex Contaminated Groundwater Sites*. URL: https://www.nap.edu/catalog/14668/alternatives-for-managing-the-nations-complex-contaminated-groundwater-sites
National Ground Water Association (NGWA). 2016. Twenty-First Century Infrastructure – the Groundwater Industry's Role [NGWA Position Paper]. https://www.ngwa.org/docs/default-source/default-document-library/advocacy/position-papers/twenty-first-century-infrastructure-the-groundwater-industrys-role.pdf?sfvrsn=71004501_2
Raucher, B. 2005. The value of water: What it means, why it's important, and how water utility managers can use it. *Journal of the American Water Works Association*, 97 (4), 90–98.
Rea, K. 1999. Glossary [of economic terms]. URL: http://www.chass.utoronto.ca/~reak/eco100/glosslist.htm (accessed November 12, 2005).
Resources for the Future (RFF). 2002. *Glossary*. URL: http://www.weathervane.rff.org/glossary/index.html (accessed November 12, 2005).
Sabalow, R., Kasler, D., Reese, P. 2016. California farmers say 'no apologies,' as well drilling hits record levels. Sacramento Bee, September 26. URL: https://www.sanluisobispo.com/news/state/california/article104251276.html
Samuelson, P.A. 1964. *Economics: An Introductory Analysis*. McGraw-Hill Incorporated, New York, 838 p.
Smith, A. 1976. (originally 1776). *An Inquiry into the Nature and Causes of the Wealth of Nations*. The University of Chicago Press, Chicago, IL, 568 p.
The Economist Newspaper, Inc. (ENI). 2006. *Research Tools; Economics A-Z*. URL: http://www.economist.com/research/Economics/alphabetic.cfm (accessed April 19, 2006).
Turner, K., Pearce, D., and Bateman, I. 1993. *Environmental Economics: An Elementary Introduction*. The Johns Hopkins University Press, Baltimore, MD, 328 p.
United Nations Educational, Scientific and Cultural Organization (UNESCO). 2003. *Water for People, Water for Life*. The United Nations Water Development Report prepared by the UN World Water Assessment Programme, 575 p.
United States Geological Survey (USGS). 2009. *Ground Water Watch; Active Ground-Water Level Network*. URL: http://groundwaterwatch.usgs.gov/default.asp (accessed January 25, 2009).
United States Geological Survey (USGS). 2018. *Irrigation Water Use; 2015 Water Use*. URL: https://water.usgs.gov/watuse/wuir.html
Uranium Information Centre, Ltd. (UICL). 2002. *Radioactive Waste Management*. Melbourne, Australia. URL: http://www.uic.com.au/wast.htm (accessed July 4, 2006).
Worldwatch Institute (WI). 2006. *Household Water Use, Select Cities and Countries*. Washington, DC. URL: www.worldwatch.org
Young, R.A. 2005. *Determining the Economic Value of Water: Concepts and Methods*. Resources for the Future Press, Washington, DC, 357 p.

ADDITIONAL REFERENCES CONSULTED

Agency for Toxic Substances and Disease Registry (ATSDR). 2005. Public health statement for arsenic. URL: http://www.atsdr.cdc.gov/toxprofiles/phs2.html (accessed May 13, 2006).

British Geological Survey (BGS). 2006. Groundwater studies for arsenic contamination in Bangladesh. URL: http://www.bgs.ac.uk/arsenic/bphase1/B_find.htm#_Toc442156256 (accessed May 13, 2006).

Custodio, E. and Gurgui, A. (eds.). 1989. *Groundwater Economics: Selected Papers from a Symposium Held in Barcelona, Spain*. Elsevier Science Limited, Amsterdam, the Netherlands, 625 p.

Docking Institute of Public Affairs (DIPA). 2001. *The Value of Ogallala Aquifer Water in Southwest Kansas*. Fort Hays State University, Hays, KS, 41 p. URL: http://www.gmd3.org/pdf/ogal1.pdf (accessed May 13, 2006).

Downey, J.S. and Dinwiddie, G.A. 1988. *The Regional Aquifer System Underlying the Northern Great Plains in Parts of Montana*. Summary. U.S. Geological Survey Professional Paper 1402-A, North Dakota, South Dakota, and Wyoming, 64 p.

Driscoll, F.G. 1995. *Groundwater and Wells*. Johnson Screens, St. Paul, MN, 1089 p.

European Environment Agency (EEA). 1999. Groundwater quality and quantity in Europe. Environmental Assessment Report No 3. Copenhagen, Denmark.

Guru, M.V. and Horne, J.E. 2006. *The Ogallala Aquifier*. The Kerr Center for Sustainable Agriculture Inc., Poteau, OK.

International Groundwater Resources Assessment Centre (IGRAC). 2018. Managed aquifer recharge portal. URL: https://apps.geodan.nl/igrac/ggis-viewer/viewer/globalmar/public/default

Jarvis, T., Giordano, M., Puri, S., Matsumotos, K., and Wolf, A. 2005. International borders, ground water flow, and hydroschizophrenia. *Ground Water*, 43 (5), 764–770.

Jorgensen, D.G., Helgesen, J.O., Signor, D.C., Leonard, R.B., Imes, J.L., and Christenson, S.C. 1997. Analysis of regional aquifers in the Central Midwest of the United States in Kansas, Nebraska, and Parts of Arkansas, Colorado, Missouri, New Mexico, Oklahoma, South Dakota, Texas, and Wyoming. U.S. Geological Survey Professional Paper 1414-A, 67 p.

Koussis, A.D., Kotronarou, A., Destouni, G., and Prieto, C. 2003. Intensive groundwater development in coastal zones and small islands. In *Intensive Use of Groundwater*, Llamas, R. and Custodio, E. (eds.). A.A. Balkema Publishers, Lisse, the Netherlands, pp. 133–155.

Mondal, N.C. and Singh, V.S. 2004. *Integrated approach to delineate the contaminated groundwater in the tannery belt: A case study*. In *Proceedings of the Asia Pacific Association of Hydrology and Water Resources (APHW) Second Conference*, Singapore, July 5–8.

Murawski, J. 2018. *The Amount of E. coli and Fecal Matter in NC Wells has Spiked Since Hurricane* Raleigh News and Observer, Florence October 24.

National Research Council (NRC). 1997. *Valuing Ground Water: Economic Concepts and Approaches*. National Academy Press, Washington, DC, 189 p.

Ontario Ministry of the Attorney General (OMAG). 2002. Report of the walkerton inquiry. Prepared by Justice Dennis R. O'Connor. Queen's Printer for Ontario, Toronto, ON, Canada. URL: http://www.attorneygeneral.jus.gov.on.ca/english/about/pubs/walkerton/ (accessed May 13, 2006).

Queensland Department of Natural Resources, Mining and Water (QDNRMW). 2006. *Facts; Water Series; The Great Artesian Basin*. URL: http://www.nrm.qld.gov.au/factsheets/pdf/water/w68.pdf (accessed May 14, 2006).

Reilly, T. 2005. Personal communication by e-mail with the United States Geological Survey, December 19, 2005, concerning documented distances of long ground water flow paths.

Schafer, K.W. 1996. *International Experience and Expertise in Registration Investigation, Assessment, and Clean-Up of Contaminated Military Sites*. URL: http://www.umweltbundesamt.de/altlast/web1/berichte/mooreeng/dmeng_t.htm (accessed May 19, 2006).

Sophocleous, M. 2003. Environmental implications of intensive groundwater use with special regard to streams and wetlands. In *Intensive Use of Groundwater*, Llamas, R. and Custodio, E. (eds.). A.A. Balkema Publishers, Lisse, the Netherlands, pp. 93–112.

Tsakris, G. 2003. *Editorial: Groundwater Management in Transboundary River Basins*. May 2003. URL: http://www.waterinfo.gr/pages/editorial_new.html (accessed September 6, 2003).

United Nations (UN). 1996. *United Nations Declaration on World Food Security*. http://www.fao.org/3/w3613e/w3613e00.htm

United Nations (UN). 2000. *United Nations Millennium Declaration*. September 8. https://www.un.org/en/development/devagenda/millennium.shtml

United Nations (UN). 2006. *The UN Works for Freshwater*. URL: http://www.un.org/works/sustainable/freshwater.html (accessed May 19, 2006).

United States Department of Agriculture (USDA), Water Quality Information Center. 1997. *Nebraska Management System Evaluation Area*, Shelton Site. URL: http://www.nal.usda.gov/wqic/wgwq/neb1.shtml (accessed May 19, 2006).

United States Geological Survey (USGS). 1999. *Ground Water and Surface Water, A Single Resource*. U.S. Geological Survey Circular 1139, 79 p.

United States Environmental Protection Agency (USEPA). 2000. EPA/ROD/R09-00/114 EPA Superfund Record of Decision: San Gabriel Valley (Area 1) EPA ID: CAD980677355 OU 05, El Monte, CA09/29/2000.

United States Environmental Protection Agency (USEPA). 2018. 2015 Drinking water infrastructure needs assessment and survey. List of managed aquifer recharge projects.

United States Environmental Protection Agency (USEPA). 2006. *Drinking Water State Revolving Fund*. URL: http://www.epa.gov/safewater/dwsrf/index.html (accessed May 19, 2006).

United States Geological Survey (USGS). 2002. *Bibliography on the Occurrence and Intrusion of Saltwater in Aquifers Along the Atlantic Coast of the United States*, Barlow, P.M. and Wild, E.C. Open-File Report 02-235, 30 p.

United States Geological Survey (USGS). 2006. *Water-Quality: Saltwater Intrusion in California*. URL: http://ca.water.usgs.gov/water_quality/salt/index.html (accessed June 30, 2006).

Vrba, J. 2003. The impact of aquifer intensive use on groundwater quality. In *Intensive Use of Groundwater*, Llamas, R. and Custodio, E. (eds.). A.A. Balkema Publishers, Lisse, the Netherlands, pp. 113–132.

Walkow, F., Enders, K., and Peklo, P. 1999. The pollution of soil and water in Bitterfield. In *Abstracts; Workshop on the SAFIRA Project*, November 17–18, Bitterfield, Germany, 78 p. URL: http://www.ufz.de/data/ufz-bericht23-00523.pdf (accessed May 19, 2006).

Washington (State) Department of Ecology (WDE). 2005. *Cleaning Up Hanford's Groundwater*. Ecology Publication Number 02-05-015.

West Bengal and Bangladesh Arsenic Crisis Information Centre (ACIC). 2006. *The Arsenic Crisis*. URL: http://bicn.com/acic/ (accessed May 13, 2006).

Wilhite, D.A. 2006. *Ogallala Aquifer Depletion. Center for Agricultural Meteorology and Climatology*, University of Nebraska, Lincoln, NE. URL: http://www.iitap.iastate.edu/gcp/issues/society/ogallala/ogallala. html (accessed May 13, 2006).

Wilson, E.J. 2006. *Carbon Capture and Sequestration: Context and Considerations for Deployment*. Humphrey Institute of Public Affairs, University of Minnesota, Minneapolis, MN. May 11, 2006. URL: http://www1.umn.edu/iree/pdfs/canada_wilson_5_06.pdf (accessed June 30, 2006).

Worm, K. 2004. Groundwater drawdown. *Water Is Life*. University of Wisconsin-Eau Claire, Eau Claire WI. URL: http://academic.evergreen.edu/g/grossmaz/WORMKA/ (accessed May 13, 2006).

ADDENDUM TO CHAPTER 1

Economic Concepts Reference

To understand the economics of groundwater, some initial economic concepts are useful. These concepts are those that we hear in news media at times. It is important to have a commonly accepted definition for these concepts. For the purposes of this text, the following terms will be useful to keep in mind:

Cost is the expense of producing and providing one more unit of a good or service (Raucher, 2005, p. 93) which derives its existence from changes in the resources of the ecosystem specified in monetary terms for quantitative economic comparison or in descriptive terms for qualitative consideration. Producing and delivering one more (or many more) unit(s) of a good or service may also cause a loss or a disadvantage to society or the ecosystem, which is described quantitatively or qualitatively and referred to as an external cost (ILRT, 2005) or "externality."

One person's use of the groundwater may indicate that it was not available for the neighboring water company to produce and sell. In this case, the groundwater may have a use value that was foregone. The economist says that this "opportunity cost" of the person's use is the value in the next alternative use that could not occur.

The concept of cost also includes disadvantages or damages to people or to the ecosystem from some action or project. Economists attempt to quantify these costs wherever possible based on economic relationships that have currently been monetized. Examples include the health effects of human consumption from contaminated water or the loss of wetland habitat of fish and other wildlife. Often these costs may not be monetized and can only be counted or qualitatively described.

Price is the monetary amount that a purchaser or consumer pays for a unit of a good or service. Prices are usually based on costs (Raucher, 2005, p. 93). The price charged for a good or service may not be equal to the expense of producing or providing it; the price of a good or service in the market reflects its demand by consumers as well as their tastes and preferences, its relation to other goods and services including substitutes and complements, "distribution of resource endowments," and other factors (Mishan, 1982, p. 28). For example, if a drought has occurred, water may become scarce, and people may have to pay more than $0.37/L for groundwater, bidding against other thirsty human beings. In paying more, the producer's cost may not change, but the exchange value (price) in the market may go up.

Benefits refers to improved welfare or satisfaction from individual or societal advantages or gains owing to a change in the amount or condition of a good or service that derives its existence from changes in the resources of the ecosystem specified in monetary terms (Freeman, 1993, p. 8; RFF, 2002). If insufficient information on the monetary value of benefits exists, then they may be expressed in other units, such as cubic meters of groundwater, number of people affected and tons of subterranean biomass. Benefits also include the value of services or objects that may be counted in other ways, such as the savings of time or the avoidance of costs. Advantages or gains to those individuals or groups not directly targeted to receive them are referred to as external benefits (ILRT, 2005). In some instances, benefits may have to be qualitatively described if the ecosystem or economic relationships are not or poorly understood. A significant point is that benefits may be based on purposes that are not well defined or described in monetary units but may still have substantial value in decision-making processes.

Value refers to the worth of a result, service, object, or good, the extent of well-being that a person obtains through exchange from having one more unit of a good or service (Freeman, 1993, p. 19; Raucher, 2005, p. 91) of a particular quality (which for groundwater might be based on location, purity, timing and reliability) (Raucher, 2005, p. 91), as well as sustainability (Daly and Farley, 2004, p. 373). Values may be "extrinsic," derived from human economic and social welfare attainment, or "intrinsic," which are inherent in the object or condition and separate from any human attainment (Young, 2005, p. 24).

Value may be more than the price. If one needs to consume some amount of groundwater to avoid dehydration and becoming ill, then the price one has to pay for the quantity of water consumed may be substantially less than its value. The value of additional groundwater flow carrying important nutrients to a coral reef may be more than the price of that groundwater multiplied by the volume of the maintained flow but may include the survival of an important species of fish and even aesthetic characteristics of the coral and the habitat that it provides, which everyone can appreciate in the future.

The *economy* is the sum of goods and services that are exchanged between individuals as well as among individuals and small and large corporations. Economies can be local, such as within a town, regional as in a state or nation, or can be global. The economy relies on resources provided by nature which people and corporations transform to serve the immediate (ranging from a few seconds to many years) needs and are subsequently returned to nature as used up or worn-out materials and wastes. Nature must somehow deal with these byproducts over time, which may sometimes take centuries and longer (the half-life of radioactive waste ranges from seconds to billions of years depending on the isotope [UICL, 2002]). The economy refines, concentrates and transforms resources to products useful to people and distributes them through exchange among individuals and corporations.

Economic *efficiency* results when a good or service is produced at a minimum cost. An Italian economist, Vilfredo Pareto, proposed an allocative efficiency criterion that "it must not be possible to change the existing resource allocation in such a way that someone is made better off and no one worse off, since, if this were possible, the existing resource allocation must involve a 'welfare waste'" (Bannock et al., 1979, p. 144). Under this paradigm, economic benefits must exceed the costs for some decision or action without a further change in the amount or condition of the good or service making someone else better off.

Supply refers to the amount of a good or service of a certain quality offered by a seller at a particular price (Bannock et al., 1979, p. 430). For example, a water company may be able to sell a certain number of units of water at $1.00, or a person may be willing to work for a set wage at a well-drilling company.

Demand is "the willingness or ability to pay a sum of money for some amount [and quality] of a particular good or service" (Bannock et al., 1979, p. 119). If water is in short supply, people might be willing to pay more for it. If the quality of water is not high, then people might not want to pay much for it.

A *market* is "an institution in which buyers ['demanders'] and sellers ['suppliers'] ... carry out mutually-agreed on exchanges" (Field, 1994, p. 68) and hence is considered to be the convergence of buyers and sellers of goods and services or the services of productive factors (e.g., labor, materials, or capital) for the purpose of economic exchange (Rea, 1999). A "perfect market" is defined as having perfect communication between buyers and sellers, accomplishing instantaneous equilibrium from inputs of all traders' supply and demand functions to obtain "market-clearing" prices for their exchanges and completing costless transactions with no "middlemen" or brokers (Hirschleifer, 1976, p. 200). In reality, no perfect market exists for economic exchanges. The exchanges in the market usually take place through the use of money paid for a good or service. In the ideal setting, the market provides sufficient information about the good or service exchanged, and the buyers and sellers are mutually satisfied with the value of the exchange.

The *ecosystem* is the sum of all the biotic and abiotic features of the earth, including human beings, fish and wildlife, plants, microorganisms, sunlight, soil and geologic matrix, atmospheric gasses, WATER and the processes – hydrologic cycle, atmospheric and oceanic circulation, heat exchange, biochemical degradation, gravitational pull, etc. – to which they inherently contribute, some known and others unknown, which are being understood with the help of science. The ecosystem supplies all the resources, available in finite quantity, used in the economy by people and their corporations.

Sustainability, in both an ecosystem and an economic sense, is the capability of a resource to have a portion removed and yet maintain itself indefinitely, which means that on average, what is removed is replenished over the long term; this is an inherent capability of the resource. For groundwater, if the removal is greater than the natural or artificial replenishment, the aquifer will be depleted. Sustainability in a financial sense is the ability of a firm or individual to have means of obtaining sufficient revenue or support to cover all costs or requirements of operation on a continuing basis; if costs or requirements exceed revenue or support, then a water company is not sustainable.

Partial equilibrium analysis is an evaluation of a small part of the economy, such as one market or industry, without considering changes in the rest of the economy. Such analyses are useful for expositional purposes and show only a static or short-run position in the market.

General equilibrium analysis provides evaluation of conditions and consequences throughout the economy from changes in a market or set of markets (Bannock et al., 1979, p. 200). General equilibrium analyses rely on the results from empirical studies to establish relationships among the markets and may use input–output modeling to help describe outcomes of changes. The effects of increasing wheat production, farm implements and agricultural extension agents may be examined relative to expanding groundwater production, well drillers and pump manufacturers.

Box 1.1.5 provides an example of a partial equilibrium analysis of supply and demand relationships for a marketed good or commodity at a point in time. The economic theory behind such an example is summarized in a later chapter. In this example, the cost of supply increases (it costs more to produce an added unit of a hypothetical commodity in this case) as more units are produced, which is reflected in the supply curve, *S*. The supply curve reflects the producers' costs per unit at

BOX 1.1.5 SIMPLE EXAMPLE OF PARTIAL EQUILIBRIUM ANALYSIS OF SUPPLY AND DEMAND

a = benefit to consumer at price P resulting from market process
$b + c$ = total amount paid by consumer for quantity Q and revenue to supplier at demand Q
c = cost of production to the supplier

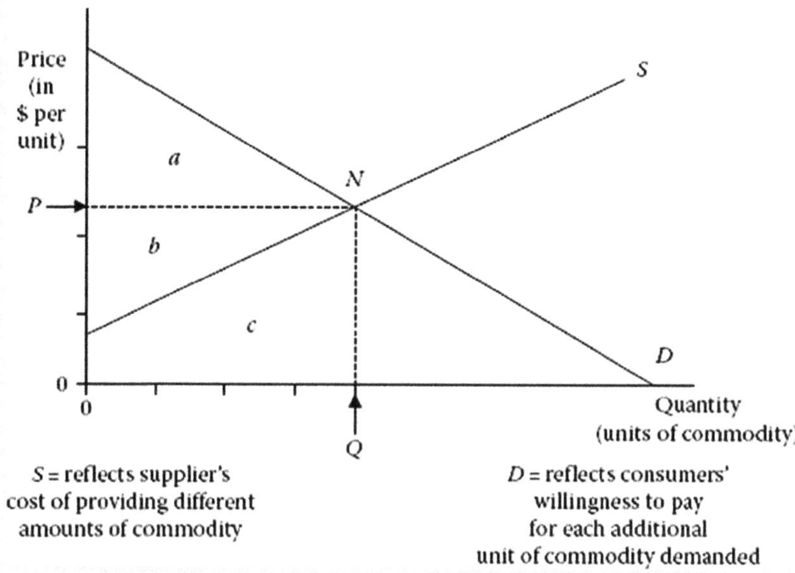

each quantity made available. The demand curve, *D*, gives the consumers' willingness to pay for one more unit at each increment of quantity. As explained in more detail in a later chapter, the demand curve portrays the consumers' willingness to pay for each additional unit, such that it derives a consumer value for a certain quantity equal to the price ($/unit) multiplied by the number of units demanded. The graph indicates that the consumer has a high benefit for the first units (the difference between the demand curve, *D*, for any quantity above its intersection with the horizontal axis) but less so as more units are supplied. At the intersection of the supply and demand curves, the market is considered to be in equilibrium and sellers are just willing to accept the amount that consumers are just willing to pay. As the consumers only pay the price in the market and not a different price at each quantity, the area above the price line, *P*, up to the demand curve, *D*, is considered to be, in economic terms, the consumer's "surplus," a benefit received but not paid for.

In the past, generally, water, or, specifically, groundwater, was not typically marketed as a commodity in humid regions but was treated as a public good with no cost other than to access it. This suggests that many people used groundwater for their individual benefit, rather than for the society, or, where controlled, it was accessed by a central producer and distributed to consumers. Groundwater's low access cost in many situations has contributed to significant and often unconstrained use. Similarly, as groundwater is often viewed as WIDELY AVAILABLE and naturally made pure at NO OR MINIMAL COST, the subsurface has been used for waste disposal, either by directly draining or injecting waste into the underground environment or allowing it to slowly leach into it. It is now recognized that there is a cost of waste disposal to adjacent users as groundwater migrates across boundary lines and to future users as groundwater moves slowly and contaminants may degrade slowly in it or perhaps not at all. As groundwater becomes scarce in many locations, and already scarce in drier climates, it is treated as a market commodity, being offered for sale to the customer/bidder paying the highest price or in smaller quantities, such as bottled water, in markets with many water suppliers of a low-cost convenience product.

While other key concepts are introduced in later chapters, it is important to consider that the discipline of economics attempts to explain exchanges of goods and services among people and corporations, as well as provide information for organized analyses to decision makers. This function of contributing to decision-making at all levels and to policy development, thereby influencing the process of transactions in the economy, makes the field of economics a catalyst for change, facilitating the potential and nature of exchange between individuals as well as within and among societies. Part of what follows examines the economic effects on those who were previously considered as nonusers of immediate local groundwater resources – the effects on adjacent and even distant water users and the ecosystem that supplies the water for those users as well as the natural purposes on which we rely.

REFERENCES

Bannock, G., Baxter, R.E., and Rees, R. 1979. *The Penguin Dictionary of Economics*. Penguin Books Ltd., Harmondsworth, Middlesex, U.K., 467 p.

Daly, H.E. and Farley, J. 2004. *Ecological Economics: Principles and Applications*. Island Press, Washington, DC, 454 p.

Field, B.C. 1994. *Environmental Economics*. McGraw-Hill, Inc., New York, 482 p.

Freeman, A.M. 1993. *The Measurement of Environmental and Resource Values*. Resources for the Future, Washington, DC, 516 p.

Hirschleifer, J. 1976. *Price Theory and Applications*. Prentice-Hall, Inc., Englewood Cliffs, NJ, 506 p

Institute for Learning and Research Technology (ILRT). 2005. *Economics Glossary*. University of Bristol, Bristol, UK URL: http://www.bized.ac.uk/cgi-bin/glossarydb (accessed November 12).

Mishan, E.J. 1982. *Cost-Benefit Analysis*. George Allen and Unwin (Publishers) Ltd., London, UK, 447 p.

Raucher, B. 2005. The value of water: What it means, why it's important, and how water utility managers can use it. *Journal of the American Water Works Association*, 97 (4), 90–98.

Rea, K. 1999. Glossary [of economic terms]. URL: http://www.chass.utoronto.ca/~reak/eco100/glosslist.htm (accessed November 12, 2005).

Resources for the Future (RFF). 2002. Glossary. URL: http://www.weathervane.rff.org/glossary/index.html (accessed November 12, 2005).

Uranium Information Centre, Ltd. (UICL). 2002. *Radioactive Waste Management*. GPO Box 1649N, Melbourne, Australia. URL: http://www.uic.com.au/wast.htm (accessed July 4, 2006).

Young, R.A. 2005. *Determining the Economic Value of Water: Concepts and Methods*. Resources for the Future Press, Washington, DC, 357 p.

2 Groundwater in the Ecosystem

PHYSICAL SIGNIFICANCE OF GROUNDWATER

To understand the role of groundwater in the economy and its production cost, it is first necessary to understand the existence of groundwater in the ecosystem. Groundwater exists as the subterranean component of the hydrologic cycle. Small organisms that are a part of the biochemical web inhabit the dark, low-oxygen underground environment. Even though we cannot see or observe groundwater easily, it exists in a complex subsurface environment that affects its accessibility and use, its quality and its sustainability as a resource in the economy. What is the magnitude of groundwater in the ecosystem?

Groundwater is the second smallest of the four "pools" of water in the earth's ecosystem. Figure 1.2.1 shows the major "pools" and "fluxes" of water in the global water balance. Aquifers hold 10,530,000 km^3 (2,526,000 cubic miles) of groundwater, about 0.8% of all the earth's water (USGS, 2016). However, groundwater comprises 99% of the earth's freshwater, excluding the frozen water of the polar ice caps (USGS, 2016). Figure 1.2.1 and Box 1.2.1 provide the estimates of the water supply of the world. The subsurface space exists under most of the world's land, in which the pores of the geologic structure – whether minute fractures or microscopic spaces among silt and sand grains – are filled and saturated with water (USGS, 1984). Groundwater interacts with surface water in watersheds that run from mountains to the oceans (USGS, 1999b). Once thought to be inexhaustible, this resource has reached its limit in many places to supply safe water of adequate quality and quantity to meet human requirements (Danielopol et al., 2003) and is now recognized as mostly finite (Gleeson et al., 2015). Groundwater occurs near and at the ground surface, in shallow aquifers

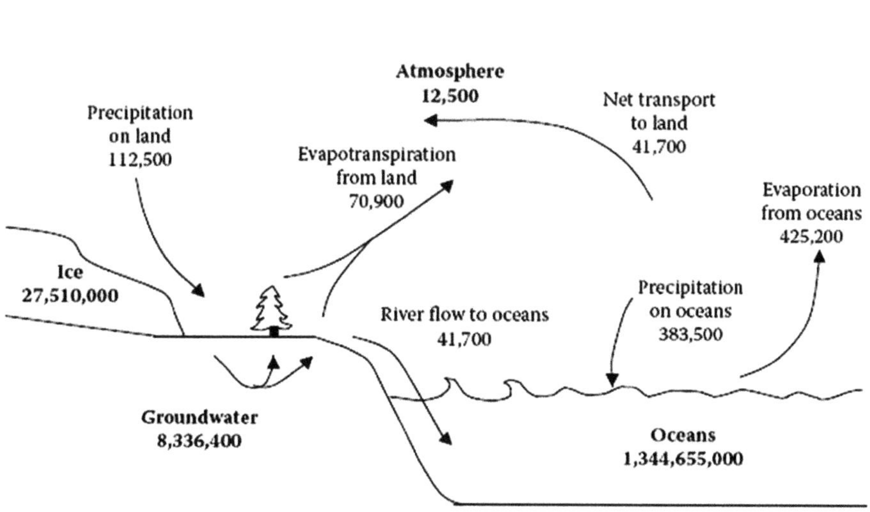

FIGURE 1.2.1 Groundwater in the global water balance. Pools in bold, units in cubic kilometers: fluxes with arrows, units in cubic kilometers per year. Groundwater is the second smallest pool water in the global water balance. It is assumed that with global climate changes, the ice–ocean relation is changing, with ice becoming a smaller pool and oceans becoming still larger. Sources: United States Geological Survey (USGS), Ground Water and Surface Water, A Single Resource, U.S. Geological Survey Circular 1139, Denver, CO, 1999b, 2. Adapted from Schelesinger, W.H., *Biogeochemistry – An Analysis of Global Change,* Academic Press, San Diego, CA, 1991.

BOX 1.2.1 WORLD'S ESTIMATED WATER SUPPLY

Water source	Water volume, in cubic kilometers	Percent of freshwater	Percent of total water
Oceans, Seas, & Bays	1,338,000,000	—	96.54
Ice caps, Glaciers, & Permanent Snow	24,060,000	68.6	1.74
Groundwater	23,400,000	—	1.69
Fresh	10,530,000	30.1	0.76
Saline	12,870,000	—	0.93
Soil Moisture	16,500	0.05	0.001
Ground Ice & Permafrost	300,000	0.86	0.022
Lakes	176,400	—	0.013
Fresh	91,000	0.26	0.007
Saline	85,400	—	0.007
Atmosphere	12,900	0.04	0.001
Swamp Water	11,470	0.03	0.0008
Rivers	2,120	0.006	0.0002
Biological Water	1,120	0.003	0.0001

Source: US Geological Survey. 2016. How much water is on Earth? URL: https://water.usgs.gov/edu/gallery/global-water-volume.html

(citing Gleick, Peter H. (editor). (1993). Water in Crisis: A Guide to the World's Fresh Water Resources. Oxford University Press, New York.)

and wetlands, respectively, and in deep aquifers hundreds and even thousands of meters below. Groundwater can migrate into rivers maintaining their flow for downstream users. The biodiversity of the groundwater portion of the ecosystem is significant (Daneilopol et al., 2003).

THE HYDROLOGIC CYCLE

Groundwater is a fundamental component of the hydrologic cycle. Indeed, hydrology is recognized as a driving variable for habitats and the ecosystem (Urbanska et al., 1997). As described in Figure 1.2.2, water evaporates from bodies of water and from the soil, as well as being transpired by plants. In this vapor phase, it rises and condenses to form clouds. As the clouds become heavily laden with water, they release it as precipitation (rain, sleet, hail, or snow). When the precipitation reaches the ground, it infiltrates the soil zone first. After the soil zone has reached its limit to allow water to infiltrate at the current rate of precipitation, water runs across the surface, becoming overland flow or "runoff." Overland flow fills the depressions on the land surface and moves toward streams and lakes. During the precipitation event, infiltration water that is not held in the soil zone percolates through the subsurface (called the unsaturated or "vadose" zone, with the pore space taken up by both water and air) to the water table (the upper plane of the saturated zone) of the "first" or "shallow" aquifer as well as its "saturated zone," referred to as groundwater, in which all the pore space is filled with water. An aquifer is a subsurface geologic formation, part of a formation or a series of formations, which are saturated with and hold water. Groundwater moves through the small, microscopic spaces between rock, sand or silt particles, and depending on the subsurface slope of the water table, percolation water reaching the saturated groundwater zone may be carried laterally by the local hydraulic gradient (slope of the water table), or move more deeply into the aquifer, and then laterally, as it moves toward a discharge zone, such as a stream or lake, ocean or pumped well.

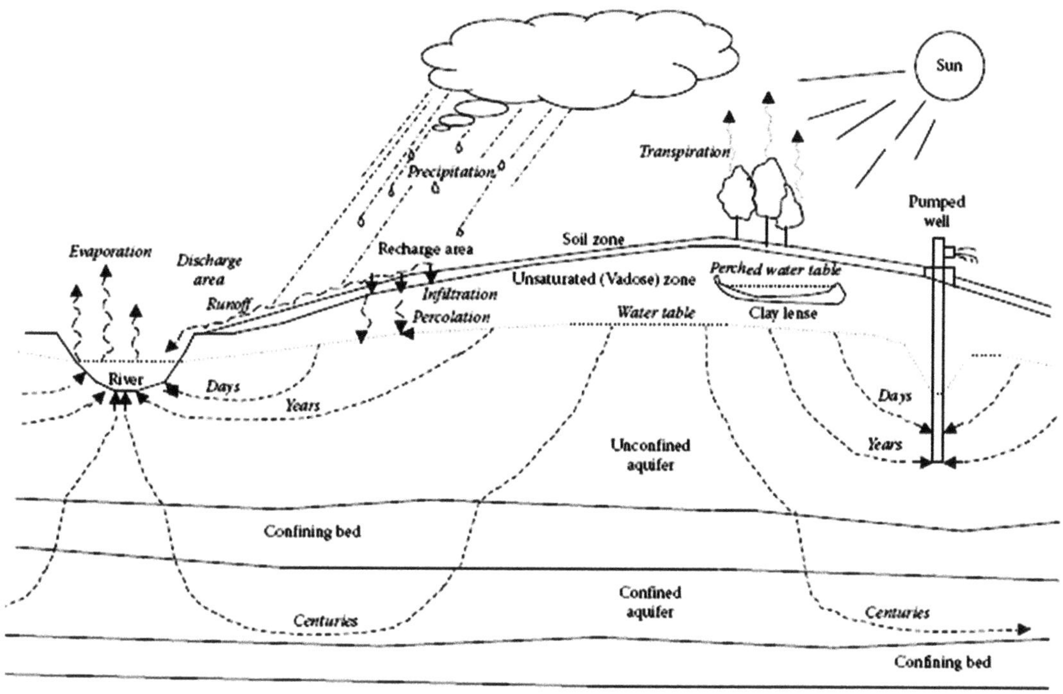

FIGURE 1.2.2 The hydrologic cycle emphasizing groundwater flow. Source: Adapted from United States Geological Survey (USGS), *Ground Water and Surface Water, A Single Resource*, U.S. Geological Survey Circular 1139, Denver, CO, 1999, p. 5.

Groundwater can flow to streams in a watershed discharging to them through the streambed. A watershed is an area of land, often referred to as a "catchment basin" (because it "catches" the precipitation that falls on it), within which the surface water and the shallow groundwater discharging to the streams have a common outflow point to a larger stream or to an ocean. During storms, the water levels can rise in the streams and allow additional water to be transmitted into the channel banks along them referred to as "bank storage" water that is slowly discharged back to the stream.

Groundwater moves from high energy (high "hydraulic head") to lower energy (lower "hydraulic head"). When water accumulates below the ground, it can create pressure from its aggregating mass in the subsurface pore spaces. This pressure provides energy to move the water, transmitting the energy through its subterranean travel to the discharge points (a stream, lake, ocean, spring, or well) that release the accumulated pressure through the flow of water. These circumstances explain the movement of groundwater even through minute pore spaces in the ground.

Additionally, some water is held by leaves, stems and branches of vegetation, called interception, which may be evaporated. A significant amount of moisture is also returned to the atmosphere by plants through transpiration, which draws water from beneath the land surface through the plants' root system. Literally, the plants can "pump" water from the ground through the processes of root uptake and transpiration.

Impervious surfaces can affect infiltration and recharge. In locations where bedrock is at the ground level creating an impervious surface, overland flow may predominate initially in a precipitation event. In these cases, rain or melting snow can only enter the rock at fractures, often too small to see. Similarly, in urban areas, rainwater may runoff from streets, sidewalks and rooftops, creating stormwater which can be routed to and concentrated in sewer lines or detention ponds. If cracks exist in the sewer lines, then the leakage from these sewers can recharge groundwater with contaminated water. Stormwater detention ponds are often designed to allow much of their detained water to percolate into the ground over longer times than the storm.

In Figure 1.2.2, the upper or shallow aquifer's water table exists typically above the elevation of the stream, indicating groundwater discharge (GWD) to the stream – flow from high hydraulic head to lower hydraulic head. In the western United States, it is common to find the water table below the elevation of the stream. In this case, water from the stream "recharges" the groundwater, transmitting water into the subsurface. Such circumstances exist in the United States Central Plains, where snowmelt maintains the flow of the Platte River that recharges the upper aquifers over which it crosses. In many locations, streams may recharge groundwater, and in alluvial (sand and gravel) valleys, an exchange between groundwater and stream and back to groundwater may occur many times (USGS, 1999), based on the geology through which the stream flows and in which the aquifer exists.

The U.S. Geological Survey has estimated that 40% to 50% of the average annual streamflow (the flow of all the water in streams in an average year) is from the GWD to streams (USGS, 2018) and may be as much as 90% in some streams (USGS, 1999). Groundwater is also discharged to oceans and maintains water elevations in most lakes. In the western United States, streamflows may also be maintained by irrigation return flow, where water is pumped from the ground and used for irrigation and then allowed to flow to the stream, either overland or underground. As stream water is used many times, the economic interests of many landowners and managers are vested in insuring that they receive their share of the use of this water.

Figure 1.2.2 highlights several key aspects of groundwater that affect its presence in the ecosystem and its use in the economy. First, the dashed lines below the line labeled "water table" indicate the relative travel times for groundwater in the saturated zone. Groundwater nearest a stream or a pumped well may reach those discharge points in a matter of days. Water further away may take years, even centuries or millennia before finally flowing through a streambed or a well. Groundwater away from the discharge points generally flows downward and the pressure from the hydraulic head can push it back up to points of discharge.

Second, deeper groundwater may be found in "confined aquifers" often lying hundreds or thousands of meters below the ground surface between two confining beds of less permeable geologic material. Confined aquifers may often be the sources of water penetrated by deep wells. These waters may have been deposited in the prehistoric geologic matrix and may be very old and are often referred to as "ancient" groundwaters. Very little or no water from recent precipitation ever reaches these deep, confined aquifers in the timeframes of people's lifetimes.

A third point is more subtle, with regard to the direction of groundwater flow in and under a watershed. Figure 1.2.2 shows the land surface sloping downward to the left from the trees and downward to the right beginning with the trees. Beneath the ground surface is the "water table," which is the groundwater level "high" point, sloping downward to the left and then downward to the right. Usually, the water table slightly mirrors the land surface, but not always. The topographic high of the ground surface separates one watershed from another. Beneath the watershed, the "groundwater-shed" may not have its high at exactly the same vertical transect. This circumstance may exist for a variety of reasons, including the location of points of discharge.

Fourth, sometimes, the subterranean features may catch and hold groundwater in places where we do not anticipate. Such is the case in Figure 1.2.2 of the clay lense located to the left of the pumped well. In this hypothetical situation, a past geologic event is considered to have left a layer of clay of limited extent that stops the water from percolating to the larger unconfined aquifer below it. This condition is referred to as "perched groundwater" and has its corresponding "perched water table," which is also an unconfined aquifer, albeit a small one. If the pumped well had been installed to the left of its location in the figure, it would have drawn groundwater at a much shallower elevation. Perched groundwater is not usually a large, dependable water supply.

Fifth, the pumped well affects the elevation of the water table around it. From the figure, it can be observed that the water table dips around the pumped well, and this condition is called the "cone of depression." The cone of depression exists around a pumped well because the pumping pulls the water to the well screen that allows water into the well and keeps out larger material, thereby reducing the pressure in the aquifer at that point and lowering the water table around the well. As the rate and volume of groundwater pumped increases, the cone of depression can grow deeper and outward from the well. For example, many wells pumped groundwater from underground in the area of Chicago, Illinois, for such a long time that the combined cone of depression could be measured north into the state of Wisconsin, a distance of well over 83.3 km from the center of the pumping area. (Further useful references for groundwater in the hydrologic cycle are Freeze and Cherry, 1979; Driscoll, 1986; USEPA, 1993; USGS, 1999; and USGS, 2020.)

NATURAL FACTORS AFFECTING GROUNDWATER RESOURCES

Nature is a major factor affecting the quantity and quality of the groundwater resource. The hydrologic cycle interacting with the earth's geology is a vital relationship in the ecosystem that influences the presence and quality of groundwater. These factors and others, not as well understood, affect the economics of groundwater. Several key concepts are important in understanding the presence of groundwater in the ecosystem on which we rely for survival and economic exchange:

Precipitation

An area must receive sufficient precipitation – rain, snow, sleet and other forms of atmospheric water that fall on the land – to allow water to saturate the soil and infiltrate the subsurface, migrating to the water table.

Aquifer

An *aquifer* is a water-bearing geologic formation, sequence of formations, or a portion of a formation capable of supplying sufficient water for particular purposes (Driscoll, 1986; USEPA, 1993). Aquifers exist in a range of geologic matrices from sand or gravel to bedrock, in which water is held in minute fractures. Groundwater also flows through solution channels in carbonate rocks, which have been created by water dissolution along the fractures. An important point to note is that for an unconfined aquifer, the saturated portion will vary based on precipitation, discharge and use. That is, the water table will fluctuate (rise and fall) depending on the volume of the water available to it. During long precipitation events or floods that cause the volume of the subsurface to be more completely saturated, it is possible for the water table to rise to the ground surface.

Deep Ancient Aquifers

In many areas of the world, water held in ancient geologic strata far below the earth's *surface* – thousands of meters deep in some cases – may be reached by wells. These deep aquifers do not receive water infiltrating from the ground surface and percolating through the subsurface. These aquifers may be large in areal extent and the only source of water in some locations. When waters in these aquifers are consumed, they disappear in perpetuity from a human-use perspective.

Evaporation

If the temperature is too warm and the precipitation is too minimal, then evaporation of the water – moisture given up because of the heat – in the soil can preclude the water from reaching the water table. In the arid southwestern United States, evaporation can be as much as 3.7 m or more per year in locations that may not receive more than 15–30 cm of precipitation each year.

Permeability and Soil Texture

The ability of soil to allow water to penetrate and move through it is called *permeability*. Soils with higher clay content are more impermeable to infiltrating water. Caliche soils in the southwestern United States also tend to cause water to run off, rather than enter the subsurface. Highly compacted soils anywhere encourage runoff. Sandy or loose soils permit water to infiltrate and, thus, are more permeable. Furthermore, clay soils with macropores – holes that worms, insects and burrowing animals create – and cracks can increase permeability.

Geologic Matrix (Or Aquifer Media)

The geologic matrix or aquifer medium that serves to hold water acts as the "container" for groundwater to exist in it. Groundwater may be contained in minute pore spaces of the geologic matrix. Underground channels for groundwater do exist in carbonate geologic formations, because water may have dissolved the formation along the fractures, creating the channels over time. Groundwater can flow from one geologic "container," or formation, to another, with the rate of flow depending on a range of factors, including the amount of pore space and the pressure under which it exists. The generic hydrogeologic settings found in the United States and elsewhere in the world are listed in Table 1.2.1. This list of hydrogeologic settings reinforces the concept of highly complex conditions under which groundwater occurs in the subsurface environment. In the United States, based on their inherent characteristics, there are 48 common settings, but the natural variation within them is great so as to produce wide ranges of specific settings found in the ecosystem.

The geology also affects the accessibility to groundwater. Clays and sands are relatively easy to drill through or *dig*. However, sand with large cobbles or boulders can damage drilling equipment and slow down the drilling progress. In some cases, more expensive and sophisticated drilling equipment may be necessary. Drilling in bedrock requires durable equipment that is more costly to operate and maintain, as well as requiring specialized training to use. These factors affect the cost to obtain groundwater.

Subsurface Water Migration

Once the water moves from the soil to the vadose or unsaturated zone, its *rate* of migration depends on the size of the particles and pore spaces making up the geologic matrix and the pressure of the water above it, creating a hydraulic head. This is also true for the saturated zones of water called the aquifer, or the groundwater at and below the water table. In clay, groundwater may move as slowly as 0.00000001 m/day (Driscoll, 1986). Attempting to pump water from such a strata would be fruitless. On the other hand, sand and gravel aquifers may allow water to travel at rates equivalent to thousands of meters per day. These aquifers are very productive; that is, they can usually provide large quantities of groundwater in relative short timeframes and can continue to do so as long as drought is not a problem for their recharge. Figure 1.2.3 portrays the typical range of hydraulic conductivity for a range of aquifers, both consolidated and unconsolidated (Driscoll, 1986, p. 75).

Water traveling in underground rivers is an interesting concept based on fact. In fractured limestone, water, over time, can move along the fractures and dissolve the calcium carbonate that makes up the limestone.

As this *dissolution* occurs over time, large channels develop, called solution channels. In some locations, the dissolution has been so extensive, forming caves and caverns in which people have observed "underground streams." As fractures may intersect, the solution channels may also intersect. Thus, in an area with limestone close to the surface and in which water has been dissolving the calcium carbonate, this underlying geology may have the rough appearance of stormwater collection lines. These areas also typically have "sinkholes," where a cavern or channel has caved in from above (Figure 1.2.4). This type of carbonate geology is called "karst." Groundwater movement is very rapid in solution limestone and karst areas, traveling as fast as 3.3 km/day (2.1 miles/day)

TABLE 1.2.1
Generic Hydrogeologic Settings of the United States (Listed Alphabetically)

1. Alluvial basins (internal drainage)
2. Alluvial fans
3. Alluvial mountain valleys
4. Alluvium
5. Alluvium and dune sand
6. Alternating sandstone, limestone, shale with deep regolith
7. Alternating sandstone, limestone, shale with thin soil
8. Beaches and bars
9. Beaches, beach ridges and sand dunes
10. Bedrock of the uplands and mountains
11. Braided river deposits
12. Buried valley
13. Coastal beaches
14. Coastal lowland deposits
15. Confined regional aquifers
16. Consolidated sedimentary rocks
17. Glacial and glaciolacustrine deposits of the interior valleys
18. Glacial lake deposits
19. Glacial till over bedded sedimentary rock
20. Glacial till over crystalline bedrock
21. Glacial till over outwash
22. Glacial till over solution limestone
23. Glacial till over sandstone
24. Glacial till over shale
25. Glaciated mountain valleys
26. Lava flows hydraulically connected
27. Lava flows not hydraulically connected
28. Moraine
29. Mountain crests
30. Mountain flanks
31. Mountain slopes
32. Ogallala
33. Outwash
34. Outwash over bedded sedimentary rock
35. Outwash over solution limestone
36. Playa lakes
37. Resistant ridges
38. River alluvium
39. River alluvium with overbank deposit
40. River alluvium without overbank deposit
41. Sand dunes
42. Solution limestone
43. Swamp
44. Thin till over bedded sedimentary
45. Triassic basins
46. Unconsolidated and semiconsolidated shallow surficial aquifer
47. Wide alluvial valleys (external drainage)
48. Volcanic uplands

Source: United States Environmental Protection Agency (USEPA), *DRASTIC: A Standardized System for Evaluating Ground Water Pollution Potential Using Hydrogeologic Settings*, EPA/600/2-85/018, Washington, DC, 1985, 163.

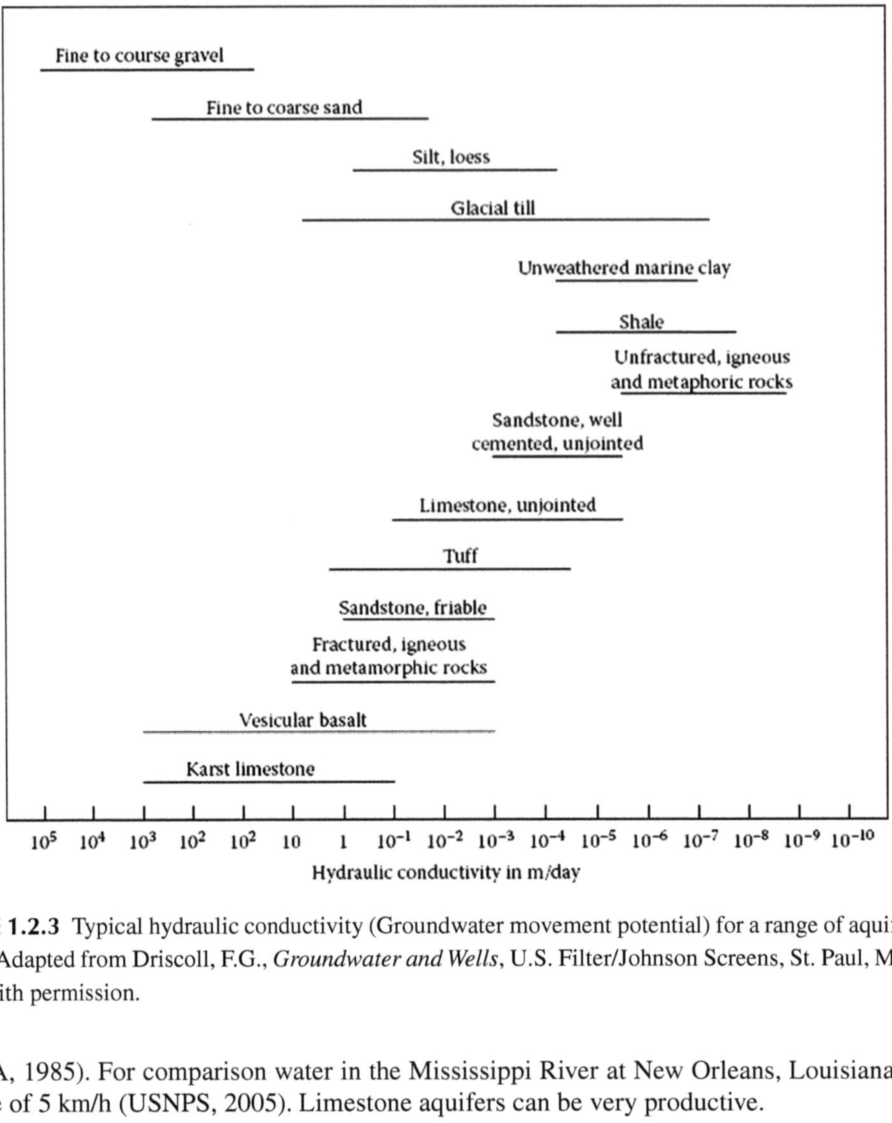

FIGURE 1.2.3 Typical hydraulic conductivity (Groundwater movement potential) for a range of aquifer types. Source: Adapted from Driscoll, F.G., *Groundwater and Wells*, U.S. Filter/Johnson Screens, St. Paul, MN, 1986, 1089. With permission.

(USEPA, 1985). For comparison water in the Mississippi River at New Orleans, Louisiana, travels at a rate of 5 km/h (USNPS, 2005). Limestone aquifers can be very productive.

Typically, groundwater moves slowly and each particle in the ground becomes an obstacle to movement in a straight line. This is the reason for groundwater contamination typically being localized around the source of contamination, in contrast to traveling tens or hundreds of kilometers in a matter of days or weeks in a river.

Subsurface water migration in an economic context means that groundwater can move from place to place, crossing political or property boundaries, carrying natural or anthropogenic (human-caused) contaminants in the process. Locations in which groundwater moves at faster rates are places where the resource can be produced more easily, minimizing the financial costs of withdrawal, which probably do not include the ecosystem costs of that production. Over long periods of time, groundwater is able to travel in the subsurface very substantial distances (Jorgensen et al., 1993; also see Box 1.2.3)

Recharge Zones

Locations that allow water to move with relative ease from the ground surface to the water table are called "recharge zones." Protecting these zones is important in areas of groundwater use because

FIGURE 1.2.4 Sinkhole collapse near Ocala, Florida. *Source*: U.S. Geological Survey.

they are the principal locations where water enters the subsurface to supply the aquifer. Rates of recharge are highly dependent on precipitation rates and the geologic matrix. In arid alluvial basins bordered by mountains, recharge rates may be no more than 0.03–30 mm/year (0.001–1.2 inches/yr). In very humid regions, such as Pacific volcanic islands overlain with alluvium, recharge rates may be as high as 1000 mm/year (40 inches/yr) (USEPA, 1985).

Discharge Zones

Areas through which groundwater moves from the subsurface environment to the surface are referred to as "discharge zones." These discharge zones can be stream channels, lake and ocean beds, springs and wetlands. In some cases, GWD to a stream may constitute most of its flow (USGS, 1999b).

Productivity or Yield

The rate at which groundwater can be pumped from an aquifer or discharged through a spring on a continuing basis is the yield of an aquifer. Bedrock aquifers have varying productivities or yields depending on the grain size, space between the grains and fractures (also referred to as secondary porosity). For example, granite with minimal fracturing and little pore space may provide almost no water. Highly fractured sandstone with pore space between sand grains is very likely to be a good source of water if it is saturated. Sand and gravel aquifers that may have porosities of 20% or more are highly productive. These may exist along stream valleys, glacial outwash areas and the base of mountainous regions. Groundwater can easily move through these geologic materials, providing reliable, economical water supply. Yields may range from 0.01 to 0.02 m^3/min (2.5–4.0 gallons/min) in the shale aquifers of Ohio to 4–76 m^3/min (1,000–20,000 gallons/min) in the limestone and dolomite aquifers of Florida (USGS, 1975, 1984).

Depth

Depth of the water table, the surface of the saturated zone at the top of an aquifer, is an important element in determining the accessibility of an aquifer for groundwater use. Depth of the water table determines a portion of the cost of drilling and installing a well. Depth also affects the amount and therefore, the cost of energy to lift the water out of the ground by pump. Depths may range from 1 m to over 2000 m (3–6,560 ft) (USGS, 1984).

Managed Yield and Sustainability

The concept of "managed yield" of an aquifer is important in the supply of groundwater and applies to unconfined aquifers, typically, the first aquifer below the ground surface, but may also apply to deeper aquifers. Managed yield is the production capacity of an aquifer with a margin of safety to maintain aquifer levels and quality for critical receptors such as minimum well depths or survival of endangered species (Meyland, 2011). Sustainability related to sustainable development of the resource, focuses on improvement in the quality of use, provides the benefit to this and future generations of a long-term water source where managed yield is practiced. The managed yield approach ensures that a range of future groundwater uses are preserved while allowing present uses to occur. Preserving future use while continuing present use is accomplished through a consciously applied, objective-driven decision-making framework of policies designed to supply current needs and preserve the services of natural capital providing for critical receptors. In an example of such a framework, groundwater pumping may be guided by maintaining a volume of withdrawal to meet current needs that enables continuance of stream flow from GWD for aquatic life survival or downgradient springs to flow.

Surface Water and Groundwater Interaction

In locations where streams flow over sand and gravel or fractured bedrock, water from the streams may move into the saturated zone of the ground, recharging the aquifer and later back to the stream depending on the water table levels and hydraulic gradients from place to place along the stream. The nearby wells in this same lithology may typically be shallow and may be drawing water that was previously in the stream. In these situations, the stream is the source water for the wells, and the aquifer acts as a conduit, with the streambed filtering out large sediment particles carried by the stream.

Wetlands

Wetlands are lands in which water saturation is the dominant factor determining the nature of the soil development and the types of plant and animal communities living in the soil and on its surface (USEPA, 2005b). Groundwater can facilitate the movement of nutrients in wetlands and influence the local microclimate by supporting plant transpiration, thereby sustaining unique flora and fauna (Harvey et al., 2007b). "Wetlands can receive groundwater inflow, recharge groundwater, or do both." (USGS, 1999b) Many wetlands are places of GWD that maintains the wetlands when no rainfall occurs and, consequently, no overland flow and runoff are generated.

OTHER FACTORS AFFECTING GROUNDWATER SUPPLY

Several other factors make groundwater supply a complex issue. While wells can be costly to install in many lithologies, harder bedrock and boulders in sediments and glacial drift make drilling difficult and more expensive. Larger diameter, higher capacity wells are more expensive to drill and install. Larger pumps are more expensive. If water is to be supplied for human consumption, it may need treatment if the well supplies a public water system. Even water from private wells may need treatment for iron, calcium, or other minerals and compounds in groundwater. Water softeners can remove minerals, while carbon filters can remove other chemical constituents, either natural or man-made, from the water. These factors are covered in subsequent chapters.

GROUNDWATER AVAILABILITY

In terms of extent, the groundwater resource is vast, the largest source of freshwater, but less than 1% of all the water in the ecosystem. The earth contains 10,523 trillion m^3 (2,780,000 trillion gallons) of groundwater (NGWA, 2020). In most of the world, groundwater is a ubiquitous resource (USGS, 1999), but access to it may be limited by depth or property rights (and technical and legal factors). In most locations, groundwater is accessible through wells. The cost to install a well takes into account the type of geology – for example, sand and gravel, fractured bedrock, or other geologic setting – and the depth to the water table and the diameter of the well, considering the volume of water needed at a site.

In more humid areas, groundwater typically may be easily accessible by shallower wells, because precipitation fills up much of the pore space closer to the ground surface. In drier climates, groundwater may only be available from deeper wells. However, this is not universally true and depends on the regional geology (USGS, 1984).

The use of groundwater has affected its availability. Importantly, the amount of groundwater in the world is reasonably fixed and with aquifers being depleted on all the continents (Danielopol et al., 2003, p.111), less groundwater is available now. In the midwestern United States, the water table of the vast Ogallala Aquifer has dropped over 30 m in some locations owing to irrigation pumping; in the northcentral United States around Chicago by 274 m (900 ft) owing to pumping for water supply; and in the southwestern United States in many locations from 15 to over 30 m (50 to over 100 ft), primarily because so much water has been pumped out for irrigation, reducing the nearby streamflows where they are linked to groundwater (USGS, 2003b). While some of the irrigated water might return to the aquifer through subsurface infiltration, most is used by plants and then transpired or evaporated from the soil zone. All these factors affect where and how groundwater is used. Subsequently, groundwater becomes more of a "stock" resource, with real limits to its use in the indefinite future.

Impermeable area, that is, surfaces which do not allow water from precipitation to infiltrate the ground surface and ultimately percolate to the water table, affects groundwater availability in the ecosystem. As much of the world's population live in rural areas that produce crops for food and the expansion of urban areas is typically into the rural areas; much of the expanding impermeable area is encroaching into former agricultural zones. In the United States, an estimated 16 million hectares (40 million acres) have been converted to roads and parking areas, at the rate of 0.07 ha/vehicle (0.17 acre/vehicle). Elsewhere in the world where the population density is higher, such as in Europe and Asia, the rate is less, but the conversion is taking place (Brown, 2003). Additionally, the area in buildings, houses and other structures is also impermeable. These areas cause water to runoff faster, resulting in less groundwater recharge, and, therefore, reduce groundwater availability in the ecosystem.

Relative to the GWD being available to support streams and other surface waters, groundwater migration and travel through the subsurface affect the delivery of water to streams. Soil infiltration of precipitation and percolation to the water table delays the effect of water transmission to streams. The result of subsurface movement of this water on streamflow is to defer and extend the delivery of this water to the stream, over time (Jha et al., 2003, p.17). This soil infiltration and percolation of water to aquifers reduces runoff and flooding potential.

The availability of groundwater for the nonhuman part of the ecosystem is of increasing interest. As groundwater is a significant source of water for rivers and wetlands, wildlife depends on it extensively. As noted previously, groundwater is the habitat for fauna that requires this environment.

Furthermore, many of these fauna migrate to surface waters for a part of their lives (Gibert et al., 1994a). Thus, availability and access to groundwater is essential for these species. However, the full relationship of this biota to the larger ecosystem is not well understood (Gibert et al., 1994a, p. 22).

GROUNDWATER QUALITY

Groundwater quality also has a significant effect on where and how groundwater is potentially used. Aquifers that contain only slightly mineralized water will readily be tapped for domestic and municipal uses. In some areas, where groundwater is from ancient deposits, brines and even radioactivity may be observed in it. Increasing depth in a geologic formation is observed to increase the temperature of the subsurface [0.6°C] per 30 m (100 ft). Subsequently, with greater depth, the solubility of many minerals rises, resulting in more mineralized groundwater (Driscoll, 1986, p. 89). In these cases, the groundwater may not be useable, or at least may require treatment. If more recent freshwater is situated over such brines and production is increased to remove (or mine) the newer water, then the brines or other mineralized water may move toward the wells, reducing its quality for certain uses over time. These changed conditions may demand treatment of the water or finding a new water source, both of which have costs associated with them. Similarly, human-caused contamination can also have the same result in affecting a well that supplies drinking water. Interestingly, brines can have beneficial uses. For example, subsurface brines provide the basis for a chemical industry and are still produced for that purpose (also see Chapter 3). More recently, brines containing lithium carbonate provide an economical supply of lithium to support battery production (Deberitz, 2006).

Natural groundwater quality is a major factor in the supply of groundwater. Table 1.2.2 identifies the common dissolved inorganic minerals in groundwater, but their presence depends on the chemical composition of the rock with which the groundwater is associated. Groundwater low in dissolved solids, mineral content and nitrate and without harmful concentrations of other chemicals

TABLE 1.2.2
Principal Naturally Occurring Dissolved Inorganic Constituents in Groundwater (Categorized by "Typical" Concentrations)

Major Constituents (Greater than 5 mg/L)	Trace Constituents (Less than 0.1 mg/L)		
Bicarbonate	Aluminum	Indium	Silver
Calcium	Antimony	Iodide	Thallium
Chloride	Arsenic	Lanthanum	Thorium
Magnesium	Barium	Lead	Tin
Silicon	Beryllium	Lithium	Titanium
Sodium	Bismuth	Manganese	Tungsten
Sulfate	Bromide	Molybdenum	Uranium
	Cadmium	Nickel	Vanadium
Minor Constituents (0.01–10.0 mg/L)	Cerium	Niobium	Ytterbium
Boron	Cesium	Phosphate	Yttrium
Carbonate	Chromium	Platinum	Zinc
Fluoride	Cobalt	Radium	
Iron	Copper	Rubidium	
Nitrate	Gallium	Ruthenium	
Potassium	Germanium	Scandium	
Strontium	Gold	Selenium	

Source: Driscoll, F.G., *Groundwater and Wells*, U.S. Filter/Johnson Screens, St. Paul, MN, 1986, 97.

may be deemed safe for use as a drinking water source. In the western United States and now increasingly in the eastern United States, groundwater is used for irrigation, which cannot support agricultural productivity if too brackish. Groundwaters with high dissolved-solids concentration may be too brackish for irrigation; however, where higher quality is scarce, it may be used for cattle watering. Extremely brackish water (total dissolved solids in excess of 10,000 ppm) has very few uses. Brackish groundwater is also found in association with oil and in areas of excessive pumping of groundwater along the coasts, allowing intrusion of ocean's salt water into what was previously a freshwater aquifer.

In major coastal cities that have relied on groundwater for water supply, groundwater production has been so substantial that saltwater intrusion has occurred (Driscoll, 1986, p. 773). This is a worldwide problem (Bear et al., 1999, cited in USGS, 2005a). Saltwater is defined as water having a total dissolved-solids concentration greater than 1000 milligrams per L (mg/L). Seawater has a total dissolved-solids concentration of about 35,000 mg/L, of which dissolved chloride is the largest component (about 19,000 mg/L). Concentrations of chloride in fresh groundwater along the Atlantic coast [of the United States] are typically less than about 20 mg/L [the drinking water standard], so there is a large contrast in chloride concentrations between freshwater and saltwater." (USGS, 2005a) In situations of high levels of saltwater in groundwater, cities need to seek alternative water sources, including other wells outside the zone of saltwater intrusion.

The United States Geological Survey documented the background concentrations of certain naturally occurring minerals in groundwater across the United States. Some of these minerals have human health effects of concern when they are consumed at higher concentrations. These groundwaters require treatment for safe human consumption. This subject is addressed further in Volume 1, Chapter 8. The USGS found that natural background levels of nitrate, ammonium and phosphorus in groundwater were less than 2 mg/L, less than 0.1 mg/L and 0.1 mg/L, respectively. Table 1.2.3 gives a perspective of nitrate concentrations in different settings.

Other natural minerals listed in Table 1.2.2 that occur in groundwater and may be harmful to human health include arsenic, asbestos, barium, cadmium, chromium, copper, fluoride, lead, mercury, nitrate, radium, selenium and uranium (USGS, 1998a, 1998b, 2000; WHO, 2004; USEPA, 2005a). As noted previously, these elements are found in the earth's crust and dissolved by water (USEPA, 2000, 2005a; WHO, 2004). Considering minerals of recent concern, higher levels of arsenic tend to be found in groundwater than in surface water (USEPA, 2005a). The health-based standard for arsenic set by the World Health Organization (WHO, 2005), the European Union and the United States is 0.010 mg/L. In the United States, locations in western states and parts of the midwestern and northeastern states have arsenic concentrations higher than the health-based standard, but the majority of the locations have concentrations below the standard in their water (USEPA, 2005a).

TABLE 1.2.3
Relative Occurrence of Nitrate in Different Settings

Settings with Lower NO_3 Concentrations	Settings with Higher NO_3 Concentrations
High rainfall, snowmelt and/or excessive irrigation	Low rainfall, snowmelt and/or irrigation
Well-drained and permeable soils underlain by sand and gravel or karst	Poorly drained soils
Areas where crop management practices slow runoff and allow more time for water to infiltrate into the ground	Cemented sandstones and crystalline rock
Low organic matter content and high dissolved oxygen that minimize NO_3 transformation to other forms	Urban, forest and range lands

Sources: United States Geological Survey (USGS), Nutrients in the nation's waters, too much of a good thing? *Circular* 1136, 1996, 16; United States Geological Survey (USGS), The quality of our nation's waters – Nutrients and pesticides, *Circular* 1225, 1999, 13.

However, locations in India, Bangladesh, Vietnam, Taiwan and Argentina have arsenic concentrations in groundwater up to 50 times the standard (EU, 2005). Health effects of high arsenic levels include skin damage or problems with circulatory systems and increased risk of cancer (USEPA, 2005c). Health-based standards for radium and uranium are 5 pCi/L and 0.030 mg/L, respectively, and health effects of high levels include increased risk of cancer.

Increasingly, the presence of man-made chemicals is observed in the ecosystem. As the U.S. Geological Survey notes, "for nearly every type of water use, water has increased concentrations of dissolved constituents" (USGS, 1999a, p. 77). The U.S. Environmental Protection Agency has recorded over 86,000 chemicals in its Toxic Chemical Substance Inventory (USEPA, 2019). As discussed in the following paragraph, with groundwater in the first or shallow aquifers vulnerable to chemical or microbial residuals applied on the ground surface or under it, the application of nutrients and pesticides has been widespread and has resulted in their presence in groundwaters around the United States (USGS, 1999a, 1999b) and worldwide (Sampat, 2000; UNESCO, 2003). The U.S. Geological Survey conducted 36 studies in 21 large river/aquifer systems in the United States during the 1990s and found nutrients and pesticides in shallow and deep groundwater, as described in Box 1.2.2 (USGS, 1999a). One conclusion based on current standards is that aquatic life may be at a greater risk from pesticides than human health.

BOX 1.2.2 FINDINGS OF NUTRIENTS AND PESTICIDES IN GROUNDWATERS OF THE UNITED STATES

Main findings of a study of groundwater quality related to nutrients and pesticides in 21 river basins in the United States include:

- Fifty-three percent of the shallow groundwater studies in agricultural and urban areas had moderate nitrate concentrations above background levels.
- Three of thirty-three major, generally deeper, aquifers – often a source of drinking water – had elevated nitrate concentrations.
- Nutrient concentrations vary seasonally because of precipitation and chemical application patterns.
- More than 50% of the wells in the agricultural areas had detectable herbicides above the drinking water standards. Atrazine, its breakdown product diethylatrazine, metochlor, prometon and simazine were the main herbicides detected.
- Insecticides were detected in less than 10% of the wells, but more wells had insecticides exceeding the health standard. Dieldrin and diazinon were the principal insecticides found.
- Deeper wells in major aquifers had substantially lower frequencies of herbicides and none exceeded the health standards.
- Contaminants in major aquifers are more likely to occur in vulnerable settings that allow rapid migration of groundwater from the shallow to the deeper aquifer.
- Groundwater can be a major nonpoint source of nutrients and pesticides to streams.
- Aquatic life may be at a greater risk from pesticides than human health. While no United States criteria existed at the time of the studies of major herbicides, 17 out of 40 streams exceeded the Canadian guidelines in the farming areas studied. Similarly, insecticides exceeded aquatic life guidelines in 18 of 40 streams examined in agricultural lands.

Source: United States Geological Survey (USGS), The Quality of our Nation's Waters – Nutrients and Pesticides, Circular 1225, Reston, VA, 1999a, 82.

Fuel and heating oil stored in above- and underground tanks, and the use and disposal of household products disposed in septic systems that are or contain volatile organic compounds (VOC), are additional sources of human-produced contaminants in the ecosystem. The U.S. Geological Survey evaluated for 55 VOC in nearly 1926 domestic wells in 39 states from 1986 to 1999 (USGS, 2002). It found at least one VOC in 12% of the samples. The solvents were detected most frequently in 4.6% of the samples. Mixtures of VOCs (more than one) were found in half of the samples tested. Only 1.4% of the samples exceeded drinking water standards or health criteria.

More recently, a group of man-made chemicals, per- and poly-fluoroalkyl substances (PFAS), used in firefighting, stain resistance, water repellants, airport and other industrial applications, have been found in groundwater. These chemicals bioaccumulate in the environment and in people and have been associated with liver and kidney damage, certain cancers and other health effects (NGWA, 2017).

Generally, groundwater quality in the ecosystem is affected by the geologic conditions in which groundwater resides, by human withdrawal – particularly, along coastal zones and by residuals from wastes and products.

GROUNDWATER FLOW

Local/On-Site Flow

Water in the ground must first flow through the soil zone and unsaturated zone (also called the "vadose zone" with the space between the subsurface particles being occupied by both water and air) below it. The type of soil – sandy or clayey – will affect the extent of precipitation – which may be more or less, respectively – that enters the ground, as well as its rate of travel through the soil zone. The water can carry nutrients and contaminants at or near the land surface into the ground as it successively fills lower pore spaces between the soil particles and migrates downward toward the water table. In coarse soils made up of sand and silt, the larger pore spaces between the sand grains may also contain silt which can slow the water movement, depending on the percent of silt.

Water in the saturated zone may flow at very different rates depending on the type of aquifer material it is being transmitted through. These conditions are described for selected aquifer types in Figure 1.2.3. The important point of focusing on groundwater flow is that beyond the local flow system, groundwater can travel great distances, first in the watershed and then beyond the watershed within the larger region.

Watershed Flow

Groundwater moves within a watershed, a typical management unit for water. A watershed is the area within a ridge of high land that drains into a stream or other water body at a particular point, usually at the intersection with a larger stream or the ocean. Watersheds vary in size depending on the scale of interest. Within a watershed, groundwater typically moves toward the points of discharge, which are streams and wetlands as well as springs and wells. If the elevation of the groundwater table is above the stream or other surface discharge point, then the hydraulic head of the groundwater will "push" it to a location of lower hydraulic head, such as the stream. A stream receiving groundwater as part of its baseflow (i.e., the ongoing movement of groundwater through a streambed to the stream, an important source of water for streams between precipitation events) is said to be a "gaining stream." If the level of groundwater is below a stream or other water body, the stream will recharge the groundwater within the watershed. The stream is then referred to as a "losing stream." In a watershed, the interaction between groundwater and surface water may be such that some portions of the stream are "gaining" and others are "losing." Watershed flow may be measured in months, years, or decades, depending on the hydraulic head and the geologic matrix.

Regional Flow

As the subsurface geology can be highly variable and does not necessarily match the other surficial features, groundwater can actually flow from one watershed to another. This condition is typical of deeper groundwater flow. Thus, the quantity and quality of groundwater at a site tens of kilometers away may influence the flow of a stream or its chemical constituents in an adjacent watershed. Regional groundwater flow from one watershed to another may take place over centuries, owing to the distance and complexities of the subsurface (USGS, 1999a, 1999b, p. 5). Box 1.2.3 describes long-distance regional flow.

BOX 1.2.3 GROUNDWATER FLOW – SHORT AND LONG

Groundwater flow in geologic formations, such as unweathered clay, shale and unfractured igneous rock may only be on the order of 10 millionths of a meter per day, almost imperceptible movement. On the other hand, groundwater in karst limestone can travel several kilometers per day, measured as what is referred to as "hydraulic conductivity" (Driscoll, 1986).

The flow system of the confined groundwater system of the Northern Great Plains of the United States extends more than 966 km (600 miles) from recharge areas in the mountains of Montana, Wyoming and South Dakota to discharge areas in the eastern Dakotas and the Canadian Province of Manitoba (Downey and Dinwiddie, 1988). The regional confined aquifers of the central midwestern United States extend on the order of 1287 km (800 miles) (Jorgensen et al., 1993).

The Great Artesian Basin in Australia is one of the largest groundwater basins in the world. It was formed from interbedded sand and clay sediments, with the sandy sediments forming permeable sandstone aquifers between the clay zones. The basin covers an area of 1.7 million km^2 (656,370 square miles) and its longest groundwater flow path is on the order of 1800 km (1,120 miles). Age dating indicates that some of the Basin's groundwater may be 2 million years old (QDNRMW, 2006).

Sources: Downey, J.S. and Dinwiddie, G.A. The Regional Aquifer System Underlying the Northern Great Plains in Parts of Montana, North Dakota, South Dakota, and Wyoming – Summary, U.S. Geological Survey Professional Paper 1402-A, 64, 1988.

Driscoll, F.G., *Groundwater and Wells*, Johnson Screens, St. Paul, MN, 1986, 1089.

Jorgensen, D.G., Helgensen, J.O., and Imes, J.L. 1993. Regional Aquifers in Kansas, Nebraska, and Parts of Arkansas, Colorado, Missouri, New Mexico, Oklahoma, South Dakota, Texas, and Wyoming Geohydrologic Framework. Regional Aquifer-System Analysis-Central Midwest. US Geological Survey Professional Paper 1414-B. URL: https://pubs.usgs.gov/pp/1414b/report.pdf (accessed March 1, 2019).

Queensland Department of Natural Resources, Mining and Water (QDNRMW), Facts; Water Series; The Great Artesian Basin, 2006, URL: http://www.nrm.qld.gov.au/factsheets/pdf/water/w68.pdf (accessed May 14, 2006).

NATURAL SENSITIVITY AND VULNERABILITY

Hydrogeologists have developed systematic approaches to identify the natural sensitivity of areas relative to their potential to allow contamination. Factors include thickness of the unsaturated zone, the existence of a confining unit (or "aquitard"), permeability of an aquifer's geologic matrix, recharge rate of the aquifer and location within the flow system (proximity to recharge and discharge zones) (NRC, 1993). Groundwater sensitivity is defined as "the relative ease with which a contaminant applied on or near the land surface can migrate to an aquifer" and is a function of the intrinsic characteristics of an aquifer (USEPA, 1993, p. 128). Many locations around the United States and around the world are naturally sensitive to contamination.

REGIONAL SENSITIVITY

Groundwater is known to exist in many types of geologic environments: unconsolidated sand and gravel and fractured bedrock, shallow or deep, with thin or thick soil layers above it, as well as confined and unconfined. All these and other factors can contribute to the sensitivity of groundwater to contamination. A study of aquifer vulnerability to disposal of liquid wastes at or just below the ground surface found that 42% of the continental United States (exclusive of Alaska, Hawaii and territories) has groundwaters that exist in a geologic environment that is highly susceptible to contamination and another 16% that are moderately vulnerable (Pettyjohn et al., 1991). Figure 1.2.5 indicates the locations of increasing risk of groundwater contamination across the United States relative to nitrogen input to land typically with well-drained soils and less-extensive woodland (USGS, 1988). About 40% of the European Union area (in 2001) has "highly sensitive soils" relative to nitrate leaching (Notenboom, 2001, p. 253). The implication is that users of chemicals and fertilizers, and disposers of waste on or in the ground, must be very careful about chemical and waste management. The costs of carelessness are high. The costs to ensure that the proper steps are taken in areas of sensitivity are much less than those to remediate the contamination later, especially if it has spread, as will be discussed later.

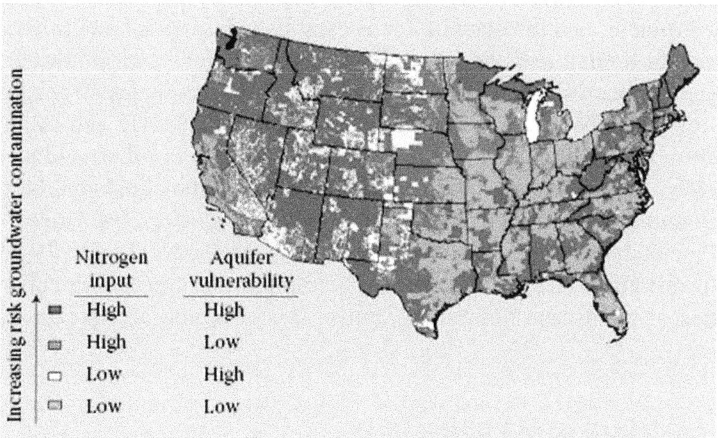

FIGURE 1.2.5 Increasing risk of contamination. Areas on the map below have the highest risk of nitrate contamination of surficial aquifers that receive high nitrogen input, have well-drained soils and have less-extensive woodland. Source: USGS, 1988. A National Look at Nitrate Contamination of Groundwater. Reston, VA.

Local Sensitivity and Vulnerability

The sensitivity of groundwater in a particular area, such as the immediate vicinity of a wellfield, is influenced by many factors. Conditions of large areas cannot always be generalized to specific locations, as circumstances change over time and space. Furthermore, specific human activities related to chemical use increase the risk of contamination, making a particular location more vulnerable. Groundwater vulnerability is defined as "the relative ease with which a contaminant applied at or near the land surface can migrate to the aquifer or wellhead of interest under a given set of contaminant source management practices, well construction characteristics, and groundwater sensitivity conditions." (Wireman and Job, 1997) These factors include the sensitivity of the ground surface (or soil zone) and subsurface environment (local geology) to contaminant migration, land uses in the area and the condition of the well(s). These specific factors that affect the local vulnerability can include (Jorgenson et al., 1998):

1. Results of past monitoring – Previous contaminant occurrence may indicate vulnerability of the groundwater resource.
2. Contaminant source risk – The existence (past or present) of a contaminant source may affect the chance of contaminant occurrence in a local groundwater supply. Furthermore, proper management of the potential contaminant sources, such as through a wellhead protection program, can also affect this risk.
3. Groundwater sensitivity – As noted earlier, the ability of the local hydrogeology to prevent the movement of contaminants through the soil or vadose zone is a critical factor in contaminant occurrence and may provide a barrier to or, conversely, a ready pathway for contaminants to reach the aquifer.
4. Well construction and condition – An inadequately constructed or deteriorated well provides pathways for contamination to enter the subsurface and contaminate groundwater.

The economic consequences of past contaminant occurrence, high risk from contaminant sources, high sensitivity to groundwater movement and poor well construction or deterioration may include lack or loss of a water supply, the need to develop a new water supply, reduction in the development opportunities for otherwise usable land and destruction of an important habitat.

Notably, the existence of unique habitats and natural sensitivities of groundwater and lands associated with them have resulted in protection of aquifers and their related land areas. In south central Texas, the Edwards Aquifer has been designated as a sole source aquifer by the United States federal government, and the state of Texas established a special administrative authority to oversee the balance of human and endangered species needs for its groundwater, with 14 threatened and endangered species identified. Box 1.2.4 lists these species in the Edwards Aquifer. Similar findings of groundwater dependence for endangered species and unique habitats relative to both quantity and quality of groundwater can be found in other studies as well (Harvey et al., 2007a, 2007b). To protect the habitat of the New Jersey pinelands on the sand plain above an extensive sand and gravel aquifer (Kirkwood-Cohansey aquifer), the United States Congress established the Pinelands National Reserve in 1978. This area covers over 404,690 ha (1 million acres) of farms, forest and wetlands, including 56 communities and ecological settings with over a thousand species of plants and animals – almost 100 of which are threatened or endangered (USNPS, 2005).

Human Activity and Ecosystem Vulnerability

Factors that make groundwater easily accessible and productive as a source of water supply also make it vulnerable to contamination from human activity. Historically, people settled in places that

BOX 1.2.4 THREATENED AND ENDANGERED SPECIES OF THE EDWARDS AQUIFER, TEXAS

All the forms of life living in the Edwards Aquifer have adapted to this existence by being colorless and eyeless. They often have other senses that are highly developed. These species rely on the water levels in the aquifer being maintained to support discharge to the springs and their substrate or spring runs in which they live. All the information currently available indicates that the subterranean aquatic ecosystem of the aquifer is the most diverse groundwater ecosystem in the world.

The threatened and endangered species of the aquifer are:

1. The Texas Blind Salamander, Typhlomolge rathbuni. Listed by both the state of Texas and the federal government as endangered.
2. The Blanco Blind Salamander, Typhlomolge robusta. Listed by the state of Texas as endangered.
3. The Cascade Caverns Salamander, Eurycea latitans. Listed by the state of Texas as threatened.
4. The Comal Blind Salamander, Eurycea tridentifera. Listed by the state of Texas as threatened.
5. The Toothless Blindcatfish, Trogloglanis pattersoni. Listed by the state of Texas as threatened.
6. The Widemouth Blindcatfish, Satan eurystomus. Listed by the state of Texas as threatened.
7. The Comal Springs Dryopid Beetle, Stygoparnus comalensisis. Listed as endangered by the federal government.
8. Peck's Cave Amphipod, Stygobromus pecki. Listed by both the state of Texas and the federal government as endangered.
9. The Fountain Darter, Etheostoma fonticola. Listed by both the state of Texas and the federal government as endangered.
10. The San Marcos Gambusia, Gambusia georgei. Listed by both the state of Texas and the federal government as endangered. It is possibly extinct.
11. San Marcos Salamander, Eurycea nana. Listed by both the state of Texas and the federal government as threatened.
12. Texas Wildrice, Zizania texana. Listed by both the state of Texas and the federal government as endangered.
13. The Barton Springs Salamander, Eurycea sosorum. Listed by the federal government as endangered.
14. The Comal Springs Riffle Beetle, Heterelmis comalensis. Listed by the federal government as endangered.

Source: Edwards Aquifer Research and Data Center (EARDC). 2019. Threatened and Endangered Species in the Edwards Aquifer System. URL: https://www.eardc.txstate.edu/Aquifer-Info/endangered.html#top (Accessed April 25, 2020).

had many natural amenities: fertile soil for crops, easy access for transportation and a reliable water supply. Shallow depth to groundwater allowed hand-dug wells. Places where water could be produced easily and in large quantities attracted larger populations. Furthermore, locations that allowed groundwater to be replenished readily – usually in sand and/or gravel or river valley alluvium – or large deep aquifers, also attracted industry. Human activity that generates wastes and excess residuals (such as from the use of chemicals that are not entirely used by plants or degraded promptly in the environment) may pose vulnerabilities to groundwater, especially when concentrated on or under the land in application areas or subsurface zones.

Industries use water for cleaning and cooling processes and to carry wastes away. Liquid wastes may be sent back into the ground via disposal wells or seepage lagoons. Industries also use solvents, petroleum products and other chemicals. Inadvertently, these chemicals may be spilled. Historically, when waste disposal or spillage occurred, business managers believed that the ground would cleanse whatever went into it. Many of these locations, because of their hydrogeology, allowed contaminants to be carried by groundwater easily into the subsurface. Until the mid-twentieth century, science did not suggest that this circumstance was problematic. As populations grew and more industries went to locations where the water was available, more waste disposal and chemical spills occurred in densely populated and used areas. Enough time elapsed that past practices that were once acceptable finally started causing serious contamination problems. For example, New Jersey is one of the early industrialized states in the United States with extensively developed industry and substantial waste disposal in areas of sand and gravel aquifers that local populations used for drinking water. New Jersey has more abandoned uncontrolled hazardous waste sites than any other of the United States, while only being 46th in size. While we understand how to control many of the wastes and handle the chemicals today, past practices have shown that we must be careful about considering the location of the industry or any activity (e.g., use of lawn pesticides) in the immediate vicinity of wells and also in the larger subsurface zone that contributes groundwater to the wells.

Another example is the use of nitrate as a fertilizer. Nitrate is soluble in water and is easily transported by groundwater into the subsurface. Nitrate is associated with many human activities, domestic waste disposal and septic systems, manure disposal, fertilizer applications on farms and residential property and golf courses. As it easily moves with groundwater and is relatively inexpensive to test its presence, nitrate is often used as an indicator of human activity affecting groundwater quality when its concentration exceeds about 3 mg/L, considered as the natural background concentration in much of the United States. As noted in Table 1.2.3, the U.S. Geological Survey found higher concentrations of nitrate in the agricultural areas and lesser concentrations in urban and forested areas (USGS, 1996).

Other chemicals that are now recognized as indicators of impacted groundwater quality include trichloroethane (TCE) and tetrachloroethene (PCE), which are industrial solvents indicating industrial activity and the potential for other industrial chemicals to be present (USEPA, 2005c). The presence of atrazine, the most widely used pesticide, may suggest that other pesticides from past agricultural operations are in the groundwater at a particular location. The occurrence of chlorofluorohydrocarbons (CFCs, such as freon) with other contaminants, such as pesticides, can be used to age-date groundwater, as it was first used in the 1940s and its use steadily increased from that time. The CFC can be used as an indicator of younger water, providing evidence of precipitation infiltrating near the sampling point in recent times (USGS, 2004).

Significantly, the demands of our human population on groundwater, specifically, and the ecosystem, more broadly, are considerable. From 1950 to 2000, significant demands have been placed on the fixed set of resources that can be obtained only from the ecosystem. Specifically, during those 50 years (WI, 2005)

- World population grew from 2.56 to 6.08 billion, nearly 2½ times
- Life expectancy went up from 46.3 to 64.7 years, an increase of 28%

From 1961 to 2000 (WI, 2005)

- Irrigated land expanded from 139.1 to 272.5 million ha, nearly doubling
- Fertilizer use grew almost 4½ times from 28 to 122 million metric tons

All these and many other factors contribute to the human demands on groundwater in the ecosystem, while this resource has declined because of depletion of aquifers worldwide.

BIODIVERSITY IN GROUNDWATER

As groundwater environments are at least as complex as surface water environments, they may have a similarly complex biological realm. Groundwater ecosystems typically rely on a flow of water and matter. Since the early 1990s, more research has been undertaken to understand the groundwater component of the ecosystem (Stanford and Simons, 1992), which, previously, had been mostly overlooked and disregarded (Gibert et al., 1994b, p. 22), except for concerns about microorganisms that could cause acute illness if they occurred in the groundwater used for drinking. Notably,

> [B]iological distributions and their controls in groundwater are likely to be as diverse and complex as those in surface waters. Just as in surface waters, the biological communities of groundwaters are highly varied, ranging from rich communities with dense populations of bacteria, fungi, protozoans and hundreds of species of invertebrates to sparse communities of almost inactive bacteria. The distribution of the groundwater biota is certainly controlled by historical factors, physicochemical variables, biological interactions and interactions among these broad classes of elements ... though, we still know little about the extent of biological distributions in groundwater and the specific factors that control these distributions.
>
> (Strayer, 1994, pp. 287–288)

> It has been estimated that the actual number of extant epigean species *[organisms still existing and living on the ground]* in the tropics is 10–20 times greater than the number currently known to science. There is no denying that tropical epigean organisms are poorly known, but it is likely that on a global scale the groundwater biota is virtually unknown. It appears that groundwater biotopes [*environments*] contain highly diverse faunas with high levels of local and regional endemism [*organisms native to those* areas]. [M]any truly epigean species colonize, or even depend upon, near-surface hypogean *[underground]* waters, especially when immature, which dramatically increases groundwater biodiversity.
>
> (Gibert et al., 1994b, p. 22)

We will consider the range of groundwater environments in the ecosystem, but only in limited ways, as so much about them is still unknown, particularly their larger function in the ecosystem. Specifically, the part of the ecosystem that comprises groundwater, covered in this chapter, includes (1) shallow, near-surface, (2) karst, (3) first aquifer, (4) littoral zones and (5) deep aquifer environments. In addition, a note is also included on biodegradation.

Ground and Near-Surface

A vast array of flora and fauna rely on shallow, near-surface groundwater conditions. These locations include wetlands, marshes and swamps, as well as many riparian areas in which tree and plant roots reach down to the water table. The total number of species and their function in wetlands are still being intensively examined (USGS, 2005b). Wetlands serve to remove nutrients and pollutants, provide spawning areas for fish and other wildlife, are wintering areas for migratory birds and moderate coastal wave action and on-land flooding (USGS, 1997). Groundwater supporting spring-fed streams provides important minerals and nutrients as well as temperature control to promote healthy aquatic life in riverine environments (Born et al., 1997, pp. 42, 75, 115). Groundwater quality and flow to coastal estuaries support highly productive coastal aquatic zones by supplying nutrients needed by phytoplankton and fauna (Bate et al., 1992).

Karst

Karst is a particular landscape created through dissolution of soluble rocks (USGS, 2003a), such as chalk, dolomite and limestone. Karst environments are characterized by solution channels in the rock with water in the channels and pore space. "In the United States, 20% of the land surface is karst.... Natural features of the landscape such as caves and springs are typical of karst regions" (USGS, 2003a) as well as sink holes (see Figure 1.2.4). Bacteria densely populate the water and sediment in caves and carry out various functions, such as oxidation of sulfur compounds, iron and manganese; nitrification; denitrification; and nitrogen fixation. Karst environments are subjected to quick changes in flow rates at different levels affecting the biota. Shallow karst has simpler assemblages of fauna, whereas deeper water has more diversity (Gibert et al., 1994b). In some locations, organisms existing on the ground actively migrate and colonize a karst aquifer during parts of their lifecycles (Gibert et al., 1994b). Research is being carried out on the ecological relationships and functions in karst (EC, 2001; UMM, 2005).

First Aquifer

The first aquifer may be an extension of the groundwater in a wetland at the ground surface or near surface, or may be tens of meters below the surface serving as the habitat for microorganisms that process energy and matter, form minerals and store biomass – offering the potential to naturally remediate polluted groundwater (Danielopol et al., 2003, p. 108). These aquifers were originally thought to be biologically homogenous and are now recognized to be heterogeneous (Gibert, 1992, p. 7). Groundwater ecosystems have significant crustacean (animal) species known as stygobites that have adapted to low-energy environments, in addition to bacteria and other microorganisms (Danielopol et al., 2003, p. 108). The first aquifer may have dissolved oxygen varying from high to low concentrations, "depending on the hydrologic and geologic conditions and the activity of the biota," but sufficient to allow aerobic processes to occur (Strayer, 1994, p. 295). Other factors affecting groundwater biota include supply of organic matter, temperature, sediment texture, salinity, low metabolic rates, long prereproductive periods, small broods and long intervals between broods; however, predation is poorly understood (Strayer, 1994, pp. 297–304).

In the subsurface zones around the streams extending out under the floodplain where significant groundwater–surface water interaction occurs, referred to as the "hyporheic zone," significant movement of water, energy and organisms transact to create a highly productive subterranean ecosystem. The hyporheic zone allows important microbial and nutrient interaction between ground and surface waters along the streams. This zone may laterally extend 2 km (1.2 miles) or more from a stream, as well as 10 m (33 ft) or more below a stream. The macroinvertebrates may spend a part of their lifecycle in groundwater before emerging to the stream. Bacteria provide nutrient cycling (Gibert, 1992, p. 7). The "standing crop biomass in the hyporheic zone could easily exceed benthic biomass in the river" (Gibert, 1992, p. 7). Thus, the significance of this zone for purifying water for streams and wells producing groundwater from it is now being recognized (Danielopol et al., 2003). Unfortunately, the effects of contaminants, such as pesticides, on hyporheic organisms is not well understood, but is of concern because the higher trophic levels are characterized by longer life spans and low reproductive rates (Notenboom and van Gestel, 1992, p. 316). "The potential for ecological recovery is low in groundwater; small effects might therefore have long-lasting consequences" and the effects may be more severe for microbial processes in anaerobic conditions in deeper groundwater (Notenboom, 2001, p. 257; 259).

Overpumping can change the structure of the groundwater habitat. In one case, a "reduction in the self-purification potential and therefore water quality was observed within the bank filtration area along the Danube [River] at Vienna [Austria] after the major part of the sediment-associated microbial biomass was removed together with the habitat, namely the fine sediment fraction, by overpumping of the water" (Danielopol et al., 2003, p. 116).

Littoral Zones

As noted earlier, groundwater contributes to the ecological productivity of estuarine wetlands. Groundwater is also discharged along lakes and the coasts of the continents and is subjected to tidal influences (USGS, 2005a). The contribution of high levels of nutrients to littoral environments is a concern because of the extensive nutrient pollution of the near-surface aquifers in these zones (USGS, 1999a, 1999b). Additionally, groundwater- and surface-water flux through streambeds and wetlands in riparian zones influence the functioning of the biochemical reactions to affect the exchange of nutrients and organic material between the subsurface and surface components of a watershed system (USEPA, 1996; Tremolieres, 2001) and change the chemistry of the water as it moves through the sediments (Hoehn, 2001). Furthermore, maintaining vital, hardy wetlands significantly affects the regional biodiversity (USEPA, 1996). Research is demonstrating the significant effects of changes in the chemistry of these lake and ocean waters on the aquatic organisms because of the changes in groundwater quality (Stanford and Simons, 1992; USGS, 2005a; Lecher and Mackey, 2018).

Deep Aquifer Environments

Deep aquifer environments typically are characterized by microscopic interstitial spaces and anaerobic conditions. Microorganisms – bacteria and archaea – have been found 4 km below the continents and 7.2 km (4.5 miles) under the ocean floor (Monastersky, 1997). "They use inorganic chemicals, such as hydrogen and hydrogen sulfide, rather than organic matter for their energy and carbon dioxide as their source of carbon. Geothermal, rather than solar, energy catalyzes chemical reactions that generate these life-sustaining chemicals from rocks and … [w]ater … the only absolutely essential ingredient" (Wirsen, 2005). Scientists found a diversity of microorganisms in the groundwater from gold mines in South Africa in fractured bedrock 3 km (2 miles) deep (Wirsen, 2005). Other scientists, who carried out research on a salt deposit 610 m (2,000 ft) beneath the ground surface in New Mexico, discovered what may be the oldest living microorganism found to date, surviving in a minute pocket of brine within a salt crystal formed over the last 250 million years (Wirsen, 2005). Additionally, discharge of groundwater from below the continental shelves is observed to add minerals to saltwater environments (Church, 1996; Moore, 1996), sustaining their aquatic life. Thus, deep aquifers may hold a vast array of microorganisms, the role of which in the ecosystem and the cycling of nutrients is not well understood.

Note on Biodegradation

With the diversity of the organisms and the geologic matrix in the subsurface and, specifically, in groundwater, there exists a potential for biochemical degradation of contaminants reaching the aquifers. The perspective of approaches to this capability in the United States is to provide natural attenuation; while in Europe, the focus has been to reduce the contaminant mass (Danielopol et al., 2003, p. 120). Danielopol (2003, p. 120) cited various results relying on "indigenous subsurface microorganisms" that "degraded" or "immobilized" a range of contaminants such as "monoaromatic benzol-toluene-ethylene-xylenes (BTEX), polycyclic aromatic compounds (PAHs), solvents and other chlorinated organic compounds (for example TCE, PCE, and vinyl chloride [VC]), nitroaromatic compounds (dyes, explosives, pesticides, and pharmaceuticals), heavy metals like Hg, Pb, Cd, Se, and Cr, and radionuclides." Notenboom (2001, p. 259) cautioned that the use of pesticides may pose risks to the "intrinsic sensitivity of … groundwater organisms," which are problematic because these organisms have low reproduction rates. Furthermore, Notenboom *et al* noted (1992, pp. 301-307) that organic contamination can destroy subsurface organisms. Thus, human action affecting the subterranean ecosystem may eliminate useful biological production, as well as rely on it for long-term (years, even centuries) remediation if conditions are right.

GROUNDWATER OCCURRENCE AND HUMAN INTERACTION

As indicated earlier, the complex subsurface environment presents many different and unique hydrogeologic settings in which groundwater occurs. Even given the complexity, these subsurface settings usually extend over large areas and are replicated in nature that has allowed systematic development of the resource. The occurrence of groundwater through wells, springs, wetlands and groundwater–surface water interaction affects our use of it.

WELLS

Typically, groundwater comes from wells installed with human action – hand-or machine-dug holes in the ground to and through the water table into the aquifer. These holes may be a meter to hundreds of meters deep. Once the hole is created, a container can be lowered to withdraw the groundwater, or a pump can be used to lift it to the surface. More details on wells are presented in Volume 1, Chapter 6. As groundwater is nearly ubiquitous, its occurrence allows inhabitation of areas that otherwise would not foster concentrated human activity and economic development.

SPRINGS

Springs are places where groundwater flows out onto the ground surface, down a cliff face, or through a river, lake, or ocean bed. They can be artesian, that is, the groundwater may be under pressure and may flow up to the surface. Springs are usually the result of water flowing along a fracture in the bedrock or along the plane of bedrock strata. Some small towns rely on springs for their source of drinking water without treatment. When springs appear without sources of contamination affecting them and are protected, their waters may be imbibed without treatment and sometimes considered clean enough to be bottled and sold as an alternative drinking water source, although regular testing is important to ensure its safety.

Springs are typically considered as either a "concentrated" or "seepage" spring. A concentrated spring flows from a point in a hillside, at which groundwater is naturally discharged from an opening in the rock. Seepage springs exist where groundwater flows or "seeps" from the ground covering an extensive zone, rather than one site (Jennings, 2005).

WETLANDS

Wetlands are typically lower lying areas in which groundwater maintains a saturated condition in the soil, often times with water from seasonal precipitation occurring in pools and over large areas. These are mainly locations of GWD (Winter, 1989) and are affected by the quality of the groundwater underlying the larger area around the wetland. Wetlands are considered to be significant locations for aquatic spawning and production (USEPA, 1996). "Many wetlands are among the most productive of natural ecosystems, exceeding the best agricultural lands and rivaling the production of tropical rain forests" (USEPA, 1996, citing NRC, 1991).

Groundwater occurrence in wetlands is critical for the natural balance in the ecosystem, providing water, minerals and critical temperature variations, as well as a zone of natural decomposition and nutrient recycling (USEPA, 1995):

- Over one-third of the threatened and endangered species in the United States have only wetlands as their habitat.
- Almost half of the species require wetlands during their maturing stage.

- A wide range of other animals and plants rely on wetlands as a source of food, water and shelter. Additionally, 50% of the endangered animal species inhabit wetlands and 28% of the endangered plant species require wetland environments in the United States (USEPA, 1996).
- Estuarine and marine fish and shellfish, certain bird species and some mammals require coastal wetlands for their survival.
- Most of the commercial and game fish reproduce and nurture their offspring in coastal marshes and estuaries.
- Nearly 35% of the rare and endangered species in the United States rely on wetlands habitat (USEPA, 1996).

Wetlands are a manifestation of groundwater occurrence. Most wetlands are maintained by groundwater or have a significant groundwater component (Gibert et al., 1994a, p. 531, citing Winter, 1989). Consequential ecological interactions exist between groundwater-supplied wetlands and wildlife (Gibert et al., 1994a, p. 387). Wetlands are complex, biologically productive zones in the ecosystem (Cowardin et al., 1979), and groundwater flow to wetlands during drought is critical to their ecological maintenance (Gomez et al., 2001, pp. 318–319).

Loss of wetlands has occurred over time with the development of land-based activities. In the seventeenth century, the coterminous 48 states of the United States are believed to have had over 89 million ha (220 million acres) of wetlands. As of 1997, this same area had approximately 42.7 million ha (106 million acres) of wetlands. Alaska had an estimated 68.8–80.9 million ha (170–200 million acres) of wetlands in 1980 (about half of the state) (USEPA, 2005d). From 1986 to 1997, the United States lost 23,674 ha (58,500 acres) of wetlands each year (USEPA, 2005d). A number of wetlands in the United States have been lost or are threatened by groundwater pumping (USEPA, 1996). Wetland loss is also an international concern, with 11 (out of 37) European Union countries reporting in 1999 that of the 420 wetlands, 153 were not endangered and 11 were endangered from groundwater mining, with 46 endangered owing to other factors (EEA, 1999, p. 102). Loss of wetlands and their functions in the ecosystem are due to a range of natural and human factors, and some major ones are presented in Table 1.2.4.

Groundwater–Surface Water Interaction

Groundwater–surface water interaction is increasingly recognized as an important part of the hydrologic cycle that affects both the quantity and quality of water in either source. As noted earlier, in

TABLE 1.2.4
Major Factors Contributing to Wetlands Loss

Human Actions		Natural Threats
-Drainage	-Construction	-Erosion
-Dredging and stream channelization	-Runoff	-Subsidence
-Deposition of fill material	-Air and water pollutants	-Rise in sea level
-Diking and damming	-Changing nutrient levels	-Droughts
-Tilling for crop production	-Releasing toxic chemicals	-Hurricanes and other storms
-Levees	-Introducing nonnative species	
-Logging	-Grazing by domestic animals	
-Mining		

Source: United States Environmental Protection Agency (USEPA), Wetlands: Status and Trends, 2005d, URL: http://www.epa.gov/OWOW/wetlands/vital/status.html (accessed August 23, 2005).

the hydrologic cycle description, water can flux or move from a surficial aquifer to a stream, lake, or coastal zone and vice versa. This circumstance is controlled by the geologic material constituting the stream, lake or ocean bed and the hydraulic gradient of those bodies of water and surficial aquifer at different times. In humid areas, streams often act as GWD points. Locations of GWD and exchange with surface water – fresh and marine – are highly productive ecological zones and are well documented (e.g., see Stanford and Simons, 1992, and Gibert et al., 1994b). In some locations, snowmelt water serves to recharge groundwater, for example, along the Platte River, as it crosses Colorado and Nebraska. Lincoln, Nebraska, gets its drinking water from wells along the Platte River, which draw water from the river through the streambed and into the subsurface. When the concentration of pollutants, such as pesticides, is too high in a river, the wells near the river are shut down so as to reduce the possibility of pumping contaminated water from the river through the subsurface during the spring after pesticide application.

As noted previously, groundwater may be discharged to streams and wetlands and maintain their water regime. For example, in a study of a small forested watershed northwest of Baltimore, Maryland, with no source of water other than precipitation, calculations indicated that the baseflow of the stream could be maintained from the GWD for 25–349 days. Groundwater was observed to sustain the stream for 187 days during a drought in the watershed (Cleaves et al., 1970). "Streamflow during a drought period originates almost entirely from subsurface [groundwater] storage that seeps into the streambed," such as the one that occurred in the Upper Mississippi River Basin, United States, during the drought of 1988–1989 (IDENR, 1994).

Groundwater–surface water interaction also points to another interesting environmental problem: shifting effects in the ecosystem to surface water in time. The control of nonpoint sources of pollution (such as widespread application of fertilizer and pesticides on lawns and in agricultural areas) is usually implemented to reduce overland flow of sediment and contaminants to streams. To reduce this flow, tillage practices are changed to allow water to reside on the ground for a longer duration and infiltrate. When this occurs, groundwater and its subsurface environment is the recipient of a substantial portion of the contamination. If those particular groundwaters are closely connected to the surface waters, much of the contamination may eventually reach the stream or lake which the nonpoint source controls are designed to protect. For example, GWD to the nontidal tributaries to Chesapeake Bay (United States) account for a median value of 54% of their streamflow and a median of 56% of the streams' nitrate loading to the bay (USGS, 1998a, 1998b). Thus, the role of groundwater–surface water interaction in the ecosystem must necessarily address this interaction in the context of the larger hydrologic cycle.

During times of drought and flooding, groundwater–surface water interaction is important for stream recovery. This recovery is relatively quick because of the relationships of the stream, the watershed and the hyporheic zone, in moving and storing nutrients and interstitial ecological communities (Valett et al., 1992, p. 396). The value of this interaction in the ecosystem to the productivity and sustainability of streams is receiving more attention (TNC, 2014) and requires further research (Valett et al., 1992).

THE WATER BUDGET

In its simplest form, the water flow of the hydrologic cycle can be expressed as a water budget. A water budget can exist for a community, a watershed, an aquifer, a region or a country, as long as the parameters can be reasonably estimated. A water budget is an accounting of the inflow, outflow and storage changes of water in a hydrologic unit. The equation for a water budget is expressed as inputs and outputs:

$$\text{Inputs} = \text{Outputs} \tag{1.2.1}$$

In the hydrologic cycle, the input is precipitation (P) and the output is equal to evaporation (E) and transpiration (T) from surface water bodies, land surface and vegetation, total surface water and groundwater flow (SW and GW) to oceans and consumptive use (CU) of water by human activity in manufacturing products or for processes such as evaporative cooling in houses and power plants. This relationship may be summarized as

$$P = E + T + \mathrm{SW} + \mathrm{GW} + \mathrm{CU} \tag{1.2.2}$$

For example, the water budget for the United States is as follows:

- Precipitation, 15.9 billion m^3/day (4.2 trillion gallons/day) coming to the earth's surface in the coterminous United States
- Consumptive use, 378.5 million m^3/day (100 billion gallons/day)
- Evaporation and transpiration, 10.6 billion m^3 (2.8 trillion gallons) and
- Total surface water and groundwater flow to oceans, 4.9 billion m^3 (1.3 trillion gallons) (USGS, 1999a, 1999b).

The U.S. Geological Survey has estimated that 40% to 50% of the average annual streamflow of the United States is the GWD to streams. From this information, GWD in the United States can be derived: the average annual streamflow is 4.7 billion m^3 (1.24 trillion gallons); if multiplied by 0.4, the calculation results in 1.9 billion m^3 (502 billion gallons). This net GWD is equal, for example, to about 1.25 times the flow of the Mississippi River measured at its mouth below New Orleans, Louisiana. The GWD is greater in the humid eastern United States, where GWD may maintain as much as 90% of the baseflow of smaller streams (USGS, 1999a, 1999b, p. 40).

Can a water budget be developed for an aquifer? For a groundwater system budget, it is expressed differently from the one mentioned earlier, as follows (Dunne and Leopold, 1978, p. 219):

$$\text{Inputs} - \text{outputs} = \text{Change in Storage}\left(\text{of the a quifer}\right) \tag{1.2.3}$$

Using some of the information from Figure 1.2.3 and further knowledge of groundwater flow, the following water budget equation can be developed for an aquifer underlying a basin or watershed:

$$\mathrm{GWS} = I + \mathrm{INJ} + \mathrm{EL} + \mathrm{SWR} - \mathrm{SMV} - E - T - \mathrm{CU} - \mathrm{GWD} \tag{1.2.4}$$

where

GWS = net groundwater storage in the aquifer
I = infiltration water
INJ = water injected into the subsurface through drains or injection wells
EL = other emplaced liquid wastes (e.g., from landfills) from which the liquid portion has leached to the water table
SWR = surface water recharge to the aquifer
SMV = soil moisture and water in the vadose zone
E = evaporation, which may be important where the groundwater table is close to or at the earth's surface, such as in wetlands
T = transpiration by vegetation
CU = consumptive use from human activity, not returned to the subsurface
GWD = net groundwater discharge to streams

Several other relationships are important to fully understand this water budget equation:

$$I = P - R, \tag{1.2.5}$$

where
I = infiltration water
P = precipitation
R = runoff or overland flow

Also,

$$\text{PER} = P - R - \text{SMV}, \tag{1.2.6}$$

where
PER = percolation water transmitted to the water table
P = precipitation
R = runoff or overland flow
SMV = soil moisture and water held in the vadose zone

This equation can also be written as

$$\text{PER} = I - \text{SMV}. \tag{1.2.7}$$

Rewriting the aquifer water budget through substitution gives

$$\text{GWS} = \text{PER} + \text{INJ} + \text{EL} + \text{SWR} - E - T - \text{CU} - \text{GWD}. \tag{1.2.8}$$

This equation shows that the net groundwater storage (GWS) in an aquifer is equal to the inputs of percolation water from precipitation transmitted to the water table, water injected into the subsurface through drains or injection wells, other emplaced liquid wastewater and surface water recharge to the aquifer, minus evaporation, transpiration, consumptive use and GWD. The equations for GWS and PER contain parameters important to managing human activities on the land surface that affect the amount of water reaching the aquifer, as well as the quality of that water.

Key points related to the groundwater budget include, but are not limited to the following:

- I (infiltration) is affected by the area of impervious surface in the watershed; more paved or roofed surfaces reduce natural opportunities for precipitation to reach the water table.
- Storm drains, septic systems and other unsewered drains from commercial and industrial operations also may collect and inject (INJ) water of varying quality back into the ground.
- Stream channelization can reduce the ability of streams to provide recharge to groundwater, and, conversely, groundwater to discharge to streams.
- In some places, evaporation rates may be considerably higher than precipitation rates, necessitating groundwater irrigation for crop growth.
- Communities that rely on groundwater for their water supply often treat the resulting wastewater and discharge it to a stream. As a result, these locations may experience long-term lowering of the water table and the need to replace wells with deeper ones.
- Recharge of groundwater by untreated stormwater may have mixed results: water quantity is increased, but water quality may be negatively affected, depending on the conditions in the zone of capture of the recharge site and the geochemistry of the geologic formation. This circumstance is highlighted in Box 1.2.5, which summarizes the results of investigations of stormwater infiltration ponds on groundwater quality.

BOX 1.2.5 STORMWATER RECHARGE EFFECTS ON GROUNDWATER QUALITY

A 5-year study of three representative stormwater catchment and infiltration ponds in Maryland, assessed the quality of groundwater and the infiltration of stormwater. The findings included the following:

- Contaminants reached groundwater indirectly because of (1) biochemical reactions in the impoundments that ultimately affected unsaturated-zone and groundwater pH and redox or that caused solubilization of metals from time to time and (2) dissolution of the rock aggregate used in subsurface impoundments that released trace metals to solution.
- Infiltration rates during dry weather were less than 0.01 m/day (0.4 inch/day) and during storm events ranged from 11.7 m/day (38 feet/day) at one site to 65.8 m/day (216 feet/day) at another site.
- The trace-element composition of groundwater beneath stormwater impoundments contained heavy metals; stormwater infiltrate was the sole or predominate source. Concentrations were modest and usually did not exceed USEPA drinking water regulations; nevertheless, concentrations of cadmium, chromium and lead exceeded USEPA [maximum contaminant levels] MCLs in some groundwater samples, and concentrations of barium, copper, nickel, strontium and zinc commonly were elevated above background concentrations.
- Concentrations of anthropogenic organic compounds were usually below detection limits in stormwater and groundwater.

Source: Abstracted from Maryland Geological Survey (MGS), Geochemistry and Factors Affecting Ground-Water Quality at Three Strom-Water-Management Sites in Maryland, Report of Investigations No. 59 (Authored by Wilde, F.D.), Baltimore, MD (1994), 201.

GROUNDWATER AND CLIMATE CHANGE

Will climate change affect groundwater quantity or quality? Climate change has a range of effects. An increase in carbon dioxide and/or other greenhouse gases raises temperatures, shifts precipitation patterns, increases evapotranspiration rates and raises ocean levels, among other things (IPCC, 2001, Chapter 1).

Effects on groundwater can vary from region to region and aquifer to aquifer, but can be noticeable and predictable, particularly on shallow aquifers (IPCC, 2001, Chapters 1 and 4; ECan, 2004). The principal impacts on groundwater are changes in aquifer recharge, groundwater–surface water interaction and groundwater use (Loaiciga et al., 1998, p. 5; ECan, 2004, Chapter 10). As the result of changing climatic conditions, if an upland region receives less precipitation, then less water will infiltrate and recharge aquifers, which will also affect groundwater quality and reduce GWD and their baseflow (ECan, 2004, Chapter 10). Conversely, if more precipitation is received, a greater amount of recharge will be available. As temperatures rise, more precipitation may be offset by higher evaporation rates. Higher temperatures in areas of already sparse rainfall may create greater soil moisture deficits than those that previously existed. Small changes in precipitation could have large changes in groundwater recharge. Modeling rainfall and groundwater levels for hypothetical climate change in central Tanzania indicated that a 15% decrease in precipitation could lead to a 40%–50% reduction in the recharge to groundwater, at a steady temperature (i.e., no change in evapotranspiration) (IPCC, 2001, Chapter 4). A study in 2008 by the Massachusetts Institute of Technology indicated a similar finding: a 20% rise in precipitation levels might raise aquifer recharge by up to 40% and, conversely, a 20% decline in the precipitation could reduce recharge by as much as 70%, which is of great significance to arid regions (MIT, 2008).

Two factors are considered to be important: timing of recharge and character of the aquifer (ECan, 2004, Chapter 10). The extent of response of aquifers to changes in the precipitation may be slower depending on the location. Fractured bedrock and unconsolidated aquifers nearer the ground surface will be more responsive to changes in precipitation. However, the effects on deeper aquifers may be different: deeper aquifers can probably be relied on to make up for deficiencies that shallow aquifers cannot supply. Thus, deeper aquifers may be depleted in situations where shallow aquifers cannot meet the water demand.

Rising sea levels may also impact the groundwater along the coasts of continents. Higher sea levels may raise the hydraulic head along the coastline. Groundwater tables of surficial aquifers may also rise and saltwater may advance inland, increasing the salinity of groundwater used for drinking water, as well as harming aquatic animals and plants and deeper-rooted vegetation. This change in groundwater quality may also affect coastal wetlands and the use of coastal groundwater for irrigation (USEPA, 1989, pp. 122–123).

SUMMARY: GENERAL OBSERVATIONS ON NATURAL FACTORS AFFECTING GROUNDWATER IN THE ECOSYSTEM

Groundwater has a strong three-dimensional underground aspect that makes it difficult to observe and sample (ITFM, 1997) – as well as making it either less or more expensive to access, depending on the location, geology and depth. As groundwater is out of sight beneath the ground surface, there is a cost to access it, usually by a well. Some general observations about groundwater in the ecosystem and their implications for its economic use include the following:

Groundwater exists under continental land masses. As groundwater exists in the vast volume of subterranean space nearly everywhere, maintaining critical wildlife habitats while supplying half the world's population with water and increasingly relied on for irrigation to grow food.

Groundwater is a finite resource. Deeper ancient aquifers that supply irrigation water for many arid regions are being depleted. Drought and weather extremes are affecting groundwater recharge.

The environment in which groundwater exists is very complex. USEPA (1985) has identified 48 different typical groundwater environments in the United States, which have analogs on other continents. Each hydrogeologic setting has a chemistry that has been affected by the geologic environment within which it exists. The type of geologic environment influences the cost of producing groundwater from it and its use.

Furthermore, ecologically, the fact that the subsurface is alive with organisms has been given much recognition only since the 1990s, and we still do not understand their function in the ecosystem. More research is needed to comprehensively define the many subsurface organisms and their roles in the ecosystem.

Groundwater can be near the ground surface or deep. A shallow water table (e.g., 3–10 m) means that a well may not need to be too deep and may cost less than a deeper well (e.g., 30–150 m). Groundwater is also contained in bedrock aquifers hundreds of meters deep.

Groundwater is more difficult to sample. Typically, a well must be drilled and casing installed to sample groundwater. The technology for real-time monitoring on a continuous basis is being refined. As microorganisms live in groundwater and could be indicators of its quality, technology for sampling this biota may also receive attention.

Groundwater chemistry and quality are affected by the chemistry of the geologic formation in which it exists, as well as its associated microbiological community, related to land use in the watershed above the aquifer. While groundwater typically moves through the interstitial pore space slowly, it reacts with the geologic formation through which it moves, carrying water-soluble molecules (as ions) with it. If removal of any of these elements or contaminants is needed, this treatment adds cost to making the water drinkable or usable in other applications.

Groundwater moves more slowly than surface water. Typically, groundwater moves in the subsurface in fractions of centimeters to meters per day, depending on the geological matrix in which it resides. If groundwater is contaminated, then slowly moving groundwater is likely to ensure that the problem will be localized. Slow movement of groundwater may mean long times for remediation of contamination.

Pollution can fan out in a plume or follow a particular geologic feature, for example, an esker. As noted earlier, pollution may spread slowly and in doing so fan out in a plume. Particular geologic features, such as fractures or a buried alluvial channel, can facilitate movement to be more rapid and move in a particular direction. This feature and the contamination being transported along it could potentially be difficult to determine. Large areas of contamination as well as contamination traveling along paths difficult to track may be expensive to remediate, or they may be abandoned as a resource because of that substantial expense.

Areas of more rapid infiltration and recharge of an aquifer are areas which are both highly productive and are easily contaminated. The same characteristics that make certain aquifers very productive (rapid infiltration, faster times of travel, larger hydraulic conductivity) for groundwater also allow contaminants to move more easily into the aquifer.

It is easier to produce groundwater in some geologic environments than others. The U.S. Geological Survey (1985) has documented the range of yields of major aquifers in each state. Some aquifers can only supply up to 38 L/min (10 gallons/min) from a well. Others are able to produce 7,570–18,930L/min (2,000–5,000 gallons/min) and are the source of high-production wells used by public water systems and private irrigators. The complexity of the aquifer or aquifer system increases the cost of understanding whether these aquifers are good long-term sources of supply or will be difficult and costly to remediate if they become contaminated.

The quality of groundwater and adjacent surface water is related. Groundwater and surface water flow and interact in watersheds as a single resource and component of the hydrologic cycle and ecosystem. While they can be viewed distinctly, they merge and intersect when near-surface, shallow groundwaters discharge to or are recharged by the surface waters. Their biota are related and may have a significant effect on their collective water quality in a watershed.

FOR FURTHER CONSIDERATION

1. Do you live in a humid or arid region? Do streams flow when there is no precipitation? To what do you attribute this "stream flow in dry weather" in your area?
2. What is the average annual precipitation for your area? What is the evaporation rate? If one-fourth of the precipitation infiltrates to the soil and subsurface, what portion of the precipitation remains as runoff?
3. What happens to stormwater in your area? Is it collected and sent to a stormwater retention pond? Are stormwater ponds in your area designed to allow water to infiltrate? Or is stormwater allowed to percolate into the soil without collection? Are there water wells nearby the stormwater pond? How far are they from the pond? Based on what you understand, is it possible that the stormwater in the pond affects the quality of the groundwater?
4. Obtain a copy of your water system's (or for a groundwater-supplied water system near you) Consumer Confidence Report (possibly also referred to as the Annual Water Quality Report). What contaminants were tested in the drinking water? Does this system, relying on groundwater, test for any of the naturally occurring constituents listed in Table 1.2.2? What does this suggest to you about the water and the geologic matrix in which it resides?
5. Why is it important to understand the current status of the groundwater environment for an economic analysis of a proposed project using or affecting groundwater?

REFERENCES

Bate, G.C., Campbell, E.E., and Parker-Nance, T. 1992. Coastal aquifers in southern Africa: Their ecological importance. In *Proceedings of the First International Conference on Ground Water Ecology*. Stanford, J.A. and Simons, J.J. (eds.). American Water Works Association, Bethesda, MD, pp. 377–384.

Bear, J., Cheng, A.H.-D., Sorek, S., Ouazar, D., and Herrera, I. (eds.). 1999. *Seawater Intrusion in Coastal Aquifers – Concepts, Methods and Practices. Kluwer Academic Publishers, Dordrecht, the Netherlands*, 625 p.

Berkey, E., and Zachry, T. (eds.). 2005. *Subsurface Contamination Remediation; Accomplishments of the Environmental Management Science Program*. ACS Symposium Series 904. American Chemical Society, Washington, DC, 396 p.

Born, S., Mayers, J., Morton, A., and Sonzogni, B. 1997. *Exploring Wisconsin Trout Streams; The Angler's Guide*. The University of Wisconsin Press, Madison, WI, 287 p.

Brown, L.R. 2003. *Plan B: Rescuing a Planet under Stress and a Civilization in Trouble*. W.W. Norton & Co., New York. URL: http://www.earth-policy.org/Books/PB/PBch3_ss4.htm (accessed August 5, 2005).

Brown, L.R. 2004. *Outgrowing the Earth*. W.W. Norton & Company, New York, 239 p.

Church, T.M. 1996. An underground route for the water cycle. *Nature*, 380, April 18, 579–580.

Cleaves, E.T., Godfrey, A.E., and Bricker, O.P. 1970. Geochemical balance of a small watershed and its geomorphic implications. *Geological Society of America Bulletin*, 81, 3015–3032.

Cowardin, L.M., Carter, V., Golet, F.C., and LaRoe, E.T. 1979. *Classification of Wetlands and Deepwater Habitats of the United States*. U.S. Department of the Interior, Fish and Wildlife Service, Washington, DC, Northern Prairie Wildlife Research Center Online, Jamestown, ND. URL: http://www.npwrc.usgs.gov/resource/1998/classwet/classwet.htm (Version 04DEC98).

Danielopol, D.L., Griebler, C., Gunatilaka, A., and Notenboom, J. 2003. Present state and future prospects for groundwater ecosystems. *Environmental Conservation*, 30 (2), 104–130.

Deberitz, J.. 2006. *Lithium: Production and application of a fascinating and versatile element*. S.V. Corporate Media GmbH, Munich, Germany, pp. 14–21.

Downey, J.S. and Dinwiddie, G.A. 1988. The Regional Aquifer System Underlying the Northern Great Plains in Parts of Montana, North Dakota, South Dakota, and Wyoming – Summary, *U.S. Geological Survey Professional Paper* 1402-A, 64.

Driscoll, F.G. 1986. *Groundwater and Wells*. U.S. Filter/Johnson Screens, St. Paul, MN.

Dunne, T. and Leopold, L.B. 1978. *Water in Environmental Planning*. W.H. Freeman & Company, San Francisco, CA, 818 p.

Edwards Aquifer Research and Data Center (EARDC). 2019. *Threatened and Endangered Species in the Edwards Aquifer System*. URL: https://www.eardc.txstate.edu/Aquifer-Info/endangered.html#top (accessed April 25, 2020).

Environment Canada (ECan). 2004. *Threats to Water Availability in Canada*. National Water Research Institute, Burlington, Ontario, Canada, NWRI Scientific Assessment Report Series No. 3 and ACSD Science Assessment Series No. 1. 128 p. URL: http://www.nwri.ca/threats2full/ch10-1-e.html (accessed February 4, 2007).

European Commission (EC). 2001. *Groundwater Ecology: A Tool for Management of Water Resources*. European Commission (EC), Luxembourg, 413 p.

European Environment Agency (EEA). 1999. *Groundwater Quality and Quantity in Europe*. European Environment Agency (EEA), Copenhagen, Denmark, 123 p.

European Union (EU). 2005. *French Researchers May Have Solution to Water-Born Arsenic*. Europa-Research. URL: http://europa.eu.int/comm/research/headlines/news/article_05_02_28_en.html (accessed August 13, 2005).

Forest Conservancy District Board for Baltimore County (FCDB). 2005. *Tree Facts*. Reference Item 26. URL: http://www.bcfb.sailorsite.net/HelpfullHints/Trees_and_Shrubs.htm (accessed August 24, 2005).

Freeze, R.A. and Cherry, J.A. 1979. *Groundwater*. Prentice Hall, Upper Saddle River, NJ, 604 p.

Geotechnical, Rock and Water Resources Library (GROW). 2005. URL: http://www.grow.arizona.edu/images/water/panevap.gif (accessed August 24, 2005).

Gibert, J. 1992. Groundwater ecology from the perspective of environmental sustainability. In *Groundwater Ecology*, Gibert, J., Danielopol, D.L., and Stanford, J. (eds.). Academic Press, San Diego, CA, pp. 3–13.

Gibert, J., Danielopol, D.L., and Stanford, J. (eds.). 1994a. *Groundwater Ecology*. Academic Press, San Diego, CA, 571 p.

Gibert, J., Vervier, P., Malard, F., Laurent, R., and Reygrobellet, J.L. 1994b. Dynamics of communities and ecology of karst ecosystems: Example of three karsts in eastern and southern France. In *Groundwater Ecology*, Gibert, J., Danielopol, D.L., and Stanford, J. (eds.). Academic Press, San Diego, CA, pp. 425–450.

Gleeson, T., Befus, K.M., Jashenko, S., Luijendijk, E., and Bayani Cardenas, M. 2015. The global volume and distribution of modern groundwater. *Nature Geoscience*, 9, 161–167.

Gleick, P.H. (editor). 1993. *Water in Crisis: A Guide to the World's Fresh Water Resources*. Oxford University Press, New York.

Gleick, P.H. 2000. *The World's Water 2000–2001: The Biennial Report on Freshwater Resources*. Island Press, Washington, DC, 310 p.

Gomez, R., Vidal-Abarca, M.R., and Suarez, M.L. 2001. Importance of the subsurface-surface water interaction in the wetland structure and dynamic in arid and semiarid areas. In *Groundwater Ecology: A Tool for Management of Water Resources*. European Commission (EC), Luxembourg, pp. 317–322.

Harvey, F.E., Ayers, J.F., and Gosselin, D.C. 2007a. Ground water dependence of endangered ecosystems: Nebraska's eastern saline wetlands. *Ground Water*, 45 (6), 736–752.

Harvey, F.E., Swinehart, J.B., and Kurtz, T.M. 2007b. Ground water sustenance of Nebraska's unique sand hills peatland fen ecosystems. *Ground Water*, 45 (2), 218–234.

Hoehn, E. 2001. Exchange processes between rivers and ground waters: The hydrological and geochemical approach. In *Groundwater Ecology: A Tool for Management of Water Resources*. European Commission (EC), Luxembourg, 413 p.

Illinois Department of Energy and Natural Resources (IDENR). 1994. *The Changing Illinois Environment: Critical Trends*, Vol. 2. Water Resources, Drought Impacts on Water Resources. URL: http://www.dnr.state.il.us/orep/ctap/ctapvol2/pp85-100.htm (accessed August 13, 2005).

Intergovernmental Panel on Climate Change (IPCC). 2001. *Climate Change 2001: Working Group II: Impacts, Adaptation and Vulnerability*. URL: http://www.grida.no/climate/ipcc_tar/wg2/index.htm (accessed February 3, 2007).

Intergovernmental Task Force on Monitoring Water Quality (ITFM). 1997. *Conceptual Frameworks for Ground-Water Quality Monitoring*. (August 1997) Denver, CO, 94 p.

Iowa State University (ISU). 1994. *Summer Tree Maintenance: Watering*. Iowa State University, Department of Horticulture, Ames, IA. URL: http://www.ipm.iastate.edu/ipm/hortnews/1994/8-12-1994/hohtree. html (accessed August 24, 2005).

Jennings, G.D. 2005. *Protecting Water Supply Springs*. Cooperative Extension Service, Raleigh, NC. URL: http://www.bae.ncsu.edu/programs/extension/publicat/wqwm/ag473-15.html (accessed August 23, 2005).

Jha, M., Pan, Z., Takle, E.S., and Gu, R. 2003. The impacts of climate change on stream flow in the upper mississippi river basin: A regional climate model perspective. Working Paper 03-WP 337. Center for Agricultural and Rural Development, Iowa State University, Ames, IA. 26 p. URL: http://www.econ.iastate.edu/research/webpapers/paper_10664.pdf (accessed August 13, 2005).

Jorgensen, D.G., Helgensen, J.O., and Imes, J.L. 1993. Wyoming Geohydrologic Framework. Regional Aquifer-System Analysis-Central Midwest. US Geological Survey Professional Paper 1414-B. URL: https://pubs.usgs.gov/pp/1414b/report.pdf (accessed March 1, 2019).

Jorgenson, D., Wireman, M., and Olson, D. 1998. Assessing the vulnerability of public water supply wells to microbial contamination. *Ground Water Monitoring and Remediation*, 18 (2), 60–66.

Lecher, A.L., and Mackey, K.R.M. 2018. Synthesizing the effects of submarine groundwater discharge on marine biota. *Hydrology*, 5 (4), 60. URL: https://www.mdpi.com/2306-5338/5/4/60/htm (accessed Apri 25, 2020).

Loaiciga, L.C., Maidment, D.R., and Valdes, J.B. 1998. *Climate Change Impacts on the Water Resources of the Edwards Balcones Fault Zone Aquifer, Texas*. ASCE/USEPA Cooperative Agreement CR824540-01-0. American Society of Civil Engineers, Reston, VA, 72 p.

Mangin, A. 1994. Karst hydrogeology. In *Groundwater Ecology*, Gibert, J., Danielopol, D.L., and Stanford, J. (eds.). Academic Press, San Diego, CA, pp. 43–67.

Maryland Geological Survey (MGS). 1994. Geochemistry and factors affecting ground-water quality at three strom-water-management sites in Maryland. Report of Investigations No. 59 (Authored by Wilde, F.D.). Baltimore, MD, 201 p.

Massachusetts Institute of Technology (MIT). 2008. Water supplies could be strongly affected by climate change. URL: http://web.mit.edu/newsoffice/2008/agu-groundwater-1218.html (accessed February 1, 2009).

Meyland, S.J. 2011. Examining safe yield and sustainable yield for groundwater supplies and moving to managed yield as water resource limits become a reality. In *Water Resources Management*, WIT Press. pp. 813–823. URL: https://www.witpress.com/Secure/elibrary/papers/WRM11/WRM11073FU1.pdf (accessed January 18, 2020).

Monastersky, R. 1997. *Life in Rock – Deep Dwellers: Microbes Thrive far Below Ground*. Newsletter of the Wisconsin Water Well Association, Waupaca, WI, pp. 1–5 (reprinted from *Science News*, March 29, 1997, Science Service).

Moore, W.S. 1996. Large groundwater inputs to coastal waters revealed by ^{226}Ra enrichments. *Nature*, 380, 612–614.

National Ground Water Association (NGWA). 2017. *Groundwater and PFAS: State of Knowledge and Practice*. NGWA Press, Westerville, OH.

National Ground Water Association (NGWA). 2020. *Groundwater Facts*. URL: https://www.ngwa.org/what-is-groundwater/About-groundwater/groundwater-facts (accessed April 25, 2020).

National Research Council (NRC). 1991. *Restoration of Aquatic Ecosystems: Science, Technology, and Public Policy*. National Research Council (NRC), Washington, DC, 485 p.

National Research Council (NRC). 1993. *Ground Water Vulnerability Assessment; Predicting Relative Contaminant Potential under Conditions of Uncertainty*. National Academy Press, Washington, DC, 224 p.

Notenboom, J. 2001. Managing ecological risks of groundwater pollution. In *Groundwater Ecology: A Tool for Management of Water Resources*. European Commission (EC), Luxembourg, pp. 247–262.

Notenboom, J. and van Gestel, K. 1992. Assessment of toxicological effects of pesticides on groundwater organisms. In *Proceedings of the First International Conference on Ground Water Ecology*. Stanford, J.A. and Simons, J.J. (eds.). American Water Works Association, Bethesda, MD, pp. 311–317.

Pettyjohn, W.A., Savoca, M., and Self, D.. 1991. *Regional Assessment of Aquifer Vulnerability and Sensitivity in the Conterminous United State*. U.S. Environmental Protection Agency Publication EPA/600/S2-91/043. URL: https://nepis.epa.gov/Exe/tiff2png.cgi/30003U3Q.PNG?-r+75+-g+7+D%3A%5CZYFILES%5CINDEX%20DATA%5C91THRU94%5CTIFF%5C00000147%5C30003U3Q.TIF (accessed April 25, 2020).

Queensland Department of Natural Resources, Mining and Water (QDNRMW). 2006. Facts; Water Series; The Great Artesian Basin. URL: http://www.nrm.qld.gov.au/factsheets/pdf/water/w68.pdf (accessed May 14, 2006).

Sampat, P. 2000. *Deep Trouble; The Hidden Threat of Groundwater Pollution*. Worldwatch Paper 154. Worldwatch Institute, Washington, DC, 55 p.

Schelesinger, W.H. 1991. *Biogeochemistry: An Analysis of Global Change*. Academic Press, San Diego, CA.

Stanford, J.A. and Simons, J.J. (eds.) 1992. *Proceedings of the First International Conference on Ground Water Ecology*. American Water Works Association, Bethesda, MD, 419 p.

Strayer, D.L. 1994. Limits to biological distributions in groundwater. In *Groundwater Ecology*, Gibert, J., Danielopol, D.L., and Stanford, J. (eds.). Academic Press, San Diego, CA, pp. 287–310.

The Nature Conservancy (TNC). 2014. *Groundwater and Stream Interaction in California's Central Valley: Insights for Sustainable Groundwater Management*. URL: https://www.scienceforconservation.org/products/groundwater-and-stream-interaction (Accessed December 15, 2020).

Tremolieres, M. 2001. Functioning of interfaces surface water/ground water/forest in a fluvial hydrosystem: The case of the rhine floodplain. In *Groundwater Ecology: A Tool for Management of Water Resources*. European Commission (EC), Luxembourg, pp. 143–164.

United Nations Educational, Scientific and Cultural Organization (UNESCO). 2003. Water for people, water for life. The United Nations Water Development Report. UN World Water Assessment Programme, 575 p.

United States Environmental Protection Agency (USEPA). 1985. *DRASTIC: A Standardized System for Evaluating Ground Water Pollution Potential Using Hydrogeologic Settings*. EPA/600/2-85/018. Washington, DC, 163 p.

United States Environmental Protection Agency (USEPA). 1989. *The Potential Effects of Global Climate Change on the United States; Report to Congress*. EPA-230-05-89-050. Washington, DC, 457 p. URL: http://www.epa.gov/climatechange/effects/downloads/potential_effects.pdf (accessed February 3, 2007).

United States Environmental Protection Agency (USEPA). 1993. *Ground Water Resource Assessment*. Office of Water. EPA 813-R-93-003. Washington, DC, 232 p.

United States Environmental Protection Agency (USEPA). 1995. *America's Wetlands: Our Vital Link Between Land and Water*. EPA843-K-95-001. Washington, DC. URL: http://www.epa.gov/owow/wetlands/fish.html (accessed August 24, 2005).

United States Environmental Protection Agency (USEPA). 1996. *An Assessment of the Ecological Impacts of Ground Water Overdraft on Wetlands and Riparian Areas in the United States*. EPA 813-S-96-001 (July 1996). Washington, DC, 103 p.

United States Environmental Protection Agency (USEPA). 2000. *Technical Fact Sheet: Final Rule for (NonRadon) Radionuclides in Drinking Water*. Washington, DC, EPA 815-F-00-013 (November 2000). URL: http://www.epa.gov/safewater/rads/technicalfacts.html (accessed August 13, 2005).

United States Environmental Protection Agency (USEPA). 2005a. *Arsenic in Drinking Water*. Washington, DC. URL: http://www.epa.gov/safewater/arsenic.html (accessed August 13, 2005).

United States Environmental Protection Agency (USEPA). 2005b. *Wetlands Definitions*. Washington, DC. URL: http://www.epa.gov/owow/wetlands/what/definitions.html (accessed July 31, 2005).

United States Environmental Protection Agency (USEPA). 2005c. *Current Drinking Water Standards*. Washington, DC. URL: http://www.epa.gov/safewater/mcl.html#mcls (accessed August 13, 2005).

United States Environmental Protection Agency (USEPA). 2005d. *Wetlands: Status and Trends*. Washington, DC. URL: http://www.epa.gov/OWOW/wetlands/vital/status.html (accessed August 23, 2005).

United States Environmental Protection Agency (USEPA). 2005e. *NPL Fact Sheets for Minnesota: New Brighton/ Arden Hills*. Washington, DC. URL: http://www.epa.gov/R5Super/npl/minnesota/MN7213820908.htm (accessed August 24, 2005).

United States Environmental Protection Agency (USEPA). 2019. *About the TSCA Chemical Substance Inventory*. URL: https://www.epa.gov/tsca-inventory/about-tsca-chemical-substance-inventory#howare (accessed April 25, 2020).

United States Geological Survey (USGS). 1975. Estimated Yield of Fresh-Water Wells in Florida. URL: https://pubs.usgs.gov/of/1974/0256/report.pdf (accessed April 24, 2020).

United States Geological Survey (USGS). 1984. *Water of the World*. U.S. Government Printing Office, 1984-421-618/107, 19 p.

United States Geological Survey (USGS). 1985. National Water Summary 1984. USGS Water Supply Paper 2275. United States Government Printing Office, Washington, DC, 467 p.

United States Geological Survey (USGS). 1988. A National Look at Nitrate Contamination of Ground Water. Prepared by Nolan, B.T., Ruddy, B.C., Hitt, K.J., and Helsel, D.R. URL: http://water.usgs.gov/nawqa/nutrients/pubs/wcp_v39_no12/ (accessed March 13, 2009).

United States Geological Survey (USGS). 1995. Surficial Aquifer System of the New Jersey Coastal Plain: Significance to Resource Management; Fact Sheet. *FS* 086-95 (February 1995), 2 p.

United States Geological Survey (USGS). 1996. Nutrients in the nation's waters, too much of a good thing? *Circular* 1136, 24 p.

United States Geological Survey (USGS). 1997. *Global Change and Submerged Aquatic Vegetation Research*. USGS FS-09097 (June 1997). URL: http://www.nwrc.usgs.gov/about/5-year-plan.htm (accessed August 13, 2005).

United States Geological Survey (USGS). 1998a. *Radium-226 and Radium-228 in Shallow Ground-Water, Southern New Jersey*. Fact Sheet FS-062-98 (June 1998).

United States Geological Survey (USGS). 1998b. Ground-water discharge and base-flow nitrate loads of non-tidal streams, and their relation to a hydrogeomorphic classification of the chesapeake Bay Watershed, Middle Atlantic Coast. Water-Resources Investigations Report 98-4059. United States Geological Survey (USGS), Baltimore, MD, 71 p.

United States Geological Survey (USGS). 1999a. The quality of our nation's waters: Nutrients and pesticides. *Circular*, 1225, 82.

United States Geological Survey (USGS). 1999b. *Ground Water and Surface Water: A Single Resource*. U.S. Geological Survey Circular 1139. Denver, CO, 79 p.

United States Geological Survey (USGS). 1999c. *Sustainability of Ground-Water Resources*. U.S. Geological Survey Circular 1186, Denver, CO.

United States Geological Survey (USGS). 2000. Arsenic in Ground-Water Resources of the United States. Fact Sheet FS-063-00 (May 2000).

United States Geological Survey (USGS). 2002. Occurrence and Status of Volatile Organic Compounds in Ground Water from Rural, Untreated, Self-Supplied Domestic Wells in the United States, 1986–1999. Water-Resources Investigations Report 02-4085. Rapid City, SD, 51 p. (authored by Moran, M.J., Lapham, W.W., Rowe, B.L., and Zogorski, O.S.).

United States Geological Survey (USGS). 2003a. *Earth's Water: Runoff*. Water Science. URL: http://ga.water.usgs.gov/edu/runoff.html (accessed August 24, 2005).

United States Geological Survey (USGS). 2003b. Ground-Water Depletion across the Nation. Fact Sheet 103-03. (November 2003) (Bartolino, J.R. and Cunningham, W.L.) URL: http://pubs.usgs.gov/fs/fs-103-03/ (accessed September 2, 2005).

United States Geological Survey (USGS). 2003c. Karst. Fact Sheet. URL: http://water.usgs.gov/ogw/karst/index.htm (accessed August 25, 2005).

United States Geological Survey (USGS). 2004. *Resources on Isotopes*. URL: http://wwwrcamnl.wr.usgs.gov/isoig/period/cfc_iig.html (accessed August 13, 2005).

United States Geological Survey (USGS). 2005a. Ground Water in Freshwater-Saltwater Environments of the Atlantic Coast. Circular 1262 (authored by Paul, M.B.).

United States Geological Survey (USGS). 2005b. *National Wetlands Research Center, Strategic Plan 2005–2009*. URL: http://www.nwrc.usgs.gov/about/5-year-plan.htm (accessed August 13, 2005).

United States Geological Survey (USGS). 2016. *Groundwater Storage – The Water Cycle*. URL: https://water.usgs.gov/edu/watercyclegwstorage.html

United States Geological Survey (USGS). 2018. *What is Drought?* URL: https://ca.water.usgs.gov/california-drought/what-is-drought.html

United States Geological Survey (USGS). 2020. Groundwater: What is Groundwater? URL: https://www.usgs.gov/special-topic/water-science-school/science/groundwater-what-groundwater?qt-science_center_objects=0#qt-science_center_objects (accessed April 25, 2020).

United States National Park Service (USNPS). 2005. *New Jersey Pinelands (factsheet)*. URL: http://www.nps.gov/pine/ (accessed August 5, 2005).

United State National Park Service (USNPS). 2018. Mississippi River Facts. URL: https://www.nps.gov/miss/riverfacts.htm (Accessed December 14, 2020).

University of Illinois (UI). 2005. Precipitation. Weather World 2010. URL: http://ww2010.atmos.uiuc.edu/(Gh)/guides/mtr/hyd/prcp.rxml (accessed August 24, 2005).

University of Montana at Missoula (UMM). 2005. *Research at Flathead Lake Biological Station*. URL: http://www.umt.edu/flbs/Research/default.htm (accessed August 13, 2005).

Urbanska, K.M., Webb, N.R., and Edwards, PJ. (eds.). 1997. *Restoration Ecology and Sustainable Development*. Cambridge University Press, Cambridge, U.K. https://www.worldcat.org/title/restoration-ecology-and-sustainable-development/oclc/48105481

Valett, H.M., Fisher, S.G., Grimm, N.B., Stanley, E.H., and Boulton, A.J. 1992. Hyproheic-surface water exchange: Implications for the structure and functioning of desert stream ecosystems. In *Proceedings of the First International Conference on Ground Water Ecology*, Stanford, J.A. and Simons, J.J. (eds.). American Water Works Association, Bethesda, MD, pp. 395–405.

Walton, B. 2015. *Groundwater Depletion Stresses Majority of World's Largest Aquifers*. Circle of Blue, June 16. URL: https://www.circleofblue.org/2015/world/groundwater-depletion-stresses-majority-of-worlds-largest-aquifers/

Whitman, W.B., Coleman, D.C., and Wiebe, W.J. 1998. Prokaryotes: The unseen majority. *Proceedings of the National Academy of Sciences of the USA*, 95, 6578–6583

Winter, T. 1989. *Hydrologic function of wetlands. U.S. Geological Survey Yearbook*. Reston, VA.

Winter, T. 1994. *Personal communication*.

Wirsen, C. 2005. Is life thriving deep beneath the seafloor? Recent discoveries hint at a potentially huge and diverse subsurface biosphere. In *Oceanus* (online research magazine). Woods Hole Oceanographic Institution, Vol. 42, No. 2. URL: http://oceanusmag.whoi.edu/v42n2/wirsen.html (accessed August 23, 2005).

Wireman, M. and Job, C. 1997. The use of ground water sensitivity assessments for the purpose of the ground water disinfection rule. *Ground Water Monitoring and Remediation*, 17 (4), 63–66.

Wolock, D.M. U.S. Geological Survey Scientist. *Personal Communication. June* 27, 2005.

World Health Organization (WHO). 2004. Guidelines for Drinking Water Quality. Chapter 8, Chemical Aspects. URL: http://www.who.int/water_sanitation_health/dwq/gdwq3/en/index.html (accessed August 13, 2005).

World Health Organization (WHO). 2005. *Arsenic in Drinking Water*. URL: http://www.who.int/mediacentre/factsheets/fs210/en/index.html (accessed August 13, 2005).

Worldwatch Institute (WI). 2005. *Worldwatch Global Trends*. URL: http://www.worldwatch.org/pubs/globaltrends/ (accessed August 25, 2005).

3 Groundwater in the Economy

GROUNDWATER'S INFLUENCE IN THE ECONOMY

Water affects every aspect of the economy. Obviously, it is essential for life and we could not live without it. We require water for food production. Water is used in manufacturing processes as a raw material and as a medium for removing wastes. It also influences places where development occurs. If water is not easily available, people may not live in that particular location and industries will not be established there. However, what about groundwater, in particular? A foremost consideration is that, in economic terms, groundwater currently produced at a site is likely to be considered the most economical source of water supply – being available in sufficient quantity and quality for a significant period of time at a location convenient for its intended or primary use at the least cost, when compared with other options for water supply and use. Groundwater has influenced the development of cultures and communities around the world.

Groundwater serves and benefits the economy in five major ways:

1. As a necessity – people and all living things need water to live, for drinking, to satisfy their thirst and remain healthy (UNESCO, 2003, p. 5; Daly and Farley, 2004, p. 87), and enable people to work in the economy by supplying labor.
2. As a commodity – people derive benefit from groundwater, especially if it is their only water source for drinking, for food preparation, as well as for bathing, laundering and cleaning; and industries and farms use it as an input to produce manufactured goods and crops for food (UNESCO, 2003, pp. 8, 326; Daly and Farley, 2004, p. 87).
3. For ecological system services – water is the fundamental component of the hydrologic cycle which people rely on to provide the balance in nature to transmit nutrients, moisture and energy; sustain all living organisms, even in ways that we have not yet understood; and provide necessary habitat for water-based food sources, a required input to fish, fowl and game consumption (UNESCO, 2003, pp. 330–331; Daly and Farley, 2004, p. 87).
4. For residual absorption, that is, a specific and significant ecological system service – water dilutes, decomposes within its natural limits and transports human wastes and agricultural and industrial residuals elsewhere as a service to the user and disposer (UNESCO, 2003, pp. 81, 86; Daly and Farley, 2004, p. 75).
5. For aesthetic and recreational purposes – people prefer good quality water environments for leisure pursuits, educational interests and tourist attraction, which, in some respects, appear as a commodity (FAO, 1993; UNESCO, 2003, p. 16), such as hot springs, geysers and streams, lakes and wetlands maintained through groundwater discharge.

Clearly, some groundwater uses have the "private good" characteristics of a commodity when one use precludes the water for other (rival) purposes. Obversely, groundwater moves, serves as a residual conveyance and sink and sustains natural wonders, such as hot springs, and one person's use may detract from but not eliminate another's enjoyment of its existence or consumption of its service. In this latter case, groundwater is a "public good," as it would require substantial means to prevent others' use of it (FAO, 1993). As groundwater is mobile, it provides both aerobic and anaerobic contaminant degradation potentials, cycles nutrients and holds and releases energy while supporting all organisms including human beings; its ecosystem services have a public good characteristic that has no market-based value, but rather is invaluable.

First, we will explore the significance of groundwater availability, use and quality relative to the economy. Specific examples, such as the extensive use of groundwater for irrigation in the process of food production, will be used to highlight the resource's importance in the economy. Second, we will develop a model to understand the economic role of groundwater more generically.

GROUNDWATER AVAILABILITY

Economic Effects of Geology

In most locations, groundwater is economically accessible through wells. These wells are usually deep and typically narrow holes in the ground, usually installed by a well driller using a drill rig, but can be put in by hand techniques for shallow groundwaters (The subject of "Groundwater Access" is addressed in more detail in Chapter 4). Wells usually have a metal casing; however, recent wells have plastic (PVC or polyvinyl chloride) casing, especially in developing countries. Depending on the depth from the ground surface to the water table, wells may be just several meters deep to hundreds of meters deep, even over 1000 m deep (Brown, 2004, p. 104), using oil-well technology. The cost associated with well installation takes into account the type of geology (e.g., sand and gravel, fractured bedrock or other geologic setting) and the depth to the water table and the diameter of the well, considering the volume of water needed at a site or in a community.

The groundwater resource is extensive, as noted in Chapter 2. Approximately half of the groundwater is shallow (Driscoll, 1986, p. 55) and, therefore, potentially accessible for use through wells. About half of this larger resource "is considered extractable if no consideration is given to changes in streamflow, the effect on the environment, and the costs of extraction" (USWRC, 1978, p. IV–18).

Groundwater lies beneath the entire surface of the continents, but the type of geologic formation in which it resides affects its accessibility and its economic production of usable quantities. Formations with a substantial percentage of larger connected open pore space may produce considerable quantities of groundwater. The best example is an aquifer of sand and gravel. Some are commonly capable of producing 1.9–3.8 m^3/min and more. Other formations composed of compacted clay and other rock may hold considerable water, but as the pore spaces are small and not well connected, very little water can be produced, in the range of 0.004–3.8 m^3/min (USGS, 1985). Thus, geology affects resource availability. The geology that allows development to occur away from streams and farms to have wells by their houses rather than long pipelines from distant streams with sufficient flow represents an economic advantage for using accessible groundwater: an individual can live almost anywhere. In the western United States, drier range lands can be used for cattle grazing, because wells pumped by windmills and solar power make groundwater available where no other water is visible. The irrigated agriculture of the midwestern United States from the Mississippi River to the Rocky Mountains has depended on large volumes of groundwater, with wells in some locations from 3 m to as much as 670 m deep (USGS, 1985).

Economic Impacts of Groundwater Use

Economic use of groundwater has affected availability. In the midwestern United States, the water table of the vast Ogallala Aquifer has dropped up to 30 m and more in some locations, such as in areas of Nebraska, because so much water has been pumped for irrigation. While some of the irrigated water might return to the aquifer through subsurface infiltration, most of it is used by the plants and then transpired or evaporated from the soil zone. Similarly, Monmouth County, New

Jersey has used the Englishtown Aquifer whose water level has fallen about 30 m as a result of pumping for drinking water supply (USGS, 1985). Because of the fact that groundwater is the only principal water source in the northern region of Jordan, the water table has historically fallen at the rate of 1 m/year, mainly owing to irrigation (Schiffler, 1998, pp. 198–201). Falling water tables have been witnessed in every continent (UNEP, 1999; Sampat, 2000; Narayan et al., 2002; EEA, 2003). Falling water tables affecting groundwater availability have reduced grain harvest in some countries, such as Saudi Arabia, Yemen, India and China (Brown, 2004).

As groundwater tables drop, wells may go dry and costs of operating wells may rise because of increased energy costs to pump the water up a longer column. A liter of water weighs about 1.2 kg. Pumping 1136 L/day (300 gallons/day) to an individual home from a shallow well may not be a major cost. However, pumping tens of thousands of cubic meters from 100 to 200 m (325–650 feet) from below the earth's surface is a considerable cost. If a well is older and not deep enough, a new well may need to be installed to a deeper part of the aquifer to obtain adequate supply. Furthermore, if the rock that must be drilled through is harder, then the cost to install the well will be higher. All these factors affect where and how groundwater is used in the economy.

Conjunctive Management of Water Sources

Recognition of conjunctive management of groundwater and surface water together in watersheds holds significant potential to improve water resource availability. As the renewable portion of groundwater (the flow of groundwater that is naturally or artificially recharged) may interact with surface waters (USGS, 1999), incorporating this relationship into utilizing the water resource optimally can facilitate maximum water use for the expanding needs in the economy. Artificial recharge of groundwater (through septic systems, retention ponds, or porous surfaces) keeps water near the points of use (Fields, 1994, p. 24), allowing nature to cycle it within a locality rather than hastening its movement to streams and beyond economic capture.

The International Groundwater Resource Assessment Center identifies 262 managed aquifer recharge sites in the United States on its interactive map (IGRAC, 2018). Of these sites, 36% have river water as their source, 26% have a lake water source, 14% use wastewater as a source and 6% capture stormwater. An important advantage of subsurface storage of water is elimination of evaporation that causes major water losses from surface reservoir storage. Water quality policy in the United States promotes infiltration of stormwater to reduce contaminant loads in streams, a significant conjunctive management approach needing further research on groundwater quality impacts for future groundwater users (Brumley et al., 2018).

Land Value Effects

Groundwater availability also influences land values. For irrigated agriculture, land prices are affected by the amount of irrigation infrastructure in place (Tsur et al., 2004, p. 104). In the drier climatic areas of the western United States, groundwater has been the source for over 47% of irrigation water, as indicated in Table 1.3.1, in part reflecting significant drought in many states in that region (USGS, 2014, 2018b). Table 1.3.2 gives the area of irrigation using groundwater in the western United States. In the more humid eastern United States, groundwater accounts for 65% of irrigation water (USGS, 2018). In particular, water availability affects agricultural land values because of the increased value of production on irrigated lands. From 2008 to 2013, farmland area irrigated by groundwater in the United States increased by 1.8% to 15,054,162 hectares (37,199,644 acres) with an average use of 0.15 hectare-meters (1.2 acre-feet per acre) (USDA, 2018). In the United States, an average value

TABLE 1.3.1
Groundwater Used in Irrigation in the 17 Western United States Plus Arkansas, Louisiana and Missouri – 2010–2015 (Units in Million Gallons per Day)*

	Total Irrigation Water Use (mgd)		Groundwater Irrigation Use (mgd)		Percent Groundwater of Total Irrigation (%)	
States	**2010**	**2015**	**2010**	**2015**	**2010**	**2015**
South Dakota	362	211	198	139	55.6	65.9
North Dakota	165	233	77.5	102	47.0	43.8
Nebraska	5,660	6,090	4,300	5,420	76.0	89.0
Texas	6,830	5,490	5,100	4,480	74.7	81.6
Kansas	3,040	2,680	2,880	2,560	94.7	95.5
Oklahoma	564	931	429	787	76.1	84.5
Colorado	9,710	9,000	1,300	1,310	13.4	14.6
Idaho	14,000	15,300	3,820	4,900	27.3	32.0
Oregon	5,260	5,160	1,910	1,220	36.3	23.6
Montana	7,160	9,450	127	59.6	1.8	0.6
Wyoming	4,370	7,790	437	537	10.0	6.9
Washington	3,150	2,520	798	720	25.3	28.6
New Mexico	2,700	2,370	1,240	1,050	45.9	44.3
California	23,100	19,000	8,690	13,900	37.6	73.2
Utah	3,220	3,030	494	537	15.3	17.7
Nevada	1,570	2,070	653	972	41.6	47.0
Arizona	4,570	4,530	1,690	1,970	37.0	43.5
Arkansas	8,720	11,600	7,380	9,280	84.6	80.0
Louisiana	928	1,050	670	720	72.2	68.6
Missouri	1,400	1,370	1,350	1,300	96.4	94.9
Totals	106,479	109,875	43,543.5	51,963.6	40.9	47.3

Sources: U.S. Geological Survey (c). 2010. Estimated use of water in the United States in 2010. *Circular*, 1405. URL: https://pubs.usgs.gov/circ/1405/pdf/circ1405.pdfU.S. Geological Survey (USGS). 2015. Estimated use of water in the United States in 2015. *Circular*, 1441. URL: https://pubs.usgs.gov/circ/1441/circ1441.pdf

* 1,000,000 gallons/day = 3,785 cubic meters/day

(unweighted) of agricultural land that is irrigated is nearly 166% greater than that of unirrigated land in the 20 states surveyed, with a range of –2.1% to 528.6% (USDA, 2018). Table 1.3.3 provides details on the agricultural land values in 20 states of the United States. Agricultural reporting for Canada also shows irrigated farmland having increased value (FCC, 2020).

Climatic Effects

Climatic effects can significantly affect groundwater availability and ultimately its use. The drought in California from 2014 to 2017 had the effect of farmers shifting from surface water supplies to groundwater. Irrigated agriculture in California reduced surface water use by 64% from 2010 to 2015 and increased groundwater use by 60% during that time. Groundwater was available but future availability may be constrained by the significant loss because of use of 10–11 cubic kilometers of groundwater from 2012 to 2016 (Wolf, 2017). Twenty years of precipitation deficit in California suggest that many decades may be needed of greater precipitation for groundwater levels to recover (Harter and Brewster, 2018).

TABLE 1.3.2
Area of Groundwater Irrigation in the 17 Western United States

	Total Irrigated Area*		Percent of Area Using Groundwater†	Principal Irrigation Method‡
	Hectares	Acres	Percent	(Type/% irrigated land) (G = Gravity; S = Sprinkler)
Arizona	344,736	851,861	37	G/89
California	3,055,037	7,549,161	51	G/60
Colorado	934,639	2,309,543	43	S/64
Idaho	1,421,190	3,511,839	41	S/88
Kansas	1,153,887	2,851,317	98	S/100
Montana	757,729	1,872,389	3	G/61
Nebraska	3,357,903	8,297,560	92	S/89
Nevada	279,214	689,953	52	G/68
New Mexico	281,108	694,632	58	S/66
North Dakota	86,493	213,728	72	S/91
Oklahoma	172,639	426,602	88	S/100
Oregon	628,951	1,554,173	33	S/73
South Dakota	149,679	369,864	65	S/88
Texas	1,817,842	4,491,987	90	S/89
Utah	455,314	1,125,106	25	S/59
Washington	656,985	1,623,389	32	S/87
Wyoming	573,959	1,418,284	10	G/74

Source: U.S. Department of Agriculture. 2012. Farm and Ranch Irrigation Survey.

* Total Irrigated Area includes area irrigated by groundwater only, by surface water only and by both water sources and may include other sources such as reclaimed irrigation water.

† Twenty-seven percent of area irrigated by groundwater was irrigated with both groundwater and surface water or another water source.

‡ G = gravity or flood irrigation, S = sprinkler system; some irrigated areas may use more than one irrigation method, including drip irrigation; Kansas and Oklahoma reported area of principal irrigation method exceeding Total Irrigated Area.

GROUNDWATER QUALITY

Groundwater quality also has a significant effect on where and how groundwater is used. Aquifers from which only slightly mineralized water can be produced will readily be tapped for domestic and municipal uses. In some areas, where groundwater is from ancient deposits that are now lithified (turned to rock) and fractured, allowing water to flow through the fractures, its quality may include higher concentrations of dissolved solids and even radioactivity. In these cases, the groundwater may not be useable, or at least may require treatment. In places where more recent freshwater is situated over brines and where production has increased to remove (or mine) the newer water, the brines or other mineralized water may move toward the wells, reducing its quality for certain uses over time. Significant and increasing withdrawal of groundwater along the United States Atlantic and Gulf coasts, where 60 million people resided and worked in coastal counties in 2016 (US Census Bureau, 2018), has induced saltwater intrusion resulting in closure of private and municipal wells and has resulted in the use of other wells and water sources (Barlow and Reichard, 2010). The consequences of these changed conditions are treatment of the contaminated water or finding a new water source, both of which have costs associated with them. Similarly, human-caused contamination can have the same effect on drinking water supply wells.

TABLE 1.3.3
Average Value per Hectare for Irrigated and Nonirrigated Cropland for 20 Selected U.S. States

	2018 Cropland Prices per Hectare (US$)*		Percent Irrigated Greater (Less)
State	Irrigated	Nonirrigated	than Nonirrigated
Missouri	12,899	9,316	38.5
Kansas	7,166	4,720	51.8
Nebraska	15,197	8,772	73.2
South Dakota	N.A.	8,031	N.A.
Florida	19,052	14,703	29.6
Georgia	11,243	7,438	51.2
Arkansas	8,352	5,066	64.9
Louisiana	6,894	7,042	(2.1)
Mississippi	8,204	6,227	31.7
Oklahoma	N.A.	4,423	N.A.
Texas	5,931	5,066	17.1
Arizona	20,880	N.A.	N.A.
Colorado	11,367	3,089	268.0
Idaho	13,220	3,707	256.7
Montana	7,413	2,076	257.1
Nevada	N.A.	N.A.	N.A.
New Mexico	9,711	1,087	793.2
Utah	13,838	2,965	366.7
Wyoming	5,362	1,878	185.5
California	34,101	12,108	181.6
Oregon	12,479	5,313	134.9
Washington	21,745	3,459	528.6

Source: USDA. 2018. *Census of Agriculture*, Washington, DC.

* 1 ha = 2.471 acres N.A = not available.

Contamination of groundwater has not only caused the resource to be abandoned, but also the land under which it lies to be removed from use in some cases. For example, in Love Canal outside Buffalo, New York, waste chemicals were disposed and buried, and a community was built on it later. After some years, it was discovered that the residents living above the old chemical waste disposal area (canal) were becoming sick because of the chemicals in the groundwater and soil. For over 20 years, the site was abandoned while being remediated. Groundwaters must be intercepted for treatment in such cases rather than being allowed to naturally discharge to the local stream and contaminate it (USEPA, 1987). In other locations, such as Shenyang, China and Jaipur, India, and other large cities with industries using toxic chemicals and storing solvents in underground tanks, previous groundwater sources could no longer be used and alternate water supplies have been tapped (Sampat, 2000, p. 19). Currently, government and private sector programs all over the world are seeking to redevelop these locations, referred to as "brownfields," with an estimated 450,000 brownfields sites in the United States (Meyer et al., 1995; USEPA, 2018a). Half of the brownfields sites in the United States exist because of petroleum products that leaked into the subsurface contaminating groundwater (USEPA, 2018b). Many other hazardous and toxic contaminants are found at brownfields sites (Campbell, 2017). Remediating and investing in these sites improves the local tax base,

BOX 1.3.1 UNDERGROUND BRINES AS THE RAW MATERIAL FOR A CORPORATION

The Dow Chemical Company was established with the fascination and interest in the chemicals in underground brines by a chemist, Herbert Dow, in the 1880s, who observed them in samples from northeast Ohio natural gas-well operations. The brines were a waste by-product of gas production. Mr. Dow refined the method of using electrolysis for extracting chemicals from brines, producing bromine in the 1890s at a cost lower than other producers of that time. Dow and his associates used wells to produce brine for commercial chemical production in Michigan, Louisiana and California. One of the first brine wells used by the predecessor to Dow Chemical Company was 914 m deep near Canton, Ohio, in 1889 (Brandt, 1997, p. 14). From these subterranean aqueous sources, the company has grown to one of the largest chemical companies in the world with annual revenues of $20 billion.

Source: Brandt, E.N., *Growth Company: Dow Chemical's First Century*, Michigan State University Press, East Lansing, MI (1997), 649.

provides jobs, uses local infrastructure already in place and reduces demand for undeveloped land while improving environmental quality.

Interestingly, waters that were once considered as nonpotable groundwaters because of higher concentrations of certain naturally occurring chemicals, such as brines (brackish groundwater), can have beneficial uses in the economy. In Michigan and other locations, the subsurface brines, often, but not always, in association with oil and natural gas, have provided the basis for a chemical industry and are still produced for that purpose. The ancient waters held by the subsurface were first used for producing commercial chemicals over 100 years ago. Early products from the brines which were used in the manufacture of other chemicals include bromine for the pharmaceutical and oil markets, chlorine for cleaning and salts, magnesium for the metals industry and iodine for medical and pesticide applications (Brandt, 1997). Box 1.3.1 highlights the early development of underground brines as economic deposits with commercial applications.

MAJOR ECONOMIC GROUNDWATER USES IN NATIONAL AND INTERNATIONAL CONTEXTS

International

Globally, while difficult to accurately calculate, groundwater provides approximately 50% of potable water supplies, 40% of self-supplied industrial demand and 20% of irrigated agricultural water use (UNESCO, 2003, p. 78). Worldwide groundwater withdrawals (GWW) are estimated between 600 and 700 km^3 (UNESCO, 2003, p. 78). Most of the world's cities and towns rely on groundwater as a source for municipal water supplies (Shah et al., 2000). By far, the largest use of groundwater is for agriculture, with 44 countries from all the continents using 74% of their GWW for agricultural use (including livestock watering), as reported in a statistics from 1973 to 1996 (WRI, 2005). Furthermore, groundwater is "the world's most extracted raw material … forms the cornerstone of the Asian 'green agricultural revolution,' provides about 70% of piped water supply in the European Union, and supports rural livelihoods across extensive areas of sub-Sahara Africa" (UNESCO, 2003, p. 78).

The uses of groundwater reported earlier have generated real products and benefits for the national economy of the United States, and similar uses around the world contribute similarly to the world economy. Groundwater is a significant economic resource and a basic requirement for

44% of the population in the United States who rely on it as a source of drinking water and for other household purposes (NGWA, 2019). As streamflow in the United States derives 40% to 50% of its volume from groundwater discharge (USGS, 2018), it can be stated that groundwater supports at least half of the labor supply supporting the economy of the United States. In the European Union, as 75% of the drinking water supply comes from groundwater, this resource similarly maintains a large proportion of the labor supply in the EU economy (Sampat, 2000). Table 1.3.4 indicates that up to as many as 2 billion people worldwide rely on groundwater as a source of drinking water. Table 1.3.5 shows the percentages of the drinking water supply in countries of the European Union which are derived from groundwater. In the industrial and power generation sectors, principal water utilization is for the production of food, paper, chemicals, refined petroleum and primary metals, as well as for cooling (USGS, 2003). Groundwater in the United States accounted for 29% of all freshwater withdrawals, 48% of irrigated agricultural water supplies in 2015 (USGS, 2018) and was the source of water for 25% of irrigated cropland (Brown, 1996). Box 1.3.2 highlights the key groundwater irrigation costs in the United States, showing the farmers' high capital and energy expenditures. In India, groundwater irrigates lands that produce 60% of the grain production (Shah et al., 2000), doubling yields that occurred solely from rainwaters, while farmers' incomes have expanded by 80%–100% (FAO, 1993). Thus, groundwater is very significant to national economies around the world.

TABLE 1.3.4
Proportion of Drinking Water from Groundwater Based on World Region

Region	Proportion of Drinking Water from Groundwater (%)	People Served (Millions)
Asia-Pacific	32	1000–1200
Europe	75	200–500
Latin America	29	150
United States	51	135
Australia	15	3
Africa	NA	NA

Source: Reprinted from Sampat, P., Groundwater Shock, *World Watch*, January/February, 12, 2000, www.worldwatch.org. With permission.

TABLE 1.3.5
Percentage of Drinking Water Supplied by Groundwater in Selected European Countries

Country	Percentage (%)	Country	Percentage (%)	Country	Percentage (%)
Austria	99	Germany	72	Portugal	80
Bulgaria	60	Greece	50	Slovak Republic	80
Czech Republic	43	Hungary	95	Spain	21
Denmark	98	Italy	80	Sweden	49
Finland	57	Netherlands	68	Switzerland	83
France	56	Norway	13	United Kingdom	28

Source: European Environmental Agency (EEA), *Groundwater Quality and Quantity in Europe*, EEA, Copenhagen, Denmark, 1999, 123.

BOX 1.3.2 GROUNDWATER IRRIGATION COST AND PRICING IN THE WESTERN UNITED STATES

A perspective on costs of groundwater in irrigation development in the western United States, particularly in the High Plains region:

- The region of the western United States drawing on Ogallala Aquifer groundwater has experienced rising costs to irrigate agricultural fields. Factors affecting the cost to irrigate crops using groundwater include:
    - Depth to water
    - Energy prices
    - Water scarcity
    - Public policy affecting water allocation for competing uses
    - Legislative budget concerns
    - Awareness of impacts of climate change on surface water availability
- Increasing energy expenditures and falling water tables prompted a large number of irrigators to employ better water use techniques, even substituting center-pivot supply, low-pressure sprinklers and below-ground drip application.
- Variable irrigation cost for a well of 200-feet deep was $95 per acre-foot in 2008 ($114 per acre-foot in 2018$US) in the High Plains. Other average variable groundwater irrigation cost examples include:
    - National average cost of pumping in 2013 Farm and Ranch Irrigation Survey - $17/acre-foot ($18/acre-foot in 2018$US) – 60% higher than for surface water
    - Nebraska – $15/acre-foot (in 2013; $16/acre-foot in 2018$US)
    - Kansas – $17/acre-foot (in 2013; $19/acre-foot in 2018$US)
    - California – $39/acre-foot (in 2013; $42/acre-foot in 2018$US)
    - Arizona – $38/acre-foot (in 2013; $42/acre-foot in 2018$US)
    - Compared with a more humid eastern state: Maryland – $8/acre-foot (in 2013) – 20% lower than for surface water
- A farmer's cost to pump groundwater does not include the effects of water table or water quality changes on other nearby irrigators or having groundwater abstraction greater than recharge.
- Surface water and groundwater irrigation pricing policy can vary among states:
    - States may control surface water allocations closely
        - Typically farmers pay the cost of channeling or piping the surface water to the fields, including pumping which is affected by distance and energy prices
    - Groundwater abstraction may not be as intently overseen in some states while other states may assign annual permitted groundwater pumping quantity
        - Well and pump operating and maintenance costs may be the only costs farmers consider
        - Energy costs increase with increasing depth
        - Energy prices and depths to water table differ from region to region
    - Enterprises supplying water to farm fields may charge on a volume basis and on an irrigated area or delivery duration method

(*Continued*)

BOX 1.3.2 (Continued)

- Recovering costs of building and using an agricultural irrigation undertaking may be accomplished by implementing water pricing to finance construction, operation and maintenance, while public policy goals may focus on charges and fees to use water more efficiently for resource conservation or to meet environmental objectives. A range of water pricing methods may include:
  - No charge – with no effect on water demand
  - Self-provision cost – affecting demand
  - Per unit price – affecting demand and potentially meeting public objectives
  - Indirect charge on land area or time duration basis – affecting demand and potentially meeting public objectives
  - Income-based fee based on land area – potentially meeting public objectives
- Charges for irrigation water supply can be fixed – based on land area, crop, or water access – or volumetric – based on per unit use that must be measured.
  - Volumetric charges may be uniform rates, increasing block rates, decreasing block rates, or seasonal rates.

Sources: Christian-Smith, J., and Kaphiem, C. (2011). *Volumetric Water Pricing and Conjunctive Use: Alta Irrigation District.* Pacific Institute. URL: https://www.pacinst.org/reports/success_stories/volumetric_water_pricing_and_conjunctive_use.pdf

Merrett, S., (2002). *Water for Agriculture: Irrigation Economics in International Perspective.* Spon Press, London.

Seo, S., Segarra, E., Mitchell, P.D., and Leatham, D.J. (2008). Irrigation technology adoption and its implication for water conservation in the Texas High Plains: A real options approach. *Agricultural Economics*, 38 (1): 47–55.

U.S Department of Agriculture. (2013). *Farm and Ranch Irrigation Survey.* URL: https://www.nass.usda.gov/Publications/AgCensus/2012/Online_Resources/Farm_and_Ranch_Irrigation_Survey/fris13_1_012_012.pdf

Wichelns, D. (2010). Agricultural Water Pricing: United States. Organisation For Economic Co-Operation and Development, 27p. URL: https://www.oecd.org/unitedstates/45016437.pdf

United States

A wide range of uses of groundwater has been observed in certain regions in the United States, where large surface water supplies are not available or may only be a seasonal resource. These uses are described in Table 1.3.6 for the year 2015, with percentages based on the total groundwater volume of 84,700 million gallons per day (320.6 million m^3/day) used in the United States.

The USGS has estimated that groundwater withdrawal in the United States in 2015 was nearly 7% more than that in 2010 and 26% of all the water withdrawals in 2015 (USGS, 2018). During the same time period between 2010 and 2015, surface water withdrawals declined in the United States by 14% (USGS, 2014, 2018). Thus, at the margin, groundwater provides an increasing volume to meet the demands of the economy in the United States. Table 1.3.7 indicates the United States' national projection from its last in-depth national study of water use. Interestingly, the projection for the year 2000 and the actual population were surprisingly close; however, greater efficiencies were

TABLE 1.3.6
Percent of Groundwater Use in the United States during the Year 2015

Category	Purpose	Percentage of Fresh Groundwater Use*
Public drinking water	Life requirement for humans	17.9%
Domestic water	Private water supplies	3.8%
Commercial and industrial	Washing, processing and cooling	3.2%
Thermoelectric	Power plant cooling	0.7%
Irrigation	Agricultural plant life requirement	67.5%
Livestock watering	Life requirement for farm animals	1.5%
Aquaculture	Life requirement for farmed fish	1.9%
Mining	Conveying mined materials and wastes and processing	3.4%
Wildlife watering	Life requirement for wildlife	Not measured, not reported
Waste disposal	Sink for residuals and wastes	Not reported
Feedstock (brines)	Raw material for chemicals and processes	Not reported

Source: United States Geological Survey (USGS). 2018. Estimated Use of Water in the United States in 2015. USGS Circular 1441, USGS, Reston, VA, 2003.

* Except where noted.

TABLE 1.3.7
Past Projection of Population and Water Use (Withdrawals) in the United States

Population (Millions)			Fresh Groundwater Use (Million m³/Day)			Total Freshwater Use (Million m³/Day)		
1975	2000			2000			2000	
Estimation	Projection	Actual	1975	Projection	Actual	1975	Projection	Actual
216.4	268.0–282.2	281.4	310.4	NE	315.3	1336 (USGS)	1160	1307
						1507 (USWRC)	(USWRC)	(USGS)

Sources: United States Water Resources Council (USWRC), *The Nation's Water Resources 1975–2000*, Vol. 2, U.S. Government Printing Office, Washington, DC, 1978.U.S. Geological Survey (USGS). 2004. *Estimated Water Use in the United States in 2000.* USGS Circular 1268. United States Government Printing Office, Washington, DC.

Note: USWRC reported a 79.1 million m³ overdraft of fresh groundwater in 1975, which was 25.4% of the total groundwater withdrawn in that year. Most of this overdraft (the respective percentages of overdraft in parentheses) was in the Missouri (24.6%), Arkansas-White-Red (61.7%), Texas-Gulf (77.2%), Lower Colorado (48.2), Great Basin (41.5%) and California (11.5%) water resources regions. Overdraft is defined in the USWRC report as groundwater use exceeding the natural recharge (USWRC, 1978, p. II-11).

NE = not estimated.

expected with regard to water use than those that occurred during 1975–2000. Young (2005, p. 71) noted the problems in making technology forecasts for water as an input to production. Notably, total water use has declined by 21% from 2000 to 2015. Table 1.3.8 shows a comparison between groundwater and surface water use in the United States from 2010 to 2015. Groundwater use was relatively flat for many years and rose in 2015, while total freshwater use is declining significantly.

Irrigation

Notably, the use of groundwater as a source of irrigation more than tripled by 17% from 1950 to 2015 and has resulted in the reduction in the use of surface water for that purpose. The national

TABLE 1.3.8
Comparison of Fresh Groundwater and Surface Water Use in the United States – 2010–2015

Table A [In Million Gallons per Day (MGD)]

	Groundwater				Surface Water			
	2010 (MGD)	**2015 (MGD)**	**Change (MGD)**	**Percent (%)**	**2010 (MGD)**	**2015 (MGD)**	**Change (MGD)**	**Percent (%)**
Public supply	15,700	15,200	(500)	−3.2%	26,300	23,800	(2,500.00)	−9.5%
Domestic	3,540	3,210	(330)	−9.3%	64	49	(14.80)	−23.2%
Irrigation	49,500	57,200	7,700	15.6%	65,900	60,900	(5,000.00)	−7.6%
Livestock	1,200	1,240	40	3.3%	797	760	(37.00)	−4.6%
Aquaculture	1,820	1,600	(220)	−12.1%	7,610	5,950	(1,660.00)	−21.8%
Industrial								
Fresh	2,900	2,670	(230)	−7.9%	12,100	11,300	(800.00)	−6.6%
Saline	48	43	(6)	−11.4%	937	743	(194.00)	−20.7%
Mining								
Fresh	1,120	1,010	(110)	−9.8%	1,130	877	(253.00)	−22.4%
Saline	2,790	1,860	(930)	−33.3%	282	256	(26.00)	−9.2%
Thermoelectric								
Fresh	587	425	(162)	−27.6%	116,000	94,700	(21,300.00)	−18.4%
Saline	134	172	38	28.4%	43,800	37,600	(6,200.00)	−14.2%
Total								
Fresh	76,000	82,300	6,300	8.3%	230,000	198,000	(32,000.00)	−13.9%
Saline	3,290	2,340	(950)	−28.9%	45,000	38,600	(6,400.00)	−14.2%

Table B* [In Million Cubic Meters per Day (Mm³/d)]

	Groundwater				Surface Water			
	2010 (Mm³/d)	**2015 (Mm³/d)**	**Change (Mm³/d)**	**Percent (%)**	**2010 (Mm³/d)**	**2015 (Mm³/d)**	**Change (Mm³/d)**	**Percent (%)**
Public supply	59	58	(2)	−3.2%	100	90	(9.46)	−9.5%
Domestic	13	12	(1)	−9.3%	0	0	(0.06)	−23.2%
Irrigation	187	217	29	15.6%	249	231	(18.93)	−7.6%
Livestock	5	5	0	3.3%	3	3	(0.14)	−4.6%
Aquaculture	7	6	(1)	−12.1%	29	23	(6.28)	−21.8%
Industrial								
Fresh	11	10	(1)	−7.9%	46	43	(3.03)	−6.6%
Saline	0	0	(0)	−11.4%	4	3	(0.73)	−20.7%
Mining								
Fresh	4	4	(0)	−9.8%	4	3	(0.96)	−22.4%
Saline	11	7	(4)	−33.3%	1	1	(0.10)	−9.2%
Thermoelectric								
Fresh	2	2	(1)	−27.6%	439	358	(80.63)	−18.4%
Saline	1	1	0	28.4%	166	c	(23.47)	−14.2%
Total								
Fresh	288	312	24	8.3%	871	750	(121.13)	−13.9%
Saline	12	9	(4)	−28.9%	170	146	(24.23)	−14.2%

Sources: United States Geological Survey (USGS), *Estimated Use of Water in the United States in 2010,* USGS Circular 1405, United States Government Printing Office, Washington, DC (2014). URL: https://pubs.usgs.gov/circ/1405/ United States Geological Survey (USGS), *Estimated Water Use in the United States in 2015,* USGS Circular 1441, Washington, DC, 2018, URL: https://pubs.er.usgs.gov/publication/cir1441.

Note: MGD = 1,000,000 gallons per day = 3785.4 cubic meters per day.

* All values converted from original MGD of Table A to Mm³/d

factors and costs affecting irrigation use in the United States are highlighted in Box 1.3.2. Notably, groundwater accounts for 48.5% of total irrigation water in the United States (see Table 1.3.8); more arid states that have higher groundwater application rates tend to rely on surface or flood irrigation, than on the more efficient methods, such as sprinkler or microirrigation systems. However, arid states have clearly improved water efficiency through the use of sprinkler systems, in particular, in Texas and Colorado. In 2000, groundwater used for irrigation in the 17 western U.S. states with typically drier climates accounted for 71% of all groundwater used nationally for irrigation, 48% of all groundwater used nationally for all uses and 13% of all water used in the United States (USGS, 2018). Clearly, groundwater used for supplying food is significant.

Drinking Water

Drinking water supply is the second largest usc of groundwater in the United States and a primary use in other countries as well (Sampat, 2000) and has a significant role in the economy. In 2000, in the United States, 39,063 community water suppliers used groundwater as their principal source out of the 50,132 community water utilities that serve 25 or more people (the largest, serving 2.3 million people). However, small water systems that serve transient, nonresidential populations in schools, hospitals and other facilities, such as roadside rest areas, and have their own water source are not included. At the fundamental level of providing a commodity that is essential to the local economy in the United States, these water utilities operate as an economic monopoly (they are organized to be the only water deliverer in their communities and as such have no significant competition), and supply water that meets health-based standards to domestic, commercial, industrial and thermoelectric power users that are not self-supplied (USGS, 1998). These water utilities have a large investment in their treatment and distribution systems, and often finance the upgrading and maintenance of these systems, issuing municipal or private bonds, depending on whether they are public- or investor-owned, respectively.

Several trends in water use and infrastructure are evident in groundwater systems in the United States. Table 1.3.9 gives the population served by and the per capita use of public water suppliers for drinking water and related uses in the United States from 1975 to 2015. The per capita use rose and then declined during this period. Notably, the population relying on groundwater from public water

TABLE 1.3.9
Drinking Water Use from Public Water Suppliers in the United States, 1975–2015

	Public Water Supply From				Per Capita Use Of		
	Groundwater		Surface Water			Surface	
Year	Population (in 1000s)	Supplied (Mm^3/d)*	Population (in 1000s)	Supplied (Mm^3/d)	Groundwater (L/day)†	Water (L/day)	Total (L/day)
1975	64,700	41.6	110,000	71.9	644	655	647
1995	91,200	57.2	137,000	95.0	628	708	678
2015	87,400	67.0	216,201	104.8	659	416	477

Sources: United States Geological Survey (USGS), (1977) *Estimated Use of Water in the United States in 1975*, USGS Circular 765, United States Government Printing Office, Washington, DC, 39.United States Geological Survey (USGS), *Estimated Use of Water in the United States in 1995*, USGS Circular 1200, United States Government Printing Office, Washington, DC (1998), 71.United States Geological Survey (USGS), *Estimated Use of Water in the United States in 2015*, USGS Circular 1441, United States Government Printing Office, Washington, DC (2018a), 65.

* Mm^3/d = 1,000,000 cubic meters = 264,172,052 gallons or 1 million m^3 = 264.172 mgd

† 1 L = 0.264 gallon or 1 gallon = 3.7854 L

supplies increased by 35%, and public water supply demands rose to 61%. In addition, the population using surface water supplies grew by 96% and the quantity used by this population increased by 46%. This relationship suggests that, at the margin of use, a greater pressure on groundwater sources for public supply existed than for surface water sources. Table 1.3.10 portrays the statistical results and a summary of a review of infrastructure for groundwater-supplied systems in the United States. Significant economies of scale accrue to large water systems, based on the data presented.

TABLE 1.3.10
Characteristics of Groundwater-Supplied Community Water Systems in the United States, 2006

	Groundwater-Supplied* Community Water System Service Population Ranges								
Category	**100 or Less**	**101–500**	**501–3300**	**3301–10,000**	**10,001–50,000**	**50,001–100,000**	**100,001–500,000**	**Over 500,000**	**All Sizes**
Primarily groundwater systems									
100% Groundwater	10,308	12,155	8408	2072	1384	164	70	10	34570
Mostly Groundwater	7	203	310	558	346	54	47	3	1527
Average number of wells at 100% GW systems	1.9	2.2	3.1	6.7	7.8	25.8	50.2	31.1	3.2
Avg. kilometers of pipe in place†	6.4	6.4	35.4	109.4	300.8	725.7	1243.8	4827.0	61.1
For Groundwater Systems									
Avg daily production (m^3/day)‡	12.5	212.0	1540.5	6813.0	30700.1	65726.5	217444.5	1021798.6	16173.3
Avg. total revenue ($ in 1000s)	19	35	207	853	2840	9730	24556	142215	510
Avg. total expenses ($ in 1000s)	10	43	168	598	2674	7379	23182	99861	382
Avg. total Capital investment in the past 5 years ($ in 1000s)	75	180	598	982	3382	14420	45405	134434	1089
Type of capital expenses in the past 5 years (% of systems reporting each type of expense)									
Land	0.1	2.7	8.1	1.1	25.3	48.0	41.4	66.7	5.8
Water source	59.8	50.5	50.8	45.7	58.3	71.1	57.9	100.0	53.2
Distribution and transmission	30.5	49.0	73.254.7	88.8	84.1	87.7	83.0	100.0	52.2
Treatment	30.8	25.5	24.7	33.2	50.8	81.2	68.8	100.0	29.4
Storage	29.5	24.7	35.2	33.3	55.7	77.5	69.0	33.3	32.6
Security	14.0	3.0	4.7	40.0	42.1	63.1	44.9	66.7	12.2
Other	3.6	5.7	11.5	17.9	17.9	67.8	49.4	66.7	11.3

Sources: United States Environmental Protection Agency (USEPA), *Community Water System Survey*, EPA 815-R-09-002, (May 2009), 3 volumes.

* Except where indicated by footnote.

† Includes all groundwater- and surface water-supplied community water systems surveyed.

‡ Includes systems that are 100% and primarily groundwater-supplied.

The selected characteristics of groundwater-supplied community water utilities (or "groundwater systems") provided earlier are derived from a statistical study, and the analysis of these utilities in the United States, completed in 2006, included

Ownership. Sixty-two percent of the groundwater systems in the United States are privately owned, with 87% of these serving populations of 500 or fewer persons.

Average Daily Production. Average daily production ranges from 0.0001 m^{3}day for systems serving 100 or fewer persons to 1021.80 m^3/day for the largest groundwater systems.

Average Number of Wells. Small systems serving 100 or fewer persons have an average of 1.9 wells per system. The largest systems serving more than 500,000 persons have an average of 31.1 wells per system. Overall, the average wells per system is 3.2.

Treatment. Because groundwater is typically considered as a cleaner water source, many systems do not treat it before delivering it to the consumers, to save cost. Twenty-eight percent of systems surveyed serving 100 or fewer persons did not treat groundwater before its use by the consumers. However, the largest systems in the survey all provided treatment.

Pipeline in Place. The small systems averaged 6.4 km of pipeline to deliver groundwater. The largest systems utilized 4827 km of pipeline. The average distance of the pipeline in place was 61.1km per system, with a mean of 63.4 connections per kilometer. While the largest systems replaced an average of 12.9 km per system in the prior 5 years, the smallest systems in the survey did not replace any pipeline during that time.

Revenues. The average annual revenue for the smallest systems was $19,000, while the largest systems had an average revenue over $142 million. Most of the water systems used a uniform or declining block rate, but about 1/3 of the large systems serving more than 50,000 people used an increasing block rate.

Expenses. The survey found that expenses for groundwater systems ranged on an average from $10,000 per year for the smallest to over $100 million for the largest. Examining expenses on a water-unit basis indicated significant economies of scale for groundwater systems: $1.19/m^3 produced for systems serving 100 or fewer people down to $0.35 for systems serving more than 500,000 people, a factor of just over 3.

Capital Investment. The average 5-year capital investment ranged from $75,000 for systems serving 100 or fewer people to $134.4 million for systems serving over 500,000 people. The four major types of capital expenses across most system size categories were distribution and transmission systems, water source, treatment and storage (Job, 2004).

Industrial

Industrial use, including manufacturing, thermoelectric cooling and mining, constitutes 5.7% of all groundwater use in the United States. Fresh groundwater use declined from 1995 to 2000 by 17%, to 18 million m^3/day, while saline groundwater use increased by 23.6% to 4.8 million m^3/day over the same time, primarily because of increased mining demands. The total industrial groundwater use was 2.8% of the total industrial water demand in 2000.

Per Capita Use

Considering all types of water use withdrawals from both the fresh ground water and surface water sources, the per capita use in the United States has changed over time. Table 1.3.11 documents the changes in the water use per person in the United States from 1900 to 2010. In this period reported, the per capita use peaked around 1970. Since that time, the per capita use has declined, most probably owing to the increased demand from a larger population that made the relatively fixed total freshwater resource scarcer. During the 1970s and 1980s, sources of groundwater were found to be contaminated at some municipal well fields, requiring new sources to be found and stream habitat for wildlife were recognized as a required allocation for surface waters, and these were the factors that potentially caused increased costs of supply and actions to conserve water.

TABLE 1.3.11
Per Capita Water Use in the United States, All Uses, 1900–2010

Year	Population	Total Freshwater Use[a] (Million m³/day)[b]	Per Capita Use (L/day)[c]
1900	76,212,168	151.4	1987.3
1910	92,228,496	249.8	2710.4
1920	106,021,537	348.3	3285.7
1930	123,202,624	416.4	3380.4
1940	132,164,569	514.8	3895.2
1950	151,325,798	772.2	5102.7
1960	179,323,175	1029.6	5742.5
1970	203,302,031	1404.4	6908.4
1980	226,542,199	1430.9	6317.8
1990	248,709,873	1279.5	5144.4
2000	281,421,906	1306.0	4640.9
2010	308,745,538	1063.7	3450.0

Sources: USCB, 2008; 2010. *Census of Population and Housing, Washington, DC.*United States Water Resources Council (USWRC), *The Nation's Water Resources 1975–2000*, Vol. 2, U.S. Government Printing Office, Washington, DC, 1978.United States Geological Survey (USGS). 2014. Estimated Use of Water in the United States in 2010, USGS Circular 1405, Reston, VA, 2003.

[a] Includes all fresh water uses of both groundwater and surface water sources: public supply, domestic supply, irrigation, livestock, aquaculture, commercial, industrial, mining and thermoelectric power.

[b] 1,000,000 cubic meters = 264,172,052 gallons or 1 million m^3 = 264.172 MGD

[c] 1 L = 0.264 gallon or 1 gallon = 3.7854 L

All Groundwater Uses

Considering all groundwater uses, the resource supports a significant portion of the water-use economy of the United States with 29% of the direct freshwater uses supported by groundwater, as identified in Table 1.3.8 for the year 2015. This calculation of groundwater use does not include the maintenance of 40% of the annual streamflow, then the proportion becomes 99% of all the fresh groundwater uses (USGS, 1999, 2018).

Groundwater Services Sector and Employment

Groundwater contributes considerably to the economies of the world's nations, underlying that its support is a substantial service sector of which a portion focuses on groundwater and its use. This sector provides groundwater assessments and surveys; installs wells; manufactures drill rigs, hand drills, well casing and pumps; gives technical consultation; remediates groundwater contamination; and provides groundwater for use. Table 1.3.12 and Box 1.3.3 provide a perspective on the groundwater services sector in the United States and China. The groundwater services sector in the United States has many companies that deal with large and small wells as well as simple and complex conditions. The well and treatment services, environmental consulting and remedial services firms most closely related to the groundwater industry numbered over 30,000 in 2018–2019, employing over 170,000 people and having sales exceeding $40 billion. This sector that is briefly described for China (People's Republic) is focused on extensive small-scale farm irrigation equipment and support.

In Asia, expanding groundwater use for irrigation as well as other purposes has caused a proliferation in the local manufacture of less expensive, but durable water pumps. In India, this industry has increased at a 20% annual rate since 1982. This pump production is a competitive business and has significant economies of scale, with the advantage of pumps selling at low cost for the poor rural

TABLE 1.3.12
Firms, Employment and Sales Most Closely Related to the U.S. Groundwater Industry in 2018–2019

Services	Firms/Establishments	Employment	Sales
Water and Well Services Contractors drill and service water wells, install and service pumps, install and service point-of-use water treatment devices	5,272 firms	52,000 people	\$13.537 billion
Environmental Consulting Establishments provide advice and assistance on water supply, environmental issues and management of contamination, toxic substances and hazardous materials	20,820 establishments	59,254 people	\$11.019 billion
Remediation Establishments address remediation and cleanup of contaminated buildings, mine sites, soil, or groundwater	4,521 establishments	59,004 people	\$15.670 billion

Source: National Ground Water Association. (2020). America's Groundwater Industry Employment. Fact Sheet. URL: https://www.ngwa.org/docs/default-source/default-document-library/groundwater/usa-groundwater-use-fact-sheet.pdf?sfvrsn=5c7a0db8_4 (Accessed April 17, 2020).

BOX 1.3.3 THE GROUNDWATER PRODUCTS AND SERVICES SECTOR IN CHINA: A PARTIAL OVERVIEW OF THE SMALL FARM IRRIGATION MARKET IN 1995

- China's population, 1995: 1.2 billion people with 80% living in rural areas
- Irrigated area of China: 53,790,000 ha with 24.8% irrigated by groundwater
- Irrigation equipment enterprises, 1995, included 166 irrigation equipment enterprises employing 65,000 workers with output of 2620 million RMB yuan (US\$310.2 million)
- Domestic market for irrigation and drainage equipment annual sales income in 1995, of 2384 million RMB yuan (US\$282.3 million)
- Production of total units of 2,605,316 (an increase of over 40% from the previous year) included
- Sprinklers
- Farm water well drills
- Large pumps
- Small and medium
- Mini-pumps
- Deep-well pumps (long-spindle)
- Submersible pumps
- Small submersible pumps
- Water-turbine pumps

Source: Weiping, Z., Irrigation technology transfer in support of food security, (Water Reports-14) *Proceedings of a Subregional Workshop*, Harare, Zimbabwe, April 14–17, Food and Agriculture Organization of the United Nations, Rome, Italy (1997).

populace (Shah et al., 2000). Box 1.3.3 (China) documents the intensely developed groundwater irrigation business sector in China.

UNINTENDED EFFECTS OF GROUNDWATER USE IN AN ECONOMY

As noted earlier, the use of the subsurface as a chemical waste sink and groundwater as a conveyance to move it away, neglecting the public good aspect of groundwater, ignores the significant cost on the economy of degrading its benefits. Releasing wastes instead of treating them produces costs and/or damage upon people – and the ecosystem that they rely on – in the future and potentially at other locations, considering the mobile nature of the resource (Young, 2005, p. 278). The company that caused the contamination problem of Love Canal, New York, had to pay an enormous sum, $129 million, in 1995, to compensate the government and others for the site cleanup – monies that could have been used for other productive activities and services, not including the costs to the lives of the families who lived and suffered there (USDOJ, 1995). Elsewhere, groundwaters applied for irrigation to feed many more people than before are being used up faster than replenished naturally, resulting in falling water tables, increased pumping and maintenance costs and dry wells. This condition was observed in the Great Plains of the United States, Saudi Arabia, Libya and northeastern China (including Beijing) (Brown, 1996). The effects on the local and national economies can be itemized, but in all cases, they are still not fully recognized in an economic sense, as in the following examples:

- Reduction in the area farmed and the potential return to dryland farming in parts of the Great Plains of the United States because of extensive aquifer depletion (USWRC, 1978).
- Urban infrastructure damage from land subsidence in and around Mandan, China (Brown, 1996) and disrupted land use in Arizona, United States (Gelt, 1992).
- In Saudi Arabia, 75% of water used in the economy is from nonrenewable groundwater resources, stemming in part from significant subsidies and aids for groundwater-irrigated agriculture, resulting in aquifer depletion, water quality degradation and falling agricultural productivity (Darghouth, 2002). About 80% of Saudi Arabia's fossil groundwater beneath the desert has been used up to produce food (National Geographic, 2019).
- In India, 40% of total annual water use of 600 billion cubic meters are from groundwater. Eighty percent of the population of 1.35 billion people rely on groundwater for drinking water and food production. Polluted and depleting groundwater jeopardize India's agricultural production and the safety of its principal drinking water source. Ninety percent of groundwater in India is used for irrigation (Shah et al., 2000; Schneider, 2019).
- In China, over 80% of wells used for agricultural, industrial and household purposes are not safe for human consumption due to contamination (The Guardian, 2016).
- Water tables of the North China Plain are falling at 1 meter per year from irrigation of growing two crops each year: wheat in winter and maize in summer (Yang, et al., 2017).
- In the South Korean island of Cheju, groundwater overdrafting of the coastal aquifer resulted in saltwater intrusion (Kim et al., 2003).
- About 11 of 37 European countries reported overexploitation of aquifers, including 33 cases of endangered wetlands and 53 situations with saltwater intrusion (EEA, 1999).
- Small island nation-states in the Caribbean Sea have experienced seawater intrusion because of overpumping their aquifers (IPCC, 2001).

ECOLOGICAL AND AESTHETIC USES OF GROUNDWATER

A less obvious, but significant, natural use of groundwater is to maintain stream baseflow (USWRC, 1978; USGS, 1999) that far exceeds all measured groundwater uses in 2015 in the United States, providing 45,699 million m^3/day to streams (USGS, 1999, p. 2). Groundwater discharges through stream channels allow additional water uses downstream beyond that which runoff as overland flow to streams could sustain. This condition is more obvious during the drier times of the year and in irrigated areas in the western United States, and elsewhere in the world with groundwater return-flow to streams. As indicated in the *National Water Quality Inventory, 1994 Report to the United States Congress*

> The importance of groundwater flow into streams and other surface waters cannot be underestimated. Groundwater can transport contaminants to streams and affect surface water quality and quantity, which may impact drinking water supplies drawn from surface waters, fish and wildlife habitats, swimming, boating, fishing, and commercial navigation. Modifications to the quantity or quality of groundwater discharged into surface water ecosystems can also have major economic repercussions as a result of adverse impacts on recreation, public health, fisheries, tourism, and general ecosystem integrity.
>
> (USEPA, 1996)

An example is instructive. In a more humid area, such as central Wisconsin, the Little Plover River baseflow is maintained by groundwater discharge to the stream. The Central Wisconsin Groundwater Center has predicted that due to the irrigation pumping near the Little Plover River, at maximum pumping rates, some sections of the river would dry up and eliminate its trout fishery (Anderson, 1998). This latter groundwater use to maintain a fishery is typically not quantified and not valued as a product or service of groundwater.

The groundwater in the subsurface environment plays a significant role supporting surface waters, such as wetlands, ponds, lakes and streames, as well as estuaries and near coastal zones, thereby enriching these water bodies with nutrients for flora and fauna that inhabit them. More animals and plants per unit area exist in a wetland than in any other kind of habitat (USEPA, 2004b). These waters may be used for recreation, sports, research and instruction. Importantly, we have not fully incorporated these processes, as well as other processes that we still do not fully understand, into the economic exchange of goods and services from which we benefit everyday. Research in subsurface processes including groundwater may lead to many other discoveries about their values, even though they are typically out of sight below our homes and cities. The costs of disrupting or replacing these subterranean services are difficult to estimate, but should not be ignored in evaluating any economic activity affecting groundwater entering the subsurface and moving through it or residing in it, as these costs could be incurred onsite or by adjacent property owners or groundwater users and significantly affect their livelihood.

RESIDUAL ABSORPTION AS A USE OF GROUNDWATER

Chemical Wastes

One use of groundwater typically ignored in most analyses and transactions is its value as a conveyance and treatment mechanism for wastes as well as chemical and biological residuals. The value of this mechanism should be determined, at a minimum, as the cost of the alternative piping and treatment technologies that would need to be applied to obtain a comparable level of service. While using groundwater and the subsurface to transport and dispose residuals might be considered efficient, as it is done at no apparent cost to the disposer, the actual cost may be high and unknown, as was found with abandoned waste sites that included groundwater contamination, with average remedial costs of $12.4 million per site in the United States (Anderson, 2017). Furthermore, underground injection and landfilling of chemical wastes is the disposal method of choice for nearly one-fourth, about 531.7 million kg, of these wastes in the United States (USEPA, 2006). The subsurface environment is expected to hold these wastes so that they do not migrate, as groundwater is increasingly relied on for water supply. Significant volumes of hazardous or toxic substances, about 102.6 million kg, disposed by industries are required to be injected deep below groundwaters used and separated by a confining zone from water supply. Another 0.136 million kg are disposed in shallower injection wells. Ecosystem damage and costs to the economy beyond those evaluated as alternative disposal mechanisms require further research.

Pesticides

In 1997, the United States users of pesticides purchased 544.3 million kg for application, most of which was land applied. At that time, 20,700 pesticide products were produced having 890 active ingredients. These pesticides were sold for approximately $11.9 billion (NAS, 2000). The extent to which these chemicals leached to groundwater is still not known, but their presence in groundwater has been well documented (Barbash and Resek, 1996).

COMPETITION FOR THE SUBSURFACE ENVIRONMENT

Approximately 15 million wells produce groundwater in the United States for public and private use (Westwater Resources, 2006). In the future, disposers of carbon dioxide (carbon sequestration) in response to climate change and wastes from desalinization of ocean water and inland brackish water may use the subsurface environment for sinks to hold these residuals. For the economy to consider hidden or unidentified costs, actions in the subsurface which appear to have no costs and benefits that are difficult to quantify, should be reevaluated for their full scope and be assigned appropriate values, so that later users do not bear the costs of the needed remedies without being informed. Research into the nature and magnitude of these benefits and costs will become a priority with greater competition for the subsurface environment.

HEALTH AND ECONOMIC PRODUCTIVITY

Economic productivity is related to the health of the populace and water can affect that health (UNESCO, 2003, pp. 101–125). Chapter 4 will address the issues of health and groundwater; however, it is important to note that significant analyses have been conducted which address health as well as clean and safe water. The economic benefits of good health from safe water directly stem from greater productivity in the home and at the workplace, due to reduced illness and death (Hutton, 2001). Furthermore, macroeconomic effects (e.g., employment, exports, tourism) of poor drinking water can be significant (Hutton, 2001).

OTHER SOCIOECONOMIC FACTORS IN OVERVIEW

"The earnings, incomes, and lifestyles of the people significantly affect priorities on use of the Nation's [and world's] water resources. For example, a major change in population growth, a shift in its geographical location, or a change in the composition (age, family size, occupation, etc.) will affect the level, distribution, and mix of water requirements both nationally and regionally. Increasing real incomes and leisure time tend to place higher priorities on instream [surface] water uses – recreation, fish and wildlife, natural areas – rather than on the traditional offstream uses. Many traditional economic and water-use relationships change substantially as prices, incomes, asset values, and productivities adjust to changing energy and trade relationships" (USWRC, 1978, p. III-9). In 1950, three of the largest water-using sectors – agriculture, mining and manufacturing – contributed 40% of the total earnings in the United States (USWRC, 1978, III-13), while using 70% (490.6 million m^3) of the fresh groundwater and surface water withdrawn (696.5 million m^3) (USGS, 2003). Earnings in these industries were predicted to be 25% of the total earnings in 2000 (USWRC, 1978, p. III-13), with 41% (640.9 million m^3) of the freshwater withdrawals (1544.4 million m^3) (USGS, 2003). Agricultural and economic policies, including import tariffs, subsidies, access to credit and exchange rates, affect groundwater withdrawals in irrigated farming and raising livestock (Schiffler, 1998, pp. 148–152). Clearly, water use plays a significant role in the overall economy – increasingly in nonproduction and service sectors.

ECONOMIC MANAGEMENT CHARACTERISTICS OF GROUNDWATER

Economic management characteristics of groundwater comprise a mix of physical and socioeconomic attributes affecting the market and open access public use of the resource. The purpose of markets is to efficiently allocate resources through prices resulting from the balancing of supply and demand for labor and capital on the supply side, as well as products and services on the demand side by human beings and the ecosystem supporting us. Groundwater has features that make it both a capital (fixed) and variable input in supply (Young, 2005, pp. 51–52), as well as product and service in response to demand. As groundwater can move in the subsurface interstitial space, but may also be contained in deep reservoirs, the results on the ecosystem of its use in the economy and, therefore, its value, vary from place to place (Young, 2005, p. 37). To provide an economic understanding that can link its natural occurrence to economic use, several concepts have evolved to better portray a range of important characteristics of natural resources, described in Box 1.3.4 (following Daly and Farley, 2004). These characteristics provide a basis for considering policies that might best be applied to groundwater for its range of uses over time.

Given the characteristics of natural resources, how do they apply to groundwater? Because of the range of physical conditions of groundwater, it takes on seemingly contradictory characteristics. These characteristics are displayed in Table 1.3.13. This table shows that the economic management characterization of groundwater and its major uses is complex. Importantly, there exists no substitute for water for its uses of greatest significance (Daly and Farley, 2004, p. 88), and where it is the only source, no substitutes for groundwater may exist in many locations (McCabe et al., 1997). Groundwater may be found as "renewable" in situations where it is only used at rates of replacement and replenishment, often referred to as "safe yield," or "nonrenewable" where groundwater is

BOX 1.3.4 NATURAL RESOURCE CHARACTERISTICS

Resource Characteristic	Characteristic Description
Stock-flow	Material of the resource is incorporated into products and used at any rate (subject to natural limits of availability); it can be stored and consumed
Fund-service	The resource is not incorporated into a product, but is used at rate limited by natural processes; it cannot be stored, but becomes "worn out" rather than consumed
Excludable	Ownership of the resource precludes others from using the resource, usually the result of institutional legal rights or technical features of the resource. This characteristic can be equated to property rights
Nonexcludable	A resource that cannot be precluded from use by anyone through institutional or technical factors
Rival	The physical aspect of a resource that allows it to be used by only one person. All stock-flow and some fund-service resources are rival resources
Nonrival	One person's use of this resource does not limit other people's use of it. Any nonrival resource is a fund-service resource
Rival between generations	Resources that can only be used by one generation
Renewable	Resources that reproduce or are replenished naturally and sustainably
Nonrenewable	Resources that are sources of materials that are consumed or used up
Congestible	Nonrival resources that have so many users at particular times or over time, that they appear to become rival resources and their quality declines
Substitutable	One resource can replace another but is not typically a perfect replacement
Complementary	Use of one resource causes other resources to be used

Source: Daly, H.E. and Farley, J., *Ecological Economics*, Island Press, Washington, DC (2004), 454.

TABLE 1.3.13
Characteristics of Groundwater by Major Use Categories

Major Use	Stock-Flow or Fund-Service	Excludable	Rival	Between Generations	Substitutable
Drinking water (for people and livestock)	Depends on rate of use	Yes	Yes	Depends on ground surface management (infiltration/ percolation potential), recycling and rate of use	No, except if surface water is available
Irrigation	Depends on rate of use and residual absorption	Yes	Yes	Depends on rate of use	No, except if surface water is available
Industrial	Depends on rate of use	Yes	Yes	Depends on ground surface management, recycling and rate of use	No, except if surface water is available
Mining	Depends on rate of use	Yes	Yes	Typically yes	No, except if surface water is available
Thermoelectric cooling	Depends on rate of use	Yes	Yes	Typically yes	No, except if surface water is available
Residual absorption	Funds-service	Depends on access	Yes	Yes	Moderate at margin, otherwise, no
Aesthetic/recreation	Funds-service	Depends on access	No[a]	No	No
Surface Water Supply Support					
Streams/wetland	Stock-flow	No	No	No	No

Source: Based on Daly, H.E. and Farley, J., *Ecological Economics*, Island Press, Washington, DC, 2004, 91, Table 1.

[a] Except if access allows congestion to occur.

contained in fossil aquifers or used at rates in excess of replenishment. Where groundwater is nonrenewable, the resource may be facing "congestion" because of the large scale in the number of users demanding such a collectively large volume of water. Even where groundwater is renewable, if the scale of use is large, this congestion may be evident. In both these cases of fossil aquifers with no renewability and congested "potentially renewable" surficial aquifers used beyond replenishment, the groundwater resource is a rival resource between generations. In those situations, the groundwater consumed is lost for use to the next generation.

The finite stocks of groundwater set a limit on the amount of its economic production, with the constraints not obvious until the resource is significantly depleted; a condition from which it may not recover (Daly and Farley, 2004, p. 91). In this regard, it shares characteristics with a range of other resource types, including petroleum, mineral resources, fisheries and wetlands. Notably, through the application of technology, water-use efficiency can increase (Schiffler, 1998, pp. 28–31). However, even with improved efficiency and because water is not substitutable and is essential for life, distribution of water to all people and livestock represents a significant ethical issue for market allocations of water (Daly and Farley, 2004, p. 196).

Based on these characterizations, scale of use significantly affects other people's use of groundwater. If large volumes are used for many or multiple purposes in a location (reflecting stock-flow, nonexcludable and rival characteristics), then an aquifer may become depleted because of massive water consumption. Likewise, extensive or concentrated residual or waste disposal can preclude groundwater from being used from a qualitative standpoint, being "worn-out" from this use (reflecting fund-service, nonexcludable, nonrival characterization). It may take decades, even generations, before natural processes can breakdown residuals and wastes to "renew" the aquifer. These circumstances – stress on groundwater because of demands on quantity and quality – point to problems in the market for this good.

ECONOMIC CHALLENGES AND MARKET FAILURE

Before developing a model of groundwater in the economy, it is important to ask about the circumstances in which the market has failed to address the economic needs relative to groundwater. Identifying market failure along with a specific model of groundwater in the economy will serve as a guide for policies that could respond to real challenges in water allocation and use in the future. Clearly, groundwater policy should not be developed in isolation. However, both groundwater existence and water demand are sufficiently ubiquitous to allow a focus on these elements of the resource together.

Market failure for groundwater relates to several problems associated with excludable and rival resources:

1. Affordability
   Distribution of water at an affordable price for all who need it is a serious problem (Schiffler, 1998; UNESCO, 2003; Daly and Farley, 2004, p. 197). Groundwater is in limited supply and being depleted on every continent. Over 1 billion people are not served by adequate water supply worldwide (UNESCO, 2003, p. 109). In the United States, the government has established plans to provide safe water supply to at least 50% of the people on the tribal lands who are underserved, by 2008 (USEPA, 2003, p. 34). These people are primarily served by groundwater. Affordability of water supplies for economically disadvantaged populations is a factor around the world, with poorer people paying twice to nearly 500 times the price paid by water users with household connections to a water supply (UNESCO, 2003, p. 341).
2. Monopolies Not Efficient
   The market for water is not efficient (Daly and Farley, 2004, p. 198). Supplying water is typically carried out by a provider that has a natural monopoly for this commodity, whether publicly or privately owned. Monopolies have little incentive to control costs and do not deal with market share, since they have most, if not, all of it. With no competition, and unless economically regulated, private water suppliers may be less efficient than public suppliers (Daly and Farley, 2004, p. 198). The opportunity cost – the value in the best alternative use – must be accounted for to provide efficient allocation (Schiffler, 1998, p. 40), but may not be considered in a monopoly supply situation.
3. Future Value
   Intergenerational fairness should be a factor, as water is essential and groundwater can be mined like petroleum or other minerals, even when it may be renewable. When a resource like groundwater is a rival resource between the generations, the market will not operate efficiently to allocate it across generations (Daly and Farley, 2004, p. 192). This also relates to maximizing annual net profit, as opposed to the net present value of the resource.
4. Ecosystem Services and Pollution
   For renewable resources such as groundwater in more humid climates, the ecosystem services that they provide are nonexcludable and do not have institutions to make them excludable (Daly and Farley, 2004, p. 211). These services may include groundwater supply to wetlands and maintenance of streamflow for drinking water, wildlife and the web of natural connections, which are not clearly understood but are necessary. Such values are difficult to determine (Schiffler, 1998, p. 40). Therefore, the economy will not invest in these services. Degradation of these services is an "externality," that is, an incidental and unplanned loss of welfare to other people who will not receive reparation for their altered welfare condition

(Daly and Farley, 2004, p. 175). Furthermore, pollution itself introduced into this natural capital is a negative externality or "public bad," being nonrival, nonexcludable and not in demand. Pollution actually diminishes the natural capital available for ecosystem services that support life (Daly and Farley, 2004, p. 215). These services include residual absorption, cycling of nutrients, flood storage and local and even global climate regulation. As we do not know all the nonmarket services that groundwater may provide, we cannot price them appropriately. Many investments are only profitable because they do not account for all costs; in some cases, for depleting or destroyed ecosystem services, substitutes may not exist (Daly and Farley, 2004, p. 212). These market failures will be addressed further in Volume 2, Chapters 3 and 4 on policy. Their significance is that the individual transactions that support the overall economy do not adequately address these factors in pricing products and services.

5. Resolution in the Market Place
   Where water is scarce or groundwater is the main or sole source of water supply, markets for groundwater have evolved. In the western United States, the rule of capture (pumping whatever amount of water capable) often applies and can result in mining and depletion of the aquifer. However, in areas with some level of control of amount pumped, rights to or permits for groundwater can be purchased. Often, a municipality may need more water and will purchase rights from a nearby irrigator not using their full right (Hardberger, 2016). Resource management districts have also provided water markets by setting up exchanges enabling water rights sellers to offer their irrigation water for bid to prospective buyers for a growing season (CPNRD, 2019). Other forum for bidding also can occur (Krieger, 2016). While markets may induce efficiencies in water use, locations with scarce water and no limits on pumping to capture groundwater may experience aquifer depletion.

A MODEL OF GROUNDWATER IN THE ECONOMY

To this point, we have observed that groundwater is significant to the economy and our survival, which means that it is a critical raw material for our existence and our economic exchanges as social beings. Table 1.3.8 shows the major uses of groundwater in the economy. Daly and Farley (2004, p. 229) indicated that under current accounting practice, the sum of all goods and services multiplied by their price equals the "country-level" economic measure, its gross national product. For groundwater, this is the value added in terms of labor and capital to produce it. This measure does not take into account "noneconomic welfare," such as a depleted or polluted resource or associated effects. Nor does it account for unmeasured services, such as flood and climate control and the natural productivity of wetlands. Groundwater is an essential substance for the ecosystem to function, a necessity for life and a critical raw material for industrial activity. Following from the first law of thermodynamics (neither matter nor energy can be created or destroyed) inputs to the economy must equal the outputs. This condition for groundwater is described by the equation

$$\mathrm{GWW} + \mathrm{NME} = \mathrm{R_p^c} + \mathrm{R_p^d} + \mathrm{R_i^d} + \mathrm{ET} + \mathrm{L} \qquad (1.3.1)$$

This equation shows that GWW by the economy along with other natural materials carried by groundwater, extracted and transformed to products (NME) are equal to the groundwater commodities produced $\left(\mathrm{R_p^c}\right)$ plus the residuals from groundwater producers $\left(\mathrm{R_p^d}\right)$, discharges back to the environment by domestic and industrial $\left(\mathrm{R_i^d}\right)$ users, groundwater lost or wasted (L) and evapotranspiration (ET) resulting from domestic and agricultural/industrial uses of groundwater. The residuals from groundwater producers as well as domestic and industrial users are often returned to the ground or subsurface, either through use (such as for pesticides) or disposal (such as liquid or hazardous waste). These residuals have a potential to contaminate and damage groundwater resources, possibly removing a portion of groundwaters from future economic withdrawal and use. This model will be further refined after the concepts of groundwater function and service have been developed. This depiction is a neoclassical environmental economic model of groundwater in the economy, focusing on the residuals produced and required to be reduced to mitigate damages to health, people, or property.

Scale of withdrawal or residual release into groundwater is a minor, if not nonexistent, problem when the population density of water users is very low. As a result, the factors in the equation presented earlier for any particular location may be small. However, with population growth and high density of people and their activities, demands on resources, generally, and groundwater, particularly, are overwhelming its sustainability (Daly and Farley, 2004, p. 12). The depletion of aquifers and the closing of wells from groundwater contamination are cases in point.

GROUNDWATER FUNCTION AND SERVICE

As discussed previously in Volume 1, Chapter 2, groundwater is both a stock and a flow resource. Stock resource indicates that in some locations, groundwater is of some known extent, is used or managed as a discrete reservoir, cannot be easily replaced or replenished and is, therefore, nonrenewable (USWRC, 1978, p. IV-18). A flow resource is renewed regularly and moves through the environment, with its use incorporating this movement or flow function, which may result in offsite services. Locations exist where groundwater may be identified with both of these categorizations. As a stock resource, groundwater provides many services in the economy that can be identified and valued. These are described in Table 1.3.14. These services range from provision of drinking water and irrigation water, to provision of a medium for wastes and support of living organisms. Our valuing of the services provided by the ecosystem functions of groundwater is less than perfect because of our ignorance of their inherent importance and connections in nature that support us, except at a basic level (if we fully understood their significance, we would be able to design and implement solutions to completely resolve the ecosystems issues now assuming sufficient fiscal resources). These services generate effects on or for people, such as a change in welfare from an increase or decrease in the availability of drinking water or a change in human health or health risks attributable to a change in water quality. These effects can be evaluated in monetary and nonmonetary terms. Approaches to attempt to value them in economic terms are described in Bergstrom et al. (1996).

Groundwater is also a flow resource that, in many cases, has a movement that may be almost imperceptible, and, in other situations, very obvious, such as in springs from the ground. While many of the services that groundwater provides may be the same for both its stock and flow characteristics, the flow component offers several services that do not exist when considering it only as a stock resource. Specifically, groundwater may provide transport and treatment of wastes and discharge to surface waters. This discharge to surface waters is significant, because it incorporates services of groundwater that were heretofore rarely considered and certainly not evaluated in an economic context, such as support of recreational swimming, boating, fishing, hunting, trapping and plant gathering when groundwater serves to maintain surface water (i.e., baseflow of streams year-round and critically in dry seasons), including wetlands. Table 1.3.14 describes the functions and services that groundwater provides to the economy.

Some of the services that groundwater provides to the economy might be considered as "ecosystem services" (Daly and Farley, 2004), as previously noted, and are not well understood and not monetizable, at least easily. These services are taken for free and may be used up irreplaceably. These services include natural provision of

- Habitat support for flora and fauna in wetlands, estuarine and coastal systems through nutrient cycling by groundwater discharge in the hydrologic cycle.
- Erosion and flood control through percolation of precipitation not becoming surface runoff, but transported to groundwater (stock) in the aquifer.

TABLE 1.3.14
Groundwater Functions and Services

A. Function – Storage of Water Reserve (Stock): Groundwater stored in an aquifer provides a reserve (stock) of water which can be directly used to generate services, including	B. Function – Discharge to Streams, Lakes and Wetlands (Flow): Groundwater contributes to the flow or stock of water in streams, lakes and wetlands. A portion of surface water and wetlands services are therefore attributable to the groundwater resource from its flow, including
1. Provision of water for drinking water	1. Provision of water for drinking water through surface water supplies
2. Provision of water for crop irrigation	2. Provision of water for crop irrigation through surface water supplies
3. Provision of water for livestock	3. Provision of water for livestock through surface water supplies
4. Provision of water for food product processing	4. Provision of water for food product processing through surface water supplies
5. Provision of water for other manufacturing processes	5. Provision of water for other manufacturing processes through surface water supplies
6. Provision of heated water for geothermal power plants	6. Provision of water for medium of heat exchange for cooling water of power plants
7. Provision of water for medium of heat exchange for cooling water of power plants and for heat pumps	7. Provision of erosion and flood control through percolation of precipitation not becoming surface runoff, but transported to groundwater (stock) in aquifer
8. Provision of water/soil support system for preventing land subsidence	8. Provision of transport and treatment (degradation) through storage for wastes and other by-products of human economic activity
	9. Maintenance of navigable streamflow
9. Provision of erosion and flood control through percolated water increasing groundwater accumulation (stock) in aquifer	10. Support of recreational swimming, boating, fishing, hunting, trapping and plant gathering
10. Provision of medium of storage and treatment (degradation) for wastes and other by-products of human economic activity	11. Support of commercial fishing, hunting, trapping and plant gathering
11. Provision of clean water for support of living organisms	12. Support of onsite observation or study of fish, wildlife and plants for leisure, educational, or scientific purposes
12. Provision of passive or nonuse services (e.g., Existence or Bequest Motivations)	13. Support of indirect, offsite fish, wildlife and plants uses (e.g., viewing wildlife photos)
13. Provision of unknown ecosystem services, critically important to the balance of life to enable economic activity	14. Provision of clean air through support of living organisms
	15. Passive or nonuse services (e.g., existence or bequest motivations)
	16. Provision of clean water through support of living organisms
	17. Regulation of climate through support of plants
	18. Provision of nonuse services (e.g., existence services) associated with surface water body or wetlands environments or ecosystems supported by groundwater
	19. Provision of unknown ecosystem services of critical importance to the balance of life to enable economic activity

Source: Adapted and modified from Bergstrom, J.C. et al., *Water Resour. Bull.*, 32, 279–291, 1996.

- Water/soil support system for preventing land subsidence.
- Transport, treatment (degradation) and storage for wastes and other by-products of human economic activity.
- Clean air and water through support of living organisms.
- Regulation of climate through support of plants.
- Other services that have not yet been identified in the hydrological and terrestrial realms that support important human needs and balance for life sustenance.

Currently, these are considered free services, but as the economy grows and more groundwater is used and/or more wastes and residuals are released to the ecosystem, these services may become threatened and their value to the economy and the well-being of humankind may be more completely recognized as they become scarce. Once identified as significant or essential, the value of these services may be acknowledged through policy or lifestyle changes, or commodity or service prices in the market place. Initial recognition of these ecosystem services is being incorporated in a renewed thinking on the function and management of watersheds and the role of groundwater in their water balances and ecological support.

RECYCLING AND REUSE

In areas where groundwater and surface waters are limited, recycling and reuse of groundwater may provide the additional supply (WSAA, 2004; Yasmeen, 2006). Recycling water means having a process to collect the water used and then treating the water for human consumption and returning it to drinking water distribution systems, or providing less treatment and redistributing it for nonpotable uses (USEPA, 1994). The water could also be returned to the subsurface environment and aquifer for storage and use at a later time, or to maintain important ecosystem functions (USEPA, 1994). A system for collecting the water is typically the sewer line already supplying the local wastewater treatment plant. In locations having industries connected to the sewer system, these companies are usually required to treat their processing water before releasing it to the sewer. The local wastewater treatment plant will remove pollutants to the level of the environmental standard set previously for the discharge point.

The water could be directly used after treatment in some applications, such as lawn watering and other kinds of outdoor or industrial washing and cleaning. However, for human consumption it would need to be chlorinated and meet the other maximum contaminant levels described in Chapter 6. The city of El Paso, Texas, United States, obtains 55% of its water supply from the Mesilla and Hueco Bolson aquifers, 3% from desalination of brackish water and the remainder from the Rio Grande. To meet water demand, the city reclaims nearly 6 million gallons per day of wastewater for irrigation at parks and schools, construction projects and industrial sites, as well as recharge its aquifers (EP Water, 2018). By 2020, the city plans to treat wastewater for direct potable use with the goal of using that treatment technology for 10% of its water supply by 2030 (Kounang, 2018).

In other less humid and arid locations, such as Orange County, California and Scottsdale, Arizona, after treatment at the wastewater plant, treated wastewater has been pumped back into the aquifer for many years to recharge aquifers. Recharge zones may also be cisterns in the ground, wells to reinject the recycled water back into the ground or retention and infiltration ponds that allow stormwater as well as treated wastewater to percolate through the soil and ground. This action would permit microbes in the ground to remove any remaining bacteria or contaminants that had survived the wastewater treatment process or remain from stormwater catchment (NRC, 1994: Brumley et al., 2018). Groundwater use, treatment, recharge and recycling are also the principal features of a septic system that relies on natural cycling of wastewater after septic treatment.

A Simple Model

From both business and public policy standpoints, groundwater can be viewed in a simplified model as a part of the natural environment–economy relationship (modified from Fields, 1994), as presented in Figure 1.3.1, a residual-based model. While groundwater can supply many functions and services that can be valued in the economy, groundwater and the geologic matrix in which it exists have characteristics that make groundwater a commodity, a medium for residuals and waste conveyance and a residuals and waste sink.

This model has the economy as a dynamic part of the environment, drawing resources – in this case, groundwater – from the larger environment. A water producer extracts groundwater from the environment and supplies domestic consumers and industrial or agricultural users with the good (water), G. Both domestic consumer and industrial user may recycle and discharge water. The water producer may also discharge a residual, R_p^d, back to the environment. The water producer's residual may be a sludge from the treatment process, containing metals, for example, which may be landfilled.

The domestic consumer recycles residuals, R_i^r, which might represent water after septic system treatment that is subsequently cycled back to the water table. Or, the residual, R_i^r, might be water treated in a wastewater treatment plant and returned to use through the water producer after further treatment. The domestic consumer also discharges residuals, R_i^d, which might be water with solids to be removed from domestically consumed groundwater by wastewater treatment and must be landfilled or land applied, and treated wastewater discharged to a stream.

Likewise, the industrial or agricultural water consumer may recycle residuals, R_i^r, which might be process water not consumed in production which is treated onsite for certain pollutants from production. Such pollutants could be reused, depending on the production process, or disposed and returned to the environment. The R_i^d, can also represent the industrial water consumer's residual discharge to the environment. This discharge could be water containing waste, which is released to the sewer system connected to the local wastewater treatment plant, land applied if the waste is appropriate to that use, transmitted to the subsurface by injection well, or irrigation waters percolating down below the root zone and carrying unused nutrients and pesticides. In all the industrial situations, the water consumer should have the applicable permit from the government authority to

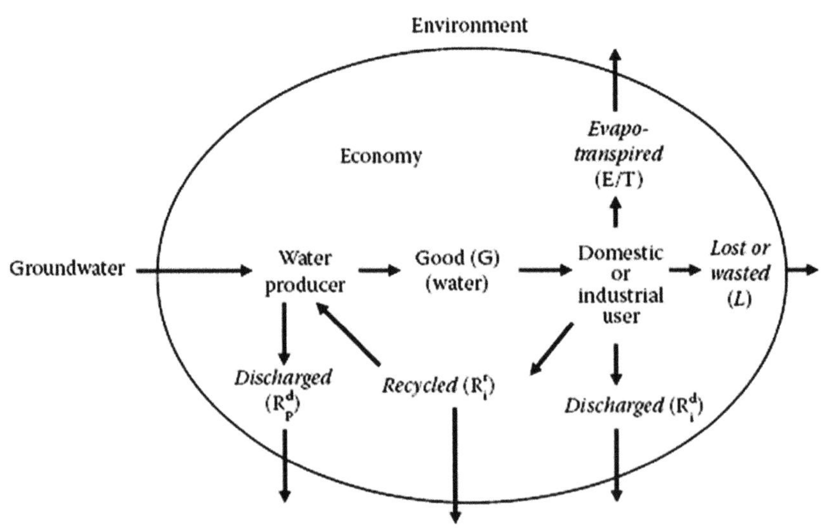

FIGURE 1.3.1 A Residual-Based Model of Groundwater in the Economy. *Source:* Adapted from Field, B.C., *Environmental Economics: An Introduction,* McGraw-Hill, New York, 1994, 24.

release the residual water with its pollutants. In the agricultural settings, central authorities may take steps to control or reduce applications of fertilizers and pesticides.

In the model, L signifies water lost or wasted. This water that does not reach the domestic or industrial consumer may be substantial. Schiffler (1998, p. 37) indicated that unaccounted-for water may be in excess of 50% of water produced in developing countries, but often may be between 20% and 30% in those countries, and 10% and 20% in developed countries. Interestingly, this is a loss to the economy, but is not a loss to the ecosystem, as it may be "recycled" back to an aquifer from a leaking pipe. However, in such a circumstance, the water quality may have changed significantly because of treatment, or lack of treatment and hence, it may not have its original quality.

EXPANDED ECOSYSTEMS SERVICES MODEL

Considering the total function of groundwater providing an array of services, which may include known and unknown services from which we benefit everyday, a different equation and model has evolved. The equation might incorporate earth structure, flood water mitigation, wetlands maintenance, nutrient cycling, streamflow maintenance, climate balance, estuary and fisheries support and wildlife sustenance, among many other possible features. Furthermore, the model may have a natural, and perhaps unknowable, inherent, constraint – carrying capacity for human interactions – beyond which the groundwater environment could not sustain those interactions. Thus, the model requires uses for groundwater withdrawn for human purposes, net recharge of residual liquids and the earth structure to be less than the groundwater carrying capacity constraint. This relates to the scale of the economy and human uses of groundwater. If they are smaller than the carrying capacity, then the inherent sustainability is preserved. If it is exceeded, then sustainability is not possible, even though all flows and residual quantities may be balanced in the equation. An equation that would recognize the ecosystem aspects of groundwater might conceptually be given as

$$\mathrm{NBGWQQ} = f\left[\mathrm{GWR,GWD,NBAC,GWWH,RRH,}(\mathrm{E/T})\mathrm{H,}(\mathrm{E/T})\mathrm{N,ES,OES}\right] \tag{1.3.2}$$

where the net balance of ambient groundwater quantity and quality (NBGWQQ) is a function of

GWR = groundwater recharge from precipitation, streams, lakes, wetlands and oceans, including erosion and flood control

GWD = groundwater discharge to streams, lakes, wetlands and oceans, for estuary and fisheries support and wildlife sustenance

NBAC = net natural biotic and chemical process substances

GWWH = groundwater withdrawn for human activities

$\mathrm{RRH} = \mathrm{R}_p^c + \mathrm{R}_p^d + \mathrm{R}_d^d + \mathrm{R}_i^d$, recharge of residual liquids from human activities, including nutrients, contaminants and substances injected into the subsurface

(E/T)H = (E/T)D + (E/T)I, evapotranspiration from human activities

(E/T)N = evapotranspiration from nature's activities (other than from human activities)

ES = earth structure (maintenance of ground surface rather than land subsidence)

NHS = net habitat support

OES = other ecosystem service flows of and support for groundwater

Subject to

$$\mathrm{GWCC} \geq f(\mathrm{GWWH,RRH,ES,NHS})$$

where GWCC is the carrying capacity of groundwater in an aquifer.

The carrying capacity of groundwater may be defined in several ways. The term could mean the quantitative use of groundwater such that only water that is replaced is consumed – the concept of "safe yield." It could also be defined in terms of an aquifer's ability to degrade only a limited volume or concentration of a contaminant or mix of contaminants. A third notion of carrying capacity might include a combination of quantity and quality factors, a range of water use that still allows the aquifer to absorb a certain amount of chemical or biological residual.

While it may seem odd to include groundwater quantity and quality in the same equation, at a conceptual level and in ambient conditions, they are considered or taken into account together in a particular location or for an aquifer. The services that groundwater provides, whether to human activities or through other ecosystem services, usually factors them together both in terms of available supply and preferred demand.

This model is depicted in Figure 1.3.2, portraying the complex and little understood interplay of groundwater ecosystem services with the economy. In the economy, each of these services would have a price associated with it, reflecting its value in the market. Typically, this value is actually the cost (including a profit to the producer) of producing the service and not the inherent value of the inputs, such as a value for the groundwater itself (e.g., existence value). Prices may probably be location- specific or dependent, based on the relative availability and quality of groundwater there and the interaction of that location or local market with other locations and their markets. It must be noted that the prices could be assigned to both "goods," such as groundwater used in wheat production, and "bads," such as pollution of groundwater. In neither of these examples is groundwater actually priced in the market directly. Groundwater pricing will be addressed in the later chapters on micro- and macroeconomics. "Ecosystem prices" may be much different than the "prices in a market economy," as the economy would price the value added for the throughput of groundwater, whereas with regard to groundwater mining (groundwater use greater than managed yield), an ecosystem price may be actually negative, that is, a cost to the ecosystem. The model with prices associated with it might look like the following function:

$$\begin{aligned}(\mathrm{a})\mathrm{NBGWQQ} = f\big[&(\mathrm{b})\mathrm{GWR} - (\mathrm{c})\mathrm{GWD} \pm (\mathrm{d})\mathrm{NBAC} - (\mathrm{e})\mathrm{GWWH} + (\mathrm{f})\mathrm{RRH} + (\mathrm{g})\\ &(\mathrm{E/T})\mathrm{H} + (\mathrm{h})(\mathrm{E/T})\mathrm{N} + (\mathrm{i})\mathrm{ES} + (\mathrm{j})\mathrm{OES}\big]\end{aligned} \tag{1.3.3}$$

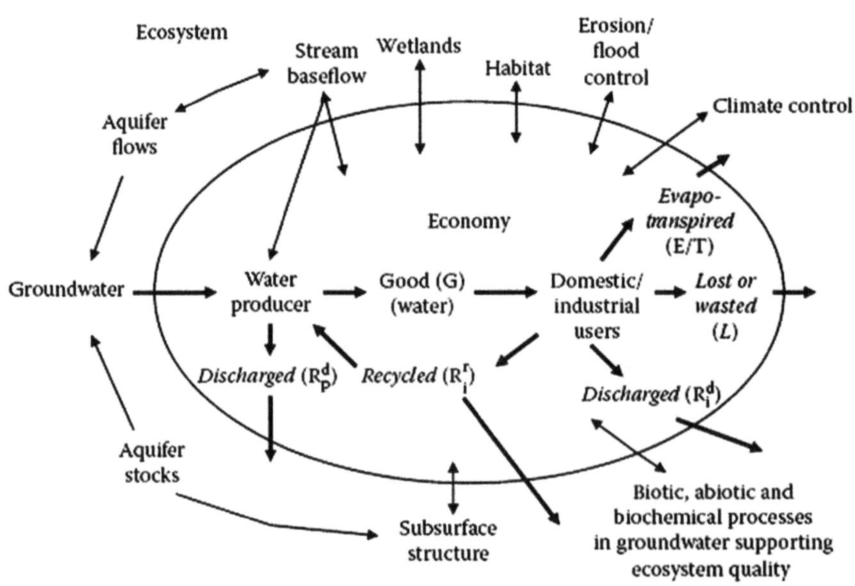

FIGURE 1.3.2 Ecosystem-Based Model. *Source:* Adapted from Field, B.C., *Environmental Economics: An Introduction,* McGraw-Hill, New York, 1994, 24.

where a, b, c, d, e, f, g, h, i and j are the prices of the factor in the equation, reflecting its value to the ecosystem within which the economy exists. Considerable research would be needed to derive these values, if they could be determined at all, reflecting our fundamental lack of knowledge of these relationships. If incorporated into the economy from a neoclassical economics perspective, they may most probably only reflect some value added that could be recognized by the market in reflecting its scarcity and not its true worth in the natural balance. Our inability to price these factors owes to the high level of uncertainty associated with their true function which gives evidence of our real ignorance of nature and its value to all our economic processes and throughput.

ECONOMIC EFFECTS ON GROUNDWATER FROM CLIMATE CHANGE

Drought and Pumping

Reduced aquifer recharge, deteriorated quality and increased pumping are the major anticipated effects of climate change on groundwater resources. Future drought may accentuate the competition between human and ecological uses of water (Loaiciga et al., 2004), in particular, with regard to drawing deeper groundwater resources. This circumstance may focus greater attention on human uses of groundwater for drinking water supply and irrigation, even encouraging more efficient water use. During the drought in California from 2014 to 2017, farmers installed thousands of new wells going as deep as 366 meters (1200 feet) to reach and pump groundwater, mining and lowering the water tables of aquifers of the San Joaquin Valley (AP, 2016).

Saltwater Intrusion

A further key concern related to groundwater directly is sea-level rise that may threaten the advance of saltwater intrusion into coastal aquifers. Water supply wells relied on by people and their local economies may have to be abandoned. Search for new fresh sources of water may include drawing on groundwater further inland or using other sources. Another option is desalination of seawater, which then involves disposal of wastes with high salt content that could include injection into deep geological strata with expectation of remaining in place and not migrating to drinking water sources. Saltwater intrusion, CO_2 sequestration and desalination waste disposal on larger scales should be factors in modeling pressures on groundwater and demands on the economy.

Flood Effects

The effects of back to back hurricanes on North and South Carolina in 2018 demonstrated the effects of flooding from heavy, long-duration rains on already saturated ground to impact groundwater wells. Wellheads could be damaged by floating debris allowing contaminated floodwater down the well. Septic systems could also be connected in the saturated vadose and soil zones to household wells, bringing unwanted wastewater into wells. Potentially thousands of wells were affected and state health laboratories reported household wells sampled testing positive for microbes indicating possible pathogenic contamination (Beitsch, 2018).

Carbon Sequestration

Other effects that are beginning to be recognized may result from carbon (dioxide) sequestration and desalinization waste disposal. As concern grows for taking action to reduce carbon dioxide (CO_2) emissions from economic activities that involve combustion of carbon-based fuels, one disposal sink being actively considered is deep rock capped by less permeable rock to reduce the potential for upward migration of these wastes. This option offers environmental benefits of reducing CO_2 emissions to the atmosphere. However, the costs to capture and store CO_2 are significant, but technological research is indicating a downward trend (USDOE, 2005a, 2005b). Another concern is the potential environmental damages from injection not addressed by technology, including acidification and loss of integrity of rock structures, accidental leakage back to the atmosphere and rapid, catastrophic "blowouts" of stored CO_2 (Smekens and van der Zwaan, 2004, pp. 4–6).

NATURAL CAPITAL CONSIDERATIONS

The economic functions and services of groundwater described previously infer that groundwater and its subsurface environment provide results as might also be derived from the application of human capital. This performance may be considered "natural capital." These functions and services may most obviously include water collection and storage, pollution degradation and distribution of water underground to points of use. These services are provided at an economic cost for surface water supplied to water customers. The capital items associated with that water supply include collection pipes, reservoirs, treatment plants and distribution pipes. Larger groundwater systems may also have some of these features. Other services were identified previously in this chapter. The point is that the physical capital aspects of this water provision process are provided by groundwater in the subsurface environment as natural capital functions. When alternatives are evaluated for water supply, this natural capital should be accorded as a factor in the options being analyzed. Legislation in the United States, the 2019 Water Supply Cost Savings Act, gives support to this consideration. This law fosters evaluation of local water source alternatives (including groundwater wells) compared to regional pipeline connection when communities apply for financial assistance to improve their water service. For small communities, the difference in human-made capital demands, energy and environmental effects and cost may be significant.

IMPLICATIONS FOR SUSTAINABLE GROUNDWATER IN THE ECONOMY

Sustainable resource policy for groundwater has multiple economic dimensions. Using up groundwaters that cannot be easily replaced and the natural unrestricted flow of groundwater transfers economic effects to other actual or potential users. Some of these users will have reduced or increased supply, while others will receive a resource of changed quality. Still others may not recognize that their groundwater use has influenced others' use of a surface water resource, whether a stream, lake, or wetland. These transfers of effects could be accounted for in the economy if costs of use are not ignored but incorporated as fully as possible in transactions involving groundwater, and tracked separately in terms of macroeconomic transactions. This will be addressed further in later chapters.

From an ecological standpoint, as water is central to the very function of societies and as groundwater use is expanding as the next most viable and economic source following exhaustive use of surface waters, economic policy development must consider groundwater's role in both the hydrologic cycles of watersheds and as a conveyance and sink for wastes and residuals in the economic processes that occur in those watersheds. By extension, these hydrologic and economic processes must subsequently be incorporated into state, national and international policy more proactively (UNESCO, 2003).

This simple ecosystem model can be a central concept from which solutions will evolve to address the critical issue of groundwater management. The model represents the integration of the natural environment and the socioeconomic realm through which we must make choices in the future. A point of significant attention for future policy is the sustainability of groundwater as a private commodity to meet the daily human needs of water and food as well as a public good to maintain habitat, geologic structure under places prone to subsidence and baseflow of streams and other ecosystem processes, including mitigation of effects of wastes and residuals. Sustainability of groundwater requires this integration to meet the high objective of the Brundtland Commission to "meet [] the needs of the present without compromising the ability of future generations to meet their own needs" (UNWCED, 1987). Clearly, groundwater is such a resource which is nearly consumed in both physical and economic terms in some locations in less than a generation, and yet is so relied on at the margin of expanding water demands that it is the future resource of necessity in many places. Accordingly, we must find ways to manage the physical and economic resources for the future.

SUMMARY

Groundwater provides many water uses, some of which are not obvious. This is because it has the characteristics of both stock and flow resources. It is a resource that is vital to the economies of many countries and localities. Groundwater as a raw material of the environment which is captured and used in the economy has long been recognized. The underlying policy implications of groundwater's movement or storage in the environment could be accounted for in transactions that affect the economy and even more essentially, the ecosystem about which we have incomplete information to address in market transactions. The simple models reflect different worldviews, one of groundwater use and residuals management in an economy that does not recognize other groundwater values, and the other incorporating ecosystem services difficult to measure, monetize and price in the market but essential to the economy. As a result of the challenge to recognize the ecosystem services in the economy, market failures include distribution inequity, inefficient allocation, intergenerational fairness and inability to price ecosystem services. Relating these issues to the economic management characteristics of groundwater indicate that the resource is difficult to own, challenging to control access and easily depleted or contaminated.

The scale of use affects the extent of depletion and contamination of groundwater, even considering its renewability in many situations. Groundwater affects the economy at the individual level, for example, water for a particular consumer, as well as on large scale, watershed or national level, such as widespread depletion of an aquifer or as an input to national production of crops. As we do not understand all the ways in which groundwater supports the economy because of the inherent interrelations of the ecosystem, it is difficult to value groundwater in the economy other than through the market, which sets prices based on the value added for the throughput of groundwater, even though this may have negative results for the ecosystem. The groundwater services sector contributes at least $45 billion to the US economy. Prices for the factors that we can define in the ecosystem could not be determined, because we are too ignorant of the intricacies of the balance of nature. The models can be used as a guide to the development of sustainable groundwater policies, taking into account issues such as distributional and intergenerational equity. Future models may need to consider saltwater intrusion, flood effects, disposal of CO_2 and desalination wastes on a larger scale. The performance of the natural capital of groundwater in providing many services including water collection, storage, treatment and distribution should be recognized in economic evaluations of alternative water provision.

FOR FURTHER CONSIDERATION

1. Can you identify the major uses of groundwater in the economy of your area, region, or state? On a relative basis, which groundwater uses would you say are most significant? Compare your answer to the information published for states at https://pubs.usgs.gov/circ/1441/circ1441.pdf, page 16.
2. To support the uses identified in (1) above, what kind of projects have been constructed or may be needed? How do these projects provide services to people or support the environment beyond the project sites? Can you identify any unintended effects of the projects beyond the project sites?
3. Considering the hydrologic cycle, how do the projects draw on groundwater? Is any groundwater reused by returning the water to the hydrologic cycle? What kind of projects would be needed to return the water to the aquifer from which it was withdrawn in the hydrologic cycle?

4. When groundwater is withdrawn from an aquifer or its quality changed because of contamination, what potential costs to other users of the aquifer may occur? What measurements should be taken before groundwater is withdrawn or water of a potentially different quality allowed to enter in the aquifer to assist management of the aquifer?
5. Using data from the U.S. Geological Survey or a local source for groundwater use and data on annual precipitation from the U.S. Weather Service with an assumption that one-fourth of precipitation infiltrates to reach groundwater, construct a model or water budget for your area or state. Do uses exceed water infiltrated, based on the infiltration assumption? If you find that infiltration is different for your area, use that fraction or percentage in your infiltration assumption.

REFERENCES

Anderson, B. 2017. Taxpayer dollars fund most oversight and cleanup costs at Superfund rites. *The Washington Post*, September 20, 2017.

Anderson, S. 1998. Groundwater declines raise concerns. In *Wisconsin Water Well Association Newsletter*. Wisconsin Water Well Association, Madison, WI, pp. 1–4.

Associated Press (AP). 2016. *Central Valley Farmers Drill More, Deeper Wells As Drought Limits Loom*. September 25. URL: https://sanfrancisco.cbslocal.com/2016/09/25/central-valley-farmers-drill-more-deeper-wells-limits-loom/

Barbash, J.E. and Resek, E.A. 1996. *Pesticides in Ground Water; Distribution, Trends and Governing Factors*. Ann Arbor Press, Inc., Chelsea, MI, 588 p.

Barlow, P.M. and Reichard, E.G. 2010. Saltwater intrusion in coastal regions of North America. *Hydrogeology Journal*, 18 (1), 247–260.

Beitsch, R. 2018. Few wells tested for contamination after major flooding from hurricanes. *Pew Stateline*, December 14. URL: https://www.pewtrusts.org/research-and-analysis/blogs/stateline/2018/12/14/few-wells-tested-for-contamination-after-major-flooding-from-hurricanes

Bergstrom, J.C., Boyle, K.J., Job, C.A., and Kealy, M.J. 1996. Assessing the economic benefits of ground water for environmental policy decisions. *Water Resources Bulletin*, 32, 279–291.

Brandt, E.N. 1997. *Growth Company: Dow Chemical's First Century*. Michigan State University Press, East Lansing, MI, 649 p.

Brown, L.R. 1996. *Tough Choices: Facing the Challenge of Food Scarcity*. W.W. Norton Publishing, New York, 159 p.

Brown, L.R. 2004. *Outgrowing the Earth: The Food Security Challenge in an Age of Falling Water Tables and Rising Temperatures*. Earth Policy Institute, W.W. Norton & Co., New York, 239 p.

Brumley, J., Arks, C., Chau, A., Lowrance, R., Huang, J., Richardson, C., Acree, S., Ross, R., and Beak, D. 2018. *The Influence of Green Infrastructure Practices on Groundwater Quality: The State of the Science*. U.S. Environmental Protection Agency, Washington, DC, EPA/600/R-18/227.

California Coastal Commission (CCC) 1993. *Seawater Desalination in California*. URL: http://www.coastal.ca.gov/desalrpt/dtitle.html#TOCDesalination (accessed February 9, 2007).

Campbell, K. 2017. Types of pollution found in brownfields. *Sciencing*. April 24, 2017. URL: https://sciencing.com/types-pollution-found-brownfields-22631.html

Christian-Smith, J., and Kaphiem, C. 2011. *Volumetric Water Pricing and Conjunctive Use: Alta Irrigation District*. Pacific Institute. URL: https://www.pacinst.org/reports/success_stories/volumetric_water_pricing_and_conjunctive_use.pdf

Central Platte Natural Resources District (CNPRD). 2019. *The Central North Platte Groundwater Exchange Program*. URL: https://cpnrd.org/groundwater-exchange-program/

Daly, H.E. and Farley, J. 2004. *Ecological Economics*. Island Press, Washington, DC, 454 p.

Darghouth, S. 2002. *Water Resources Management and Agriculture: Issues and Options*. World Bank presentation in Riyadh, Saudi Arabia. October 22. URL: http://www.planning.gov.sa/PLANNING/ALL%20PRESENTATION/Day4/4B-agriculture%20and%20water/22 (accessed February 25, 2005).

Driscoll, F.G. 1986. *Groundwater and Wells*. U.S. Filter/Johnson Screens. St. Paul, MN, 1089 p.

El Paso Water (EP Water). 2018. *Reclaimed Water*. URL: https://www.epwater.org/our_water/water_resources/reclaimed_water

European Environmental Agency (EEA). 1999. *Groundwater Quality and Quantity in Europe*. EEA, Copenhagen, Denmark, 123 p.

European Environment Agency (EEA). 2003. *Indicator Fact Sheet (WQ03b): Saltwater Intrusion.* (Author: Scheidleder, A.) EEA, Copenhagen, Denmark, 4 p. URL: http://themes.eea.eu.int/Specific_media/water/indicators/WQ03b,2003.1001/Fact_sheet_16_Saltwater_Intrusion_final.pdf (accessed February 25, 2005).

Farm Credit Canada (FCC). 2020. 2019 FCC Farmland Values Report. URL: https://www.fcc-fac.ca/fcc/resources/2019-farmland-values-report-e.pdf (accessed December 15, 2020).

Field, B.C. 1994. *Environmental Economics: An Introduction.* McGraw-Hill, New York, 482 p.

Foldager, R. 2003. Economics of desalination concentrate disposal methods in inland regions: Deep-well injection, evaporation ponds, and salinity gradient solar ponds. Thesis. New Mexico State University, Las Cruces, NM, 71 p. URL: http://wrri.nmsu.edu/research/rfp/studentgrants03/reports/foldager.pdf (accessed February 9, 2007).

Gelt, J. 1992. Land subsidence, earth fissures change Arizona's Landscape. *Arroyo (Newsletter of the University of Arizona Water Resources Research Institute)*, 6 (2), Summer. URL: http://ag.arizona.edu/azwater/arroyo/062land.html (accessed February 26, 2005).

Gleik, P. 2000. *The World's Water; The Biennial Report on Freshwater Resources, 2000–2001.* Island Press, Washington, DC.

Hardberger. 2016. *Texas Groundwater Markets and the Edwards Aquifer: A Case Study for the Political Economy of Water Markets Project.* URL: https://static1.squarespace.com/static/56d1e36d59827e6585c0b336/t/5805468415d5dbb1ab59a3a9/1476740751543/Texas-Groundwater-Markets-Hardberger.pdf

Harter, T., and Brewster, B. 2018. *Groundwater Recovery in California – Still Behind the Curve.* California WaterBlog. April 8. URL: https://californiawaterblog.com/2018/04/08/groundwater-recovery-in-california-still-behind-the-curve/

Hutton, G. 2001. Economic evaluation and priority setting in water and sanitation interventions. In *Water Quality – Guidelines, Standards and Health: Assessment of Risk and Risk Management for Water-Related Infectious Disease.* Lorna, F. and Jamie, B. (eds.). World Health Organization IWA Publishing, London, UK

Intergovernmental Panel on Climate Change (IPCC). 2001. *Climate Change 2001: Impacts, Adaptation, and Vulnerability.* Cambridge University Press, New York, 1042 p. Hydrology and Water Resources Chapter. URL: http://www.grida.no/climate/ipcc_tar/wg2/159.htm (accessed January 15, 2007).

International Groundwater Resource Assessment Center (IGRAC). 2018. *Managed Aquifer Recharge Portal.* URL: https://apps.geodan.nl/igrac/ggis-viewer/viewer/globalmar/public/default

Job, C.A. 2004. Selected characteristics of ground water-supplied community water systems. ground water monitoring and remediation. *National Ground Water Association*, 24 (2) (Spring), 48, 50, 52.

Kim, J.-H., Yum, B.-W., Kim, R.-H., Koh, D.-C., Cheong, T.-J., Lee, J., and Chang, H.-W. 2003. Application of cluster analysis for the hydrogeochemical factors of saline groundwater in Kimje, Korea. *Geosciences Journal*, (December) 7 (4), 313–322.

Kounang, N. 2018. El Paso to drink treated sewage water due to climate change drought. *CNN*, December 5. URL: https://www.cnn.com/2018/11/30/health/water-climate-change-el-paso/index.html

Krieger, L. M. 2016. California drought: High-bidding farmers battle in water auctions. *The Mercury News*, August 12. URL: https://www.mercurynews.com/2014/07/19/california-drought-high-bidding-farmers-battle-in-water-auctions/

Manning, F.S. and Thompson, R.E. 1995. *Oilfield Processing: Crude Oil.* PennWell Books, Tulsa, OK, 400 p.

McCabe, W.J., Job, C.A., Simons, J.J., Graves, J.S., and Terada, C.J. 1997. History of the sole source aquifer program: A community-based approach for protecting aquifers used for drinking water supply. *Ground Water Monitoring and Remediation*, 17 (3), 78–86.

Merrett, S., 2002. *Water for Agriculture: Irrigation Economics in International Perspective.* Spon Press, London.

Meyer, P.B., Williams, R.H., and Yount, K.R. 1995. *Contaminated Land: Reclamation, Redevelopment and Re-Use in the United States and the European Union.* Edward Elgar Publishers, Cheltenham, UK

Narayan, K.A., Schleeberger, C., Charlesworth, P.B., and Bristow, K.L. 2002. Modeling saltwater intrusion in the lower Burdekin Delta, North Queensland, Australia. In *2002 Denver Annual Meeting.* October 27–30, Geological Society of America, Denver, CO.

National Academies of Science. 2000. *The Future Role of Pesticides in US Agriculture.* National Academies Press. URL: https://www.nap.edu/read/9598/chapter/4

National Geographic. 2019. *Saudi Arabia's Great Thirst.* URL: https://www.nationalgeographic.com/environment/freshwater/saudi-arabia-water-use/

National Ground Water Association (NGWA). 2001. *U.S. Ground Water Industry Market Backgrounder.* Columbus, OH.

National Ground Water Association (NGWA). 2019. *Groundwater Fundamentals*. URL: https://www.ngwa.org/what-is-groundwater/About-groundwater

National Ground Water Association (NGWA). 2020. *America's Groundwater Industry Employment*. Fact Sheet. URL: https://www.ngwa.org/docs/default-source/default-document-library/groundwater/usa-groundwater-use-fact-sheet.pdf?sfvrsn=5c7a0db8_4 (accessed April 17, 2020).

National Research Council (NRC). 1994. *Ground Water Recharge Using Waters of Impaired Quality*. The National Academies Press, Washington, DC. https://doi.org/10.17226/4780.

Sampat, P. 2000. Groundwater shock. *World Watch*, January/February, 13 (1), 10–22.

Schiffler, M. 1998. *The Economics of Groundwater Management in Arid Countries; Theory, International Experience and a Case Study of Jordan*. Frank Cass Publishers, London, UK, 394 p.

Schneider, K. 2019. Groundwater scarcity, pollution set India on perilous course. *New Security Beat*, January 15. URL: https://www.newsecuritybeat.org/2019/01/groundwater-scarcity-pollution-set-india-perilous/

Seo, S., Segarra, E., Mitchell, P.D., and Leatham, D.J. 2008. Irrigation technology adoption and its implication for water conservation in the Texas High Plains: A real options approach. *Agricultural Economics*, 38 (1): 47–55.

Shah, T., Molden, D., Sakthivadivel, R., and Seckler, D. 2000. *The Global Groundwater Situation: Overview of Opportunities and Challenges*. International Water Management Institute. http://www.iwmi.cgiar.org/pubs/WWVisn/GrWater.htm (accessed February 24, 2005).

Smekens, K. and van der Zwaan, B. 2004. *Environmental Externalities of Geological Carbon Sequestration Effects on Energy Scenarios*. The Fondazione Eni Enrico Mattei Note di Lavoro Series. URL: www.feem.it/Feem/Pub/Publications/WPapers/default.htm (accessed February 4, 2007).

Southern California Coastal Water Resources Project (SCCWRP). 1993. Toxic effects of elevated salinity and desalination waste brine. In 1992–1993 Annual Report. URL: http://www.sccwrp.org/pubs/annrpt.htm (accessed February 9, 2007).

Stowell, C. 2005. Solution to Warming: Go Underground; Researchers Consider Storing Carbon Dioxide beneath Earth's Surface. *Daily Texan Online*. February 22, 2005. URL: http://www.dailytexanonline.com/news/2005/02/22/Focus/Solution.To.Warming.Go.Underground-872198.shtml (accessed February 4, 2007).

Texas Water Development Board (TWDB) 2007. *Desalination – Frequently Asked Questions*. URL: http://www.twdb.state.tx.us/Desalination/Desal/Frequently%20asked%20questions.asp (accessed February 9, 2007).

The Guardian. 2016. "Four-fifths of China's water from wells unsafe because of pollution." April 12. URL: https://www.theguardian.com/environment/2016/apr/12/four-fifths-of-chinas-water-from-wells-unsafe-because-of-pollution

Tsur, Y., Roe, T., Doukkali, R., and Dinar, A. 2004. *Pricing Irrigation Water; Principles and Cases from Developing Countries*. Resources for the Future. Washington, DC, 319 p.

United Nations Educational, Scientific and Cultural Organization (UNESCO). 2003. *Water for People, Water for Life The United Nations Water Development Report*. UN World Water Assessment Programme UNESCO, Barcelona, Spain, 575 p.

United Nations Environment Programme (UNEP). 1999. *Global Environment Outlook 2000*. URL: http://www1.unep.org/geo-text/0088.htm (accessed February 25, 2005).

United Nations Food and Agriculture Organization (FAO). 1993. *The State of Food and Agriculture 1993*. FAO Agriculture Series No. 26. FAO, Rome, Italy.

United Nations Food and Agriculture Organization (FAO). 1997. Review of the irrigation equipment manufacture and supply sector in China. In *Proceedings of a Subregional Workshop*. April 14–17, Harare, Zimbabwe. (Author: Zhou, W.) URL: http://www.fao.org/docrep/W7314E/w7314e00.htm#Contents (accessed February 25, 2005).

United Nations World Commission on Environment and Development (UNWCED). 1987. *Our Common Future*. Oxford University Press, Oxford, UK

U.S. Census Bureau (USCB). 2008. *Census of Population and Housing*. Washington, DC. URL: http://www.census.gov/prod/www/abs/decennial/index.htm (accessed August 28, 2009).

U.S. Census Bureau (USCB). 2018. *Coastline County Population Continues to Grow; 60 Million Live in the Path of Hurricanes*. URL: https://www.census.gov/library/stories/2018/08/coastal-county-population-rises.html

U.S Department of Agriculture (USDA). 2013. *Farm and Ranch Irrigation Survey*. URL: https://www.nass.usda.gov/Publications/AgCensus/2012/Online_Resources/Farm_and_Ranch_Irrigation_Survey/fris13_1_012_012.pdf

U.S. Department of Agriculture (USDA) Economic Research Service. 1997. Data: Western Irrigation. URL: http://www.ers.usda.gov/data/westernirrigation/ShowTables.asp?tabList=1#flag (accessed March 24, 2005).

U.S. Department of Agriculture (USDA) 2018. *Census of Agriculture, Land Values*, 2018 Summary Washington, DC. URL: https://downloads.usda.library.cornell.edu/usda-esmis/files/pn89d6567/qb98mj07s/rv042w81z/AgriLandVa-08-02-2018.pdf

U.S. Department of Agriculture (USDA) Economic Research Service. 2002. *Agricultural Resources and Environmental Indicators: Water Use and Pricing in Agriculture*. Publication No. AH722 (April). (Authors: Gollehon, N., Quinby, W., and Aillery, M.), Washington, DC.

U.S. Department of Energy (USDOE). 2005a. *Carbon Sequestration (Website)*. URL: www.netl.doe.gov/technologies/carbon_seq/index.html (accessed February 4, 2007).

U.S. Department of Energy (USDOE). 2005b. *Carbon Sequestration R&D Overview*. November 6, 2005. URL: http://www.fossil.energy.gov/sequestration/overview.html (accessed February 4, 2007).

U.S. Department of Justice (USDOJ). 1995. Press Release: Occidental to Pay $129 Million in Love Canal Settlement. December 21, 1995.

U.S. Environmental Protection Agency (USEPA). 1987. Press Release: Love Canal Record of Decision Signed. October 26. URL: http://www.epa.gov/history/topics/lovecanal/04.htm (accessed February 26, 2005).

U.S. Environmental Protection Agency (USEPA). 1994. *Guidelines for Water Reuse*. EPA/625/R-04/108. Washington, DC, USA. URL: http://www.epa.gov/nrmrl/pubs/625r04108/625r04108.pdf (accessed October 20, 2006).

U.S. Environmental Protection Agency (USEPA). 1996. National Water Quality Inventory, 1994 Report to Congress. EPA-813-R-96-001. 63 p.

U.S. Environmental Protection Agency (USEPA). 1997. Pesticide Annual Sales and Usage; 1994 and 1995 Market Estimates. 733-R-97–002. 35 p.

U.S. Environmental Protection Agency (USEPA). 2002. *Community Water System Survey 2000*. EPA 815-R-02–005B. December, 180 p.

U.S. Environmental Protection Agency (USEPA). 2003. 2003–2008 EPA Strategic Plan. EPA-190–03–0032003. September 2003. 239 p. URL: http://www.epa.gov/ocfo/plan/2003sp.pdf (accessed June 13, 2005).

U.S. Environmental Protection Agency (USEPA). 2004a. *Drinking Water Costs and Federal Funding*. EPA 816- F-04-038. June 2004. URL: http://www.epa.gov/safewater/sdwa/30th/factsheets/pdfs/fs_30ann_dwsrf_web.pdf (accessed February 9, 2007).

U.S. Environmental Protection Agency (USEPA). 2004b. *It Pays to Save Wetlands*. EPA-843-H-04-001. Fact Sheet.

U.S. Environmental Protection Agency (USEPA). 2006. Releases: Chemical Report; 2004. URL: http://www.epa.gov/tri/ (accessed October 20, 2006).

United States Environmental Protection Agency (USEPA). 2009. *Community Water System Survey 2006*, EPA 815-R-09-002, May, 3 volumes.

U.S. Environmental Protection Agency (USEPA). 2018a. *Overview of EPA's Brownfields Program*. URL: https://www.epa.gov/brownfields/overview-epas-brownfields-program

U.S. Environmental Protection Agency (USEPA). 2018b. *Petroleum Brownfields*. URL: https://www.epa.gov/ust/petroleum-brownfields

U.S. Geological Survey (USGS). 1977. *Estimated Use of Water in the United States in 1975*. USGS Circular 765. United States Government Printing Office, Washington, DC, 39 p.

U.S. Geological Survey (USGS). 1983. *Estimated Use of Water in the United States in 1980*. USGS Circular 1001. United States Government Printing Office, Washington, DC, 56 p.

U.S. Geological Survey (USGS). 1985. *National Water Summary 1984*. USGS Water Supply Paper 2275. United States Government Printing Office, Washington, DC.

U.S. Geological Survey (USGS). 1988. *Estimated Use of Water in the United States in 1985*. USGS Circular 1004. United States Government Printing Office, Washington, DC, 82 p.

U.S. Geological Survey (USGS). 1993. *Estimated Use of Water in the United States in 1990*. USGS Circular 1081. United States Government Printing Office, Washington, DC, 76 p.

United States Geological Survey (USGS). 1977. *Estimated Use of Water in the United States in 1975*, USGS Circular 765, United States Government Printing Office, Washington, DC, 39.

U.S. Geological Survey (USGS). 1998. *Estimated Use of Water in the United States in 1995*. USGS Circular 1200. United States Government Printing Office, Washington, DC, 71 p.

U.S. Geological Survey (USGS). 1999. *Ground Water and Surface Water, A Single Resource*. Circular 1139. Denver, CO, 79 p.

U.S. Geological Survey (USGS). 2003. *Ground Water in Freshwater-Saltwater Environments of the Atlantic Coast*. (Author: Barlow, P.M.) USGS Circular 1262. USGS, Reston, VA.

U.S. Geological Survey (USGS). 2004. *Estimated Water Use in the United States in 2000*. USGS Circular 1268. United States Government Printing Office, Washington, DC. URL: http://water.usgs.gov/pubs/circ/2004/circ1268/htdocs/text-do.html (accessed March 25, 2005).

U.S. Geological Survey (USGS). 2005a. *Ground Water in Freshwater-Saltwater Environments of the Atlantic Coast*. USGS Circular 1262. United States Government Printing Office, Washington, DC. URL: http://pubs.usgs.gov/circ/2003/circ1262/ (accessed February 9, 2007).

U.S. Geological Survey (USGS). 2005b. Thirsty? How 'bout a cool, refreshing cup of seawater? URL: http://ga.water.usgs.gov/edu/drinkseawater.html (accessed February 9, 2007).

U.S. Geological Survey (USGS). 2010. Estimated use of water in the United States in 2010. *Circular*, 1405. URL: https://pubs.usgs.gov/circ/1405/pdf/circ1405.pdf

U.S. Geological Survey (USGS). 2014. Estimated use of water in the United States in 2010. *Circular*, 1405. URL: https://pubs.usgs.gov/circ/1405/pdf/circ1405.pdf

U.S. Geological Survey (USGS). 2015. Estimated use of water in the United States in 2015. *Circular*, 1441. URL: https://pubs.usgs.gov/circ/1441/circ1441.pdf

U.S. Geological Survey (USGS). 2018a. Estimated use of water in the United States in 2015. *Circular*, 1441. URL: https://pubs.usgs.gov/circ/1441/circ1441.pdf

U.S. Geological Survey (USGS). 2018b. What is Drought? URL: https://ca.water.usgs.gov/california-drought/what-is-drought.html

U.S. National Oceanic and Atmospheric Administration (USNOAA). 2005. *Monterrey Bay National Marine Sanctuary Resource Management Issues: Desalination*. URL: http://montereybay.noaa.gov/resourcepro/resmanissues/desalination.html (accessed February 9, 2007).

U.S. Water Resources Council (USWRC). 1978. *The Nation's Water Resources 1975–2000*. Vol.2. U.S. Government Printing Office, Washington, DC.

Wolf, J.. 2017. UCLA-led researchers track groundwater loss during drought in California's Central Valley. USCL Newsroom; Environment+Climate, May 17. URL: http://newsroom.ucla.edu/releases/groundwatcr-loss-in-central-valley

Water Services Association of Australia Inc (WSAA). 2004. *Dual Water Supply Systems; Supplement; Melbourne Retail Water Agencies Edition of the Water Supply Code of Australia*. WSA 03-2002. WSAA, Melbourne, Victoria, Australia, 71 p. URL: http://www.citywestwater.com.au/about/docs/MRWA_Dual_Water_Supply_System_Supplement.pdf (accessed October 20, 2006).

Weiping, Z.1997. *Irrigation technology transfer in support of food security, (Water Reports-14) Proceedings of a Subregional Workshop*. Harare, Zimbabwe, April 14–17, Food and Agriculture Organization of the United Nations, Rome, Italy.

Westwater Resources. 2006. WaterBank; Frequently Asked Questions. URL: www.waterbank.com/faq1.html (accessed October 21, 2006).

Wichelns, D. 2010. *Agricultural Water Pricing: United States. Organisation For Economic Co-Operation and Development*, 27p. URL: https://www.oecd.org/unitedstates/45016437.pdf

Wolock, D.M. 2005. U.S. Geological Survey Scientist. Personal Communication. June 27, 2005.

World Resources Institute (WRI). 2005. Earth Trends. URL: www.earthtrends.wri.org (accessed February 23, 2005).

Yang, X.-L., Chen, Y.-Q., Steenhuis, T.S., Pacenka, S., Gao, W.-S., Ma, L., Zhang, M., and Suil, P. 2017. Mitigating groundwater depletion in North China Plain with cropping system that alternate deep and shallow rooted crops. *Frontiers in Plant Science*, 8, 980.

Yasmeen, A. 2006. New layouts to have dual water supply system. *The Hindu*; Online edition of India's National Newspaper. Tuesday, July 25, 2006. URL: http://www.hindu.com/2006/07/25/stories/2006072520950300.htm (accessed October 20, 2006).

Young, R.A. 2005. *Determining the Economic Value of Water Concepts and Methods*. Resources for the Future Press, Washington, DC, 356 p.

Zhou, W. 1997. *Review of the irrigation equipment manufacture and supply sector in China*. In *Irrigation Technology Transfer in Support of Food Security (Water Reports-14). Proceedings of a Subregional Workshop*. Harare, Zimbabwe, April 14–17, Food and Agriculture Organization of the United Nations, Rome, Italy.

4 Groundwater Law

A COMPLEX FRAMEWORK

A complex mix of local, state, national and international laws affects the use of groundwater as a source of water supply and as a sink for contaminants. These laws recognize groundwater as a resource fundamental in economic exchanges. In the United States, states typically have the leading legal role in controlling the use of groundwater (USEPA, 1991). In Europe, a groundwater directive identifies the role of national governments in responding to both quantity and quality issues of the resource (EU, 2006). In the United States, law basically treats groundwater as a local or state resource. The U.S. federal law does not recognize the emerging understanding that groundwater can migrate across long distances over time and may be essential to streamflow in many areas of the country, although this perception is changing with a 2020 decision of the U.S. Supreme Court. European Union law recognizes explicitly the relation of ground and surface water in watersheds (EU, 2000). Some U.S. state laws have linked ground and surface water legally to provide conjunctive management. International law dealing with transboundary (cross-border) issues embraces the need to consider the effects on neighboring countries. The framework of state, federal and case law in the United States indicates who has the property rights in groundwater, the process for dealing with damages incurred by adjacent property owners, who can dispose of contaminants in the subsurface and in what concentration or volume and who must pay for the cost of cleanup of contaminated groundwater. Most recently, extensive drought driving groundwater use in California has prompted a new law there on managing groundwater sustainably (CAL, 2014). These legal frameworks provide the basis for economic relationships and exchanges among governments, corporations and individuals which result in the need for economic information and analysis applied to groundwater.

Leading legal authorities Joseph Sax and Robert Abrams said of groundwater law:

> Water law has traditionally separated groundwater and surface water and provided independent rules of allocation and assignment of property rights. Much of the historical reason for the duality is attributable to the lack of knowledge about the movement of water beneath the earth's surface. This separation of legal regimes continues in the law of a majority of American states, even though the science of geohydrology has come of age. It is now possible to trace subterranean water movement and to establish groundwater discharge as a major component of base streamflow in most streams. Legal doctrine has not kept pace ... It is a mistake to consider groundwater as only a minimally relevant subset of water law.
>
> (Sax and Abrams, 1986, p. 786)

SOCIOECONOMIC CONSIDERATIONS

The economics of groundwater resources has, as a principal social basis, law to guide who pays and who benefits from actions taken to or through it. As economics works to bring a balance between supply and demand and monetize fairness in applying and using the resources of labor and capital, law brings a balance between the ownership of that labor and capital and the associated rights. As exchanges of ownership and rights occur, or as damage to owners of and users with rights to groundwater is incurred, law provides a part of the basis for value, costs, benefits and prices. The other major element of determining the values of water is, of course, human desire and need, fashioned by tastes and preferences of individuals and groups (NRC, 1997, p. 157), as well as imperfect but improving information on the availability and quality. Groundwater law, then, affects and is affected by the uses to which groundwater is put (Goldfarb, 1988, p. 6).

RANGE OF USES RECOGNIZED IN LAW

In the United States, nine major uses of water are recognized in law: navigation, irrigation, potable water, waste disposal, industrial process and cooling purposes, hydroelectric power, recreation, fish and wildlife maintenance and scenic/aesthetic enjoyment (Goldfarb, 1988, p. 6). A similar range of uses of water is recognized by the European Union (EU, 2000) and by international organizations (UN, 2003). Science has expanded our understanding of how surface water and groundwater interact and the significance of groundwater in all these uses is now better understood. In areas or circumstances where surface flows are minimal, groundwater may be relied on to provide all these uses, including maintaining streamflows for fish and wildlife sustenance and scenic/aesthetic enjoyment.

A principal concern for water uses in the past has been an adequate supply of water. Laws affecting the provision of safe drinking water, because of the social preference for healthful uncontaminated water, have caused additional treatment of drinking water, resulting in increased supply costs and value of drinking water. As recognition of groundwater contribution to the other water uses grows (Bergstrom et al., 1996), cost impacts of reduced quantity and quality of groundwater as well as the value of maintaining the quantity and quality of groundwater may increase. As this recognition grows, the law in the United States may eventually incorporate these considerations, as already done by the European Union directives.

Goldfarb (1988, pp. 6–7) notes that one major water use has been neglected: demand-side conservation. As he indicates, "Water is, and will continue to be, scarce in many parts of the United States. Consequently, we are learning to perceive water conservation as an end in itself, a use on a par with other uses of water." Goldfarb projects this conservation use, "intentional nonuse and use reduction," as "one of the most volatile areas of water law for the foreseeable future" (Goldfarb, 1988, pp. 6–7). Schiffler (1998, pp. 334–335 and elsewhere) indicates that investment in water-conserving techniques in water-short areas is effectively a way to supply greater needs for water. This issue relates to a topic addressed later: sustainability of the groundwater resource.

A major characteristic of water that has only received much attention in the last half of the twentieth century, and for groundwater, in the last quarter of that century, is quality. In this light, Goldfarb moves away from terms such as "consumptive" and "nonconsumptive" uses to a new terminology of "transformational" and "nontransformational" uses, in line with some current thinking on sustainability (e.g., see Daly and Farley, 2004, pp. 62–64). "A transformational use is one which causes a significant change in the existing condition of a waterbody" (Goldfarb, 1988, p. 7). "A 'nontransformational use' leaves a waterbody in a basically unaltered state. This approach has the advantage of placing all water uses on the same plane. Thus, it deals directly with the fundamental value question inherent in water resource management: Are the social and economic gains produced by water use worth the costs of ecosystem change?" (Goldfarb, 1988, p. 7). This subject was examined in Volume 1, Chapter 2.

GROUNDWATER QUANTITY AND QUALITY

Water law in the United States historically has treated quantity and quality in separate statutes. The European Union addresses all major pollutant releases affecting ground and surface water quality in one Directive, whereas the United States divides them among several statutes. When population was not as dense and concentrated, concerns about quality were less. But as the use and the range of uses increased, quality became a significant concern. While quality and quantity as well as groundwater and surface water are inextricably related, they are still usually legislated separately in the United States with potentially important economic consequences. For example, the U.S. Clean Water Act (CWA) has provided the basis for communities to treat their wastewaters at a central treatment plant with water discharge typically into a stream. The US Safe Drinking Water Act regulates injection of liquids including stormwater and wastewater into the subsurface. Communities using groundwater for water supply must evaluate the regulatory, hydrologic, engineering and economic considerations and decide whether, faster release of water to a stream or reclaiming and recycling the water by employing hydrologic or redistribution processes is in their best interests. Such an evaluation can recognize the importance of the hydrologic cycle in maintaining water supplies and not respond only to water quality effects. In areas where the water tables may be falling, causing both municipalities and residential well owners to install deeper wells or causing greater public expenditure on what might have been unnecessary public water systems expansion may be an inefficient approach in a systems analysis context.

In such cases, all costs may not be accounted for because the "hydrologic system" was not defined in the enabling legislation. The law simply did not reflect an understanding of groundwater in the hydrologic cycle and its relation to both ground and surface water quality. Similarly, many U.S. state laws address water quantity and allocation issues without relating them to water quality considerations. Some state laws maneuver around this situation with broad legislation for the conservation and protection of all waters of the state but do not explicitly acknowledge the significance of the hydrologic cycle in water management.

GROUNDWATER MOBILITY

The dynamic factor that distinguishes water from resources like land is its mobility (Hanemann, 2005, pp. 14–16). Groundwater moves at different rates depending on the geologic matrix in which it occurs. Because it moves and its other characteristics that may be changing over time move with it, groundwater quality and quantity change as well. This condition of groundwater mobility affecting quality and quantity influences the effects that one user may have on a neighboring user that may carry with it costs and/or benefits depending on the nature of the use. The other economic aspect of groundwater mobility that is not widely acknowledged yet is the infrastructure attribute of natural transmission and distribution of water underground without pipes. The cost of installing and maintaining a pipe network is significant and aquifers provide the services of water storage transmission and distribution freely – a service of its natural capital characteristic (WSC, 2018).

GROUNDWATER RIGHTS IN USE AND ACCESS

"Property rights are defined as 'a bundle of entitlements, privileges, and limitations defining the owner's rights to use a resource'" (Schiffler, 1998, p. 92; citing Tietenberg, 1994). Groundwater as property to be owned in some form is addressed in the rights: (a) to use a resource — e.g., pump and drink groundwater, (b) to the return flows from the asset — e.g., groundwater flowing across a boundary to a down-gradient property owner, (c) to change the asset's form and substance — e.g., adding pollutants to groundwater even at low concentrations and (d) to transfer a resource — e.g., selling groundwater, all of which can be held by either a central government (state or nation), a community or a person (Schiffler, 1998, p. 93). However, because groundwater moves, it is difficult to define a groundwater property right (Schiffler, 1998, p. 95). Depending on where in the world groundwater use is taking place, its consumption or use as a pollutant sink may be governed by property rights (i.e., exclusive use by the owner) or open-access to the resource (i.e., use is not restricted, controlled or managed and available to anyone for any purpose). National or state laws may control groundwater use in most cases in a legal status between these extremes.

Legal resource distinctions affect consumers' use of groundwater. An open-access resource categorization may apply to groundwater, as it would to air, in cases where drilling and production of the resource are not controlled or limited in any way. Water consumers may actually be denied the use of the resource in circumstances of (a) restricted common property, (b) communal property and

(c) private property (Schiffler, 1998, p. 93; citing Dales, 1968). Thus, rights in groundwater affect who can use it and to which uses it can be put. The three categories of restricted property rights just listed have different implications for groundwater users (Schiffler, 1998, p. 93):

1. Restricted common property might apply to groundwater in a national park or reserve, where the resource is owned by the central government in trust for anyone to use but resource use is controlled.
2. Communal property might apply if a community has exclusive rights to groundwater, such as a spring, and can exclude other users.
3. Private property has direct application in locations such as the western states where rights to groundwater stocks can be managed for exclusive use by the owner.

Whereas private property rights exclude use by others, public goods by their nature do not have restrictions on who uses them and how they are used. Depending on the location and authority over groundwater, the resource may be a private good or a more public good. This distinction affects both its use and its value. However, the more detailed the identification of rights, whether privately or publicly held, the greater the groundwater value since this definition specifies what rights are conferred with control over the resource. In cases where particular use is legally specified, groundwater value may be lower, since market forces are constrained (Schiffler, 1998, pp. 93–94). "(F)ully specified property rights are a powerful incentive to conserve resources, while their absence leads to the overutilization of resources," as would be the case if an aquifer had open access to anyone desiring to install a well and produce groundwater (Schiffler, 1998, pp. 94, 95).

Schiffler (1998, p. 94) notes some conditions under which groundwater might be considered an open-access or a private resource:

- Open access: aquifers of high porosity; no groundwater management
- Private: aquifer not porous; no interference from adjacent wells; access regulated.

Other conditions may be consistent with these categorizations. It is notable that private rights to groundwater do not allow an owner to do anything he or she may desire, since groundwater moves across property boundaries and excessive pumping or large amounts of pollutants reaching an aquifer may damage neighboring users' consumption of groundwater.

Under the legal systems of most countries, governments own the groundwater (Schiffler, 1998, p. 97) and in the United States, most states typically own and manage a water body as "trustee" for its citizens. In these states, an individual can only own the right to use the water, termed "usufructuary right." This right of use ownership is private property. This right is not absolute, and the government may determine that its use of the resource for the community is a priority and can take private property with just compensation. A major concern of the legal system is "at what point does a governmental regulation to protect the public health, safety, and welfare become a 'regulatory taking' of private property for which just compensation must be paid?" (Goldfarb, 1988, p. 11). Once this is decided, the legal system must also draw on information from the economic system to determine the appropriate level of compensation, which may vary from place to place.

Both a legal and an economic problem is the law's historical recognition of boundaries of water bodies, which do not exist in the physical environment. For example, surface water courses are not lined pipes. Groundwater of varying quality and quantity can interact with a stream and vice versa. This interaction changes with time and location; groundwater recharge from a stream and discharge to a stream vary with precipitation events and seasons of a year. However, a large body of law exists that governs rights to and use of groundwater, treating it as a "stock" resource with well-defined boundaries.

GROUNDWATER LAWS OF THE UNITED STATES

Groundwater law in the United States has aspects existing at all jurisdictional levels: local, state and federal. Much of early U.S. groundwater law was derived from English court law and then from U.S. court cases, which used common law to determine outcomes. Later, state law in the 1800s and then federal law in the 1900s, particularly in the latter years, addressed issues significant to groundwater and factors that affected its management, such as hazardous chemicals and wastes that were disposed of on or in the ground without adequate safeguards. The laws at any level were typically developed to address particular economic issues that arose, related to property use, income, health and environmental degradation, and may not have considered the effect of the issues related to the hydrologic cycle at that time. The effect of extreme weather events from changing climatic conditions may further influence the economic outcomes tied to associated changes in resource status and use under law. For example, the city of El Paso, Texas, USA, to have sufficient water supply, is producing and treating brackish water to supply a growing population in an arid climate (EPW, 2018). Wichita Falls, Texas, USA, has both direct potable and indirect potable water reuse facilities complying with drinking water standards to supply the city with water in its dry north central Texas location (Kowalick, 2018).

State Groundwater Doctrines, Law and Policy in the United States

Currently in the United States, 50 state groundwater doctrines have evolved, although some are quite similar. The doctrine related to groundwater quantity has a local focus, having evolved from English court law and ultimately into state law. These laws and court interpretations typically deal with the effects of one groundwater user on the availability and quality of water to another user. This section reviews those doctrines in the United States as they apply to groundwater and has been summarized primarily from Goldfarb (1988). Legal terms for groundwater categories are important in understanding some of the legal and hydrologic issues associated with these doctrines.

Terminology for Groundwater Categories in Case Law

In the hydrologic cycle, one cannot always distinguish between groundwater and surface water. However, because law has not historically always relied on science, some courts continue to distinguish even among artificial categories of groundwater, as well as ground and surface waters. This situation is changing slowly. Legal processes have considered the following categories.

Subflow of Surface Streams

Subflow is "the saturated zone directly beneath and supporting a river or lake in direct contact with surface water. Where subflow can be identified, it is considered as part of the water course itself" (Goldfarb, 1988, p. 19). From a hydrologic standpoint, subflow, as defined here, is highly dependent on the hydrologic head of surface and groundwater and the considerable potential variation in groundwater discharge and recharge relationships to a stream, even over short distances (e.g., less than 0.8 km). Even the contiguous saturated substratum immediately maintaining a river or a lake may now be much larger than originally understood.

Hydrology recognizes the "hyporheic zone" as "the porous areas of the stream bed and stream bank in which stream water mixes with shallow groundwater" (Runkel et al., 2003).

Underground Streams

An underground stream follows a definite channel below the ground surface. Typically, an underground stream is considered surface water unless evidence can be shown that the water flows in a "known" channel and does not "percolate" (Hutchins et al., 1971; Goldfarb, 1988). From a hydrologic perspective, "known" channels of underground streams would exist in carbonate formations (limestone or dolomite) because of solution channeling of water through what were first small fractures. As a result, underground streams do not exist in all subsurface environments, only in carbonates. In some parts of the United States (e.g., Florida, Central Texas and Northeast Iowa), these areas are extensive.

Percolating Waters

Percolating waters are waters that have infiltrated the soil and move through the ground, but not by a defined channel (Goldfarb, 1988). These waters represent what is considered to become the principal body of groundwater once they are part of the subsurface saturated zone referred to as the "aquifer." Most "percolating waters" begin as precipitation, infiltrate the ground surface becoming soil water for a time, and then, if not evapotranspired, travel (migrate) to the water table by way of the microscopic pore spaces between the soil particles, along fine soil and rock fractures, and down macropores — worm, insect and animal holes in the soil and unsaturated zone.

Wastewaters

Certain waters may escape constructed works by seepage, loss, waste, drainage or percolation. A legal and economic question is to whom the wastewater belongs when it is released (Goldfarb, 1988). Where water is scarce, it is possible that escaping wastewater may be an important water source. In some locations, escaping wastewater may be contaminating other potable waters.

While these definitions continue in legal use, many courts now attempt to incorporate modern understanding of hydrology. Because in many instances no clear distinction can be made between ground and surface waters, it is important to recognize the function of the hydrologic cycle in water quantity and quality issues.

Water Quantity – State Law

Historically, groundwater law or doctrines developed separately from those for surface water because of a lack of understanding of the relationship between ground and surface water. English court case law related to groundwater was based on the principle that groundwater movement was unknowable (NRC, 1997). Currently, a disjointed, multifaceted system of groundwater law exists. Goldfarb (1988) and the National Research Council (1997) identify similar major categories of U.S. state groundwater doctrine that are important:

1. The rule of absolute ownership
2. The reasonable use rule
3. Correlative rights rule in the eastern United States, based on the Restatement of Torts rule
4. Correlative rights rule in the western states
5. State permits
6. Prior appropriation
7. Critical management areas

These are summarized in Table 1.4.1 and the text below. Note that groundwater and surface water are managed under different legal approaches in the United States, further explained below. For example, in California, surface water is managed under riparian and prior appropriation doctrine, whereas groundwater is managed under Western Correlative rights law. In Texas, surface water is under prior appropriation doctrine with permits while groundwater is under absolute ownership. In Indiana, surface water is managed under riparian law, but groundwater is under absolute ownership doctrine.

TABLE 1.4.1
State Groundwater Production Law in the United States

Legal Approach	Groundwater Owner	Pumping/Use	Water Shortage Conditions	Pumping Zone Encroachment	Groundwater Mining	Surface Water Influence	Example States
1. Common law							
a. Absolute ownership (English rule)	Landowner	Unrestricted		Landowner held harmless for encroaching an adjacent user's capture zone	Cannot be wasteful	Courts apply "underground stream" doctrine	Texas, Indiana
b. Reasonable use (American rule)	Landowner, but no legal interest until groundwater is produced	Use on own property without waste		If use is flagrant waste, or user is not overlying, judicial remedy can be sought	If use is wasteful or adjacent, may be controlled by courts	Courts apply "underground stream" doctrine	Nebraska, Arizona, California, plus 11 eastern states
2. Correlative rights							
a.Western (overdraft situations)	All well owners have equal right to groundwater	If wasteful practice, court can order reduced use	Court can order comparable proportionate decrease among users	Users have proportional pumping interest in water shortage	Judicial action may provide proportional use or restrict production	Courts apply "underground stream" doctrine	California, Nebraska
b. Eastern (from second restatement of Torts	Landowner using groundwater	If conflicts arise, court can assign groundwater use based on "most beneficial use"	Court can reduce use	Users have proportional pumping interest in water shortage	Judicial action may provide proportional use or restrict production	"Reasonable/beneficial use" doctrine applied	Eastern United States (MI, OH, FL, WI, NJ, MO, NE)
3. Statutory approach							
a. Permits	Landowner must follow requirements of state permit	Well construction and use permits required (production may be specified)	User may be required to ration use; public water supply may have priority	Addressed in permit by well placement and use limits; Public water & domestic use have priority in water shortages	Usually seasonal issue and not long-term problem	May be addressed in permit	Florida, Iowa, Wisconsin, Minnesota
b. Appropriation	State permit holder (not necessarily landowner)	Senior permit holder has rights over junior permit holders	Senior permit holder has rights over junior permit holders	Senior user may have to install new well to greater depth		Where groundwater production reducesstream-flow, priority of appropriation rights determines use	Most western states (except Texas, Nebraska, Arizona and California)
4. Management area	Landowner/State	Use conditions specified by state or mgtdistrict; required metering and pumping reductions			Setup "mgt areas", e.g., bans on large volume wells or pumping restrictions		Calfornia and Arizona (groundwater basins)

Source: Modified from Goldfarb, W., *Water Law*, Lewis Publishers, Chelsea, MI, 1988. See also Craig, R.K.; Adler, R.W.; and Hall, N.D. *Water Law,* Foundation Press, St. Paul, MN, 2017.

Absolute Ownership

Under this doctrine, the landowner can pump as much groundwater from beneath his property as he desires, without the responsibility for the effects on neighbors. Indiana and Texas are the only remaining states with this system (Goldfarb, 1988). Economic implications are that the use can result in significant losses on adjacent property.

Reasonable Use

Under the reasonable use doctrine, a groundwater user may produce as much water as can be pumped under the conditions of water being used on overlying land without waste (Goldfarb, 1989, p. 44). The water can be transported only if it can be done so without any harm to the groundwater use of the adjacent overlying landowners (Goldfarb, 1988 p. 44). Eleven eastern states and Arizona apply the reasonable use rule. Economic implications are that this doctrine protects adjacent landowners from harm and wasteful practices of neighboring groundwater users.

Correlative Rights – Eastern United States (From the Restatement of Torts Rule)

Restatement of Torts (Second Edition), legal scholars' idealization of what law should be, defines a groundwater use rule for well owners. A well owner may produce groundwater for his and for nonoverlying land uses but cannot interfere unreasonably with other users. Any one of three factors determines "unreasonable interferences with a neighbor's use of groundwater: (1) causing unreasonable harm by lowering the water table or reducing artesian pressure, (2) exceeding the owner's reasonable share of the total annual supply, or (3) having a direct and substantial effect on surface supplies and causing unreasonable harm to surface users" (Goldfarb, 1988 p. 44).

The economic effects of the Restatement Rule are that it "promotes conjunctive management," "protects aquifers from mining," and "protects minimum flows in watercourses" (Goldfarb, 1988 p. 44). This rule is used in Michigan, Ohio and Wisconsin with laws similar to it in Arkansas, Florida, Nebraska, New Jersey and Missouri.

Correlative Rights – Western United States

In California, the State Supreme Court determines correlative rights to groundwater during water-short periods on the basis of the following (Goldfarb, 1988):

1. "Overlying owners are entitled to no more than their 'fair and just proportion' for onsite uses;
2. As between transportation out of the basin first in time is first in right;
3. Overlying users have priority over transporters" (Goldfarb, 1988 p. 45). A "fair and just proportion" is equated to the fraction of an owner's property to the total area overlying the aquifer.

Prior Appropriation

Appropriation of water rights has four main principles: (1) acquisition of a right of beneficial use, regardless of land ownership, (2) acquisition of a specific amount of water, (3) the transfer of the right separately from the land and (4) the right is indefinite as to duration of time as long as the beneficial use continues in line with the law.

Many western states have nominally adopted this legal approach, individually modifying it to fit conditions. A major economic implication of this doctrine is that the priority of use is not based on the economic value of that use necessarily.

Management Area

The most complex legal systems for groundwater management have evolved in U.S. states and for areas with recurring water shortages, primarily in the west and also in Florida and New Jersey (Goldfarb, 1988). Typically, those systems involve the conjunctive use of surface water and groundwater. In these systems, groundwater use is controlled to protect certain uses or users. In these systems, water use, integration of ground and surface water rights, measures to affect use (such as taxes or fees), education and technical assistance and control of mining are forms of management that may be exercised by the state (in Arizona, Colorado and New Mexico), by local management districts (Kansas, Nebraska and Texas) or by the courts in conjunction with water districts (California). Economic implications of the management area approach are protection of existing uses into the future and possible reduction of impacts to adjacent landowners. The management area concept progressed from Nebraska's creation of Natural Resource Districts in 1975 with authority to measure and control use and quality of groundwater, to Arizona's active management areas in 1980 for control of overdrafting of groundwater, to California's Sustainable Groundwater Management Act of 2014 establishing groundwater sustainability agencies to develop groundwater sustainability plans for their groundwater use areas to regulate groundwater withdrawals so as not to cause undesirable results. The state of Connecticut adopted in law its approach to protecting groundwater quality through wellhead protection area management. These state groundwater management area approaches are summarized in Box 1.4.2, State Laws for Sustainable Groundwater Through Management Areas. Importantly, these states' laws emphasize local area management of groundwater and utilize both similar and different methods to guide and control groundwater use. Similar methods include establishing management areas, focusing on local water use control, mitigating groundwater overdraft and groundwater use measurement. Differences are emphases on groundwater quantity versus groundwater quantity and quality, irrigation use compared to total groundwater use and ecosystem consideration. Management area approaches are combined with other legal doctrine, such as correlative rights.

State Groundwater Drainage Law

State drainage law deals with the rights of landowners to "repel" or "expel" water at their properties' boundaries (Goldfarb, 1988). The application of drainage law to groundwater covers two principal circumstances: (1) preventing groundwater from flowing into one's property and (2) lowering the water table which results in dewatering or subsidence of an adjacent landowner's property.

Preventing groundwater drainage into one's property is generally covered by the same rule as surface water drainage in those states (Goldfarb, 1988):

> Common enemy rule: A property owner can control drainage waters regardless of effects to adjacent property owners. Seven states follow this rule, some with modifications to lessen damages to neighbors.
>
> Natural flow rule: A property owner cannot obstruct the natural flow of water in situations where adjacent property owners would be affected. Agricultural drainage is an exception in some states applying this rule.
>
> Reasonable use: A property owner is able to make "reasonable use" of land and subsurface space below it even if the use causes injury to an adjacent owner. If a neighbor's damage becomes unreasonable, then the owner affecting the injurious use is liable. A majority of states follow this rule.

The cases of dewatering or subsidence typically relate to mining, and less typically to sewer construction. In "absolute ownership" states, damages from dewatering or subsidence are not recoverable. Property owners in reasonable use states are usually similarly affected, as long as use is not wasteful or waters are transported off-site. In "correlative rights" states, since neighboring landowners have common and equal rights to percolating waters, the damaged landowner generally wins. In western states with water rights appropriation, a property owner cannot interfere with the flow of a groundwater supply of prior appropriation. One prior appropriation state, New Mexico, allows mining operations to compensate adjacent owners or provide an alternate water supply.

Where subsidence may be a factor, adjacent property owners are under mutual obligations of support for the surface and subsurface. One owner's activities must not affect the subsidence of a neighboring owner's land, which breaks the obligation. If this results, the damaged owner, should he seek legal remedy, will likely win. Because courts have given mixed decisions on activities involving mining and removal of solids by water withdrawal and other means that have caused subsidence, the case law is not consistent (Goldfarb, 1988 p. 63–64).

State Laws for Well Installation

States have typically adopted laws and related regulations controlling the location and installation of water supply and monitoring wells to protect public health. A study of 12 states in the north-central United States (Job and Gabanski, 1988) indicated a wide variety of state legal approaches to allowing access to groundwater by well installation. The analysis addressed both water supply and monitoring wells. The study found that

- Some states apply water supply well installation requirements to monitoring wells, even though disinfection and yield tests are contrary to most uses of monitoring wells.
- Most states have a requirement for one or some combination of licensing water well contractors, registering water well engineers or a well permit. Typically, such requirements are intended to ensure that wells are properly installed to provide adequate groundwater supply and to reduce contamination problems. To maintain well drilling licenses and registrations, states may also require drillers and water well contractors to take continuing education courses. Organizations such as the National Ground Water Association and the Irrigation Association offer continuing education courses and examinations accepted by states (NGWA, 2019; IrrA, 2019).
- Permanent abandonment requirements for unused wells are specified in most states to reduce the potential that a well will not be used for waste disposal or contribute to groundwater contamination as it is not being maintained.
- Most states require termination of wells at least 30.48 cm (12 inches) above ground surface and 60.96 cm (24 inches) above highest known flood elevation.

A second study of all 50 U.S. states showed other important considerations in well installation (USEPA, 1996b). This study found that 48 states have well construction standards, many following standards developed by the American Water Works Association. Forty-seven states have minimum setback distances from microbial contaminant sources (such as septic systems or sewer lines), with minimum setback distances ranging from 3 to 91 m. Thirty-six percent of the states had minimum microbial setbacks of 15.2 m, with an additional 30% having minimum setbacks from 22.9 to 30.5 m. Three states had no required microbial setback.

Additionally, 29 states use hydrogeologic criteria to guide well construction. Twenty-five states base minimum casing depth on hydrogeologic factors, such as an unconfined or confined aquifer. Other hydrogeologic factors included in state law or regulation for the consideration in well installation are review for setback, grouting depth and type, karst conduit flow, groundwater flow and flooding setback. Setback distances of wells from potential sources of contamination are part of state and local water codes to ensure safe well water. The setback is a requirement that water supply wells be at least a certain distance from a septic tank or sewer line to provide adequate time for sufficient natural degradation of chemicals a die-off of harmful organisms than may endanger well water. As examples, setback distances for homeowner wells from septic leach fields in Minnesota are 50 feet (15.2 meters) except for special cases. The minimum setback for a septic field from a well in Colorado is 50 feet, but through variance the minimum setback may be 25 feet based on hydrogeologic information for the site (Alley and Job, 2020). Setbacks and other well construction requirements provide protection of groundwater quality.

An emerging issue from a hydrogeologic standpoint is the use of groundwater as a medium for heat exchange through groundwater heat pumps. States recognize that groundwater needs protection from fluids in these heat pumps and have addressed these matters in state regulation, such as in Illinois, which requires a setback from 22.9 to 60.2m depending on the type of heat pump.

Economic implications of well installation requirements include the following:

- Standards for well installation affect the costs of wells and of access to groundwater.
- Contamination can travel along the casing of a poorly installed well or from improperly located wells in relation to potential contaminant sources and potentially affect groundwater users with both acute and chronic illness depending on the contaminant source; health impacts and damage to the adjacent users are minimized by the application of the standards.

U.S. Federal Groundwater Quantity–Related Laws

Protecting Quality for Future Quantity

A range of United States federal laws affect groundwater quantity. The Safe Drinking Water Act affects groundwater quantity and availability in several ways. The underground injection control (UIC) provisions are applied to injection of nonhazardous treated wastewater meeting drinking water standards into an underground source of drinking water through managed aquifer recharge to be used subsequently for water supply. The drinking water standards that are issued as "maximum contaminant levels" to ensure delivery of safe drinking water to consumers are often applied to hazardous waste disposal site remediation. The result is that groundwater that was at one time not usable for drinking water and other purposes may now be of adequate quality to be used for those purposes. The wellhead protection program provides a framework for states and public water systems to manage contaminant sources in areas that recharge groundwater supplying public wells so that the groundwater can continue to be of adequate quality to be used for community water supply. The sole source aquifer program provides for review of projects receiving federal financial assistance in areas of approved federal designation as relying on groundwater as their principal or sole source of drinking water to ensure that the projects are protective of their aquifers' continued use for water supply.

Less Quality Protection Reducing Future Supply Potential

Under the Safe Drinking Water Act, managers of production processes generating untreated wastewaters not wanting to incur the cost of deep injection below a drinking water aquifer and impermeable strata in a "no-migration" zone can apply to the Environmental Protection Agency for a permit to dispose of the fluids into a currently "unused" portion of an aquifer that does not have groundwater of drinking water quality. Aquifers with high total dissolved solids or brackish waters may not currently be a source of drinking water but with proper treatment may be able to provide safe water supply. The UIC program may regulatorily classify such an aquifer as an "exempted aquifer" meaning that a disposer of untreated fluid does not have to meet drinking water standards for the fluid being injected into an otherwise potentially usable aquifer. Injection of untreated fluids into "exempted" groundwaters may be making them less desirable and more costly to transform with treatment to safe drinking water in arid locations with limited water sources.

Supply-Related Tax Subsidy for Irrigation

The U.S. Internal Revenue Code was modified based on a decision in 1965 by the United States Court of Appeals, Fifth Circuit, in the case of the *United States v. Marvin Shurbet* et ux, 347 Fed. (2d) 103, in Internal Revenue Service (IRS) Revenue Ruling 65–296 (November 19, 1965) to allow landowners engaged in irrigated agriculture in the High Plains to deduct from income the depreciated value of the groundwater stock that they drew on for their livelihood. This depletion allowance reduces the federally taxable income of these farmers and subsidizes groundwater depletion over a vast area of eight mid-western states: Texas, New Mexico, Oklahoma, Kansas, Colorado, Nebraska, Wyoming and South Dakota, with portions of all these states supplied by the Ogallala Aquifer (Jenkins, 1980; Smith and Wyatt, 1980). The depletion allowance taken was calculated by

1. Deriving the total value of groundwater based on the sales value of the property with and without groundwater and dividing by the number of acres of the property;
2. Dividing the water table decline for the year by the saturated thickness of the aquifer at the time of acquisition;
3. Multiplying the results of 1 and 2 to establish the per acre depletion allowance (Smith and Wyatt, 1980, p. 49).

Laws Affecting Groundwater Recharge

Several U.S. federal laws provide funding or permits for managed aquifer recharge projects, in addition to the Safe Drinking Water Act. The Water Infrastructure Finance and Innovation Act of 2014 (WIFIA) provides federal financial assistance for managed aquifer recharge projects. Under the CWA, the National Pollutant Discharge Elimination System municipal separate storm sewer system permits may be issued for stormwater runoff to be collected by storm drains releasing untreated water to the subsurface to minimize degradation of surface waters and to provide aquifer recharge. The Consolidated Appropriations Act, 2018 (Pub. L. 115-141) funded the Department of Transportation to construct ground water recharge projects in areas of water scarcity, wetlands creation or improved habitat connectivity and stormwater mitigation.

The Reclamation Act of 1902 [Public Law 161] is principally focused on providing irrigation water to western states, including assessment of the practicality of irrigation projects. The Department of the Interior, Bureau of Reclamation, oversaw funding for the High Plains States Ground Water Demonstration Program Act (1983) (Public Law 98-434). For all western states, the program identified 21 recharge projects, all of which were concluded by 2000. Techniques used were as follows: well injection using active pumping (Colorado, Idaho, Nevada, Oregon, South Dakota, Texas, Utah and Washington), natural instream recharge (Arizona), passive infiltration ponds (Idaho), surface infiltration (Kansas, Nebraska, Oklahoma), gravity-feed passive injection wells (Kansas, Oklahoma), agricultural practices to enhance surface infiltration (Montana) and inter-aquifer transfer (Oregon).

From 2016 to 2018, the Bureau of Reclamation funded several managed aquifer recharge projects including the Shafter-Wasco Irrigation District Kimberlina Groundwater Recharge Basin and Banking Project in California, the Stockton East Water District – Groundwater Recharge Basin, Southeast Recharge Basin in California; the Lower Santa Cruz River (Tucson, Arizona) Enhanced Recharge Demonstration Project; and El Paso Water Utilities Aquifer Storage and Recovery Project in Texas.

Water Conservation in Irrigation

The U.S. Department of Agriculture Environmental Quality Incentives Program under various federal farm legislation (farm bills), such as the Farm Security and Rural Investment Act (2002) and Agriculture Improvement Act (2018), provides cost-share payments, incentive payments and loans to agricultural producers to improve irrigation systems for ground and surface water conservation. In 2018, the legislation also included conservation and efficiency payments for managed aquifer recovery practices in areas of groundwater depletion.

Water Supply Equipment

Federal law also addresses equipment involved in water supply. The Motor Carrier Safety Improvement Act covers safety of operation of drill rigs that install water wells. Heavy trucks that transport drill rigs must meet certain safety requirements, including securing drill rig masts and cable tool rig derricks, for traveling on highways. SDWA has lead-free equipment requirements that prohibit use of pipe, fittings, fixtures, solder or any flux that contain lead in the installation or repair of any public water system and any plumbing in residential or non-residential facilities supplying water for human consumption.

Other Quantity-Related Federal Laws

The Surface Mining Control and Reclamation Act (SMCRA) provides for the assessment of probable cumulative hydrologic impacts before the Department of the Interior approves a mining permit. Performance standards and a reclamation plan must protect groundwater. Water supplies, if affected, must be replaced. Economic consequences of loss of water supply represent infrastructure decline for a community with potential loss of jobs and income.

The Geothermal Steam Act provides for the use of superheated subsurface water and steam for many energy applications including heating and electrical output. These applications provide economic value for groundwater which might not be captured in typical water use.

Water Quality

Concerning groundwater quality protection, most U.S. states have adopted federal approaches, such as in the Resource Conservation and Recovery Act (RCRA), UIC, Wellhead Protection and Superfund Programs, described briefly in the subsequent text. Many states have their own standards for certain uses of groundwater, some of which relate to both quantity and quality considerations, and may be enforceable standards. These are described to some degree in the Box 1.4.1, Examples of Groundwater Classification in Three U.S. States. Most frequently, states use groundwater classification systems as a basis of permitting discharges to groundwater. This use of a classification system is closely related to regulated activities affecting or potentially affecting groundwater quality and for establishing remedial objectives. States' approaches for the use of groundwater classification systems are basically (1) "a formal system with classes defined by state statute or regulation, and (2) a method which distinguishes groundwater by use, quality, or hydrogeologic characteristics without formal class designations" (Environmental Law Institute, 1990). Twenty-eight states have groundwater classification systems and the other twenty-two do not have them. However, 14 of these states protect all groundwater in the state for drinking water use. Several others have nondegradation standards to protect groundwater quality. Other states have developed similar classification systems (ICF, Inc., 1990). Examples of groundwater classification systems from three states are given in Box 1.4.1.

State Law Affecting Land Management Practices with Groundwater Quality Implications

During the period from the late 1980s to the early 1990s, states were active in enacting legislation affecting land management practices for the purposes of groundwater protection. Major legislative initiatives passed into law were summarized by the National Conference of State Legislatures (NCSL) (NCSL, 1994). Table 1.4.2 provides a brief description of these laws. During this time, most states, following U.S. federal laws, focused on protecting groundwater quality from underground storage tank leakage, pesticide use control, more comprehensive approaches to groundwater protection and protecting groundwater sources of public water systems. An underlying theme in these laws is improved information to better manage the protection of groundwater resources. From an economic standpoint, more and better information on groundwater quality may affect an individual's tastes and preferences for particular geographic area or site and its water supply.

State Groundwater Quality Protection

Management Area and Wellhead Protection

Perhaps the best current example of a comprehensive approach to groundwater protection in the United States is the local implementation of state wellhead protection programs. States are required under U.S. Safe Drinking Water Act (SDWA), Section 1428, to develop wellhead protection

BOX 1.4.1 EXAMPLES OF GROUNDWATER CLASSIFICATION IN THREE U.S. STATES

Colorado
Authority: Basic standards for groundwater, 3.11.0 (5 Colorado Code of Regulations 1002-8)

I. Domestic use and quality (less than 10,000 mg/L TDS)
II. Agricultural use and quality (less than 10,000 mg/L TDS)
III. Surface water protection (interacts with surface water)
IV. Potentially useable (less than 10,000 mg/L TDS)
V. Limited use and quality (10,000 or more mg/L TDS)

New Jersey
Authority: New Jersey Administrative Code, Title 7, Chapter 9, Subchapter 6

I. Aquifers that support special ecological systems
   A. Natural areas, high-quality surface waters and exceptional ecosystems that depend on groundwater

PL. Aquifers in the New Jersey Pine Lands

I. Aquifers that provide potable groundwater
   A. With conventional treatment
   B. Subsequent to enhancement or restoration or water quality after extended period of time

I. Aquifers that provide groundwater that is not considered suitable for human consumption owing to natural hydrogeologic characteristics or natural water quality
   A. Aquitards
   B. Groundwater with natural or regional concentrations of chloride exceeding 3000 mg/L or more than 5000 mg/L. Total dissolved solids or contaminants not subject to conventional treatment that exceed-state standards

West Virginia
Authority: West Virginia Code, Chapter 20, Article 5 M
All groundwater is protected as if it was a source of drinking water.

programs for approval by the U.S. Environmental Protection Agency based on addressing the following elements, with the expectation of local implementation required in some states:

- Specify roles and duties of state agencies, local government entities and public water suppliers;
- Delineate the wellhead protection area (WHPA) for each wellhead;
- Identify potential sources of contaminants within each WHPA;
- Develop management approaches to protect the water supply within WHPAs;
- Develop contingency plans for each public water supply system to respond to well or well-field contamination;

- Site new wells properly to maximize yield and minimize potential contamination;
- Ensure public participation.

Most states and two territories have approved WHP programs. A review of this program (Job, 2007) reveals that

1. Most states apply a mix of approaches to delineate wellhead protection areas: fixed radius (0.1–0.4 km), modeling (5 year time of travel) or hydrogeologic mapping, and recognize both microbial and chemical protection zones.
2. States identified a wide range of resources to be used to develop potential contaminant source inventories locally, including historical documents, databases, previous contaminant source inventories, land use maps, zoning records, field inspection and past aerial reconnaissance.

TABLE 1.4.2

Summary of State Law Affecting Land Management Practices for Groundwater Quality Protection, 1988–2014

Category	Brief Description	Example States
Comprehensive groundwater protection	May obligate state agencies to write a comprehensive plan addressing groundwater quality protection from contaminating sources, including classifying aquifers, setting ambient groundwater quality standards, groundwater discharge permits, nonpoint source best management practices and critical groundwater area specification	Arizona, Arkansas, Connecticut, Idaho, Indiana, Massachusetts, Minnesota, New Hampshire, North Dakota, Oregon, South Dakota, Washington, West Virginia
Groundwater monitoring	Sets up a state monitoring program to test ambient groundwater quality and potential contaminant sources and define vulnerable zones	Idaho, Missouri, Montana, Nebraska, Washington
Wellhead protection	Allows state agency to define roles of local and state governments in conducting a wellhead protection program, including delineating a protection area around public wells, locating potential contaminant sources, giving technical support to communities and formulate regulations to prevent contamination of groundwater	Nearly all states
Land use management	Allows local government to implement zoning laws and other control measures to manage land use that may affect groundwater quality	Florida, New Jersey, New York
Nonpoint source pollution	May specify that a state agency is to establish a priority list of state waters, including groundwaters, for future protection, and issue technical advice for the best management practices that will minimize contamination	Iowa, Maine, New York, Vermont
Pesticides	Controls the use, selling, labeling and discarding of agricultural chemicals and nutrients; may also address use fees, education and monitoring	Alabama, California, Colorado, Connecticut, Florida, Idaho, Illinois, Iowa, Kansas, Louisiana, Maryland, Michigan, Minnesota, Montana, New York, North Dakota, Oregon, South Carolina, South Dakota, Vermont, West Virginia, Washington, Wisconsin

(Continued)

TABLE 1.4.2 (*Continued*)
Summary of State Law Affecting Land Management Practices for Groundwater Quality Protection, 1988–2014

Category	Brief Description	Example States
Underground storage tanks	Specifies conditions for registering, building, improving, emplacing, abandoning and monitoring underground storage tanks; may also set up funds for cleanup of UST leaks and contamination of adjacent properties and provide financing for tank improvements	Nearly all states
Quantity/Drought Management	Emphasizes local area management of groundwater and utilize a range of methods to guide and control groundwater use. Similar methods include establishing management areas, focusing on local water use control, mitigating groundwater overdraft and groundwater use measurement. Different methods based on state interests are emphases on groundwater quantity versus groundwater quantity and quality, irrigation use compared to total groundwater use, and ecosystem consideration. Management area approaches are combined with other legal doctrine.	Arizona, California, Nebraska

Sources: National Conference of State Legislatures (NCSL). Groundwater Protection Legislation, Survey of State Action, 1988–1992, June 1994.Nebraska Legislature. 1975. Groundwater Management Act; 2004. Groundwater Management and Protection Act. URL: https://dnr.nebraska.gov/sites/dnr.nebraska.gov/files/doc/water-planning/2016NEWGWMgmtProtectionActStatutes.pdf (accessed May 25, 2019).Arizona Department of Water Resources. 1980. Overview of the Arizona Groundwater Management Code. http://www.azwater.gov/AzDWR/WaterManagement/documents/Groundwater_Code.pdf (Accessed May 24, 2019).California Legislature. 2014. Sustainable Groundwater Management Act. http://leginfo.legislature.ca.gov/faces/codes_displayexpandedbranch.xhtml?tocCode=WAT&division=6.&title=&part=2.74.&chapter=&article (accessed May 24, 2019).Job, C. 2007. A summary of state wellhead protection programs. Ground Water Monitoring and Remediation, 17 (2), 61–63.

3. Sharing the implementation of WHP among state government, water suppliers and other local agencies is the predominant approach for implementation. However, five states implement the program primarily by applying state law or revised regulations and four states leave implementation solely to local government.
4. The most prevalent management approaches include education, zoning, monitoring ambient groundwater quality, design/operating/performance standards, prohibition of uses/activities/facilities, technical assistance, permits, land acquisition, agricultural controls and best management practices. Typically, states and local agencies rely on existing authorities related to groundwater protection to manage contaminant sources in WHPAs.

Public input to the development and implementation of this program is required. While historically little funding was provided to implement this program, a private nonprofit approach emerged.

Some states enacted laws to implement the program. The programs of Arizona, California, Connecticut and Nebraska are summarized in Box 1.4.2. Connecticut's program includes permits that restrict certain activities around community public water wells.

BOX 1.4.2 STATE LAWS FOR SUSTAINABLE GROUNDWATER THROUGH MANAGEMENT AREAS

States in more arid regions that rely on groundwater for a significant portion of their water supply have utilized water management areas or districts to control groundwater use and quality. Three states are highlighted in summary form with differing requirements for managing groundwater through designated area management procedures. An example is also included of a state protecting groundwater quality though management of wellhead protection areas established under state law.

Nebraska

Ground Water Management Act, 1975 (and subsequent amendments including Groundwater Management and Protection Act, 2004)

- Manage groundwater to reduce nitrate and other contaminant levels to eliminate health hazards
- Preferred regulators are natural resources districts to control activities contributing to groundwater contamination
- Department of Environmental Quality to regulate contaminant sources to prevent serious groundwater quality deterioration
- Natural resource districts may
  - Issue regulations
  - Require reports from groundwater users
  - Require reporting of water uses and irrigated areas by landowners to certify uses and areas
  - Require meters on wells to acquire water use data
  - Decommission wells not properly identified as active or inactive
  - Conduct investigations to meet purposes of Act
  - Report to and consult with DEQ
  - Issue cease and desist orders to enforce the Act
  - Hold public hearings on proposed regulations
  - Issue stays on construction of new wells and on increase of areas irrigated
  - Adopt regulations to control surface runoff from groundwater irrigation
  - Maintain a groundwater management plan and consider available groundwater supplies, precipitation, crop needs, groundwater data, groundwater use, groundwater quality concerns and goals, water conservation, availability of supplemental water including groundwater recharge, coordination of different water sources, groundwater management objectives, economic value of different uses of groundwater and boundaries of a proposed management area within the district
  - Establish groundwater management areas to protect groundwater quantity and quality and resolve conflicts among groundwater users and appropriators of surface water and adopt controls for the management area
  - Develop an integrated management plan for a river basin or portion of a basin that
    - Sets clear goals and objectives to sustain balance between water uses and supplies
    - Maps area of integrated water management

(Continued)

BOX 1.4.2 (Continued)

- Controls for groundwater and surface water with protections for groundwater users dependent on recharge from streams and for surface water appropriators from overappropriated surface water or groundwater uses, including
    - Allocate groundwater to be withdrawn
    - Adopt system of rotation of groundwater use
    - Adopt well-spacing requirements
    - Require measurement of groundwater withdrawals
    - Require reduction of irrigated area
    - Limit expansion of irrigated area
    - Use of best management practices, chemical content of water and soils, educational requirement to protect groundwater quality and availability
    - Monitor water quality at wells in management area
    - Decide on requests to transfer groundwater off land where withdrawn
    - Determine new or replacement well depths
    - Close all or part of management to issuance of well permits
    - Establish different water allocations for different irrigation methods
    - Establish different provisions for different hydrologic relationships between groundwater and surface water
- Enables data collection and evaluation to implement the management plan and increase understanding of surface water and hydrologically connected groundwater
- May include incentive programs affecting water use
- Track depletions and gains to streamflow

• Develop action plan at direction of DEQ and with public input to mitigate or eliminate increase in or spread of groundwater contamination in a management area and monitor groundwater quality to provide landowner/operator with information on chemical use for specific soils and crops
• Require application for a permit to install or modify a well in a management area

Source: Nebraska Legislature. 1975. Groundwater Management Act; 2004. Groundwater Management and Protection Act. URL: https://dnr.nebraska.gov/sites/dnr.nebraska.gov/files/doc/water-planning/2016NEWGWMgmtProtectionActStatutes.pdf (accessed May 25, 2019)

Arizona

Groundwater Management Act, 1980

- Specifies five "active management areas" (AMAs) in groundwater overdraft and provides for areas of general management and irrigation non-expansion
- Establishes deadline for reaching "safe yield" (groundwater withdrawal that does not exceed annual replenishment of aquifers) of 2025
- Establishes essential requirements
    - Groundwater rights and permits identifying who can pump groundwater
    - Prohibition of irrigation on new agricultural lands in AMAs
    - Plans with targets and water management criteria for AMAs
    - Developers building in AMAs to demonstrate an assured water supply of at least 100 years

(Continued)

BOX 1.4.2 (Continued)

  - Metering/measurement of water pumped from large wells
  - Annual reporting of water withdrawal and use
- Provides for conservation requirements
- Limits land area for irrigation

Source: Arizona Department of Water Resources. Overview of the Arizona Groundwater Management Code. http://www.azwater.gov/AzDWR/WaterManagement/documents/Groundwater_Code.pdf (accessed May 24, 2019)

Connecticut
Aquifer Protection Area Program Statutes, Sections 22a-354a through 22a-354bb of Connecticut General Statutes (regulation 1988)

- Protects major public water supply wells in sand and gravel aquifers
- Designates aquifer (wellhead) protection areas around active well fields in towns serving more than 1000 people
- Restricts development of certain new land use activities that use, store, handle or dispose of hazardous materials through permits
- Requires existing regulated land uses to register and follow best management practices
- Regulates major facilities, state facilities and public service companies (utilities) located within the aquifer protection areas

Source: Connecticut Department of Energy and Environmental Protection. 2019. Aquifer Protection Area Program. https://www.ct.gov/deep/cwp/view.asp?a=2685&q=322252&deepNav_GID=1654 (accessed June 10, 2019)

California
Sustainable Groundwater Management Act (SGMA), 2014

- Mandates management and use of groundwater maintained through 2040 without causing undesirable results
- Prioritizes groundwater basins based on use and overdraft
- Establishes Sustainable Groundwater Agencies (GSAs) to develop groundwater sustainability plans (GSPs) for the basins, incorporating (partial listing here) historical data, projected supply and demand, existing and potential recharge areas, measurable objectives every five years, steps to achieve objectives, monitoring and management of groundwater levels, monitoring of other groundwater-related factors, mitigation of overdraft, contribution of recharge areas, impacts on groundwater dependent ecosystems, use of surface water for recharge and water budget and sustainable yield assumptions
- Authorizes GSAs to regulate groundwater in their plans by
  - Setting spacing requirements for new wells to minimize well interference
  - Requiring well extractors to operate on a rotational basis
  - Limiting or suspending groundwater extraction
  - Controlling construction of new wells and enlargement of existing wells

(Continued)

BOX 1.4.2 (Continued)

- Providing temporary and permanent transfers of groundwater extraction allocations
- Establishing accounting rules for carrying unused groundwater allocations into another year

- Requires balancing pumping and recharge to achieve sustainability of GSP in critically over-drafted basins by 2040
- Requires remaining high and medium priority basins to reach sustainability by 2042

Source: California Legislature. 2014. Sustainable Groundwater Management Act. http://leginfo.legislature.ca.gov/faces/codes_displayexpandedbranch.xhtml?tocCode=WAT&division=6.&title=&part=2.74.&chapter=&article (accessed May 24, 2019)

Pollutant Discharges to Groundwater

Some states control pollutant discharges to groundwater under their state implementation of the National Pollutant Discharge Elimination System or a similar permit system. In these states, facilities that discharge pollutants to ground waters of the state require a state discharge to groundwater (DGW) permit. The states establish the requirements and process for permit eligibility and general conditions for maintaining these discharge permits. The goal of a state PDES-DGW permitting program is to restore, enhance and maintain the ground water quality of the state. These states may also have groundwater quality standards used in the DGW permit that may be the same as the health-based maximum contaminant level under SDWA for the same constituent (MPCA, 2019; NJDEP, 2019; SDDENR, 2019; WDE, 2019).

The Groundwater Foundation, started in 1988 in Lincoln, Nebraska, has used the recognition of communities' groundwater protection efforts to encourage groundwater protection. This program basically encourages communities to value a protected groundwater supply, albeit valuing may not be in monetary terms. This approach has principally relied on providing information to the public to change local tastes and preferences for groundwater quality protection.

Economic implications of wellhead protection include

- Several thousand communities around the United States have fully implemented wellhead protection programs, recognizing that it is less expensive to implement protective land management practices in the area supplying the public wells than to remediate contamination (USEPA, 1996a; Job, 2007; AwwaRF, 2004).
- Furthermore, businesses see that protecting the local water supply brings more value to the infrastructure of a community which can be attractive to other businesses and promote more environmentally sound economic development (Job, 1996; AwwaRF, 2004).

States continue to enact law and develop regulation in response to economic circumstances and environmental conditions affecting groundwater.

U.S. Federal Groundwater Quality Protection–Related Laws

Since the mid-1970s, the U.S. Congress has passed a number of laws with significance to groundwater quality and management. Many of the laws have similar counterparts at the state level. While quantity is largely dealt with at the state and local levels, quality has principally been dealt with at the federal level in surface water, which has been carried over to groundwater only to a limited extent.

This circumstance is attributable to the fact that most groundwater is perceived as traveling slowly, is a local issue therefore, and does not affect interstate waters. The latter point is often not the case. Groundwater maintains the baseflow of many streams and rivers in the United States and around the world, thus providing drinking water supply, habitat for aquatic fauna and flora and a natural diversity for the food web and for hunters and nature enthusiasts, as well as navigation and recreation on those streams. Maintaining the quality and quantity of streamflow actively affects commerce among states, which is recognized in the Constitution and therefore should be a federal interest.

Most federal groundwater quality protection laws focus on eliminating, reducing or managing the sources of actual or potential contamination. Federal groundwater quality law promotes to a limited extent that an individual or business should incorporate the cost of protecting and maintaining groundwater quality as a cost of doing business – the concept that a "polluter pays" for the full cost of doing business, including waste disposal. The CWA, with many references to groundwater, does not provide a coherent framework for groundwater protection, as it exists. Under the CWA, groundwater could be considered in regional wastewater planning, but few of the regional wastewater management (Section 208) plans did this. After amendments, regulations provided for monitoring funds to be used to measure groundwater quality, and in 1987 the nonpoint source pollution control provisions were strengthened and allowed to apply to funding for projects protecting groundwater. A continuing shortcoming that has been the focus of several court suits is the lack of recognition in the act that groundwater and surface water interact extensively, with the quality of one affecting the other. This situation, in turn, may affect the amount of pollution permitted to be discharged but is usually not accounted for because of the need for more data on the variation in this interaction over time and from place to place. Related to groundwater–surface water interaction is the matter of public water systems drawing water from either source and paying for the cost to treat water to be sufficiently clean and safe to meet federal standards – a cost transfer without compensation by the polluter(s).

At further issue is the CWA municipal separate storm sewer systems (MS4) permit program that allows untreated stormwater to enter the subsurface through drains that by their placement and construction do not pass the stormwater through the soil zone. Using the soil zone as treatment (filtration, adsorption and degradation) of many contaminants before the stormwater reaches the aquifer (Brumley et al., 2018). Six thousand six hundred ninety-five (6,695) MS4 systems are required to be place but EPA has not completed its research to determine adverse effects to groundwater quality (USEPA, 2019). If impacts are problematic, costs to affected communities relying on groundwater to treat degraded water quality for safe drinking water supply could be significant.

The SDWA as amended in 1986 and 1996 principally to set standards for the safety of consumers' drinking water quality has focused more on promoting the prevention of contamination through the management and modification of potentially contaminating sources of wellhead protection areas around public water supply wells and wellfields. It also set up a control program to regulate the disposal of the largest volume of hazardous waste through underground injection into the deep subsurface zones below aquifers used for drinking water supply. SDWA also contains several consumer right-to-know provisions at the local, state and national levels so that consumers are informed about contaminants in their drinking water. Additionally, the Food and Drug Administration (FDA) is to consider whether to establish standards for bottled water quality. These sections of the law have implications for informing the larger populace about product quality, and therefore influence the public's tastes and preferences for different qualities of water.

Many U.S. federal laws and their implementing regulations affect groundwater quality:

- The Safe Drinking Water Act UIC program controls the quality of injected fluids into the subsurface, as explained above. The UIC program for controlled injection of wastewater underground is the legal complement to the Resource Conservation and Recovery Act programs for controlled land disposal of solid wastes.
- The Resource Conservation and Recovery Act (RCRA) program controls the generation, storage, transportation and disposal of chemicals and their wastes. These controls affect the amount of waste and chemicals potentially released to the ground in regulated landfills. Under

this act, underground storage tanks are also regulated to minimize the release of chemicals to the subsurface environment and groundwater.

- The Federal Insecticide, Fungicide and Rodenticide Act (FIFRA) controls the production and use of pesticides in the commerce of the United States through a registration process. Pesticide registrants must conduct research on the degradation and health effects of their products to demonstrate that damage will not be incurred or will be minimal and be balanced by the economic benefits of the pesticide.
- The Pollution Prevention Act promotes waste source elimination and reduction, and recycling and best management where necessary. Potentially, this reduces the waste requiring disposal in landfills or injected underground.
- The Toxic Substances Control Act provides for a Toxic Release Inventory, which, in combination with Wellhead Protection and Comprehensive State Groundwater Protection Programs, promotes the use of information in making sound decisions to protect groundwater and prevent its contamination. A major economic effect is to reduce the "spillover" effect of potential contaminant release to adjacent landowners.
- The Comprehensive Environmental Response, Compensation, and Liability Act (CERCLA or Superfund) provides for the cleanup of abandoned chemical waste sites. This activity has increased the information the public can use in influencing local decisions affecting groundwater.
- The Superfund Amendments and Reauthorization Act (SARA) provides for important information to be made available by commercial and government organizations to the public concerning hazardous and toxic substances used in and around communities.

Groundwater Quality Standards

While no U.S. federal groundwater standards exist, the Superfund program began applying federal drinking water standards to its contaminated sites as remedial goals. Because of the tremendous cost of remediating groundwater to safer low contaminant concentrations in many sites where groundwater was not used extensively and could be cleaned at less cost at the tap, alternative remedial concentrations were established. However, one major factor in U.S. federal environmental policy relative to the control of future releases of contaminants or remediation (or nonremediation) of existing contaminants has been the recognition of case law: a federal regulation applied to one property should not cause harm to adjacent properties. Furthermore, if natural degradation was the selected remedy, the federal government would need to maintain institutional control of the area under which the contaminant plume may move and ultimately degrade to levels at or below drinking water standards. As long as adjacent users can produce water that is drinkable by federal standards, no harm was done to that user and he/she would not have to incur the cost of additional treatment to safely consume the water. The federal drinking water standards, referred to as "maximum contaminant levels" (MCL), are presented in Volume 1, Chapter 8. Table 1.4.3 presents a summary list and description of U.S. groundwater-related law.

U.S. Federal Wetlands Laws

Wetland-related laws, while not specifically recognizing that most wetlands are locations of groundwater discharge during most of the year, protect these areas from contamination and destruction. CWA Section 404 provides a permit program to control dredging and filling of wetlands. Wetlands, as the term implies, are water-saturated soils and subsurface zones, in which the watertable is manifest near or slightly above the ground surface with vegetation growing from it that has adapted to these conditions which may, in certain times and locations, be anaerobic. During times of high precipitation, wetlands can allow excess waters to be held in wetland soils and low-lying areas and thereby not contribute to rapid overland flow. Wetlands can provide important wildlife habitat for fish, migratory animals and birds, helping to maintain natural values and biodiversity beyond the

TABLE 1.4.3
A Brief Listing of United States Federal Groundwater-Related Law and Policy

Law or Policy	Year	Brief Description/Significant Provisions
Related to groundwater quality		
Clean Water Act (CWA)	1972	Groundwater to be considered in regional wastewater planning; provides for nonpoint source control management of groundwater quality
Safe Drinking Water Act (SDWA)	1974	Established standards for contaminants in drinking water; setup regulatory program to control underground injection of wastes; established prevention program to protect groundwater as a water supply source and the Sole Source Aquifer Program
Resource Conservation and Recovery Act (RCRA)	1976	Set up program to control hazardous waste generation, storage, transportation and disposal in response to discoveries of groundwater contamination from hazardous waste disposal around the country
Low Level Radioactive Waste Policy Act	1980	States to manage low level radioactive waste within their boundaries and can enter into compacts with other states for disposal
Comprehensive Environmental Response, Compensation, and Liability Act (Superfund)	1980	Established a federal program to remediate abandoned hazardous waste disposal sites and groundwater contamination, past landowners and operators joint and severally liable for costs of remediation [Note: drinking water standards were often applied as remedial standards in this program, making it expensive to implement]
Nuclear Waste Policy Act	1982	Set procedures to select sites for highly radioactive waste disposal
Groundwater Protection Strategy	1984	Promoted development of state capacity to protect groundwater from contamination and proposed an aquifer classification method
RCRA, Amended	1984	Set up program to regulate underground storage tanks to protect groundwater
SDWA, Amended	1986	Established the wellhead protection program to prevent contamination of groundwater sources of public water supplies and authorized a critical aquifer management program under the Sole Source Aquifer Protection Program
Superfund Amendments and Reauthorization Act (SARA) and Emergency Planning and Community Right-To-Know Act (EPCRA)	1986	Provided communities with the "right-to-know" about contaminant releases to air, land and water; created the Toxic Release Inventory under Section 313 of EPCRA
CWA, Amended	1987	Provided grants to states for groundwater protection programs and groundwater monitoring; strengthened the nonpoint source program, which included funding for groundwater-related nonpoint source control projects
Federal Insecticide, Fungicide, and Rodenticide Act, Amended	1988	Streamlined the pesticide re-registration process, which included consideration of leaching of pesticides to groundwater
Pollution Prevention Act	1989	Setup program to eliminate, reduce, recycle and improve "housekeeping" of commercial and industrial wastes
Food, Agriculture, Conservation and Trade Act	1990 and later years	Provided that areas in a wellhead protection program designated under the SDWA may be eligible to allow farmers owning them to receive payments for not cultivating them and leaving them as "conservation reserves" for up to 10–15 years
Groundwater Protection Strategy for the 1990s	1991	Promoted Comprehensive State Groundwater Protection Programs, including action through use of inter-program capabilities; consideration of use, value and vulnerability of groundwater; and actions affecting water quality through ground and surface water interaction

(Continued)

TABLE 1.4.3 (*Continued*)
A Brief Listing of United States Federal Groundwater-Related Law and Policy

Law or Policy	Year	Brief Description/Significant Provisions
SDWA, Amended	1996	Provided for source water (ground and surface water) assessments as a basis for local and state protection activities and for possible reduction in chemical monitoring of public water systems; set up local water system reporting to consumers on drinking water quality and other consumer right-to-know provisions including a national drinking water contaminant occurrence database; established regulatory requirements for drinking water contaminants found in groundwater: arsenic, radon and sulfate and for a process to determine the conditions under which groundwater supplies should be disinfected; changed the technology cost-effectiveness test to benefit-cost test through the comparison of risk-reduction benefits with technology cost; established and funded the Drinking Water State Revolving Fund to capitalize state revolving funds for financing construction of water system improvements and support state drinking water program alternatives including wellhead protection
Related principally to groundwater habitat		
Clean Water Act, Section 404	1972	To fill or dredge a wetland, must apply for a permit from the Corps of Engineers
Federal Agriculture Improvement and Reform Act	1996	Under Wetlands Reserve Program allows wetlands to be restored through permanent easements, long-term easements (30 years or the maximum allowed by state law) and restoration of cost-share agreements without easements; requires one-third of area to be allocated to permanent easements, long-term easements and restoration agreements; provides payments to be commensurate with those provided to landowners who enroll filter strips in conservation reserves.
Agriculture Improvement Act	2018	Protecting source waters for drinking water in conservation reserve land and minimize nutrient losses to surface water and groundwater
Related principally to groundwater quantity		
Internal Revenue Service Ruling 65–296	1965	Allows farmers of irrigated areas in the High Plains to take a depletion allowance for depleting the regional aquifer
Surface Mining Control and Reclamation Act	1977	Requires assessment of probable cumulative hydrologic impacts before mining permit approval; performance standards to minimize effects on quantity of groundwater (and surface water) during and after mining operation; operators must replace water supplies affected; reclamation plan must act to protect ground and surface water
Geothermal Steam Act and Amendments	1970/1988	Federal government claims geothermal resource wherever it holds mineral rights; Department of the Interior authorizes leases on federal lands; prospective users must get exploratory, drilling and production permits; air impacts controlled under Clean Air Act; water discharges controlled under CWA; owners can use percentage depletion allowance
High Plains States Groundwater Demonstration Program Act	1983	Provides for the evaluation of the potential for artificially recharging the Ogallala (High Plains) Aquifer because of extensive overdrafting of the groundwater resource throughout the aquifer
Clean Water Act, Municipal Separate Storm Sewer System	1972	Permits stormwater collection and drain projects to reduce surface water runoff and to recharge groundwater
Related to financing groundwater supply		
SDWA, amended	1996	Established grants to states to capitalize state revolving funds to finance public water systems which need to return to or maintain compliance with drinking water quality standards, with a focus on small systems

(*Continued*)

TABLE 1.4.3 (*Continued*)
A Brief Listing of United States Federal Groundwater-Related Law and Policy

Law or Policy	Year	Brief Description/Significant Provisions
Farm Security and Rural Investment Act	2002	Provides loans, grants and loan guarantees for drinking water, sanitary sewer, solid waste and storm drainage facilities in rural areas and cities and towns of 10,000 or less; authorizes a grant to a nonprofit organization to assist individuals in receiving water well financing through loans of up to $8000 each at a rate of 1% for up to 20 years; provides payments for improved irrigation system efficiency.
Water Infrastructure Finance and Innovation Act	2014	Provides financing for managed aquifer recharge projects.
Agriculture Improvement Act	2018	Continues to provide payments for irrigation system efficiency improvements; includes payments for managed aquifer recovery practices.
Related to Equipment/ Infrastructure for Groundwater Production/Supply	1999	Requires heavy trucks, including drill rigs, to meet certain safety requirements, including securing drill rig masts and cable tool rig derricks
Motor Carrier Safety Improvement Act	1986	Requires lead-free pipe, fittings, fixtures, solder or any flux, in the installation or repair of any public water system and any plumbing in residential or non-residential facilities for potable supply
SDWA, amended Water Supply Cost Savings Act	2016	Requires potential Drinking Water State Revolving Fund small community loan applicants to consider local water sources, including water wells before applying for a loan for water supply

Sources: U.S. Environmental Protection Agency, Office of Water, Progress in Groundwater Protection and Restoration, February 1990 (EPA 440/6-90-001).

Campbell-Mohn, C. et al., Sustainable Environmental Law: Integrating Natural Resource and Pollution Abatement Law from Resources to Recovery, Environmental Law Institute, West Publishing Company, St. Paul, MN, 1993.

Smith, D.D. and Wyatt, A.W., Proceedings of the 7th Annual Conference of Groundwater Management Districts Association, North Platte, NE, December 8–9, 1980, 43–49.

United States Department of Agriculture (USDA), Rural Utilities Service, Water and Environment Program, 2005, URL: http://www.usda.gov/rus/water/index.htm (accessed January 29, 2005a).

United States Department of Agriculture (USDA), Economic Research Service, Farm Policy, 2005, URL: http://www.ers.usda.gov/Features/farmbill/ (accessed January 29, 2005b).

location of the wetland. Economic consequences of wetlands loss include increased flooding and reduced fishing and hunting potential, as well as wildlife enjoyment opportunities.

U.S. Federal Environmental Valuation of Groundwater-Related Actions

U.S. Federal law and executive orders provide the bases for environmental valuation of groundwater-related actions. The U.S. Federal Office of Management and Budget (OMB) is charged by Presidential Executive Order (E.O.) 12291 and 12866 to administer an evaluation process through the other Federal Departments and Agencies that causes federal programs' costs to be compared with benefits prior to decisions for program implementation. Allowances can be made for unquantifiable and unmonetizable benefits being applied to environmental and other regulatory and federal decisions affecting groundwater as well as other resources.

In U.S. Environmental Protection Agency programs, these analyses are included in Regulatory Impact Analyses. Congress, on the other hand, has prescribed how environmental programs should take into account benefits and costs. Several major federal environmental programs take different approaches in their analyses under their respective laws.

Drinking Water Standards

The 1996 Amendments to the SDWA changed this law's requirements from cost-effectiveness of available technology to a comparison of risk-reduction benefits with implementation costs.

Underground Injection Control

The only economic test in UIC is whether the injection occurs in a zone below an aquifer used for public water supply and separated from it by a lower impermeable geologic zone to avoid drinking water contamination and associated costs; otherwise, it is basically a cost-effectiveness evaluation that is required.

Wellhead Protection

This voluntary approach uses cost-effectiveness and avoided costs of remediation as its principal economic rationale.

Resource Conservation and Recovery Act

This law addresses permits for treatment, storage and disposal of solid and hazardous wastes. The law states that its "objectives are to promote the protection of health and the environment" and does not address economic considerations.

Comprehensive Environmental Response, Compensation and Liability Act

CERCLA or Superfund provides for the use of Trust Fund monies through "fund balancing" which considers the use of funds to be "balanced" among cleanup projects. Remedial design compares cost among alternatives for each cleanup.

Federal Insecticide, Fungicide and Rodenticide Act

FIFRA requires a risk-benefit analysis for special reviews of pesticides and an analysis of likely economic consequences of canceling a pesticide.

Small Business Regulatory Enforcement Fairness Act

This act requires evaluation of effects on small business and communities.

Unfunded Mandates Act

This law requires the effects of regulations on states and local governments to be evaluated to determine which costs will not be covered by the federal government.

Surface Mining Control and Reclamation Act

A cost-effectiveness assessment of performance standards to minimize effects on quantity of groundwater (and surface water) during and after mining operation; replacement of the affected water supplies; and reclamation plan to protect ground and surface water.

U.S. Groundwater Financing Laws

The United States has limited federal financing of water supply. Most water supply systems in the United States are small groundwater systems. Financing of water supply mainly occurs through the private sector or the states financing programs. Since the 1960s, U.S. agricultural legislation has been providing grants, loans and loan guarantees for rural development and established the Rural Utilities Service to manage grants and loans for homeowner and small water systems wells, water supply and wastewater treatment. More recently, farm legislation has provided subsidy funds for irrigation system efficiency improvements and managed aquifer recovery practices. Additionally, the SDWA amendments in 1996 focused on small water systems needs and established grants to states through the Drinking Water State Revolving Fund to provide financing to systems that were not in compliance or needed upgrading to maintain compliance with drinking water quality standards for public health protection. A significant economic aspect of both of these programs is a federal subsidy to targeted water systems through principal forgiveness for economically disadvantaged communities and below-market rate loans. The 2016 Water Supply Cost Savings Act requires potential Drinking Water State Revolving Fund small community loan applicants to consider cost-effectiveness information on local water sources, including water wells, before applying for a loan for water supply.

OTHER ECONOMIC IMPLICATIONS OF U.S. GROUNDWATER RIGHTS AND LAWS

Each groundwater legal doctrine in the United States establishes a set of relationships that forms the basis for economic transactions. These relationships include individual (or corporation) with individual, individual with community or state (or vice versa) and community with state (or vice versa). The doctrines describe "who can do what." In the western U.S. states, rights to use water may be purchased within the framework of the appropriate doctrine. If actions or results take place outside the doctrine, such as overpumping and wasteful mining or groundwater contamination, adversarial transactions may result in economic exchanges, such as court suits that may ultimately result in the payment of damages or penalties by one party to another.

The federal legal framework principally identifies the limits for the potentially groundwater withdrawing or contaminating activities to operate within to minimize any effects on the groundwater resource managed by states. The costs of operating within these limits must be borne by the economic activities or receive penalties that are identified in the law.

An important point to consider is that laws that limit contamination of groundwater (a quality issue) may support supply of groundwater (a quantity matter) by preserving or remediating groundwater quality and contribute to using the resource. Laws that promote groundwater conservation (reduced demand for groundwater supply) may facilitate maintenance of groundwater quality by minimizing brackish water migration and intrusion in aquifers where water quality maybe a potential or actual problem. Both circumstances have costs associated with their resolution and implementation.

GROUNDWATER CASE LAW

State Case Law

Legal interpretations of common law applied to groundwater and of state laws provide additional jurisdictional and property use rights requirements affecting the application and management of groundwater. A full review of these cases will not be covered here. These cases in which groundwater users have turned to state courts of law for decisions on groundwater pumping and contamination significantly influence economic actions related to groundwater in states of the decisions and in states with similar laws and doctrines. In humid states with reasonable use law, groundwater case law typically addresses interference of groundwater flow; interference of surface supply from groundwater use; drainage, ownership and use of groundwater; and damage to property or activities from groundwater use, including water quality impacts (Coogan, 1975). For example, the state of Ohio recognized in 1861 law the right of absolute ownership of groundwater which the state supreme court revised to reasonable use of groundwater as a duty of landowners and set a remedy of landowners damaged by unreasonable use of a nearby property owner (Hall, Undated).

In more arid states, case law often focuses on the volume pumped that damages adjacent groundwater users. In an example from Texas, in 1904, a property owner sued a railway company for pumping so much groundwater that the landowner's spring dried up. The state supreme court ruled that the "rule of capture" allowed the company to "divert, consume or cut off groundwater without regard to neighboring properties", being part of the property (Sprouse, Shrader, Smith, 2019).In 2018, the

California Supreme Court determined that a county and the state Water Resources Control Board have a duty under the public trust doctrine to manage groundwater pumping near a river that harms the uses of the river (PLF, 2018). In this latter case, surface water users were damaged by groundwater extraction. The governments controlling groundwater pumping will need to constrain pumping to rates that do not affect streamflow. The governments will incur legal and administrative costs to set and implement oversight of groundwater pumping and the wellowners will need to modify their pumping volumes correspondingly. Beneficiaries of the ruling are the stream users.

Federal Case Law

Federal case law also carries economic consequences. Most federal case law, other than dealing with interstate groundwater use issues, focuses on pollutants discharged to groundwater that are transmitted to surface water. Like state case law, federal case outcomes indicate who must take what actions, and can address who pays for damages and who benefits from the courts' decisions. Box 1.4.3, United States Federal Case Law on Groundwater, provides brief statements of some of the determinations from 1989 to 2019 that U.S. federal courts have made regarding whether and how groundwater, once polluted, is regulated under federal CWA provisions for controlling pollutants released to surface waters. Case law findings of federal courts have concluded in different ways. In 2020, U.S. Supreme Court ruled that water pollution permits apply to point source pollutant discharges to groundwater that reach waters of the United States, thereby serving as the functional equivalent of a discharge to surface water.

The federal Supreme Court has previously made determinations related to groundwater use among states. Congress had affirmed in 1943 an interstate compact that allocates the waters of the Republican River among the states of Kansas, Nebraska and Colorado. Beginning in 1998 and resolved by the Supreme Court in 2015, Kansas sued to limit the pumping of groundwater in Nebraska that was removing water from the river by underground flow to wells in Nebraska, contrary to the compact. The Court awarded Kansas payment by Nebraska of $5.5 million in damages but instructed the compact to adjust its accounting procedures and model (SCOTUS, 2015).As of this writing (2019), another case is being considered by the U.S. Supreme Court regarding the state of Mississippi suing the city of Memphis and the state of Tennessee for taking groundwater through wells near the border of the two states from the Memphis Sand Aquifer for use by Memphis

BOX 1.4.3 UNITED STATES FEDERAL CASE LAW ON GROUNDWATER

Federal case law in the United States on groundwater often focuses on water quality issues. The following information revolves around the applicability of the CWA National Pollutant Discharge Elimination System for discharges to groundwater as a conveyance of pollution to surface waters. The information is not a complete listing but provides legal perspectives from a range of federal cases. Federal courts have concluded differently depending on the circumstances of the cases being considered. The outcomes of these cases created economic consequences as to who pays, who benefits and how much.

Federal Cases	Findings
(a) U.S. Supreme Court, Cty. of Maui v. Hawaii Wildlife Fund et al., No. 18-260 (2020)	(a) The Supreme Court determined that a National Pollutant Discharge Elimination System (NPDES) "permit is required when there is a discharge from a point source directly into navigable waters or when there is the functional equivalent of a direct discharge … through groundwater".

(*Continued*)

BOX 1.4.3 (Continued)

Federal Cases	Findings
(b) Sixth Circuit, Kentucky Waterways Alliance v. Kentucky Utilities Co.; Tennessee Clean Water Network v. TVA (2018)	(b) The court found that pollutants from coal ash ponds seeped into groundwater that subsequently reached jurisdictional surface waters, and that the NPDES permitting requirements do not apply to releases to groundwater. Congress chose to prohibit only the discharge of pollutants to "navigable waters from any point source." The court reversed a district court decision that adopted the direct hydrologic theory, finding that "any alleged leakages into the groundwater are not a violation of the CWA." 905 F.3d at 444. "[a]s long as a connection [between groundwater and surface water] is shown to be real, direct, and immediate, there is no statutory, constitutional, or policy reason to require that every twist and turn of its path be precisely traced"
(c) Kinder Morgan Energy Partners, L.P., 887 F.3d 637 (4th Cir. 2018)	(c) In analyzing whether gasoline from a ruptured underground pipeline that undisputedly leached from groundwater into navigable waters required a NPDES permit, the Fourth Circuit framed its inquiry as whether, first, the discharge was from a point source, id. at 649-50, and second, whether there was a direct hydrological connection between the groundwater and jurisdictional surface water. Id. at 651. The court cited the broad purpose of the Act to restore and maintain the chemical, physical and biological integrity of the Nation's waters, asserting that "the statute established a regime of zero tolerance for unpermitted discharges of pollutants." Id. at 652. The court reasoned that "if the presence of a short distance of soil and ground water were enough to defeat a claim, polluters easily could avoid liability under the CWA by ensuring that all discharges pass through soil and ground water before reaching navigable waters." Id. The court concluded that "an alleged discharge of pollutants, reaching navigable waters located 1000 feet or less from the point source by means of ground water with a direct hydrological connection to such navigable waters, falls within the scope of the CWA."
(d) 26 Crown Assocs., LLC v. Greater New Haven Reg'l Water Pollution Control Auth., No. 3:15-cv-1439, 2017 U.S. Dist. LEXIS 106989, *24 (D. Conn. 2017)	(d) "[I]f the Clean Water Act were to apply as a routine matter to the discharge of pollution onto the ground that ends up seeping into the ground water, then Congress's purpose to limit the scope of the Clean Water Act [to point source discharges] would be easily thwarted."
(e) Sixth Circuit, Cape Fear River Watch v. Duke Energy Progress, 25 F. Supp. 3d 798, 810 (E.D.N.C. 2014)	(e) "Congress did not intend for the CWA to extend federal regulatory authority over groundwater, regardless of whether that groundwater is eventually or somehow `hydrologically connected' to navigable surface waters."
(f) Fifth Circuit, Rice v. Harken Exploration Co., 89 F.Supp.2d 820, 827 (N.D.Tex. 1999)	(f) Fifth Circuit held that "a generalized assertion that covered surface waters will eventually be affected by remote, gradual, natural seepage from the contaminated groundwater" was outside the scope of the OPA [Oil Pollution Act] in order "to respect Congress's decision to leave the regulation of groundwater to the States."
(g) Ninth Circuit, McClellan Ecological Seepage Situation v. Weinberger, 707 F. Supp. 1182, 1196 (E.D. Cal. 1998)	(g) Discharges to groundwater are subject to CWA regulation if "the groundwater is naturally connected to surface waters"

(Continued)

BOX 1.4.3 (Continued)

Federal Cases	Findings
(h) Umatilla Waterquality Protective Ass'n, Inc. v. Smith Frozen Foods, Inc., 962 F. Supp. 1312 (D. Or. 1997)	(h) "[T]he CWA's NPDES program should apply to groundwater to adequately protect surface water," but concluded that "the law as written, as intended by Congress, and as applied in Oregon for over two decades does not regulate even hydrologically-connected groundwater."
(i) Seventh Circuit, Village of Oconomowoc Lake v. Dayton Hudson Corporation, 24 F.3d 962 (7th Cir. 1994)	(i) "[n]either the Clean Water Act nor the EPA's definition [of waters of the United States] asserts authority over ground waters, just because these may be hydrologically connected with surface waters." 24 F.3d at 965. "[t]he omission of ground waters from regulations is not an oversight," as "Congress elected to leave the subject [of groundwater regulation] to state law[.]""[i]f the discharge is made into groundwater, and the pollutants somehow later find their way to navigable surface waters via a discrete hydrological connection, the CWA is still not implicated, because the offending discharge was made into groundwater, which is not subject to the CWA" Regarding potential hydrologic connection between groundwater and jurisdictional surface water, the court's determined: "Neither the Clean Water Act nor the EPA's definition [of navigable waters] asserts authority over ground waters, just because these may be hydrologically connected with surface waters."
(j) Matter of Bethlehem Steel Corp, UIC Appeal Nos. 85-8 & 86-13 (1989) 100 Stat. 7, 52	(j) EPA "declines to exercise CWA jurisdiction over injection wells (except those that inject into ground water with a physically and temporally direct hydrologic connection to surface water)."

residents. The state of Mississippi is asking for an award of $615 million in damages and that the city stop pumping from the aquifer and use an alternate water source (AP, 2019).Other levels of federal court usually adjudicate water resources issues between states.

ENVIRONMENTAL SITE ASSESSMENTS

In the United States

One of the most effective activities at the local level in the United States that has influenced the inclusion of groundwater quality in economic and financial decision-making has been the environmental site assessment for property transactions through the legal process known as "due diligence" These assessments are used to determine whether adverse environmental conditions caused by previous activities on a property could have negative economic consequences, including potential health and environmental impacts, for future buyers, financiers or users of a property (Flaherty and Laemmerman, 1992).

Beginning in the United States with CERCLA or Superfund, which considered all previous owners of a contaminated site jointly and severally liable for contamination, commercial property owners and financiers began evaluating the environmental conditions of their properties for future use, purchase or sale. Serious attention began when the state of New Jersey enacted its Environmental Cleanup Responsibility Act (ECRA) of 1983, under which all commercial and industrial property transactions were required to have an environmental site assessment undertaken by the company financing the transaction, including the testing of wells. This action informed the buyer about the condition of the property relative to contamination. While other states debated whether to pass similar legislation, this one action had such a strong positive response among the financial community in protecting their investment and managing their potential future liability that lenders around the country required sellers to conduct such an assessment for commercial and industrial property transfers, regardless of which state the property existed in. Other states have enacted similar laws. Subsequently, the SARA expanded information that commercial and public operations with hazardous and toxic chemicals in their processes must provide to the communities in which they exist. Thus, in response to these laws, the financial community changed its "tastes and preferences" toward financing environmentally clean properties.

Environmental site assessments are often driven by concerns about groundwater contamination and protection (Willard and Carroll, 1992) and typically have two major components (Martella, 1992). The first element – also referred to as "Phase I" – is a site inspection and documentation review for the property, including identification of the use of materials and substances at the site. The second element, Phase II, is the physical evaluation of environmental conditions at the site if the Phase I assessment determines that contaminant or other environmental problems exist. Phase II may involve soil borings and monitoring well installation to determine if soil and groundwater contamination have occurred.

Internationally

Governments and businesses in the United States are drawing on environmental due diligence developed at the international level. Most recently, the application of consensus-derived, internationally accepted standards are used in business and government to guide transactions. The International Organization for Standards (ISO) has existed since 1947 to develop standards covering a range of commercial and industrial activities and recently has developed standards applicable to transactions between businesses and governments relating to environmental site assessments and associated planning activities to minimize their impacts on the environment and human health. Two standards are described briefly in Box 1.4.4.

BOX 1.4.4 INTERNATIONAL QUALITY AND ENVIRONMENTAL MANAGEMENT STANDARDS ESTABLISHED BY THE INTERNATIONAL ORGANIZATION FOR STANDARDIZATION (ISO), GENEVA, SWITZERLAND

The ISO has developed generic management system standards for

- Any product or service of any organization, and
- Any commercial, industrial or governmental activity

ISO 9000 and ISO 14000 are quality and environmental management standards that are widely used around the world. ISO 9000 is an accepted global reference applied to business among corporations for quality management. Organizations use ISO 14000 to respond to their environmental requirements. Over 300,000 certifications to ISO 14001 have occurred in 171 countries.

The ISO 9000 standards are primarily concerned with quality management to address

- The customer's expectations for quality results, and
- Required regulatory outcomes

(Continued)

BOX 1.4.4 (Continued)

while in the process of

- Increasing customer fulfillment, and
- Obtaining regular progress in its conduct of achieving these objectives

The ISO 14000 standards focus on environmental management addressing

- Reduction and abatement of damages to the environment from an organization's activities, and
- Regular betterment of its environmental results

Regardless of an organization's mission and objectives, the ISO quality management system and environmental management system (EMS) under ISO 9000 and ISO 14000, respectively, have critical characteristics to address improving their performance.

ISO 9000 standards cover documentation, audits, review, professional competence and training, customer communication and feedback, product monitoring, analysis, product improvement/corrective action/problem prevention.

ISO 14000 standards address development and implementation of EMSs, principles of environmental auditing, audit of an EMS, qualification criteria for environmental auditors and lead auditors, audit program review and assessment material, labeling issues, performance targets and monitoring within an EMS and life-cycle issues.

Both of these sets of standards will apply to environmental site assessments in terms of their conduct and completeness.

Source: ISO Web site, URL: https://www.iso.org/home.html (accessed July 2, 2019).

GROUNDWATER LAWS OF OTHER COUNTRIES

Other countries have taken approaches to their legal frameworks that are different from the United States, even when they have a common legal history or similar hydrogeology. An understanding of these differences assists in evaluating the economic consequences that may also be different then, because these laws may define "who pays and how much" in distinctly varied ways. Evaluations in international groundwater economics will draw on these differences as cases arise.

First, national groundwater-related laws of the two countries of North America neighboring the United States – Canada and Mexico – will be reviewed, followed by an overview of laws of the European Union (EU) and of selected Muslim countries. A review of national groundwater laws and policies of all countries is, unfortunately, beyond the scope of this text. Those countries reviewed provide a range of national legal perspectives.

CANADA

In Canada, water law is based on English common law, under which groundwater is administered through the "rule of absolute capture." Provinces have primary jurisdiction over water resources, similar to the legal arrangement in the United States. Approximately one-third of the Canadian population relies on groundwater for supply (Nowlan, 2007).

Canada has followed both the same as well as different approaches as the United States in its national legal framework for groundwater. The Canada Water Act addresses water quality issues for both ground and surface waters. The Canadian Environmental Protection Act directs government controls on the release of substances into the environment and the protection of human health. Canadian law also indicates that the nation is committed to carrying out the "precautionary principle," which states "where there are threats of serious or irreversible damage, lack of full scientific certainty shall not be used as a reason for postponing cost-effective measures to prevent environmental degradation." Though this is not a stated principle in the laws of the U.S., the United States has followed a similar approach in some laws. Likewise, Canada explicitly cites in its laws the "polluter pays" principle: those who cause pollution pay to remediate it or compensate those affected by it. Again, while not explicit, the U.S. has incorporated this concept into its environmental laws affecting groundwater and other resources and circumstances.

In 2003, Canadian federal and provincial governments and other stakeholder groups cooperatively developed a "Canadian Framework for Collaboration on Groundwater," which serves as a guide for future groundwater use, conservation and protection policy (Government of Canada, 2003). The framework focuses on four areas: (1) coordination and collaboration mechanisms among governments and stakeholders; (2) national cooperative programs for research, inventory, assessment and monitoring of groundwater; (3) communication to raise the awareness of the public to its role in groundwater protection, to exchange data and to hold forum on groundwater issues; (4) performance standards and uniformity across Canada with respect to training and accreditation of groundwater professionals (including scientists, engineers and well installers), best management practices and technology transfer. Future implementation of such a framework through policy, law and regulation has the effect of improving information for the public's decisions about groundwater-related use and investment. Table 1.4.4 gives an overview of the legal and policy framework at the Canadian federal level related to groundwater use and protection.

TABLE 1.4.4
Overview of Canada's Major Federal Laws and Policy Related to Groundwater

Law/Policy	Date	Brief Description
Canada Water Act	1970	Establishes intergovernmental arrangements for policy development for water quality (both ground and surface water); provides for an inventory of all waters and for comprehensive water management plans; provides for grants and loans to agencies to address pollution control; establishes water quality standards for the controlled deposition of wastes into Canadian waters and effluent discharge fees for this deposition; water quality management agencies are to develop water quality plans with public input; provides for regulation of monitoring, treatment and compliance with standards for wastes and discharges of wastes to waters; establishes an inspection program
Canada Oil and Gas Operations Act	1985	Provides for regulations to prescribe measures necessary for the disposal or gathering and injection into an underground formation of water, gas, oil or other substances produced from a pool or oil deposit
Canadian Environmental Protection Act	1999	Establishes legal standing of persons to sue for damages caused by substances introduced into the environment and for compensation; provides for the gathering of information on and monitoring of substances in the environment; requires the formulation of plans for pollution prevention and the control and abatement of pollution as well as the restoration of the environment as necessary; requires the issuance of objectives, guidelines, limits and codes of practice relating to the release of substances into the environment to ensure sustainable development; mandates the issuance of objectives, guidelines and codes of practice with respect to the elements of the environment that may affect the life and health of people
Pest Control Products Act	2002	In registering pest control products, set maximum residue limits to protect health by considering exposure through dietary sources including drinking water
National Water Supply Expansion Program		Provides federal financial assistance along with provincial assistance to rural agricultural infrastructure including wells and pipelines as well as groundwater assessments and plans for safe reliable water supply

Other aspects of groundwater stewardship are managed and protected through the federal framework of laws or through provinces or other levels of government. For example, wetlands policy is based on the legal framework of the Canada Water Act, which provides for comprehensive water management plans developed through federal, provincial and other stakeholders, as well as for controlling wastewater discharges to waters, including wetlands. Nutrients and other substances that would cause their degradation are controlled under the Canadian Environmental Protection Act. Provincial governments have parallel laws controlling wastewater releases and waste disposal, and specifically address wetlands protection, such as Alberta provincial law, Environmental Protection and Enhancement Act and Water Act. Alberta's Water Act also regulates potable water supply and its allocation and well installation, maintenance and remediation as well as the qualification of well drillers. Wellhead protection is implemented in provinces largely as a voluntary matter at the local level. This voluntary approach may be changing, as suggested by the events and actions in Ontario.

The Province of Ontario conducted a comprehensive review of laws and activities affecting a "watershed-based approach" to protect water sources in light of groundwater contamination that resulted in deaths and illness from drinking water in one of its communities (O'Connor, 2002). This review recognized the "shared responsibility of all governments and stakeholders to contribute to our collective goal of ensuring a sustainable supply of safe clean drinking water" (Ontario, 2003). The Ontario Advisory Committee conducting the review recommended the watershed protection approach to be focused on managing and reducing water quality and quantity threats in vulnerable areas and for sensitive waters, including groundwater. It also called for a watershed-based source protection law and amending numerous other acts to provide consistency in water source protection, including its Conservation Authorities Act (dealing with local controls of land use), Environmental Protection Act, Municipal Act, Planning Act, Nutrient Management Act, Drainage Act, Brownfields Statute Law Amendment Act and the Mining Act, as well as other related laws. The Advisory Committee identified the following measures that were not addressed in law at that time and that may be incorporated in new laws and regulations:

- "Require routine disclosure of chemicals used or stored on-site, with appropriate confidentiality requirements.
- Require measures for the containment of chemicals, including plans for addressing leaks and spills.
- Require monitoring, including the installation of monitoring wells in specific high-risk circumstances.
- Enter into agreements with property owners and to attach relevant water protection conditions (e.g., secondary containment for chemical storage and monitoring requirements) on development applications.
- Control the drilling of new private wells and require proper plugging and sealing of unused wells in vulnerable areas where municipal drinking water supplies are potentially affected.
- Require regular maintenance and repairs and enable periodic inspection of septic systems.
- Require the effective decommissioning of septic systems prior to redevelopment.
- Require appropriate notification of contaminants associated with historic activities, as they are discovered through redevelopment.
- Require or promote conservation initiatives.
- Deal effectively with noncompliance (e.g., adding a charge to the associated property tax bill for the work done by the municipality as a way to deal with noncompliance)" (Ontario, 2003).

Throughout this review, the Advisory Committee emphasized a watershed-based source protection program as a "cost-effective" and "efficient" approach to minimize contamination of water sources (Ontario, 2003); however, the review did not include an evaluation of cost-effectiveness or efficiency. Changes in the laws to protect ground and surface water will affect the economic relationships currently influencing pollutant release and water use by individuals, businesses and governments.

Mexico

In the past, rights to water were owned by the national government exclusively based on revisions to the Mexico Constitution in 1917 as a result of land reforms at that time (NRC, 1995, p. 71), but that is changing to provide for water markets through private transferable water use concessions (Hearne and Trava, 1997, p. 12). Most of the federal law having to do with groundwater use, quantity and quality are covered in the National Water Law and the General Law of Ecological Equilibrium and Environmental Protection. Major provisions of these laws are described in Table 1.4.5. Notably, Mexican law provides for managing both ground and surface water in a watershed context and establishes a National Water Commission, regional authorities and water associations to set requirements and oversee water use and quality.

Mexican federal law affects groundwater and water resources generally in a number of ways. Significantly, changes in the Mexican federal laws allowing individuals to appropriate water use rights for the purpose of irrigation provides for the market exchange of ground and surface waters and their pricing in water transactions. Additionally, Mexican federal law allows protection zones to be established for the purposes of preserving ecosystem factors, preventing contamination and groundwater mining, thereby recognizing these values nationally. Requirements for application of benefit–cost analysis in regulatory development promote cost-effectiveness and economic efficiency to be considered. Market-based approaches as an alternative to regulatory mandates support more economically efficient outcomes where market failure is an issue in dealing with groundwater use, allocation and quality.

European Union

The EU consists of 28 member states (countries),[1] which "have set up common institutions to which they delegate some of their sovereignty so that decisions on specific matters of joint interest can be made democratically at the European level" (EU, 2005). It is a legal union largely to address common interests that have an economic basis. Those common interests include groundwater use and protection. A review of the groundwater-related laws of the EU and its member states on a comparable level with that given previously for the United States would be even more complex than the review for the U.S., if it started at the substate level (states of the member countries) and then covered each member's national laws and then the EU "Directives" as its mutual laws are known. The review here will focus on the EU level, which provides an interesting amalgam of the perspectives of the member states reflecting common concerns related to groundwater and water resources in general, and in an extensive multi-state transboundary situation.

Notably, the EU perspective embraces the "polluter pays" and "precautionary" principles explicitly and most recently consolidated several water-related directives into one "Water Framework" Directive and specifically a "Groundwater Directive" targeted at controlling pollution of groundwater. The approach of the first directive is to treat ground and surface water as one resource, except where unique features (e.g., underground injection of liquid substances) require additional steps to address its sustainability. The second directive recognizes unique features of groundwater needing regulation. These and the other groundwater-related directives are summarized in Table 1.4.6.

Of particular significance to groundwater, these directives do not treat groundwater as a separate resource except where this is important. Both ground and surface water quality must be regularly monitored and limits on their abstraction (withdrawal) set. Specific law focuses on injection of liquids into the subsurface, which is prohibited unless a permit is obtained. Groundwater monitoring must occur before any injection or indirect discharge to the subsurface. A comprehensive groundwater monitoring program is mandated. Integrated pollution prevention requires evaluation of shifting waste releases from one media to another, including soil and groundwater. Likewise, an evaluation of transboundary effects across national borders of contaminant releases must be conducted. Wellhead protection is required around wellfields that are sources of water supply, using nitrate as an indicator

TABLE 1.4.5
Overview of Mexico's Major Federal Laws and Policy Related to Groundwater

Law/Policy	Date	Brief Description
Federal Law for the Prevention and Control of Environmental Pollution	1971	(Ley Federal para Prevenir y Controlar la Contaminación Ambiental). Regulates releases of pollutants to air, water and the sea
Federal Law for the Protection of the Environment	1982	(Ley Federal de Protección del Ambiente). Provides for the protection and preservation of ecosystems through a revised approach to protect plants, animals, soil and water; establishes guidance for inclusion of ecosystem factors in economic development
National Water Law	1992; amended in 2004	(Ley de Aguas Nacionales) Creates a National Water Commission to set requirements and oversee water use and quality in watersheds; establishes irrigation districts/ water user associations to oversee and allocate ground and surface waters; establishes domestic water use as the priority water use; sets up a national hydrologic program to inventory and classify water bodies; provides for a water rights registry of users and for concessions for water use; allows the federal government to delineate protection zones to conserve water, preserve sources of water and the ecosystem, protect against contamination and control water quality; directs the federal government to set measures to control substances in wastewaters; sets fines for water polluters; requires environmental impact assessments for new permits; prescribes prevention of contamination of national waters and their return to an adequate condition for reuse; provides for water quality standards based on water body classification and assimilation capacities of the subsurface; sets discharge permit requirements, monitoring, maintenance of records and allowance of facility inspections; requires cleanup of accidental releases to waters and cost recovery by the government; allows suspension of discharges when no permit, monitoring results not available and water use fees not paid; requires violators of the law to be responsible for remediation and environmental damage and to return water to previous quality.
General Law of Ecological Equilibrium and Environmental Protection	1988; amended in 1996	(Ley General del Equilibrium Ecológico y la Protección al Ambiente). Warrants an individual's rights to a safe, healthful environment and public involvement in environmental protection; sets objective of biodiversity preservation and protection of natural areas; provides for sustainable use of water and other natural resources; controls hazardous wastes and radiological pollutant sources; provides for evaluating risks of actions affecting the environment; allows establishment of management areas to protect ecosystem function; defines enforcement of law and penalties for noncompliance
Law for Establishing Norms and Standards	1992	Establishes a "Federal Prosecutor for the Environment" to prosecute violators of environmental standards; requires use of benefit–cost analysis in future regulations and consideration of market-based approaches rather than regulatory means
General Health Law	1984; amended in 1988	Establishes standards for contaminants in drinking water; forbids releases of polluted water to sources of drinking water

Sources: El Congreso de Los Estados Unidos de Mexico, Ley de Aguas Nacionales, 1992.Gonzalez, G.R. and Gastelum, M.E., *Overview of the Environmental Laws of Mexico,* National Law Center for Inter-American Free Trade, 1999, URL: http://www.natlaw.com/pubs/spmxen13.htm (accessed February 11, 2005).National Research Council (NRC), *Mexico City's Water Supply,* National Academy Press, Washington, DC, 1995, 71–77.Nido, L.M., and Hutt, J.B., Significant changes to Mexico's water laws, 2004

TABLE 1.4.6
European Union Groundwater Related Directives

Directive	Date	Brief Description/Significant Provisions
Directive 2006/118/EC concerning groundwater	2006	Developed in response to Article 17 of the Water Framework Directive; defines groundwater areas in river basins (watersheds); requires classification of groundwater areas relative to meeting Water Framework objectives; establishes a register for groundwater areas based on uses, including protection of habitats and species; establishes groundwater monitoring based on classification; requires river basin plan to include the summary of pressures and impacts, economic analyses of water uses and the summary of programs for protection, pollution control and remedial measures; provides for cost recovery of measures based on "polluter pays" principle; establishes a program to meet Water Framework measures for groundwater extraction and pollution prevention and control, including artificial recharge to be updated every six years.
Directive COM 2003 550 concerning groundwater protection against pollution	2003	Sets a policy framework for protecting groundwater from pollution; establishes criteria for assessing groundwater chemical status and a compliance regime for quality standards; provides a process for setting pollutant threshold values (or concentration limits) in groundwater; specifies criteria for identifying significant increasing pollutant concentration trends and for determining initiation of actions to reverse trends, considering risks to aquatic and terrestrial ecosystems and human health; provides continuity with Directive 2000/60/EC by linking to its list of pollutants to be controlled; prevents or limits indirect pollutant discharge to groundwater that would affect achieving good chemical status in waters of river basins; provides for review and updating of river basin plans.
Directive 2000/60/EC establishing a framework for water policy	2000	Focuses on sustainability of water resources; recognizes interaction of ground and surface water; protects water for drinking water supply and ecological functions; specifies objectives to reverse contamination of groundwaters and coordinate measures for surface and ground quality through a river basin management approach through river basin plans; states goal for the elimination of pollution of water from man-made substances; specifies that common minimum environmental standards and emission limits must be adopted by member states; less stringent environmental objectives can only be implemented for specific water bodies for which technically infeasible and economically expensive actions can be demonstrated as needed; requires monitoring for surface and groundwater quality; directs member states to recover full costs of water service; establishes point and diffuse (nonpoint) pollution controls and measures (standards), including prohibition of injection of contaminants into groundwater; sets controls for abstraction (withdrawal) of surface and groundwaters; provides for the development of the prevention and control of pollution to surface and groundwaters; provides for ecological monitoring and economic assessments of required actions.
Directive 91/676/EEC concerning the protection of waters against pollution caused by nitrates from agricultural sources	1991	Requires designation of vulnerable zones, which have surface and groundwaters that could be affected by nitrate from agricultural sources; provides for establishing codes of good agricultural practices, voluntary action, training and information for farmers to protect these waters in vulnerable zones from nitrate; allows action to protect water in vulnerable zones to vary based on environmental conditions and activities; provides for taking additional protective actions; requires monitoring for nitrate to be repeated over time; report every four years on the implementation of the program; requires member states to bring laws, regulations and other administrative provisions into alignment with the Directive.
Council Directive 96/61/EC concerning integrated pollution prevention and control	1996	Sets out requirements to achieve integrated prevention and control of emissions to air, water and land; requires permits for pollution emission in existing and new facilities and provides for the Council to set emission limit values; mandates integrated approach to issuing permits, including evaluation of the potential for shifting pollution from one media to another, protection of soil and groundwater and monitoring of releases; requires that best available techniques be used for prevention and control of pollutants; sets conditions for facility operator to report results of emission monitoring; provides access to information and to allow exchange of information among member states; requires consideration of transboundary effects of pollution among member states.

(Continued)

TABLE 1.4.6 (*Continued*)
European Union Groundwater Related Directives

Directive	Date	Brief Description/Significant Provisions
Council Directive of April 26, 1999 on the landfill of waste	1999	Requires member states to establish a national strategy for the reduction of biodegradable wastes going to landfills; specifies that landfills will not accept liquid wastes and wastes that are explosive, corrosive, oxidizing or flammable, medical and veterinary wastes, tires and certain other wastes and mixtures of diluted wastes; identifies permit conditions for landfills, including consideration of hydrogeologic characteristics and financial security; ensures that landfill disposal prices reflect the full cost of disposal, including long-term maintenance of the landfill; provides for conditions to accept waste including appropriate documentation and for inspections; requires monitoring of landfills; establishes landfill closure and postclosure care requirements; requires member states to report on implementation of the Directive and to bring their laws, regulations and administrative procedures into alignment with the Directive.
Council Directive 80/68/EEC on the protection of groundwater against pollution caused by certain dangerous substances	1979	Establishes policy to prevent the pollution of groundwater by substances belonging to the families and groups of substances that have harmful effects in groundwater; prevents substances of a toxic, persistent and bioaccumlative nature and limits heavy metals and other pollutants from entering groundwater; authorizes reinjection into the same aquifer of geothermal, mining and construction waters; limits indirect discharge of substances on or in the ground that might cause groundwater pollution; authorizes artificial recharge if no adverse effect to groundwater; requires hydrogeologic investigations before allowing indirect discharges of substances; requires groundwater quality monitoring before indirect discharges and artificial recharge may occur; specifies authorizations of indirect discharge must include essential precautions related to the characteristics of the substances, for monitoring the substances and groundwater quality, and for proximity to other water uses such as drinking water; mandates that authorizations be reviewed every four years; requires checking compliance with the Directive; stipulates that consultation shall occur among member states for discharges that affect transboundary groundwaters; requires member states to bring their laws, regulations and administrative procedures into alignment with the Directive.
Council Directive 98/83/EC on the quality of water intended for human consumption	1998	Establishes policy to protect human health from the adverse effects of any contamination of water intended for human consumption by ensuring that it is wholesome and clean regardless of its origin and whether it is supplied from a distribution network, from a tanker or in bottles or containers; requires water to be free from any microorganisms and parasites and from any substances which, in numbers or concentrations, constitute a potential danger to human health and to meet minimum requirements; requires member states to take steps to prevent contamination of drinking water sources; sets parametric values as maximum occurrences of contaminants allowed in drinking water at points of compliance; requires regular monitoring at compliance points; specifies remedial action when parametric values are exceeded and health is endangered; provides for short derogations from the parametric values to allow time to address quality problems; mandates that member states report on drinking water quality every three years; requires member states to bring their laws, regulations and administrative procedures into alignment with the Directive.

that must be monitored around those wellfields. Wetlands are covered by the comprehensive Water Framework Directive. Member states must bring their national laws into alignment with the EU directives within a timeframe specified in the directive, if they are not already.

Muslim Countries

Groundwater law in Muslim countries derives its basis from the Quran and Laws of Mohammad. These laws were defined as applying to all waters in most cases and applied to local circumstances typically. Aspects of them have found application on a national basis. Additionally, some national water laws in Muslim countries were derivatives of the British and French influences in these regions. Caponera (1973a, b) and Schiffler (1998) have reviewed water law in Muslim countries from which this summary is taken. Water law of Muslim countries provides an interesting and different perspective on water rights that define not only interpersonal but also economic relations on a different basis. Within the Muslim world, distinctions exist between the different religious groups concerning the use of water, such as for the priority of use.

Schiffler describes Islamic law for water in these countries in the following ways (1998, p. 122):

> "Islamic law (Shari'a) is a comprehensive legal system covering all aspects of life. Its most important sources are the Quran and the sayings of the Prophet Muhammad (the Sunna). From these two sources Muslim scholars have over the centuries elaborated Islamic law through interpretation (idjtihad), analogies (qiyas) and consensus (Ijm'a)… the Quran does not lay down any clear and specific rules on [water] use, and few such rules can be found in the Sunna. Water law in Muslim countries is heterogenous. It incorporates elements of pre-Muslim traditions in informal water law and of western tradition in formal law. Rules concerning water use also differ between the two main branches of Islam (Sunni and Shi'i) and among the four main Sunni schools of law. All these differences make it difficult to speak of Islamic water law as such. The common basic trait of water law in Muslim countries seems to be the principle of the "community of interest", as derived from the saying of the Prophet."

Caponera (1973a, b) has prepared a summary and blend of water law in Muslim countries, which is abstracted here in Table 1.4.7 as it applies to groundwater. The codification of water law in the

TABLE 1.4.7
Water Law in Muslim Countries as Related to Groundwater

Traditional Concepts – based on the *Shari'a* and interpretations of it.
Water definition – water is a nonsaleable good to which all have a right (right of thirst); groundwater belongs to the community.
Water ownership – water is the property of the state (country); traditionally, a person can own the right to water use through receipt as a gift or inheritance or by occupying land to maintain it; a person may possess water if it is in a receptacle.
Water use – an owner may control access to persons who desire drinking water; if no public water source is available, the owner must provide drinking water on the condition that the person does not damage the well.
Water rights transfer through sale – water rights are rights of use; an owner may sell water rights with the land or not do so.
Protected area around a well and spring – traditionally, well owners may control areas around their wells from trespass or the installation of wells within a small area that could influence flow to their well.

Law/Policy	Date	Brief Description
Saudi Arabia		
Shari'a and local customary law	Historic	Water is God's property and available to all; ownership only occurs when water is contained in a receptacle (jar, cistern, pool, etc.); all lands adjacent to a well or spring have right to a portion of the available water; community allocates rights to water use; surplus waters should be available for use as long as not detrimental to rights of owner or user; wells and springs must not be used in a way that causes harm to adjacent land and water users; animals are to be provided from the same sources or from surplus waters, like people; communities are responsible for well operation and maintenance; a right to use water is recorded locally and controlled by the community through a water use supervisor.
Ministerial resolution No. 328 – Digging of wells for drinking water and the extension of drinking water networks	1968	Approval of applications for drinking water wells will consider site suitability, distance from settlement, water availability and source; 37.8–56.8 L/day is the minimum that must be available to each person.

(*Continued*)

TABLE 1.4.7 (*Continued*)
Water Law in Muslim Countries as Related to Groundwater

Law/Policy	Date	Brief Description
(Draft) Water code		Ground and surface waters are the property of the state; existing land and water rights are protected as long as they are in accordance with the principles of Islamic law; domestic use and household use have priority.
Royal decree on operational procedure of the distribution of public land ordinance Royal orders 17697, 8949, 2092, 7553		Permits are required for drilling wells; drillers must demonstrate qualifications; minimum distance between wells is 500 m. Well drilling is prohibited in protected zones.
Jordan		
Undergroundwaters Control Regulation	1966	Gives to the state full powers to control use and disposal of groundwater; groundwater use permit holders must take all necessary actions to preclude infiltration of saline water into usable aquifers.
Law on Settlement of Land and Water Rights	1952	Relative to groundwater, only water brought to the surface is owned by the landowner; water rights are for the registered use of water, which may be sold or leased with the approval of the Natural Resources Authority.
Water Control Law	1953	A permit for other than irrigation purposes is a personal right and may be sold separately from the land; the "right of thirst" is recognized as not needing a permit.
Law on establishment of the Natural Resources Authority	1968	Water permits are required for groundwater use, for its use in excess of 5 m^3/day; for water used for industrial and mining purposes, for new wells, for modified wells, for installation of pumping devices; for pollutant discharges; for drilling wells; water users must not waste water under the threat of limiting the amount of use by the state; well drillers must obtain a license; groundwater extraction permits may be refused in case of unreasonable use or depletion minimum distances between wells may be established.
Indonesia		
Local customary water law	Historic	Members of a community give mutual help (derived from the Hindu cosmology), a concept similar to Muslim customary water law; a water master controls water allocation.
General Water Regulation	1936	Provides for economic water use criteria and for free use of groundwater from dug wells of 15 m in depth; wells deeper than 15 m and wells of less than 15 m deep for agricultural and industrial purposes are subject to a concession from the government; requires that responsible authority allocate water for other uses, such as irrigation or emergencies; allows landowner free use of springs while requiring maintenance of their head, conservation of their flow and protection of a 20 m zone around the spring head; requires a concession for modifying, sealing, deepening and cleaning of wells; provides for agricultural uses to have priority over domestic and sewerage uses with cultivation plans being subject to approval of provincial administration; in times of water shortage, specifies that drinking water supply has priority and then sewerage; regulates the discharge of liquid and solid wastes for which a government concession is required.
Constitution of Indonesia	1945	Land and water are controlled by the state for the benefit of the people.
Law No. 11 for Public Service Undertakings	1962	Establishes the Department of Health to protect water quality for all aspects of consumption; controls water use for mineral bottling and food processing through concessions provided by the Public Health Administration.

(*Continued*)

TABLE 1.4.7 (*Continued*)
Water Law in Muslim Countries as Related to Groundwater

Law/Policy	Date	Brief Description
Presidential Instruction on Government organization and coordination in the implementation of the five-year plan	1969	Identifies the Department of Finance as establishing and supervising water charges.
Circular of the Minister of Mines	1971	Specifies groundwater use conditions such as depth and amount subject to the decision of the Department of Mines; provides for licensing of all well drillers and companies and requires them to report activity and data.
Water Resources Law	1974	Places the responsibility for beneficial use of water in Central and provincial governments; establishes an inventory of water resources; requires the Central government to formulate plans, development, use, rights and control of water treatment and pollution as well as management of harmful effects of water use and discharge; establishes the National Development Planning Board to coordinate water resource planning with other development, including river basin development; controls use of groundwater, geothermal waters and hotsprings by the Department of Mines, including drilling and development; provides for penalties for violations of concessions and regulations; delegates certain authorities to Provinces.

Sources: Caponera, D.A., *Water Laws in Moslem Countries. Irrigation and Drainage Paper 20/1,* Food and Agriculture Organization, United Nations, Rome, Italy, 1973a.Caponera, D.A., *Water Laws in Moslem Countries. Irrigation and Drainage Paper 20/2,* Food and Agriculture Organization, United Nations, Rome, Italy, 1973b.

area of Muslim countries first took place during the time of the Ottoman Empire (1300–1922 A.D.) (Caponera, 1973a, p. 36). Table 1.4.7 describes some common elements of that code and the subsequent Turkish Republic civil code, which replaced it and still guides legal actions relative to groundwater in Muslim countries, as well as gives highlights of groundwater-related laws in selected Muslim countries, including one beyond the historical Ottoman Empire, Indonesia. Indonesian water law evolved from a succession of influences: (1) traders first from India who brought Hindu and from Indochina who instilled Buddhism through the fourteenth century, (2) later Muslim traders, (3) from the seventeenth through the early twentieth centuries, European interests, (4) and then after World War II, the independent Indonesia land and water law based on the traditional *adat* law derived largely from earlier Hindu propensities (Caponera, 1973b, pp. 54–55).

COMPARISON OF APPROACHES

At the center of any comparison of approaches at national levels is the recognition of who controls the groundwater resource. In the United States and Canada, the federal government controls waters that are defined to be interstate or interprovincial, and even then in limited ways. In more arid countries, such as Jordan and Saudi Arabia, all waters are under the authority of the national government. This relation alone affects who can pursue water rights damages and in what ways resolution may be accomplished. In other words, the perspectives defined in the laws affect the transaction costs of dealing with any problems arising out of water use and whether a person or company has the standing to sue in court.

At the local level, in the United States, states principally have administrative authority over groundwater. This allows states to specify whether the rights to a certain volume can be owned by individuals, particularly in arid states, or whether those rights to volumes are managed by states.

Again, who has the rights to access and/or to obtain a volume of groundwater determines how problems in groundwater use are addressed and the extent of the transaction costs from place to place. These circumstances are also reflective of the climate and hydrology of an area: if arid, individual groundwater volume rights may be protected by law, whereas in a more humid state, individual rights may not be protected in the same way and in fact may be controlled by the state through local governments.

The legal approaches to water rights are also the result of historical, cultural and economic development that evolved based on climate and hydrology. The more heavily industrialized economies have required extensive laws and regulations on pollution control and on water uses that affect human health and the environment. These concerns are addressed at a high level in the EU, which now has laws focusing on watersheds to more fully integrate water resource management across its varying hydrologic zones and economic interests. Such an approach recognizes that groundwater migrates under political boundaries and has transitory physical boundaries depending on local geological conditions, precipitation and pumping (abstraction).

While no one perfect approach may exist, several legal themes central to groundwater use and management have emerged and have relevance around the world. First, groundwater is a limited resource. Second, groundwater should not be considered separately from surface water, but rather as one resource existing in different physical conditions. Third, what affects groundwater in one place cannot be isolated physically and so should not be isolated legally, since it is one resource with surface water. Fourth, as population and water use expand, legal institutions should be updated to recognize the current science of hydrology and specifically hydrogeology to more fully address transboundary issues, whether property boundaries at the local level or state boundaries at the international level. Fifth, water laws reflect values of and for ground and surface water in the local and national economies recognizing that they are together a fundamental resource for which controls on access to water for a range of uses should not abrogate affordable access to a sufficient and safe volume of water to sustain life for all. Sixth, as population grows, the sustainability of the resource becomes more paramount and is now being recognized in law at national and international levels such as in Canada, Mexico and the EU.

CONCLUSION

The United States laws affecting groundwater vary from state to state and from state to federal government. States hold groundwater in trust for their residents and control access, ownership, use and quality. Most western states have laws providing prior appropriation by individuals to specific quantities of water whereas eastern law typically governs groundwater through correlative rights allowing reasonable beneficial use of groundwater. In western states, groundwater management areas have emerged as an important tool for allocating scarce water. Across the country, wellhead protection areas have been adopted to manage groundwater quality. Groundwater recharge and aquifer management are emerging as a significant focus for legal attention.

The water laws of many other countries can and should be examined and understood to allow a better understanding of the range of groundwater management approaches and values among different nations. These laws reflect historical, cultural and economic development that evolved based in part on climate and hydrology. The summaries above are just a sampling. The significance of reviewing these different approaches is to underscore how groundwater is valued in various legal doctrines. The range of legal ways to address groundwater defines rights to control and to use the resource, rights to discharge pollutants into it and rights to respond to damages to the resource, all of which affect the economic and financial conditions of individuals, companies and governments in different ways depending on the organization of the law. More integrated water policies recognizing the significance of the hydrologic cycle for both water quantity and quality may facilitate having ecologically and economically sustainable water supplies for the range of their uses.

FOR FURTHER CONSIDERATION

1. How does your state or provincial law affect ownership and access to groundwater?
2. What does your state or provincial law indicate about unreasonable use of groundwater?
3. Contrast your state's or province's groundwater law with that of another state or province regarding groundwater management areas. Is your state's or province's management area law focused primarily on groundwater quantity or quality protection?
4. What groundwater issue has emerged in your state or province recently?
5. What local, state or federal laws affect the groundwater issue you identified in Question 4? How do the laws you identified affect who is affected by the issue? What effects can you identify?
6. What are the economic aspects of the issue identified in Question 4 – what are and who may receive the advantages or disadvantages of its resolution? How significant are those advantages or disadvantages? Quantify the advantages and disadvantages if information is available to do so. Who incurs the greatest cost and who receives the greatest benefit?

NOTE

1 Austria, Belgium, Bulgaria, Croatia, Cyprus, Czechia, Denmark, Estonia, Finland, France, Germany, Greece, Hungary, Ireland, Italy, Latvia, Lithuania, Luxembourg, Malta, the Netherlands, Poland, Portugal, Romania, Slovakia, Slovenia, Spain, Sweden and the United Kingdom comprise the European Union as of 2019.

REFERENCES

Alley, W.A. and Job, C.A.. 2020. *Groundwater, Wells and Coronavirus*. URL: https://ngwa.org/publications-and-news/covid-19-news-and-resources/groundwater-wells-and-coronavirus (accessed May 3, 2020).

American Water Works Research Foundation (AwwaRF). 2004. *Demonstrating Benefits of Wellhead Protections Programs*, Denver, CO: AwwaRF.

Associated Press (AP). 2019. *Judge to report to Supreme Court on Memphis vs. Mississippi water rights case*. URL: https://wreg.com/2019/06/04/judge-to-report-to-supreme-court-on-memphis-vs-mississippi-water-rights-case/ (accessed July 2, 2019).

Arizona Department of Water Resources. 1980. *Overview of the Arizona Groundwater Management Code*. URL: http://www.azwater.gov/AzDWR/WaterManagement/documents/Groundwater_Code.pdf (accessed May 24, 2019)

Bergstrom, J.C., Boyle, K.J., Job, C.A., and Kealy, M.J. 1996. Assessing the economic benefits of ground water for environmental policy decisions. Environmental Economics Report to the U.S. Environmental Protection Agency. URL: http://yosemite.epa.gov/ee/epa/eerm.nsf/vwRepNumLookup/EE-0115?OpenDocument (accessed February 16, 2009), 34 p; Reprinted in Journal of the American Water Resources Association, 32 (2), *April* 1996, 279–291.

Brumley, J., Marks, C., Chau, A., Lowrance, R., Huang, J., Richardson, C., Acree, S., Ross, R., and Beak, D. 2018. *The Influence of Green Infrastructure Practices on Groundwater Quality: The State of the Science*. U.S. Environmental Protection Agency, Washington, DC, EPA/600/R-18/227.

California Legislature (CAL). 2014. *Sustainable Groundwater Management Act; California Government Code, Water Code, Division 6, Part 2.74, 10720-10737.8*. URL: http://leginfo.legislature.ca.gov/faces/codes_displayexpandedbranch.xhtml?tocCode=WAT&division=6.&title=&part=2.74.&chapter=&article=

Campbell-Mohn, C., Breen, B., and Futrell, J.W. 1993. *Sustainable Environmental Law: Integrating Natural Resource and Pollution Abatement Law from Resources to Recovery*. Environmental Law Institute. West Publishing Company, St. Paul, MN.

Caponera, D.A. 1973a. *Water Laws in Moslem Countries. Irrigation and Drainage Paper 20/1*. Food and Agriculture Organization, United Nations, Rome, Italy.

Caponera, D.A. 1973b. *Water Laws in Moslem Countries. Irrigation and Drainage Paper 20/2*. Food and Agriculture Organization, United Nations, Rome Italy.

Coogan, A. 1975. *Problems in Groundwater Rights in Ohio*. URL: https://www.uakron.edu/dotAsset/507139e2-063a-40ea-9a5d-1177dfbbd88d.pdf (accessed July 1, 2019).

Connecticut Department of Energy and Environmental Protection. 2019. Aquifer Protection Area Program. https://www.ct.gov/deep/cwp/view.asp?a=2685&q=322252&deepNav_GID=1654 (accessed June 10, 2019)

Congreso de los Estados Unidos Mexicanos. 1992. *Ley de Aguas Nacionales*. URL: http://www.diputados.gob.mx/LeyesBiblio/pdf/16_060120.pdf (accessed December 16, 2020).

Dales, J.H. 1968. *Pollution, Property and Prices*. University of Toronto Press, Toronto, ON, Canada.

Daly, H.E. and Farley, J. 2004. *Ecological Economics: Principles and Applications*. Island Press. Washington, DC, 454 p.

El Paso Water (EPW). 2018. *Water Resources*; Desalination. URL: https://www.epwater.org/our_water/water_resources/desalination (accessed May 24, 2019).

Environmental Law Institute (ELI). 1990. *Survey and Analysis of State Groundwater Classification Systems and Program Operations*. Prepared for the U.S EPA Office of Ground Water Protection, Washington, DC.

European Union (EU). 2000. Council Directive 2000/60/EC establishing a framework for water policy.

European Union (EU). 2005. *The European Union at a Glance*. URL: http://europa.eu.int/abc/index_en.htm# (accessed February 11, 2005)

European Union (EU). 2006. Groundwater Directive 2006/118/EC. URL: http://ec.europa.eu/environment/water/waterframework/groundwater/policy/current_framework/index_en.htm (accessed March 14, 2009).

Flaherty, A.M. and Laemmerman, L. 1992. *What is the purpose of the phase I environmental site assessment? Proceedings the Conference on Environmental Site Assessments: Case Studies and Strategies*, August 5–7, Orlando, FL, pp. 3–6.

Goldfarb, W. 1988. *Water Law*. Lewis Publishers. Chelsea, MI, 284 p.

Gonzalez, G.R. and Gastelum, M.E. 1999. *Overview of the Environmental Laws of Mexico*. National Law Center for Inter-American Free Trade. URL: http://www.natlaw.com/pubs/spmxen13.htm (accessed February 11, 2005).

Government of Canada, Natural Resources Canada, National Ad Hoc Committee on Groundwater. 2003. *Canadian Framework for Collaboration on Groundwater*. URL: https://doi.org/10.4095/214620

Hall, P.K. Undated. *Understanding Water Rights in Ohio; Parts I, II, and III*. Ohio State University Extension Law Notes. URL: https://farmoffice.osu.edu/sites/aglaw/files/site-library/Understanding%20Water%20Rights%20in%20Ohio.pdf (accessed July 1, 2019).

Hanemann, M. 2005. *The Value of Water*. University of California, Berkeley, CA. URL: http://are.berkeley.edu/courses/EEP162/spring06/Readings/valuewater.pdf Accessed November 10, 2006. 35 p.

Hearne, R.R. and Trava, J.L. 1997. *Water Markets in Mexico: Opportunities and Constraints*. International Institute for Environment and Development Discussion Paper DP97-01, 27 p.

Hutchins, W.A., Ellis, H.H., and DeBraal, J.P. 1971. *Water Rights Laws in the Nineteen Western States*. 3 vols. United States Department of Agriculture, Washington, DC.

ICF, Inc. 1990. *Review of Nine Selected State Groundwater Classification Systems*. Prepared for the U.S. EPA Office of Groung-Water Protection, Washington, DC.

ISO Web site URL: http://www.iso.ch/iso/en/iso9000-14000/index.html, http://www.isoeasy.org/, and http://www.iso14000-iso14001-environmental-management.com/ (accessed January 28, 2004).

ISO Web site, URL: https://www.iso.org/home.html (accessed July 2, 2019)

Irrigation Association (IrrA). 2019. *Irrigation E3 Program*. URL: https://www.irrigation.org/IA/Workforce_Development/Irrigation_E3/IA/Workforce_Development/Irrigation_E3/Irrigation-E3-Program.aspx?hkey=4b3a4cde-2a11-4042-b51b-f6c138058420 (accessed July 15, 2019).

Jenkins, E.D. 1980. *Water depletion allowance, Kansas case. Proceedings of the 7th Annual Conference of Groundwater Management Districts Association*, December 8–9, 1980, North Platte, NE, pp. 33–43.

Job, C. 1996. Benefits and costs of wellhead protection. *Ground Water Monitoring and Remediation*, 16 (2), 65–68.

Job, C. 2007. A summary of state wellhead protection programs. *Ground Water Monitoring and Remediation*, 17 (2), 61–63.

Job, C.A. and Gabanski, G. 1988. Monitoring wells need consistent regulation. *Waste Age*, 19 (8), 164–170.

Kowalick, Claire. 2018. Indirect potable water reuse system goes online; plant renamed. *Times News Record*, February 28, 2018. URL: https://www.timesrecordnews.com/story/news/local/2018/02/28/indirect-potable-water-system-online/382297002/ (accessed May 24, 2019).

Martella, T.K. 1992. *The EAS and what it can accomplish. Proceedings the Conference on Environmental Site Assessments: Case Studies and Strategies*, August 5–7, Orlando, FL, pp. 15–19.

Minnesota Pollution Control Agency (MPCA). 2019. *Introduction to wastewater permits*. URL: https://www.pca.state.mn.us/water/introduction-wastewater-permits (accessed May 26, 2019).

National Conference of State Legislatures (NCSL). 1994. Ground Water Protection Legislation, *Survey of State Action*, 1988–1992.
National Ground Water Association. 2019. *Advance Your Career*. URL: https://www.ngwa.org/advance-your-career/certification (accessed July 15, 2019).
National Research Council (NRC). 1995. *Mexico City's Water Supply*. National Academy Press, Washington, DC, 107 p.
National Research Council (NRC). 1997. *Valuing Ground Water: Economic Concepts and Approaches*. National Academy Press, Washington, DC, 189 p.
Nebraska Legislature. 1975. *Groundwater Management Act; 2004. Groundwater Management and Protection Act*. URL: https://dnr.nebraska.gov/sites/dnr.nebraska.gov/files/doc/water-planning/2016NEWGWMgmtProtectionActStatutes.pdf (accessed May 25, 2019)
New Jersey Department of Environmental Quality (NJDEP). 2019. *Discharge to Ground Water Permitting Program*. URL: https://www.nj.gov/dep/dwq/dgw_home.htm (accessed May 26, 2019).
Nido, L.M. and Hutt, J.B. 2004. *Significant changes to Mexico's water law (March 19, 2004)*. URL: http://www.bracepatt.com/files/tbl_s16Publications% 5CFileUpload77%5C969%5CMEXICO_WATER_LAW_REFORM__ARTICLE__1.pdf (accessed February 11, 2005).
Nowlan, L. 2007. "*Out of Sight, Out of Mind? Taking Canada's Groundwater for Granted*." In Eau Canada, (ed.), Karen Bakker, UBC Press.
O'Connor, D.R. 2002. *Ontario Ministry of the Attorney General Report of the Walkerton Commission of Inquiry: The Events of May 2000 and Related Issues*. 38 p. and appendices. URL: http://www.archives.gov.on.ca/en/e_records/walkerton/report1/index.html
Ontario Advisory Committee on Watershed-Based Source Protection Planning. 2003. Protecting Ontario's drinking water: Toward a watershed-based source protection planning Framework, Final Report. URL: https://atrium.lib.uoguelph.ca/xmlui/handle/10214/15836?show=full
Pacific Legal Foundation (PLF) 2018. California Court of Appeal imposes public trust obligations on groundwater well permits. URL: https://pacificlegal.org/california-court-of-appeal-imposes-public-trust-obligations/ (accessed July 1, 2019).
Runkel, R.L., McKnight, D.M., Rajaram, H. 2003. Modeling hyporheic zone processes. *Advances in Water Resources*, 26 (9), 901–905.
Sax, J.L. and Abrams, R.H. 1986. *Legal Control of Water Resources*. West Publishing Company, St. Paul, MN.
Schiffler, M. 1998. *The Economics of Groundwater Management in Arid Countries; Theory, International Experience and a Case Study of Jordan*. Frank Cass Publishers, Portland, OR, 394 p.
Smith, D.D. and Wyatt, A.W. 1980. Groundwater depletion allowance for Texas. In *Proceedings of the 7th Annual Conference of Groundwater Management Districts Association*, December 8–9, North Platte, NE, pp. 43–49.
South Dakota Department of Environment and Natural Resources (SDDENR) 2019. *Groundwater Discharge Permits*. URL: https://denr.sd.gov/des/gw/gwdischarge/gw_discharge_permit.aspx (accessed May 26, 2019).
Sprouse, S. Smith, Attorneys at Law. 2019. *Ownership of Texas Groundwater*. URL: https://www.sprouselaw.com/ownership-of-texas-groundwater/ (accessed July 1, 2019).
Supreme Court of the United States (SCOTUS). 2015. State of Kansas, Plaintiff V. State of Nebraska, et al., Defendants. URL: https://www.supremecourt.gov/opinions/14pdf/126orig_4256.pdf (accessed July 2, 2019).
Tietenberg, T. 1994. *Environmental Economics and Policy*. HarperCollins College Div, New York.
United Nations (UN). 2003. *Water for People, Water for Life*. UN-Water, United Nations Educational, Scientific, and Cultural Organization (UNESCO), p. 576.
United States Department of Agriculture (USDA), Rural utilities service. 2005. *Water and Environment Program*. URL: http://www.usda.gov/rus/water/index.htm (accessed January 29, 2005).
United States Department of Agriculture (USDA), Economic research service. 2005. *Farm Policy*. URL: http://www.ers.usda.gov/Features/farmbill/ (accessed January 29, 2005).
United States Environmental Protection Agency (USEPA), Office of Water. 1990. *Progress in Ground Water Protection and Restoration*. February 1990 (EPA 440/6-90-001). Washington, DC.
United States Environmental Protection Agency (USEPA), Office of Water. 1991. *Ground Water Protection: Strategy for the 1990s*. Washington, DC.
United States Environmental Protection Agency (USEPA), Office of Water. 1996a. *Benefits and Costs of Prevention: Case Studies of Community Wellhead Protection*, vol. 1. March 1996 (EPA 813-B-95-005). Washington, DC.

United States Environmental Protection Agency (USEPA), Office of Water. 1996b. *Ground Water Disinfection and Protective Practices in the United States*. Washington, DC.

United States Environmental Protection Agency (USEPA). 2019. *National Pollutant Discharge Elimination System*; Stormwater Discharges from Municipal Sources. URL: https://www.epa.gov/npdes/stormwater-discharges-municipal-sources (accessed June 6, 2019).

Washington Department of Ecology (WDE). 2019. *Water quality permits*. URL: https://ecology.wa.gov/Water-Shorelines/Water-quality/Water-quality-permits (accessed May 26, 2019).

Water Systems Council (WSC). 2018. Watermark project proves cost effectiveness of wells and well systems. Well Connected Newsletter, September 2018. URL: https://www.watersystemscouncil.org/members/newsletter-archive/

Willard, J.M. and Carroll, D. 1992. *Phase I evaluation logic. Proceedings the Conference on Environmental Site Assessments: Case Studies and Strategies*, August 5–7, Orlando, FL, pp. 43–54.

5 Microeconomics and Groundwater Markets

EVALUATING ESSENTIAL SCARCE RESOURCES

Water is an essential input for human life, agriculture and manufacturing industries. In other words, life and production of such things as food and beverages, petroleum, lumber and wood products, paper, chemicals and electronic equipment (Hanemann, 2005, p. 20) would not be possible without water in certain quantities, qualities and times. Water is thus defined as an "essential" economic commodity. Therein lies the motivation to supply water to ensure production that meets the necessary requirements of life.

Consumer demand drives producers to supply goods. Requirement for food has impelled demand for water to irrigate croplands – in the United States, Mexico, South America and around the world. Demand for irrigation water stimulated water equipment producers to provide well casing and small pumps for farmers to increase food production in India, China and elsewhere. Other than information on depth of the falling groundwater tables in these areas, the impact of millions of wells on aquifer depletion was not widely recognized but could have been anticipated and projected. These circumstances in which groundwater can have substantial ecosystem, health and survival implications also affect economic relationships locally, nationally and globally.

As we have observed, groundwater supplies fundamental human, animal and plant needs to quench thirst for bodily functions. It contributes substantially to food production and foreign trade. Groundwater is also becoming scarce in many locations and so should receive attention in markets that rely on it as a critical factor of production. In the arid regions of the United States and other regions of the world with water scarcity, markets for groundwater have emerged to sell and buy groundwater as a commodity (Wheeler et al., 2016). The basic economic relationships begin at the individual level but build up to national and international considerations.

The field of microeconomics provides tools to evaluate how scarce resources might be allocated in an efficient way in the market among a range of uses, given certain assumptions and limitations. Economics tools allow the assessment of alternative policies to anticipate and project their effects. The scarcity of both water and the effective means to produce and safeguard it are growing (Young, 2005, p. 16). In this chapter, which takes a neoclassical microeconomics perspective, we first consider the "individual" demand in relation to the "product" a firm produces. Simple equations provide insight into these relationships. Second, we will contemplate basic supply and demand for groundwater. A particular focus will be on specific decisions that producers make and their "production function." The development of groundwater markets will then be addressed. Third, a look at some financial considerations that affect groundwater production will be covered. Finally, we will consider some connections to macrolevel relationships at the national and international scale.

ECONOMIC UTILITY, PRODUCT AND PRICE

The usefulness of a product or service, such as groundwater for water supply, to human beings is a driver of demand for that product or service. People have multiple needs which means that they must decide how much among a range of products and services will satisfy those needs. Economists think in terms of an individual's "utility" for a product or service and whether a person has a requirement to be satisfied by the features or qualities of the good that are useful to him or her. "Utility," when acted on, generates a demand by that person for a product or service offered by a firm. The basic economic relationship at the individual-firm level in the market for products, referred to as the "basic market equation," is (Samuelson, 1964, pp. 430–431, 518–525; Daly and Farley, 2004, pp. 126–129):

$$\mathrm{MU}_{wn} \,/\, \mathrm{MU}_{cn} = P_w \,/\, P_c = \mathrm{MPP}_{ac} \,/\, \mathrm{MPP}_{aw} \tag{1.5.1}$$

where

MU_{wn} is the marginal utility of good "w" (water, in this case) to person *n,* that is, the increment of satisfaction from having or using one additional unit of good "w" to person *n*; this added unit could be one more cup of water. Having a utility for more water is a fundamental basis for consumer demand for that additional water.

MU_{cn} is the marginal utility of good "c" (cup or container for water) to person *n*; for example, the added satisfaction of owning one more cup, and thus the demand for more cups.

P_w is the price of water in the market, which may range from \$0.0017 up to \$0.16 and more per liter (\$0.0064 to \$0.60 per gallon).

P_c is the price of a cup in the market, which could be under \$1 to several dollars and higher.

MPP_{aw} is the marginal physical product of using one more of input factor *a* to produce an extra quantity of good "w" (water). Production is the basis for supplying water in response to a demand for more water.

MPP_{ac} is the marginal physical product of applying one more input of *a* to manufacture or create more of good "c," more cups to hold water. Again, this reflects production to supply more cups because of a demand for them.

In a competitive market, such as an arid location with drought or scarce water conditions, so many people are consuming and producers are delivering the equivalent items in the market so that no one individual consumer or producer can influence the market price. Plans to buy or sell an item are made based on the prices of the items offered in the market. In a competitive market, the left-hand side of the above equation represents the consumer and his or her demand for water and cups given the price relationship holds. The right-hand side is the producers' supply capabilities for the water and cups demanded. The prices in the middle of the economic relationship should reflect the value to the consumer to meet his or her demand for the products and the value to the producer to provide the products, considering both production cost and profit.

On the basis of observation and experience, three principles affect the use of this equation (Samuelson, 1964, pp. 430–431, 518–525; Daly and Farley, 2004, p. 127):

1. Diminishing marginal utility – This principle indicates that after a person has consumed one of the items, such as a cup of water, the person has or receives less satisfaction from having another of the same item to consume again, at least at the time of consumption.
2. Diminishing marginal physical product (MPP) – As a producer adds one more input to the manufacturing process, other inputs unchanged, the additional output resulting from extra input declines. Assuming it takes a crew of two people with a drill rig to install a shallow well in 1 day (this depends, of course, on a range of conditions and may not be reality) and when an additional person is added to the crew, they may install one well and have one-fourth of another one started in 1 day. The MPP of the additional person is 0.25. The limits of the other input factors did not allow the additional person to perform to his/her potential.
3. Equimarginal principle of maximization – Consumers will cease changing their allotment of income for a range of goods, such as water and cups, when they have reached their highest degree of satisfaction or "maximized their utility" among the array of purchased goods. In that circumstance, the marginal utility of each dollar paid for every good is the same.

Applying these principles to the basic market economics equation above means:

- Prices are central in the equation to balance the consumers' satisfaction (their marginal utilities for products and services, e.g., a cup to hold water and water to satisfy thirst) and the means of allocating resources to produce the cups and water.
- In a competitive market, prices balance the left- and right-hand sides of equation 1.5.1.

The concept of "utility" assumes a certain level of information existing in the market about a product or service. To continue with the water and cups example, the consumer may need to know how much water is necessary to drink to maintain an adequate level of health. If the local water is untreated and has high levels of certain metals in solution with the water that could harm the consumer, finding an alternative source of water may be required. This latter circumstance could affect the marginal utility of untreated water and other waters, which, in turn, will affect the value of the different waters and the prices that the consumer is willing to pay for them, based on their tastes and preferences for waters of different qualities. Thus, information may affect the individual utility and therefore demand for particular products and services.

DEMAND

Utility Function

In practice, a person's utility and demand for groundwater may be a composite of factors that then responds to their tastes and preferences. Hanemann (2005, p. 23) notes that water is not a homogeneous commodity, but has a range of characteristics, including quantity, location, timing, quality and variability/uncertainty. To a groundwater user, water delivered in a different volume, place, time and with a different chemical or microbiological composition may have an entirely different usefulness than other groundwater. Untreated groundwater for irrigation on a farm has a different utility than treated groundwater delivered to an urban apartment. Because of treatment and piping, it will also have a different cost. Reliability of supply and consumption can vary with time of the year and affect supply cost and price to the consumer.

Relying on the Lancaster (1966)–Maler (1974) model of utility, Hanemann defines a utility function that takes into account the range of characteristics of water:

$$u = u\left(x1,x2,\ldots,xN,q1,q2,\ldots qN,z\right) \tag{1.5.2}$$

where

xi, separate (water) commodities, based on their characteristics

qik, the amount or level of the kth characteristic associated with one unit of consumption of commodity i

z, other consumption

In defining the utility function for groundwater in this way, one can consider pricing water relative to the user's purpose and consequent demand for volume, place, time, composition and reliability of supply. The range of characteristics of groundwater then results in the different combinations of them that become distinct commodities from the user's perspective. That is to say, a liter of cold spring water served at a dining table to a restaurant customer is distinct from a hectare-meter or three acre-feet of groundwater with nutrients and low concentrations of dissolved metals for irrigating a field of corn over a six-month growing season to a farmer.

A number of excellent texts address the theory of economic supply and demand and the role of consumer utility. These texts can be found searching online using the keyword "microeconomics."

Demand and Willingness to Pay

As consumers value different sets or "bundles" of goods and services, as suggested by the groundwater utility function just presented, they manifest their preferences for more or less of them by their "willingness to pay" for those goods and services. Willingness to pay is defined as "the maximum sum of money the individual would be willing to pay rather than do without an increase in some good" (Freeman, 1993, p. 8). A person can go to the market (store, online website, electronic bid response, etc.) where goods are exchanged for money and decide whether they are willing to pay the prices or not. If some items sell for a very low price relative to the consumer's value for that item, the consumer may actually buy more of them. On the other hand, if the price is too high in comparison with what the consumer believed was the value for the good, then the purchase may be less. In the market today, clearly many types of choices exist to purchase water and the choice does not only concern quantity (how much to buy) but also quality (e.g., untreated spring or ozonated water). A person's value for a good (such as a specific amount of water of a particular quality) is measured by the amount of money or other item(s) of worth that will be given for it (Field, 1994, p. 47). The sacrifice in this sense is the giving up of money that would enable a purchase of a good. A land purchaser may be willing to pay more for land that has groundwater close to the ground surface and therefore has lower groundwater access costs than a similar property with groundwater at much greater depth (and greater drilling and well installation cost). Figure 1.5.1 graphs an example of a hypothetical willingness to pay, in this case, for water.

A similar concept to willingness to pay is "willingness to accept." Willingness to accept is defined as the stated compensation level (price) that an individual must receive to accept a given or prescribed risk or the loss or diminution of an environmental service (NRC, 1997, p. 173; OECD, 2005). A person may be willing to accept payment for loss of a right to pump unrestricted amounts of groundwater as long as the person has access to water supply for essential needs.

In the hypothetical case of willingness to pay for water, Figure 1.5.1 allows us to consider both marginal and total amounts that the consumer might be willing to pay for more water. Focusing on the marginal willingness to pay, Graph 1 indicates that for the first unit of water (perhaps a liter or some other appropriate volume), the individual would be willing to pay $1.15, but after that the willingness drops off. For one more unit in addition to the first, the individual would pay $0.88, for the third, $0.70 and for the fourth, only $0.50 – this last value is less than half of the amount paid for the first unit. (Considering actual use, health advisories suggest that a person consume approximately

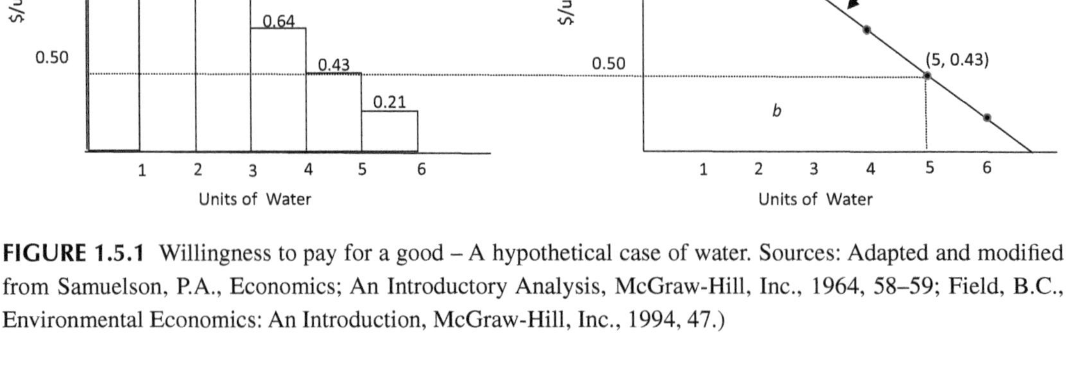

FIGURE 1.5.1 Willingness to pay for a good – A hypothetical case of water. Sources: Adapted and modified from Samuelson, P.A., Economics; An Introductory Analysis, McGraw-Hill, Inc., 1964, 58–59; Field, B.C., Environmental Economics: An Introduction, McGraw-Hill, Inc., 1994, 47.)

two liters of water per day – about half of total water consumption for an adult who also eats normal portions of other meals and snacks (receiving water from "solid" foods) and consuming other beverages (Tufts University, 2004).)

Total willingness to pay measures the complete amount a consumer would pay for a particular number of units of a good instead of fully foregoing them (Field, 1994, p. 47). In the hypothetical case of Figure 1.5.1, this person's total willingness to pay for five units of water is $4.29, which equals $1.29 for the first unit plus $1.07 for the second unit plus 0.86 for the third unit plus $0.64 for the fourth unit plus 0.43 for the fifth unit. This amount is equivalent to the area under the demand curve of Graph 2 up to five units, which is also the same as adding together areas *a* and *b* under the demand curve. In the market, the consumer pays $0.43 for each unit of water. In Graph 2, 5 units of water with a price of $0.43 per unit have a total exchange value equal to area b, $2.15.

Willingness to pay for a good or service may also be thought of as a person's demand for that good or service. Typically, in the market, many people have utility for and therefore demand for goods and services. However, their utilities and resulting demands may be different. This circumstance is reflected in Figure 1.5.2. Graphs 1 and 2 portray the marginal willingness-to-pay or demand curves for two families in a village that has a market. The quantities demanded for the same prices are not the same for each family. Their combined total demand, or "aggregate demand," is presented in Graph 3, indicating that at $4 per unit the demand in the market would be 31 units, but at $11 per unit, would be only 12 units.

In the field of groundwater, a demand for a number of goods, such as drinking water or irrigated crops, may come to mind, with more context provided in Volume 1, Chapters 7 through 9. Demand for groundwater to supply a basic need to quench thirst may be the most obvious. Agricultural irrigation is the single largest use of groundwater worldwide. Demand for groundwater to maintain streamflow and wetlands in dry seasons may be acute during a drought. Groundwater discharge to streams, wetlands and coastal areas may exceed that used in irrigation. Water for irrigation usually has a cost to produce even if the farmer owns the land – the farmer as water consumer has to install the well, pump and distribution system for the water, as well as use energy in some form to pump the water out of the ground and through the irrigation system. If the farmer had to pay the same for the groundwater as an urban water consumer, would the farmer still irrigate? Probably not, since the per unit cost for groundwater has to be a small part of the cost of producing the crops to be competitive

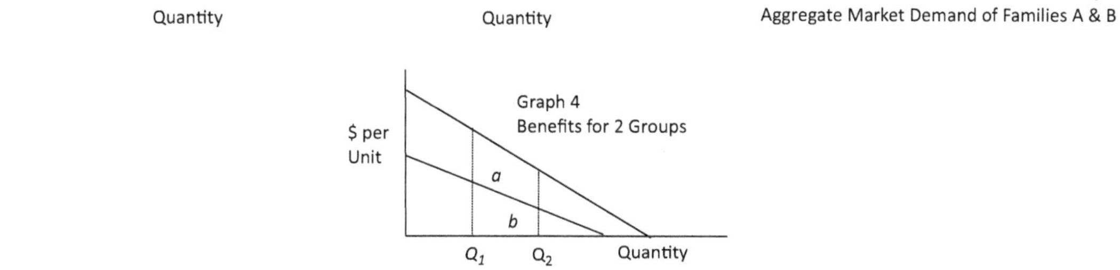

FIGURE 1.5.2 Demand and benefit curves. Sources: Adapted from Field, B.C., *Environmental Economics*, McGraw-Hill, Inc., New York, 1994, 50–51; mnmeconomics. Undated. The demand for public goods. URL: https://mnmeconomics.wordpress.com/2011/10/22/the-demand-for-public-goods/ (Accessed June 24, 2020).

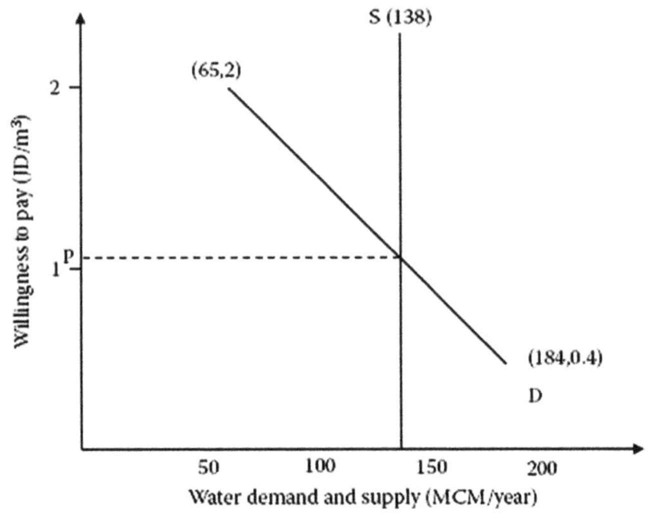

FIGURE 1.5.3 Water demand in Amman area of Jordan. Source: Adapted from Schiffler, M., *The Economics of Groundwater Management in Arid Countries: Theory, International Experience and a Case Study of Jordan*, Frank Cass Publishers, London, U.K., 1998, 271. With permission.

in the world market. Water treatment is typically part of the cost of urban water supply. If either a farmer, urban water supplier or other groundwater user pump a large volume of water that reduces the flow of an adjacent stream, the effect of the consumed stream water and impacted wildlife is also a water supply cost that these users may not have accounted for and passed on to downstream water users. These latter points get into the realm of water production being an intermediate good in the production of other goods. Water as an intermediate good may have a demand derived from a demand for other goods or services being produced or used. Before addressing that subject, we will briefly cover measurement of benefits from goods and services.

Figure 1.5.3 shows the essential information used to develop a demand curve for groundwater based on willingness to pay in the Amman, Jordan area (Schiffler, 1998, p. 271). It uses prices and estimated quantities of public and vendor-sold water in association with demographic and economic data to establish a demand curve. At Jordanian dinar (JD) 0.4/m^3, 184 million cubic meters/year are projected to be demanded based on the then current public water charge. At JD 2/m^3, only 65 million cubic meters would be expected to be demanded, derived from purchases from water vendors. An average price of slightly more than JD 1/m^3 with an inelastic supply curve [(S(138)] may be created.

BENEFITS

In environmental economics, "benefits are the gain associated with an environmental improvement" (Freeman, 1993, p. 8). From the perspective of the groundwater resource, an environmental improvement might be a good such as inducing more precipitation to infiltrate to recharge an aquifer. If people value having more groundwater, then we may be able to measure or estimate the volume and calculate its benefit to them. This result would be a gain and people would be better off. If an aquifer recharge zone is paved and built over, reducing the volume of water being able to reach the aquifer used by the population of a nearby town, then the benefit of the additional groundwater may be eliminated in whole or part with the townspeople being damaged by this result.

Benefits may be measured by various economic methods (Freeman, 1993). Figure 1.5.2 indicates conceptually how this might be done. Graph 4 has two different demand curves reflecting the willingness to pay for more quantities of some good, such as more groundwater storage in the ground; the community may have had shortages in the past. The initial quantity, Q_1, will be increased in aggregate amount to Q_2. The lower demand curve may represent apartment dwellers in the community whose demand for groundwater does not fluctuate much. If more water were available at lower price, they might wash their cars more often. The upper demand curve might be for suburbanites who have wanted to put in gardens but have had just enough groundwater to meet their typical home maintenance needs. They have a demand for and receive a specific benefit from increased groundwater availability. As long as the two groups are willing to pay for their greater water use, they benefit from more water being available. However, the benefits they derive are distinctly different. The benefit for the apartment dwellers from the increased amount is **b**, but that of the suburbanites is $b + a$. Clearly, this result mirrors the difference in values for and willingness to pay for increased water supply.

SUPPLY

Production Function

Supplying a good to the market may involve employing a range of people and resources locally and beyond the community, depending on the nature of the product or service. A basic production function is $Y = f(L, K)$, which says output (Y) is a function of labor (L) applied and capital (K) used. Furthermore, the quantity produced and supplied should be related to that demanded, as the basic market equation indicates. A supplier can build his capability to produce a good and once done will typically be able to provide many more units of it. Production of groundwater usually requires a number of input factors: installation of a well and pump, construction of a distribution line, energy source to operate the pump and staff for maintenance of the system. The well, pump and distribution system might be considered capital stock, but if financed over time may be a fixed cost each year; staff, parts for repair and energy use are variable costs related to the amount of water used for the time of day, week or season of the year. From the incidence of costs for producing and providing groundwater a cost curve can be developed.

A firm's basic decisions about supplying groundwater (or any other product or service) can be modeled in a set of simplifying equations. In practice, the equations may become more complex depending on the requirements of production, such as treatment or specific needs, and depend on assumptions about the ease of specifying relations among inputs to production. Before investing in the capacity to produce groundwater, a firm must have some sense of whether the product it wants to produce will have buyers who have demand for its output over time to warrant investment in equipment and materials for production. The firm would have done some market analysis. At this point, we will simply note that water is essential to all human beings and other living things, so there is a demand. We will also note that many firms produce groundwater, so the fundamental production relationships are known.

A basic production function for water is (adapted from Young, 2005, p. 54):

$$Y = f\left(X, G, K\right) \tag{1.5.3}$$

where

Y is the output of water and a function (*f*) of the input factors on the right-hand side of the equation
X is the variable inputs, including labor
G is the groundwater from the aquifer
K is the fixed factors of productionThe factors of production may be further specified, such as:
X, variable inputs = Labor (L) to operate and maintain a pumping well + Energy (E) to pump water from the aquifer to the ground surface = $X_L + X_E$
G, groundwater of quality 1 during normal pumping + groundwater of quality 2 during peak pumping, if greater pumping at peak demand times pulls groundwater from portions of the aquifer that have a different quality = $G_1 + G_2$

K, Capital, such as a well, pump, machinery and treatment equipment (W) + Other owned inputs (M) that could include land, management and entrepreneurship (Young, 2005, pp. 77–79) $= K_W + K_M$

Further specification is possible depending on the complexity of the situation and the need to use a more refined approach. Thus, a more fully specified equation for groundwater production might look like:

$$Y = f\left(X_L, X_E, G_1, G_2, K_W, K_M\right) \tag{1.5.4}$$

A numeric example is provided later in this section to show how to think about this economic relationship embedded in the production function.

Costs of Production

The costs of supplying groundwater to a community can be described on a "per unit" basis, such as per kiloliters or cubic meters (gallons or cubic feet). This "per unit" measurement allows both marginal and total costs (TCs) to be derived. Wells of different sizes (such as 5 cm [2 inches] diameter versus 25.4 cm [10 inches] diameter) and depths (30 m [100 ft] compared to 100 m [330 ft]) will have different costs associated with them to install the well and to pump the water (The size of the well should be based on the expected demand for water.). The production of groundwater from different wells will have different cost curves based on their respective production relationships, such as differences in depth, diameter, pump size and placement and other factors.

A set of hypothetical production cost curves is presented in Figure 1.5.4. Graph 1 shows the schedule of production costs: the first unit will cost about \$0.50 to produce, the second, \$0.35 and the seventh unit, \$1.05. Graph 2 smooths out the marginal costs (MC) into a curve. The total cost of production (TC) is the area under the curve. Graph 2, reflecting the production cost schedule, shows a production process that has falling marginal costs for the initial units of production but to produce additional units more inputs raise the cost. A rising MC curve may be typical of many goods. Such a condition may exist where low-volume demand can be satisfied by local resources, but increased output requires inputs from longer distances that cause costs to rise. In most situations, the cost to produce groundwater, typically a local resource, often has a declining cost curve as shown in Graph 3 because investment costs are spread over more and more units of water when more is demanded and supplied with low operation costs for increasing production. Notably, in these cases, no charge for the groundwater itself is made. However, where groundwater is being mined and depleted with a falling water table, the groundwater must be pumped from greater depths. In locations of groundwater depletion, production costs can increase as in Graph 2 because more energy (i.e., increased production input) is required to lift the water greater vertical distances to the ground surface. Also, where managed aquifer recharge is being conducted to restore aquifer quantity, additional supply costs will be incurred.

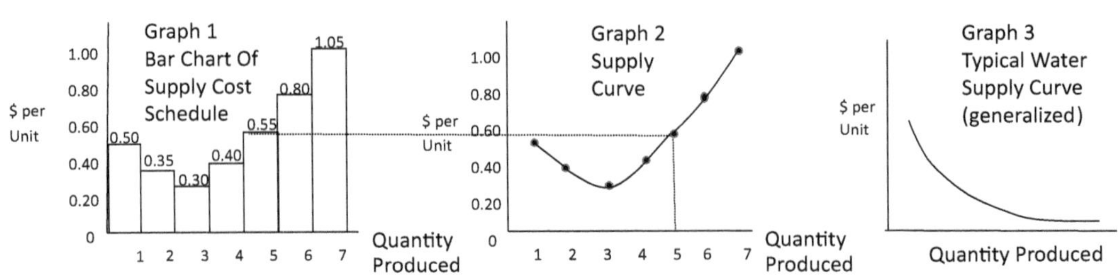

FIGURE 1.5.4 Supply curves. Sources: Adapted from Samuelson, P.A. *Economics; An Introductory Analysis.* McGraw-Hill, Inc., New York, 1964, 60–61; Field, B.C., *Environmental Economics; An Introduction*, McGraw-Hill, Inc., New York, 1994, 55.

Two types of MC curves apply to evaluating the provision of a good: short-run and long-run curves. Short-run MC curves usually fall and rise more steeply, reflecting the circumstance that, over short periods, the relation of inputs to each other are relatively fixed and unchangeable. To produce more water, a well manager may be able to increase the rate of pumping, in which case the pump will be more inefficient for the energy used, so costs will rise, but no more staff can be productively used in this case. In the long-run, another well can be installed with a more efficient pump due to technological advances, allowing the MC of producing more water to be lower for the same increased quantity. Long-run MC curves are typically more gently falling and rising than their associated short-run complements. This flatter long-run MC curve results from improvements in production not possible in the short-run.

AN EXAMPLE IN GRAPHICAL FORM

Figure 1.5.5 provides an example of a partial equilibrium analysis of supply and demand relationships for a marketed commodity. In this example, the cost of supply declined (it cost less to produce a unit of a hypothetical commodity in this case), shifting the supply curve from S_1 to S_2, with demand staying at the same level and with more units produced or demanded moving to the right on the quantity axis of the graph. The quantity available increased and the new price is lower in the market, reflecting the efficiency. The supply curves reflect the producers' costs per unit at each quantity made available. Notably, the line segment WZ indicates the decrease in supply cost, which is different than the change in price, $\text{Price}_1 - \text{Price}_2$. The demand curve, D, gives the consumers' willingness to pay for one more unit at each increment of quantity. The demand curve portrays the consumers' willingness to pay for each additional unit. The graph indicates that the consumer has a high benefit for the first units (the difference between the demand curve, D, for any quantity and the horizontal axis), but less so as more units are supplied. At the intersection of the supply and demand curves, the market is considered to be in equilibrium and sellers are just willing to accept the amount that consumers are just willing to pay. Since the consumers only pay the price in the market and not a different price at each quantity, the area above the price line is considered the "surplus" benefit, in economic terms. This surplus is area a for Price_1 and area a+b+c for Price_2. The difference in the area between the two prices is the change in benefit to consumers from the change in supply costs.

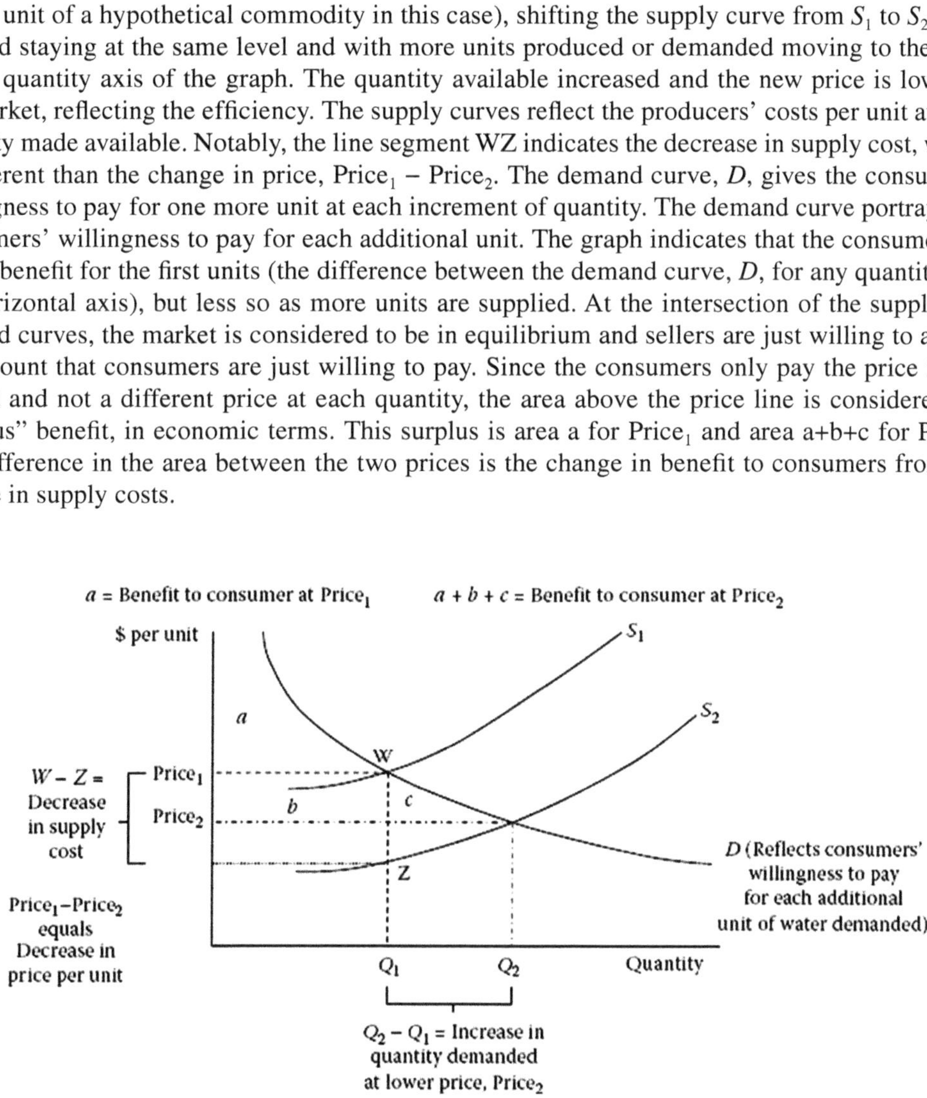

FIGURE 1.5.5 Partial equilibrium analysis of supply and demand relationships for a marketed commodity

When would we see a supply curve shift from S_1 to S_2? This could happen with a new more efficient pump being installed or after a significant extended rainfall results in a rise in the water table. If the water table rise continues, pumping costs would likely be reduced and may encourage more use.

MONOPOLIES AND SUPPLY

The examples and descriptions in the preceding sections, "Demand" and "Supply" assume a competitive market exists for factor inputs that provide the labor and capital to firms producing goods for sale. This is portrayed in Figure 1.5.6.

Graph 1, in which price equals MC, $P = \mathrm{MC}$. However, these conditions do not apply when there is a sole producer of a good, called a "monopolist." For a product, like water, a "natural" monopoly exists when both the average cost and marginal cost and their curves decline in the range of production of interest. In the relevant range of the market, the monopolist finds the MCs falling and to obtain the best profit seeks to maximize revenue while the marginal revenue (MR) is less than the competitive price (Samuelson, 1964, p. 491); that is, $\mathrm{MR_w} < P_\mathrm{w}$. Daly and Farley (2004, p. 134) indicate mathematically the circumstances that affect both consumer and monopolist. In this case, MR substitutes for the price of a good. The MR for water, $\mathrm{MR_w}$, will substitute for P_w in the right-hand side of the basic market equation:

$$\mathrm{MU}_{wn} / \mathrm{MU}_{cn} = P_\mathrm{w} / P_\mathrm{c} > \mathrm{MR_w} / P_\mathrm{c} = \mathrm{MPP}_{ac} / \mathrm{MPP}_{aw} \tag{1.5.5}$$

Rearranging the inequality:

$$\mathrm{MU}_{wn} / \mathrm{MU}_{cn} > \mathrm{MPP}_{ac} / \mathrm{MPP}_{aw} \tag{1.5.6}$$

Expressing the inequality in terms products:

$$\mathrm{MU}_{wn} \times \mathrm{MPP}_{aw} > \mathrm{MU}_{cn} \times \mathrm{MPP}_{ac} \tag{1.5.7}$$

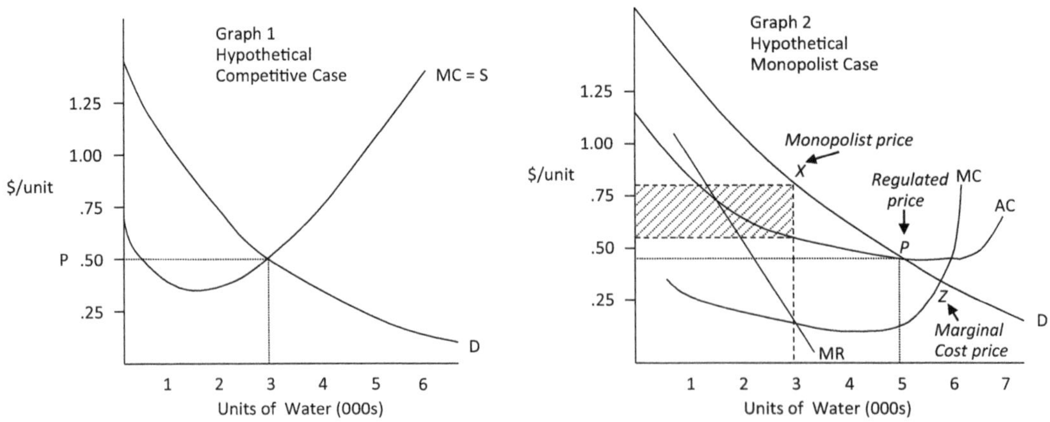

FIGURE 1.5.6 Competitive and monopolist pricing. Sources: Adapted and modified from Samuelson, P.A., *Economics: An Introductory Analysis,* McGraw-Hill Book Company, New York, 1964, 468, 491; Nicholson, W. *Intermediate Microeconomics and Its Application.* The Dryden Press, Hinsdale, Illinois, 1979, 322.

This last inequality indicates that consumer n has a higher utility for more input a to be applied to satisfy the demand for more water but the action of the monopolist is to constrain further application of input a to produce more water. To do this, the monopolist would not be maximizing profit, so the production is less than desired by the consumer. Initially, in this example, the monopolist chooses to produce 3 units, where MR = MC, but the average cost per unit is much higher at this point.

Figure 1.5.6 shows the result of the monopolist graphically (adapted from Samuelson, 1964, p. 491). Graph 2 shows the MR below the MC price at Z *where MC intersects the demand curve D, a typical competitive result, but not for monopolist, though: average costs are still not covered.* MR intersecting MC gives the amount of production in this initial case, corresponding to point X on the demand curve and a price of $0.80. The monopolist price at X is the average revenue (AR) per unit of production. Subtraction of AC from AR at X gives the profit for the monopolist, equal to the shaded (cross-hatched) area. As the monopolist tries to maximize profit, while covering average production cost per unit, the monopolist is now regulated because his/her price is too high for the few units produced. The demand curve D represents the average revenue per unit sold. The regulatory authority moves to a price along the demand curve D relative to the average cost (AC) intersection with the demand curve. This intersection might be the regulated price P (the regulatory authority will attempt to bring the price down to as close to the MC price Z as possible and eliminate as much of the excess profit as it can). If unregulated, the monopolist could capture more excess profit by charging a declining block rate for the earlier units, starting up higher on the demand curve to define the block charges. Many water utilities are monopolies and use a declining block rate structure, which encourages more water use in the large low-rate blocks.

Natural monopolies characterize most water utilities supplying communities with water. Their water pricing is not based on the efficient outcome of having the marginal cost of supply equal the consumer demand price (P = MC), but P = AC.

PRICE IN A COMPETITIVE MARKET

Price in a competitive market is obtained where the MC – or supply – curve intersects the demand curve. While water – specifically groundwater – can be offered in a competitive market (one may consider bottled groundwater and tradable water rights as examples), some of the intermediate components of water supply may also be offered competitively to producers. For example, needs for well installation may be offered for bid to a number of well drillers. For well drillers competing to install wells, their demand for well casing, grout, concrete or pumps may be such that a number of suppliers offer such products competitively. In Figure 1.5.6, Graph 1 shows a hypothetical case of competitive market pricing. In this case, the MC curve intersects the demand curve and establishes P = $0.50 as the price for three units of production. Thus, P = MC, price equals MC is the "market clearing" position for this exchange. Competitive markets exist for groundwater discussed in the section on "Groundwater Trading and Markets," later in this chapter. First, we will discuss other factors that may affect a competitive market.

Economic Efficiency

From a neoclassical microeconomic standpoint, economic efficiency should be the primary goal of a society (Young, 2005, p. 21), but not the only goal. Another goal may be the distribution of resources in the economy among individuals or groups of people, also referred to as equity. Economic efficiency is the unique allocation of resources such that no other allocation creates the circumstance that someone has a further improved condition at the same time no one else is in a worse state. Otherwise, a "welfare waste" would exist in the original allocation (Bannock et al., 1978, p. 144). This definition is called "Pareto optimality" and is further discussed in Box 1.5.1. Economists can use tools, such as incentives, taxes and subsidies, to enhance the allocation of resources. They compare the "world without" their proposed tool to the projected "world with" the tool in use.

BOX 1.5.1 PARETO EFFICIENCY CONDITIONS

The concept of economic efficiency in neoclassical economics was developed by Wilfredo Pareto (1848–1923), an Italian economist and political sociologist. Pareto's influential writing, *Manuale d'Economia Politica* (1906), provided the foundation of modern welfare economics. In it, he developed what is referred to as the concept of "Pareto optimum": "the optimum allocation of the resources of a society is not attained so long as it is possible to make at least one individual better off in his own estimation while keeping others as well off as before in their own estimation" (Weiner, 2006). This condition only applies in a "perfectly functioning competitive economy." It is "achieved when the marginal benefits of using a good or service are equal to the MC of supplying the good" (Young, 2005, pp. 17–18).
Several tenets must operate for Pareto optimality to occur (Young, 2005, 28):

1. "Individual preferences count; the economic welfare of society is based on the economic welfare in aggregate of its individual citizens"
2. "The individual is the best judge of his/her own well being"
3. "A change which makes everybody better off with no one becoming worse off constitutes a positive change in total welfare"

Since most changes in policies or decisions result in benefits to one person or group and losses to another, economists have developed a way around the third tenet: referred to as "potential Pareto improvement," "if gainers could in principle compensate losers and still be better off, the change is deemed acceptable, whether or not the compensation actually takes place" (Young, 2005, 28). This latter explanation is central to the basis for calculating benefits of actions or projects affecting water or other resources.

Sources: Weiner, S.D., *Wilfredo Pareto*, 2006, URL: http://sdweiner.home.texas.net/d/archive/theory-method/WILFREDO_PARETO.htm (accessed September 2, 2006).Young, R.A., Determining the Economic Value of Water: Concepts and Methods, Resources for the Future Press, Washington, DC, 2005, 357.

Assumptions Used to Derive "Competitive Markets"

Neoclassical economics has established a set of assumptions used in defining competitive markets, which do not apply well to groundwater. The assumptions include (Barlowe, 1972, p. 121):

1. Human beings are rational beings who behave logically and reasonably
2. Human beings act to maximize their self-interests
3. Prices allocate resources
   Other assumptions applied to balance markets (supply and demand) are (Barlowe, 1972, p. 121; Schiffler, 1998, pp. 25–39; Daly and Farley, 2004, p. 125):
4. Perfect competition
5. Perfect knowledge among buyers and sellers
6. Perfect mobility of goods and productive factors
7. Perfectly elastic supply of productive factors
8. Spontaneous order and efficiency
9. Substitutability of inputs

Texts pm microeconomics review these assumptions in detail which will not be repeated here. The assumptions as a set are affected by improved and correct information being accessible and used, understanding of the relation of different information sets and advances in technology. Use of computers to transmit information and analyze data has changed perspectives on applying some of these assumptions. From a practical "on-the-ground" standpoint, people want a reliable supply of water to meet their basic needs. If water is scarce, they may be willing to pay more for it. Water availability may be constrained by both climate and seasonal weather as well as legal factors affecting rights to water. Information about water is widely available and accessible by computer. Legal systems, rate of decision-making and groundwater flow are variable in speed and timing and may not be as efficient as we would prefer. Groundwater quality affects the uses to which it is put. Consideration of substitutability of other inputs for water may have significant, and perhaps practical, implications, such as potential health impacts of substitutes, cost and extent of availability.

SUBSTITUTABILITY OF INPUTS

Neoclassical economics assumes that capital and labor are substitutable in the production function. "Two goods are substitutes if a rise in price of one causes an increase in demand for the other because the goods perform a similar function or serve a similar taste" (Bannock, 2017, p. 428). In reality, the extent to which this assumption applies depends on the application. Some economists, however, have indicated that water may be nonsubstitutable in most uses, but this depends on context and scale (Daly and Farley, 2004, pp. 91–92). Schiffler (1998, pp. 24–25) suggests, while water may be irreplaceable in some uses, it is human use that holds the prospect for at least partial substitution and that a range of substitution is possible for water along a continuum of six levels:

1. Water users can change their behavior while relying on the same inputs.
2. Water users can adopt existing water-conserving practices (this corresponds to the classical approach of a production function).
3. Demand for water-using products can shift to less water-intensive products.
4. Water users can relocate to places in which climate contributes to less water being needed.
5. Production of goods requiring intensive water use can be shifted to regions where climate allows less water use or water is less scarce.
6. New technologies are developed to use less water in production of goods, causing the production function to shift.

While these six levels can be observed, a basic relationship is not addressed by them: hydrologic cycling and natural services that provide public goods. At the most basic level, implied in the six levels of water substitution is that water is still needed and we cannot substitute it to a zero-use condition. Furthermore, groundwater cycling of nutrients and temperature flux is required for many aquatic micro and macroorganisms to exist; their habitat would disappear if groundwater did not perform the functions it does. In a more observable circumstance, groundwater provides geologic structure and support to maintain ground surface levels rather than contributing to land subsidence and hydrogeologic pressure to preclude salt-water intrusion where groundwater levels are maintained. It would be difficult to imagine a substance other than water performing these functions.

Substitution focuses attention on defined human economic processes that create demand for water. We can partially substitute capital or labor for water in some of these processes. Water scarcity may be a significant driver for this substitution.

Decision Makers in Groundwater Production and Supply

Most economic examination focuses on the decision maker's optimization of providing or employing factor inputs to produce outputs, or the market's determination of equilibrium. The market equilibrium is defined by supply–demand analysis (Hirshleifer, 1976, p. 22). The decision maker optimizes inputs and outputs through the evaluation of total, average and marginal relationships. Actually, many people may participate in the decision to produce groundwater. These participants may be the landowner, the groundwater abstractor and seller or governmental representative trying to correct any economic issues that may arise. The natural conditions – depth to groundwater, type of geology, access to the well site, price of energy – are significant in affecting the cost of producing groundwater. A partial list of these decision makers include:

1. A land or home owner – A land or home owner deciding to develop a property and needing water must decide on the amount of water he needs over different intervals of time – e.g., per day or per year. Important factors with respect to quantity include, well depth, screened interval, pump size, diameter of casing and the number of wells.

   If the water table elevation fluctuates significantly, the well may need to be deeper and the screened interval longer to ensure access to groundwater over the entire year and to reduce the need to have additional and deeper wells drilled. Pump size and casing diameter are directly related to the volume required to be pumped at any particular instant. All of these factors affect the cost to develop and supply water. The size of the pump and the depth from which it must push or pull the water up the casing will influence operating costs for electricity.

   If the water is "hard" (higher mineral content) or has chemicals that need to be removed for human consumption or other use, groundwater treatment costs must also be included in production decisions. If the location of production is a distance from the location of use, a transmission line will be needed, usually installed below ground to be out of the way of activities on the ground surface and to avoid accidental breakage or tampering. Long distances can add large fixed costs for transmission pipes. Depending on maintenance, these costs may be considerable. Transmission losses through cracks in the line can add costs for additional pumping.
2. Well driller/installer – A well driller/installer is typically familiar with the soils and nearsurface geology and the labor and materials costs of drilling and installing wells in a locality. Geological conditions under which groundwater is found and produced vary widely, even for locations just a few tens of meters apart in many situations. The well driller/installer must decide how much to pay his drilling assistants to obtain the quality of work that will guarantee proper well construction. He must also stay abreast of materials costs for well casing, screen, grout and well protection. Equipment maintenance costs are also a significant factor. The well driller generally operates in a competitive environment and his rates (prices may be quoted on the basis of per unit of well depth and diameter, and type of casing and screen) must be adequate to cover costs but in line with other well drillers and installers. In less developed countries, millions of wells called tube wells have been installed by hand.
3. State or local technical/planning/zoning official, geologist, engineer or water master – A local planning/zoning official or state geologist can inform a land/homeowner whether his property is an area in which residential/commercial-industrial wells are allowed. Important considerations might include: location with respect to septic or waste disposal systems, limits on production over given timeframes or on pump size/capacity, depth to the top of the well screen, length of screened interval and well capping and protection requirements. A State or local official may be required to issue a permit to the landowner or driller indicating proper well construction before it can be installed or used and limits on the amount of water that can be pumped and withdrawn. The State or municipality may also require a yield or aquifer performance test. In some countries, a water master may define the volume of water that a groundwater user may produce.

   If water is to be injected in to the subsurface for aquifer storage and recovery, a state or federal underground injection control (UIC) permit will be required. Heat pump and geothermal wells may require permits, depending on the state and how groundwater is affected.
4. Water user association – Often defined in law, water user associations with responsibilities for coordinating groundwater use and management over large areas may have technical expertise that specifies the volume of water that can be pumped from an aquifer.

5. State/local public health official – A public health official may work in collaboration with a planning/zoning official to identify well location and groundwater treatment needs based on soil type and local water quality. A public health official may require that the groundwater be chemically and biologically tested before use.

   Relative to use of groundwater, state officials may determine that groundwater must be treated before human use because of naturally occurring or human-caused contamination of an aquifer or a portion of an aquifer. Local officials who are trying to promote greater groundwater availability and quality may encourage zoning to allow for stormwater retention and recharge ponds to provide for infiltration of precipitation into the ground. Local or state officials may be responsible for determining the number, density and depth of wells in aquifers to ensure a long-term supply and that adjacent users do not interfere with each other's use. They may also have responsibility through zoning and groundwater and land management regulations to prohibit or restrict certain land and subsurface uses associated with contaminants of concern that may compromise groundwater quality and use.
6. Fire chief – Fire protection is heavily dependent on a reliable supply of water that can be drawn immediately and in large volume rapidly. Fire protection needs must be balanced with regular needs of other domestic and commercial users.
7. A water utility company official – The local water utility, which may be a public or private company, and usually a monopoly organized under state law, may act in many ways like an individual landowner. The company may have or need to obtain the rights to the groundwater. The utility company decision maker, in addition to getting local and state permits, must meet federal and state drinking water quality requirements. Those standards have substantial water testing requirements and associated costs. Utility company managers can also promote conservation and water source protection within their service area to preserve the supply of groundwater available in the present and future. Importantly, water utility company officials, usually through a board or committee, set water rates or prices for the community their company serves.
8. Water resource district director/engineer – In states with arid climates and scarce water resources, an official of a water district organized under state law, may allocate water for irrigation by farmers with rights to water use. Groundwater production may also be associated with nutrient and chemical addition requirements established by the water district.
9. Groundwater testing/remediation service representatives – In locations affected or potentially affected by contamination, groundwater may not meet water quality standards for the intended use. If groundwater testing/remediation companies are hired to evaluate the extent of groundwater contamination, hydrogeologists and groundwater engineers will most likely recommend installing monitoring wells, sampling and chemical or microbiological testing of the groundwater, using reliable techniques to evaluate the presence of contamination. A monitoring well is used to produce water for chemical or microbiological analysis. A piezometer is used to measure water levels. Usually, three or more wells are required for proper sampling and establishing groundwater flow direction. Cleanup of contamination may require the installation of additional wells to produce groundwater in sufficient quantities to provide timely remediation. Large, heavily contaminated sites may have hundreds of monitoring or remedial production wells. Use of wells to inject chemical or biological treatment additions will require a state or federal UIC permit to address protection of underground sources of drinking water.
10. Finance company representative/property buyers – Before a property will be accepted for financing, many lending institutions will require buyers or sellers to have an environmental assessment of the property. This assessment entails determining whether any activity previously undertaken on the property may have contaminated the soil, groundwater or structures running through the property. If sufficient information exists to indicate that potentially contaminating activities occurred, then samples are taken and analyzed. Typically, groundwater

monitoring wells are installed and samples are analyzed. Soil samples are typically taken while installing the well. The assessment is used to describe the extent of contamination and, if appropriate, the magnitude of the activities needed to remedy it.
11. State utility rate commission – Members of state utility commissions are usually appointed by the governor of the state to set regulated prices on commodities produced and sold by utilities. These utilities are monopolies that are considered as necessarily providing a good or service to the public, which would be provided at greater social cost in a non-competitive unregulated market and not having competition to establish a market price for their good or service. Typically, groundwater is not priced at its MC. Thus, price to the consumer is based on demand and volume to maximize total revenue (TR) to the company (usually a monopoly) and not on the cost of the next available alternative, or on effects of other environmental resources that should be conserved in the long run.

Each of the decision makers, and perhaps others, in some way may be involved in establishing total, average and MCs for groundwater production and treatment. The first part of this group of decision makers was influencing the availability and production for consumption. The latter portion of the list of decision makers is involved in producing groundwater to test its quality for future use. In some situations, all of these individuals may be involved in deciding the level of resources (factors) to be applied to groundwater production and/or remediation. The rate commission seeks to minimize monopoly profits by the utility company and guide pricing to be more in line with MCs to get the outcome (price charged by a regulated utility) to be as close to a competitive price as possible to enable an adequate supply of groundwater to be produced for consumers while covering utility costs. The price set still would not take into account ecosystem value of groundwater or ecological effects of producing groundwater.

Decision Makers Affecting Demand for and Use of Groundwater

Demand for groundwater is affected by a range of considerations, among which are (modified from Foster, 1999):

- Large quantities of high-quality water are required to meet domestic and commercial needs in many cities and towns.
- Seasonally, groundwater may be more dependable in dry weather and during droughts because of the large storage potential of the subsurface environment.
- Different consumers may have different requirements for quality of water and may even have varying quality requirements for different uses. Groundwater quality is usually very steady and dependable, if found unpolluted.
- Groundwater is often viewed as a "safer" source of water because the subsurface and soil zone acts as a screen and treatment for some contaminants.
- In many locations, groundwater may be less expensive to develop, and if unpolluted, it requires little treatment.
- A major advantage is that groundwater can often be obtained where it is needed, rather than running a long pipeline to a surface water source with limited access.
- Catastrophic events may have less effect on groundwater because of its location.

Potential users of groundwater may not understand the local and even worldwide implications of their decisions to use it as their water source. Demand for groundwater stimulates demand for drilling labor and drill rigs to provide access. Companies produce pipe to transport groundwater from well to points of use. Demand for these supply factors provides upward pressure on costs. Managers of companies supplying these services and goods must make decisions on capital investment and labor rates to be paid. The key decision makers affecting the demand for groundwater may include:

1. Household consumers – Household consumers drink water and flush wastes away daily, prepare food, launder clothing, water lawns, wash cars, fill swimming pools and in hot, dry locations, such as the southwestern United States, Mexico or the Middle East, evaporate water to cool their homes and buildings.
2. Water-using appliance makers – Manufacturers of any appliance that washes anything or requires water in its process influence individual consumer's use of water. The aggregate volume of water use required by home appliances affects the volume of the water production, treatment and distribution infrastructure of a community.
3. Irrigators – Farming that uses water to irrigate the land for food production is the largest water-using activity in the world – 70% of fresh groundwater use in the United States and 74% worldwide. The extent that water-conserving irrigation methods are employed affects the amount of water used in irrigation to grow food that would not otherwise be produced.
4. Food processing – Companies that process foods and make beverages use water in cleaning as well as putting it into their consumable products.
5. Business – Cleaning activities in commercial establishments require water regularly.
6. Manufacturing and power production – These activities require water to clean and to cool processes, the latter involving evaporation of water not available for recycling. Many industrial processes have some portion that is affected by water use. Recycling and reuse of water reduces demand on water resources.
7. Any activity that results in contaminants of concern being applied and allowed to leach, buried or injected into the subsurface may reduce the quality of groundwater to a point that the resource is no longer usable in a particular location. Finding an alternate source of groundwater may require new wells to be installed at greater distance or depth to ensure an adequate water source, increasing the demand for groundwater in those other locations.

These and many other water users create the demand for water that we need every day for personal and business purposes.

An Example

Having examined some of the supporting concepts and activities affecting groundwater supply and demand, focusing on an example of a simple production process for pipe used as well casing will provide specifics that allow a better understanding of the application of economics to business and management decisions. The hypothetical example developed is for a single pipe manufacturer in a competitive market of many pipe producers. The units of production are not important. The real emphasis of the example is on the concepts of marginal analysis. The discipline of economics is largely a study of the process and results of valued exchanges at the margin of human activities: what is the cost, who pays the cost, who benefits and by how much. "At the margin" means "adding one more" of whatever is being examined, producing one more meter of pipe, for example, and evaluating whether the value or the cost of it is greater or less than that of the previous unit. Producing one more unit of pipe may have other economic and ecosystem consequences from metal mining, energy use and air emissions. The analysis may be extended to evaluate who benefits and who pays the cost of the additional production (or if applicable and appropriate, additional pollution), what contributed to any change value and whether the value of alternatives are greater or less than the unit being evaluated and who is affected by changes in the benefits and costs and how. The information derived from this analysis may be used in decisions about production processes or local decisions about type of pipe material and well depth. The economist provides the information to the decision maker.

Figure 1.5.7 (a) through (c) provides the data and graphs used in this example of a factor input approach to analyzing production. The fixed factor input in this case might be a parcel of land with a pipe production building and a metal storage area. The variable factor inputs are labor, metal and electricity, as indicated in the discussion of a production function previously. One could assume for this example that the application of a given amount of electricity to the pipe production process would also require the employment of labor. In this example, each unit increase in metal and electricity requires a similar proportional increase in labor.

Key concepts. A major focus of economic analysis is on the point of diminishing returns and the circumstances affecting and the results of actions taken at the "margin." Taking a factor input-unit approach, the following analytical concepts are important to the operator's production management and will be used in the discussion:

- Fixed inputs are the factors used in the production process that are typically permanent and not easily portable, such as land, structures and production equipment. This may also be referred to as capital.
- Variable inputs are factors that may change or fluctuate depending on production levels, such as labor, raw materials that are throughput to the production process and energy. These costs may also be similar to operation and maintenance expenditures.

(a) Water Pipe Production

Fixed Input Factors: Land and Building	Variable Input Factors: Labor, Metal and Electricity	Units of Total Production (TPP)	Average Production Per Variable Input Factor (APP)	Additional or Marginal Production (MPP)
1	1	4	4	4
1	2	12	6	8
1	3	26	8.67	14
1	4	42	10.5	16
1	5	52	10.4	10
1	6	58	9.67	6
1	7	62	8.86	4
1	8	63	7.86	1
1	9	63	7	0

(b) Net Return

Total Production[a]	Fixed Cost	Variable Cost	Total Cost	Marginal Cost	Average Cost	Price of Output	Value of Output	Marginal Revenue	Average Revenue	Net Return
	10	0	10	—		—	—	—	—	−10
4	10	10	20	2.50	5.00	2	8	2	2	−12
12	10	20	30	1.25	2.50	2	24	2	2	−4
26	10	30	40	0.71	1.54	2	52	2	2	17
42	10	40	50	0.63	1.19	2	84	2	2	43
52	10	50	60	1.00	1.15	2	104	2	2	66
58	10	60	70	1.67	1.21	2	116	2	2	67
62	10	70	80	2.5	1.29	2	124	2	2	56
63	10	80	90	10.00	1.43	2	126	2	2	33
63	10	90	100		1.59	2	126	2	2	−2

[a]Units of total production (TPP)

FIGURE 1.5.7 Example of water pipe production and net return. (a) Pipe Production, (b) Net Return, and (c) Graphs.

(c) Graphs

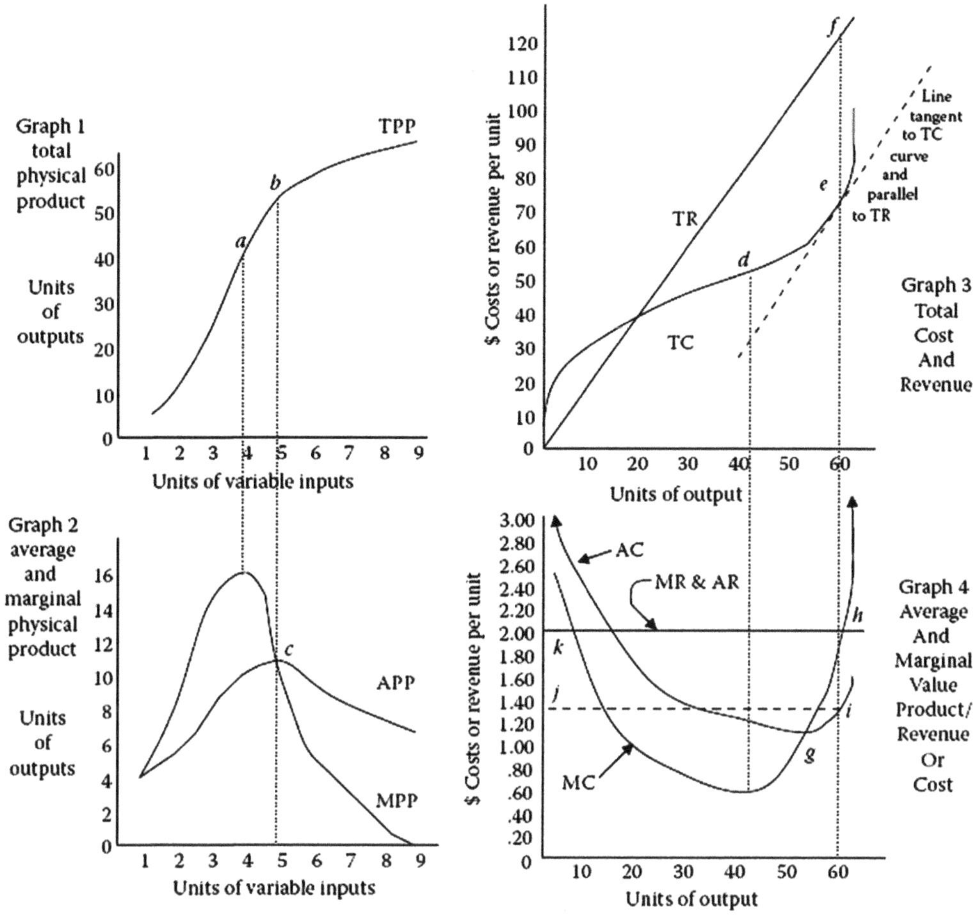

FIGURE 1.5.7 (Continued)

- Total physical product (TPP) is the total output of the production process. For water pipe, this quantity may be the total meters or feet of pipe over a specified period of time. An example is given in Figure 1.5.7a Water Pipe Production and in Graph 1.
- Average physical product (APP) is the output produced per variable input factor and is calculated by dividing TPP by variable input.
- Marginal Physical Product (MPP) is the additional production from the application of one more unit of variable inputs.
- Total Cost (TC) is the producer's or operator's cost of all input factors to provide a product or service. TC includes the fixed cost of land and structures, and the variable cost of labor and operation, including such factors as energy. TC is portrayed in Figure 1.5.7a in Graph 3.
- Average Cost (AC) is the total input cost divided by the total output. In Figure 1.5.7a, to produce 26 units of physical product requires three variable input factors with a variable cost of $6. The average variable cost (AVC) is $6 divided by three variable input factors, equaling $2 per variable input factor.
- Marginal Cost (MC) is the cost of the additional input factors to produce the level of output of interest. In the example, the marginal cost of producing more units decreases then increases.

- Total Revenue (TR) or Total value product (TVP) is the total revenue received from compensation for selling a product or service in the market or the value of production. The revenue or value is constant in the example and the TR line represents this result in Graph 3.
- Average Revenue (AR) or Average value product (AVP) is the revenue or value of a unit produced and sold. AR is calculated by dividing the TR by the total output produced and expected to be sold. In the example, the AR curve is shown in Graph 4 and is a constant amount of $2 per unit.
- Marginal Revenue (MR) or Marginal value product (MVP) is the revenue or value per unit of production for the last (or additional) output unit(s) produced. In the example, the MR curve is the same as the AR curve, since all units of output are expected to have the same revenue from sales.

Discussion. While the example here is obviously simplified, it has important features that are also common to more complex situations. As each combined unit of labor, metal and electricity is applied in Graph 1, the units of total production (TPP) increase up to a point, where, in the example, the increase is zero. The TPP curve shows how physical output increases with successive additions of variable inputs. It is also called the "production function." Point *a* on the TPP curve is the point where production changes from increasing at an increasing rate to a decreasing rate. Point *a* can also be identified by projecting a line from the highest point of the MPP curve to the TPP curve, which intersects at point *a*. The APP curve gives the relationship of the TPP divided by the variable input factor (labor and electricity), as the variable input factors increase. The APP curve is always highest where the MPP curve intersects it. To the right of this intersection, the APP curve gradually declines. Projecting a line through that point to the TPP curve, point **b**, shows where production increases at a decreasing rate. In the water pipe example, more labor, metal and electricity will not produce more pipe beyond the 63 units. In fact, more power may cause equipment to malfunction, with a loss of production.

Thus, in Graph 2, to the right of the intersection of APP and MPP, average production and marginal production both decrease and TPP in Graph 1 obviously increases at a decreasing rate. This example describes the physical law of diminishing returns.

The economic law of diminishing returns expresses this same relationship from a monetized perspective. Maximization of economic net returns is of interest in focusing on economically efficient decisions, rather than maximization of physical production. To view this maximization of economic net returns, one must place a cost on input factors and value to units produced. The point at which production managers will desire to operate is where, from an input standpoint, the MR equals or is just beyond the MC. "As long as an operator combines his variable input factors around his scarce or limiting factor, he can always expect his highest net return at this point." (Barlowe, 1972, p. 128). While the example is simple and the graph imprecise, inspecting the numbers in Figure 1.5.7b indicates that production just slightly more than 58 units with MC = 1.67 and MR = 2 generates the highest net return. However, beyond 58 units, net return diminishes.

The difference between total production costs and TR is the operator's "net return." Depending on which factors are treated as "fixed" and as "variable," this net return may also be referred to as "economic surplus," "rent," or "profit." Net return for this example can be shown graphically: it is the rectangle *hijk* in Graph 4. Notice the point *d* in Graph 3. This is the point at which the TC curve changes from costs increasing at a decreasing rate to costs increasing at an increasing rate. This point is also identified as the dotted line from the low point on the MC curve in Graph 4 projected straight up to the TC curve at *d* (called the "inflection point" of the curve because the rate of change is different above and below that point). In Graph 4, the solid line indicated as "MR & AR" gives the marginal and average revenue, $2 per unit. The dashed line *ij* shows the intersection at *i* of the average cost curve AC with the dotted line coming down from Graph 3. This dotted line intersects the TC curve at *e* and the TR curve at *f* in Graph 3. The dotted line crosses the TC curve at the point of tangency of a (dashed) line parallel with the TR curve. Mathematically (through the application of calculus), it can be shown that the length of the line segment *ef* is the greatest distance between the TR and TC curves, and therefore net revenue (TR minus TC) is maximized at this level of the firm's operation.

A principal observation about production processes is that applying more labor or resources to a fixed factor typically will cause production to increase only to a point where application of additional labor or other resources will not increase the additional physical product. The additional production in this case is called the marginal production. If the law of diminishing returns did not apply to the production of water pipe, a firm could supply all the pipe needed from one production plant, as long as an unlimited supply of labor and electricity were available. The table in Figure 1.5.7a illustrates the law of diminishing returns as it might apply to water pipe production. In the example, applying more variable input factors does not lead to an increase in Total Production in Figure 1.5.7b.

Producer/Operator Decisions. The operator of the water pipe production process would seek to employ variable inputs and produce output only up to the point where the net return was greatest. Graphically, he would choose the last whole unit of variable input before MC equaled MR that is the level of variable inputs at which the MR was still in excess of MC, providing a net return of approximately $67.00. We can see that this level is about seven variable inputs, already in the zone of diminishing returns, which begin to decrease at around five variable inputs. At approximately seven variable inputs, the operator is producing 58 units of output. The MC of one more unit is $1.67, which is close to but not in excess of the market price of $2.00.

Since the operator desires to maximize his net return, he would produce water pipe up to the point where the greatest vertical difference can be measured (or calculated) between the TR and TC curves. This point can be established as indicated above by two graphical methods of (1) drawing a line parallel to the TR curve that is just tangent to the TC curve in the right-hand side of the TR–TC graph or (2) by projecting a line from the intersection of MR–AR curve (the Price line) with the MC curve – this conforms with the Price = MC rule of economics – to the horizontal axis in Graph 4, at approximately 58 units of output. Price is the AR received in this example. The TR received is calculated by multiplying the units of output (about 60, in this case) by the price per unit expected to be sold of $2 each. The rectangle *hijk* is the amount of net return the operator receives, equaling AR minus AC, then multiplied by the production units. (Note: Do not rely on the accuracy of the graph in measuring the length and locations of the lines; these are approximate values.)

The operator would not produce beyond the point *h* on the MR and AR and MC curves because any production choice on the rising MC curve past this point means that MC of production is greater than MR.

OTHER FACTORS IN MARGINAL COSTS OF GROUNDWATER PRODUCTION

Are there production aspects that make groundwater supply costs different from those costs for surface water? In groundwater production, prices of the water sold or exchanged do not have several components of cost associated with supplying surface water in a competitive exchange (Zilberman and Schoengold, 2013). Since groundwater already exists in aquifers that typically extend under significant areas, is transmitted in the geologic matrix and flows to points of discharge, extraction and conveyance costs are not expenses to the seller or provider of groundwater in an exchange. However, environmental and third-party impact costs from pumping may be incorporated in the water price that the purchaser may pay in a water transaction.

For example, many farms in an area can be connected to the same aquifer by wells, with the aquifer serving as a conduit between pumping zones. Except for very site-specific factors, such as varying depths to the aquifer, the marginal cost to pump groundwater by any of the farms would be roughly the same. If any farmer wanted to sell some of the farm's water not needed for that growing season or year, the selling price of groundwater should be equal to the local marginal cost to produce groundwater at any of the farm wells, similar among all the farms, in this case.

GROUNDWATER MARKETS

Markets for Groundwater Supply

In areas of recurring or chronic water shortage, groundwater markets have emerged to allocate water to high value uses, principally in the western United States and in other arid regions of the world, treating it as a commodity. Groundwater markets are often thought of as neutral but efficient in approach to respond to water demand (Wheeler et al., 2016). Groundwater markets exist through a range of mechanisms including short and long-term leases, permanent transfers, exchanges, options contracts and other vehicles, including informal approaches (Wheeler et al., 2016). Nebraska had established arrangements in 1979 in its Upper Republican Natural Resource District, an irrigation district in the southwestern part of the state that enabled groundwater transfers (Wheeler et al., 2016). These arrangements included flow meters on all wells, use restrictions, 5-year allocations and controlled expansion of irrigated area. Groundwater markets started in California in the early 2000s (Evans, 2017), with a formal market established in 2017 (Hersko, 2017). In addition to California and Nebraska, water markets also exist in the states of Arizona, Colorado, Idaho, Nevada, New Mexico, Oregon, Texas, Utah and Washington. Internationally, Australia, China, Oman, India and Pakistan have water markets (Wheeler et al., 2016).

In recognition of the limits to expanding water supply, the 1992 Rio (de Janeiro) Convention set out Agenda 21 that identified important factors in addressing water supply for the future. These policy factors on the supply side supported water users paying the full cost of water, community involvement in decisions affecting water and consideration of the opportunity costs of the most valuable alternative application of water, and on the demand side more efficient water resource use and establishing water markets. These factors were qualified by major considerations of provision of safe water for all people (access to water in adequate quantity and healthful quality), recognition of communities' ability to pay for water (affordability), reconciling environmental needs with development (ecosystem as stakeholder in water demand) and resource user and polluter paying for impacts on natural resources, among other interests (UN, 1992). The needs of disadvantaged communities and groundwater-dependent ecosystems can be protected in groundwater markets when the state specifically identifies these needs as values to be addressed by reserving water for them through regulation (Wheeler et al., 2016; Babbitt et al., 2017).

Four conditions, recognized in the Rio Convention, must exist for water markets to be created and function efficiently (Wheeler et al., 2016):

1. A defined, measured and accepted extent of the limited public resource
2. Assured property rights providing access to the limited resource
3. Tradeable water rights with negligible transaction costs and market participation challenges
4. Prices for traded allocations that incorporate third-party impacts (externalities) through a market with many buyers and sellers knowledgeable in water use value

Babbitt et al. (2017) describe key considerations and steps in developing a groundwater market and trading programs. Major points include:

- Engaging stakeholders of a groundwater basin to develop their vision for the basin's social, economic and environmental well-being
- Establishing and enforcing groundwater allocations informed by water rights law and the interests and desires of the basin's people, balancing landowner, non-landowner and groundwater dependent ecosystem purposes
- Specifying clearly groundwater rights, recognizing that more than one groundwater rights doctrine may apply, such as "overlying rights" (having evolved from riparian and correlative rights doctrine) and "appropriative rights" ("first in time, first in right") and that groundwater

rights can only be acquired when surplus groundwater is available after meeting other needs and considering the results of monitoring and modeling
- Selecting the methods for quantifying and enforcing groundwater use, such as metering, exchanging irrigated areas, water volume trades or remote sensing of consumptive water use
- Establishing procedures for adjusting allocations, particularly in situations that lack historical data or have over-allocated groundwater and are unsustainable in the long term, and may include "ramp-down" for step-wise use reductions, adjudicated allocation adjustments and/or "offsets" having a specified proportion reduced in a trade to address over-allocation
- Recognizing basin hydrology, including connection to surface water, recharge, recycling and transboundary exchange, while guarding rights and needs of current users, and that may necessitate criteria or standards regarding placement of wells relative to other wells or structures or of volume of trade or adjustment
- Deciding on types of trades
  - from a range of evolving trade communication structures, including bilateral "person-to-person" contracts, formal brokerage between buyer and seller, physical or electronic bulletin boards, auctions and reverse auctions, electronic clearinghouses bringing traders together drawing on set trade rules or combinations of these or other accepted approaches
  - as to whether negotiations are public or confidential
  - allowing for flexibility in changes each year or within a growing season
  - limiting conflicts of interest in agency oversight of regulation and financial factors
  - utilizing transparent and defined steps for timely decisions
- Transferring funds and recording the transaction

Determining tradeable property use rights may have obstacles that need attention before market trading can occur. While groundwater has many characteristics of a common property resource, one wellowner's pumping may impact other users differently based on location and time of use because of the varying transmissivities and compositions of the range of geologic strata affecting groundwater movement and quality (NRC, 1994; Brozovic et al. 2010). In this case, public decision makers should determine the aquifer's sustainable volume for which groundwater use rights will be established. A related concern is the differences in hydrogeologic settings of multiple aquifers for which their groundwaters may be offered for trade as well as their interaction with surface water and a single management approach applied to all the aquifers. Management tailored to the aquifers from which the traded groundwater is withdrawn should be preferred over rigid regulatory steps to mitigate third-party and unintended ecosystem impacts (Wheeler et al., 2016). Groundwater monitoring costs can be high and serve as a constraint on use of tradeable rights (Wheeler et al. 2016). Additionally, highly detailed groundwater rights and conditions reduce flexibility in rights transfer as well as the market for the groundwater; in other words, flexibility in policy and institutions is important to respond to changes in environment, climate and socio-economic conditions (Wheeler et al., 2016). Skurray et al. (2013) suggest that steps be taken in establishing a market to preclude third-party and ecosystem impacts. Box 1.5.2 provides examples of water market attributes in Australia and Texas (USA). Characteristics of these markets are water use permits as a basis for rights to water, government conditions on production factors and changing use. Table 1.5.1 gives groundwater commodity prices in Texas from 2007 to 2016 with prices ranging from \$203 to 48,642 per hectare-meter (\$25 to \$6000/acre-foot) depending on location.

Valuing a Groundwater Use Right

In deciding to trade or purchase a groundwater use right, the purchaser must decide how best to determine the value of the right. All water rights are not the same – some are absolute, other conditional, still others may have specific use limitations or be subject to local or state use changes.

BOX 1.5.2 WATER MARKET EXAMPLES IN AUSTRALIA AND TEXAS

Western Australia

- For the season of 2011–12, groundwater users exchanged 4,225 megaliters (3,425.3 acre-feet) of allocated groundwater, 38 percent of the national total of groundwater traded (ANWC, 2013).
- Average price for water license transfers was (2012) AUS$631/megaliter (2012 US$792/acre-foot) (ANWC, 2013).
- Quantity use rights under a 10-year license may be affected by the ability of the provincial water minister to reclaim water not in use and even cancel conditional rights.
    - The water minister can amend water rights to protect third parties.
    - The government can decide to amend a license if a water use is improper.
    - Groundwater licenses have time limits.
    - Licenses are not equivalent to property but rather are water use arrangements with stipulations that may include requirements for well depth, measurement, appurtenances, recording and frequency or duration of use (Skurray et al., 2013).
- A significant limitation on license transfers: only a landowner or land occupier can receive a transfer of water to be applied to the land owned or occupied or a person with documented permission of the landowner can receive a water transfer for activities determined proper under a license.
- Water transfer applications are challenging, costly and not protective of confidentiality.
- Local regulation can nullify provincial water minister approvals of transfers (Skurray et al., 2013).

Texas USA

Edwards Aquifer (EAA, 2019)

- The Edwards Aquifer Authority (EAA) regulates groundwater underlying more than 8,000 square miles across eight Texas counties
- Groundwater use in Texas follows rule of capture: a landowner has the right to use groundwater below his or her land.
- EAA in south-central Texas operates with some restrictions on use and permits groundwater trading.
- Threatened and endangered species' habitat relies on minimum aquifer levels being maintained.
- Texas Senate Bill 1477 of 1993 changed groundwater rights structure, provided for permits giving use rights for a specific volume of water and identified the EAA to oversee the process.
- Analysis found that with no regulation, impacts to endangered species habitat were significant under low levels of groundwater (Gillig et al., 2004; Boadu et al., 2007).
- As of 2019, there were 2,036 groundwater permits to rights to groundwater use ranging from 0.004 to 22,830 hectare-meters (0.036 to 185,088 acre-feet) annually (EEA, 2019).

(Continued)

BOX 1.5.2 (Continued)

- The aquifer must be maintained at a level of 635 feet above mean sea level in the index well in San Antonio, Texas.
- Any sale or lease of groundwater rights requires an application for a permit transfer.
- Changes associated with a permit require a permit amendment, including water purpose and location of use.
- Monthly reports are required for groundwater withdrawals during times of critical aquifer water levels
- During times of water restrictions, irrigators desiring to complete a crop season must send notice to EEA within 30 days of a declared restriction.
- Irrigators planting additional crops within a declared critical restriction must send notice to EEA within 15 days of planting and the well meter reading for the planting. EEA may implement further reductions of water use in spite of permit authorized groundwater withdrawals.
- Permit holders must implement a conservation program of best management practices and document progress every three years.

Sources: Primarily abstracted from:

Wheeler, Sarah Ann; Karina Schoengold; Henning Bjornlund; Sarah Ann Wheeler. 2017. Lessons to Be Learned from Groundwater Trading in Australia and the United States. In Integrated Water Management. pp 493–517.

Australian Government National Water Commission (ANWC). 2013. Australian water markets: trends and drivers 2007–08 to 2011–12. URL: https://apo.org.au/sites/default/files/resource-files/2013/06/apo-nid34726-1230356.pdf (Accessed October 20, 2019).

Edwards Aquifer Authority (EEA). 2019. Groundwater Permit Holder. URL: https://www.edwardsaquifer.org/business-center/groundwater-permit-holder/ (Accessed October 21, 2019).

TABLE 1.5.1
Groundwater Commodity Prices in Texas

Some Available Prices for Water Commodity Prices in Texas Are

			Price	
Location	**Aquifer**	**Date**	**Per hectare-meter**	**Per acre-foot**
San Antonio	Not Known	2007	$203 to 48,642	$25 to $6000
Milan County	Carrizo-Wilcox	2014	$15,403	$1900
City of Waelder	Carrizo-Wilcox	2016	$6,242	$770
Central Gulf Coast	Gulf Coast	2016	$3,689	$455
GW Mgt Area 9	Trinity	2013	$23,283	$2,872

Source: Petrosian, Rima. 2019. Groundwater as a Commodity in Texas. URL: https://aipg-tx.org/groundwater/(Accessed October 24, 2019).

The value of a water right must reflect the specific type and characteristics to which it is subject. The use must be determined to be beneficial and reasonable.

Valuation of groundwater rights may take one or all of these approaches (TTPL, 2003; Bovee, 2017):

1. Comparable sales. Finding other sales of groundwater rights in the area can now be done through online postings of groundwater being offered for sale. States and online water markets may provide this information. Potential purchasers should obtain the details of the groundwater right, including volume, duration of use, location of use (e.g., within a particular groundwater basin) and transferability, among other factors. "Comparable sales" are probably the most common method of determining groundwater right value.
2. Potential income stream. Evaluating the income derived from purchasing a water use right provides perspective on water as an economic input to production or result planned. For agricultural purposes, computing projected income from irrigating and raising high-value crops compared to dryland cultivation provides a practical conclusion. Similar evaluations can be developed for other reasonable and beneficial groundwater uses.
3. Replacement cost. This method considers the situation in which the water right and provision are not obtainable. The potential groundwater user would examine the costs to develop the resource, obtaining the permit to install a well or wellfield and pump the water as well as the actual installing the wells and other appurtenances to have the water at the site of use. In this approach, the purchaser considers the cost of a different water right or source.

Markets for Groundwater Quality Trading

Water quality trading in the United States has existed formally since 2003 (USEPA, 2016). Water quality trading is a voluntary program operating under Clean Water Act National Pollutant Discharge Elimination System (NPDES) that enables a point source permit holder to satisfy regulatory requirements through pollutant reductions achieved by different wastewater point source or nonpoint source pollutant permit holder. At least 20 states have water quality trading programs.[1] While groundwater is not a point source of pollution, groundwater may be a nonpoint source of pollution to surface waters, carrying nutrients, pesticides, industrial pollutants and other contaminants that may infiltrate to the subsurface environment (USEPA, 2013). The U.S. Environmental Protection Agency has established a process for trading reductions of pollutants in nonpoint sources, including groundwater (USEPA, 2019).

Under the water quality trading program, a point source pollutant-generating permit holder can purchase pollutant credit, measured in pounds equivalent of the pollutant, created by a nonpoint source who implemented a best management practice that reduced its pollutant contribution beyond a pollutant baseline established under a total maximum daily load (TMDL) allocation for that pollutant discharged to a particular stream or other controlled surface water or watershed. The benefit to the point source permit holder is achieving pollution reduction at a lower cost than installing expensive treatment processes, assuming the nonpoint source pollution reduction cost per pound is less than treatment plant construction cost per pound of pollutant treated. A TMDL is the amount of pollutant allowed to be discharged to a stream, lake or other surface water.

Best management practices (BMP) are proven field controls and treatments that farmers apply to minimize soil erosion, agricultural nutrient and pesticide runoff, and animal feedlot waste transmission. BMPs may include cover crops, contour farming, no-till farming, crop rotation, fencing streams and grassed waterways, among many other steps. These practices protect stream habitats and drinking water supplies. In the context of water quality trading, BMPs frequently reduce a farmer's operating costs, providing a basis for greater implementation and attractiveness for selling credits to upstream or downstream wastewater discharge permit holders desiring to meet in-stream water quality standards. Groundwater can convey agricultural pollutants to streams and lakes, so reducing pollutants to groundwater can support achieving water quality requirements for surface

BOX 1.5.3 POINT SOURCE-NONPOINT SOURCE POLLUTANT TRADE EXAMPLE

For a water quality trade to be attractive, the point source pollutant permit holder must be able to reduce pollutant treatment costs by purchasing nonpoint source pollutant credits. A hypothetical example below involves a wastewater treatment discharger and a farmer implementing best management practices for nonpoint sources of pollution facing a watershed cap on phosphorus pollution requiring a reduction of two units across the watershed.

Assumptions for Phosphorus Pollutant Reduction

- The point source pollutant phosphorus is generated by a wastewater treatment plant that must reduce its phosphorus discharge at a cost of $150 per unit.
- In the same watershed, a farmer can reduce nonpoint source phosphorus for $100 per unit.

Case 1 – Both WWT Plant operator and farmer reduce phosphorus pollution by 1 unit:
$150/WWT phosphorus unit + $100/Farm phosphorus unit = $250
Case 2 – WWT Plant operator pays farmer to reduce phosphorus pollution by 2 units:
$100/Farm Phosphorus unit X 2 units paid by WWT Plant operator = $200
Cost Savings Benefit to WWT Plant operator = $50

In this hypothetical example, it is clearly a benefit to the WWT Plant operator to pay the farmer for the farm's capability to reduce phosphorus pollution by 2 units, rather than construct additional treatment plant capacity. The quality of groundwater, contributing nonpoint source pollution to the stream, will also improve.

Source: Pennsylvania State University Extension. 2006. A Primer on Water Quality Credit Trading in the Mid-Atlantic Region. By Charles Abdalla. URL: https://extension.psu.edu/a-primer-on-water-quality-credit-trading-in-the-mid-atlantic-region (Accessed October 24, 2019).

waters. Pollutant trade incentives encourage farmers to apply BMPs, improving groundwater quality as well as surface water quality. Box 1.5.3 gives a hypothetical point source–nonpoint source pollutant trade example that could apply to nonpoint source groundwater pollution.

CAPITAL AND FINANCING ASPECTS OF ECONOMIC PRODUCTION

CAPITAL FINANCING

To apply capital to a production process, it has to be acquired. Capital in production economics typically refers to either (Bannock et al., 1978): (1) "fixed capital" comprising durable goods such as structures and machines or (2) "circulating capital" which includes raw materials and "semifinished" goods used to process or incorporate into the output and used up in the process. These are all physical goods. The prospective owner of an enterprise may purchase capital by outright payment of cash or by financing the purchase over time. The payment of cash assumes that the owner has adequate savings and has determined that the use of the money to purchase the capital needed to produce a good is the best use of those funds. This means that it has been determined, given the preference for or avoidance of risk, that the funds will earn their highest rate of return in the new venture in which the capital will be used. In other words, he/she could invest his/her money in other ways and do no better than in his new production process. Thus, he/she has considered his/her "opportunity cost" for the funds and determined the best course of action for their productive investment. Opportunity cost is the value of the alternative foregone (Bannock et al., 1978) by choosing to use the funds in the new production venture, in this case. The evaluation of alternatives at the margin of a project or program is a significant function of economic analysis.

If the prospective owner decides to finance the purchase of the necessary capital, then he/she chooses to go the credit market, which can channel others' savings to the selected investment in the capital he/she needed for the new production process. Financial institutions that manage savings for investment establish a price for the use of the money they have for lending, referred to as the interest rate. The market interest rate is the competitive premium paid by the borrower to the lender in excess of the amount initially lent (Bannock et al., 1978). Different types of credit instruments depend on the length and nature of the loan (Gonzalez, 1989, p. 7):

- Short-term – used for short periods of months to buy circulating capital.
- Medium-term – used for periods covering several years to procure capital goods and machinery.
- Long-term – used for periods of 5 or more years to obtain durable goods such as structures.

The financial institution will set interest rates based on the loan term and type of good acquired as well as the credit-worthiness of the borrower. Goods easily or quickly used up may command higher interest rates, whereas buildings that can be sold and reused may have lower rates applied to them since the risk of loss is less. Groundwater wells are typically in place and use for many years, usually exceeding the length of a loan's term (e.g., 30 or more years for a well and 5 to 20 years for a financing term). On the other hand, equipment for on-site treatment for groundwater contamination removal may wear out more quickly and require higher interest rates to finance.

Interest payment for use of the loan funds is expressed as a percentage of the initial loan amount, referred to as the "principal," and is related to the length of the loan (e.g., 3 months or 5 years). Applicable parameters are: M_0, the initial amount of funds lent; M_t, the amount of funds to be repaid including interest; t, the time interval of the loan; r, the interest rate. Several methods of calculating interest are (Gonzalez, 1989, p. 7)

Simple interest: $M_t = M_0(1 + rt)$

Compound interest: $M_t = M_0(1 + r)^t$

Continuous interest: $M_t = M_0 e^{rt}$, where e is the natural number = 2.71828…

Discount Rate

A discount rate has similarities to an interest rate, except that the former is applied to funds or monetized equivalents (or worth) to be received or recognized in the future and stated in the value of the present time, reflecting a "time rate of preference" for money today rather than money received in the future. Discounting is applied to prospective monetized values due to uncertainty about the future, inflation expectations and changes in productivity because of technological advances, for example. Described another way, a dollar (or other monetary unit) today is worth less 1 year or 10 years or some period of time in the future. Conversely, a dollar received 10 years from now will be worth less than in the present. This concept has been applied to investments in the future.

The specific rate of discount to be applied to an investment is a significant issue. It has been asserted that for public goods a low "social discount rate" should be used so that the services and benefits of long-lived investments are not undervalued in the present by the current generation. The social discount rate is "a rate of conversion of future value to present value that reflects society's collective ethical judgment, as opposed to an individualistic judgment such as the market rate of interest" (Daly and Farley, 2004, p. 439). A low rate might be 2%–3%, and some have suggested zero and even negative rates for increasingly scarce resources in which investment could be made or value established (Daly and Farley, 2004, p. 274). A market rate of interest will be what the competitive forces at work in the financial realm indicate it should be to balance supply and demand for money.

The calculation for discounting future funds in its simplest form is

$$M_0 = M_t / (1+r)^t, \tag{8.14}$$

where

M_0 is the amount of funds in the present time
M_t is the value of future funds in time t
r is the discount rate

Other presentations of the discount rate for measuring environmental and resource values are given in the literature (for example, see Freeman, 1993, pp. 199–218). This calculation is also referred to as "present value" of future payments or values, meaning that discounting presents these future values in terms of their current monetized worth. Box 1.5.4 shows the results of applying discounting at different rates.

BOX 1.5.4 THE EFFECT OF DIFFERENT DISCOUNT RATES ON A FUTURE STREAM OF BENEFITS

Applying the formula $M_0 = M_t/(1 + r)^t$ for computing the current value of a stream of benefits nominally estimated to be $100 each year for 8 years gives the following results. The formula is applied to the $100 of benefits each year and then summed, shown as

$$M_0 = \Sigma\left[M_1/(1+r)^1\right]+\left[M_2/(1+r)^2\right]+\left[M_3/(1+r)^3\right]+\left[M_4/(1+r)^4\right]$$
$$+\left[M_5/(1+r)^5\right]+\left[M_6/(1+r)^6\right]+\left[M_7/(1+r)^7\right]+\left[M_8/(1+r)^8\right]$$

Since M_1 through M_8 are each equal to $100, r is the indicated rate and t is 8 years, the calculations provide the following results:

At Discount Rate of:		**−10%**	**−5%**	**0%**	**2%**	**5%**	**10%**
Present Value of $100	1	111.11	105.26	100.00	98.04	95.24	90.91
in Time Period	2	123.46	110.80	100.00	96.12	90.70	82.64
	3	137.17	116.64	100.00	94.23	86.38	75.13
	4	152.42	122.77	100.00	92.38	82.27	68.30
	5	169.35	129.24	100.00	90.57	78.35	62.09
	6	188.17	136.04	100.00	88.80	74.62	56.45
	7	209.08	143.20	100.00	87.06	71.07	51.32
	8	232.31	150.73	100.00	85.35	67.68	46.65
Summations		$1,323.07	$1,014.68	$800.00	$732.55	$646.31	$533.49

Notably, in the cases considered, a negative (−) 10% discount rate provides the largest present value of the sum of future benefits over 8 years, nearly 2½ times the present value of a positive 10% rate. Significantly, $100 from the 8th year in the future under the −10% rate is worth almost five times that of the 10% rate.

Under what circumstances might this occur? If population is growing and its demand for a resource, such as groundwater, is also growing and the resource stock is getting smaller through being consumed or contaminated, the resource from a societal perspective may be considerably more valuable in the future regardless of its absolute supply cost or its market price.

BOX 1.5.5 DEPRECIATION ALLOWANCE FOR A WELL

If a well has a useful life of 30 years, the annual allowance (for replacement) for constant depreciation of this capital (assuming no salvage value) may be calculated by

$$A = \left[Cr(1+r)^n\right] / \left[(1+r)^n - 1\right],$$

where

A, the annual allowance at constant depreciation (to be calculated)
C, the installed price of the well and pump = \$10,000
r, the interest rate in the market = 5% or 0.05 (for example)
n, useful life = 30 years

$$A = \left[\$10{,}000 \times 0.05(1+0.05)^{30}\right] / \left[(1+0.05)^{30} - 1\right] = \$650.51$$

Depreciation of Capital

The value of fixed capital decreases because of deterioration over time, being out-of-date due to technological advances or unneeded use. From these considerations, the concept of "useful life of a capital good evolved to recognize that these items wear out or become obsolete. "When a firm computes its annual production costs, it makes an annual allowance for the depreciation of each capital good. The purpose of these allowances is to build up a fund which allows the same item to be replaced at the end of its useful life or depreciation period" (Gonzalez, 1989, p. 8), although Bannock et al. (1978, pp. 123–124) indicate that accounting depreciation is to "ensure that the cost of the flow of services provided by capital assets is met in the price of the company's products" and not to provide for a "replacement" fund, since the funds can be used in practice to invest in other capital that will provide the highest return. The simplest form of calculation commonly used is straight-line depreciation that can be computed as an annual allowance evaluating the capital cost less its end-of-life salvage value divided by the capital's useful life:

$$A = (C - SV) / n \tag{1.5.6}$$

where

A, annual allowance
C, Initial cost (sometimes referred to as "first cost", FC, of the capital good)
SV, Salvage value of the capital good at the end of its useful life
n, useful life

Wells are capital investments that are relatively long-lived but can wear out. Depreciation of wells may be warranted in business applications. Box 1.5.5 gives an example of depreciation allowance for a well considering constant depreciation for replacement purposes and alternative investments at a particular interest rate.

Natural Capital

Natural capital equates to "stocks or funds provided by nature (biotic and abiotic) that yield a valuable flow into the future of either natural resources or natural services" (Daly and Farley, 2004,

p. 437). Natural capital available in and taken from the ecosystem is transformed by the economic system. Every raw material for industrial processes is extracted from nature and usually not assigned an economic value to be monetized except the cost of extraction and processing. The processing of raw materials to transform them into goods results in wastes and all goods themselves ultimately become wastes after they are used up or worn out, returned in both cases to the ecosystem, degrading or damaging its quality as well as its viability to provide a similar level of natural capital in the future. Planned and engineered recycling and reuse may regain some quality in future applications. Groundwater is both a final product for the satisfaction of human, animal and plant thirst and bodily requirements, as well as an intermediate good to be used in irrigated agriculture, mining and manufacturing and incorporated into other goods which provide useful products and services, such as food, paint, other objects and recreation, and used as a conveyance and sink for wastes. When groundwater is "produced" from the subsurface, it already existed there, placed in those subterranean zones through hydrologic and geologic processes, with no cost to past or current generations for its availability to be used. Natural capital, such as groundwater, can be depleted through human economic activity; it then becomes scarce and, as discussed previously, becomes commoditized and traded in the market because of its increased cost relative to alternatives.

Pollution can destroy the quality of natural capital so that it cannot be used, thereby removing it from availability for human use or assimilation by other life forms, except bacteria, which may decompose it over long periods of time to be harmless once again. Pollution can then cause previously clean, safe freshwater to be depleted. In these two cases, this water becomes scarcer. In the circumstance of degradation, groundwater does not typically receive a depreciation allowance to be replaced later by the acquisition of new freshwater.

In the case of depletion, in the United States, depreciation allowances are provided to farmers who irrigate and mine groundwater, but no new water has ever been purchased to replace the water used up. Since these groundwaters have become scarce, the value of the remaining groundwater may be greater than in the past and in the future compared to the current time. If the scarce water value were monetized, its value might be able to be calculated using a negative interest rate for its future worth to the next generations. Otherwise, its value is immeasurable and since it is getting scarcer, approaches to using it must take into account its invaluable nature. Certainly, the value of groundwater is greater than zero and more than the cost to pump it out of the ground.

This aspect of groundwater value has national and international implications well beyond the use by individuals, as each person's or firm's demand for groundwater added up has macro level consequences for the value and inherent state of this essential and ubiquitous resource.

NATIONAL AND INTERNATIONAL IMPLICATIONS OF GROUNDWATER PRODUCTION

While many texts are devoted to the development of national economies and international economics, some basic economic relationships should be highlighted relative to groundwater and macroeconomic and international considerations. National economies are built on the collective actions of individuals with preferences that can be aggregated to demand. Likewise, in response to this demand, firms acting in competitive microeconomic settings collectively provide the mechanism of transforming raw materials of the ecosystem to goods demanded by the populations served in the marketplaces and producing wastes in the process of this production. The goods, once used up or worn out, become wastes for the ecosystem to, in some way, absorb. Neoclassical economic thought assumes that the natural capital of the raw materials of the ecosystem will always be available – and the economy can grow indefinitely. At the international level, the world population was 6.1 billion in 2000 and is projected to grow to 9.4 billion by 2050, a 54% increase (USCB, 2005). During this time, total world water resources will not have grown and the usable groundwater portion, while large, will likely shrink from what it is today to meet demands for food and water, given current technology and

BOX 1.5.6 WATER FOOTPRINT FOR SOME FOODS AND COUNTRIES

- 1 L of milk needs 800 L of water.
- 1 kg of wheat needs 1100 L of water.
- 1 kg of rice needs 2300 L of water.
- 1 kg maize needs 900 L of water.
- The production of one kilogram of beef requires 16,000 L of water.
- To produce one cup of coffee we need 140 L of water.
- The water footprint of China is about 700 cubic meters per year per capita. Only about 7% of the Chinese water footprint falls outside China.
- Japan with a footprint of 1150 cubic meters per year per capita, has about 65% of its total water footprint outside the borders of the country.
- The water footprint of the United States is 2500 cubic meters per year per capita.

Sources: UNESCO, *Water Footprint*, 2004. URL: http://www.waterfootprint.org/ (accessed October 22, 2005); GDRC, *The Concepts of Water Footprint and Virtual Water*, 2005. URL: http://www.gdrc.org/uem/footprints/water-footprint.html (accessed October 22, 2005).

urgent needs for water. Just meeting human demand will take more water, not taking into account the needs of the rest of the ecosystem on which humankind rely. Thus, water will become scarcer and more valuable due to population growth and its demands (Schiffler, 1998, p. 10).

Some discussion of "virtual" water on national and international fronts seems on the surface to address the need for water, but this may be a shallow path. Virtual water suggests that one region or country can obtain water for food by importing food from other countries that have the capacity to produce food. In other words, importing countries do not have to use up their scarce water to produce food – but they must have some way to pay for it by producing other goods offered in exchange. Box 1.5.6 provides some facts about what is referred to as the "water footprint" of some foods and countries. Notably, countries that eat more foods higher on the food chain or that are produced by irrigation require more water.

Thus, countries relying on irrigated agriculture for export revenue are "sending water" out of the country, perhaps depleting aquifers that will not provide water for their future generations, in the form of food. These countries are exporting their natural capital. While the market for food may attract such economic activity, national governments may desire to consider the long-range implications for their own populations as to how increasingly scarce groundwater is being used and implement steps to use it more efficiently.

In a balanced review, Higgott (2005) notes that globalization of the world's economies, characterized by trade liberalization, financial deregulation and privatization has increased economic efficiency among trading partners, which are typically trading companies, and profited major international corporations which conduct that trade. Wealth has grown, but it is not clear that the poor of the world have benefited from the ability to move capital and operations to the lowest cost locations. Leaders of countries have expressed skepticism in this regard as well as their concern about the consolidation of capital in large multinational corporations and the need for distribution of wealth more fairly (Reel, 2005). Furthermore, some international arrangements, such as through the World Trade Organization, may not support equivalent environmental protections (Daly and Farley, 2004, p. 329) which lower production costs even more to the detriment of the country losing employment for the sake of reduced prices to consumers. Thus, countries that inadequately internalize social and environmental costs of production have a competitive trade advantage (Daly and Farley, 2004, p. 329). This point strikes at the heart of the basic market equation and thus relates macroeconomic and microeconomic consequences that require national and international attention for countries to develop policies that recognize internal and transboundary policies affecting their economies and social and intergenerational equity and justice at scales they can manage. In both the national and international arenas, an emphasis is needed to understand resource use beyond the individual and firm as products, services and information may move with greater efficiency and speed than policies can incorporate the changes in the sociopolitical fabric that could promote their wise use and stewardship of those resources, including groundwater.

SUMMARY

Human economic utility for goods is driven by tastes, preferences and needs. The market balances consumers' demand, derived from their utility for goods and services with firms' capacity to produce them, through prices. Prices of each good and service relative to each other affect firms' willingness to produce more or less of a good and consumers' willingness to pay for them. Economic production of groundwater or any other good is guided by business monetary return, which is to be maximized. Economic measurement and equations facilitate calculating the value of factor inputs to and outputs of production processes for use in decision-making. The market would indicate that groundwater should be priced at its MC, except in a monopolist's case. However, MC pricing may not reflect the full cost of producing this resource. Information may be useful or not in guiding our use of goods, depending on whether we receive the full details on the implications applying those goods to their intended purposes. Groundwater is a good with both final and intermediate uses. It can be depleted through consumption or pollution. Information about products and services that were produced through consuming or polluting groundwater may not be sufficiently complete to inform consumers about the impact of their purchases. Groundwater is a public good and natural capital provided by the ecosystem. In arid locations, groundwater is heavily relied on and has become a commodity with markets enabling purchase of use rights for allocation to what is expected to be efficient and best uses. Markets have also been established for groundwater quality trading where alternative treatment in permitted discharges results in efficient water quality outcomes. Discounting its monetized value in the future does not recognize its invaluable nature. Nations should develop an understanding of their resource use and ensure that their participation in international economic exchange does not sacrifice economic distribution concerns at the national level. Whether or not allowance is made for this removal of the groundwater resource, no one is replacing what has been taken. It is becoming scarce because of these reasons and its value may not be able to be measured, even though it exists nearly everywhere, albeit in decreasing amount.

FOR FURTHER CONSIDERATION

1. Is water sufficiently available in your community or region to satisfy all mayor needs, including drinking water to people of lesser income? What steps did your community or region take to ensure adequate water supply to all residents?
2. Does your community or one in your region purchase water through a water market? What is the price of water in the market? What factors affected that price from your perspective?
3. Compare the price of water in your community to the price of bottled water from the store for an equivalent amount of water. What factors affecting price are different in each case?

NOTE

1 States having water quality trading programs include: Arkansas, Colorado, Connecticut, Florida, Idaho, Indiana, Iowa, Kentucky, Louisiana, Maryland, Minnesota, Missouri, Montana, North Carolina, Ohio, Oregon, Pennsylvania, South Carolina, West Virginia, Wisconsin.

REFERENCES

Australian Government National Water Commission (ANWC). 2013. Australian water markets: trends and drivers 2007–08 to 2011–12. URL: https://apo.org.au/sites/default/files/resource-files/2013/06/apo-nid34726-1230356.pdf (accessed October 20, 2019).

Babbitt, C., Hall, M., Hayden, A., Garcia Briones, A.L., Young, R., and Brozović, N. 2017. Groundwater trading as a tool for implementing California's sustainable groundwater management act. *Environmental Defense Fund*. http://Edf.org/WaterTrading

Bannock, G., Baxter, R.E., and Rees, R. 1978. *The Penguin Dictionary of Economics*. Penguin Books Ltd., Harmondsworth, Middlesex, UK 467 p.

Barlowe, R. 1972. *Land Resource Economics: The Economics of Real Property*. Prentice-Hall Inc, Englewood Cliffs, NJ, 616 p.

Boadu, F., McCarl, B.C., Gillig, D.. 2007. Empirical investigation of institutional change in groundwater management in Texas: The Edwards Aquifer case. *Natural Resources Journal*, 47. pp. 117–163.

Bovee, B. 2017. *Colorado Real Estate Journal*, February 27, 2017. URL: https://crej.com/news/value-water-right/ (accessedApril 24, 2020).

Brozovic, N., Sunding, D., and Zilberman, D.. 2010. On the spatial nature of the groundwater pumping externality. *Resource and Energy Economics*, 32. Pages 154–164.

Business Trend Analysis, Inc. 1992. The bottled water market. Past Performance, Current Trends, and Strategies for the Future. A Business Information Report, p. 84.

Daly, H.E. and Farley, J. 2004. *Ecological Economics*. Island Press, Washington, DC, 454 p.

Danielopol, D.L., Griebler, C., Gunatilaka, A., and Notenboom, J. 2003. Present state and future prospects for groundwater ecosystems. *Environmental Conservation*, 30(2): 104–130.

Edwards Aquifer Authority (EAA). 2019. *Groundwater Permit Holder*. URL: https://www.edwardsaquifer.org/business-center/groundwater-permit-holder/ (accessed October 21, 2019).

Evans, I. 2017. *A New Kind of Groundwater Market for California*. In Water Deeply. URL: https://www.newsdeeply.com/water/community/2017/06/27/how-california-got-its-first-groundwater-market (accessed October 11, 2019).

Field, B.C. 1994. *Environmental Economics*. McGraw-Hill Inc, New York, 482 pp.

Foster, S. 1999. Ground for concern. *Our Planet On-line Magazine*. URL: http://www.ourplanet.com/imgversn/83/foster.html (accessed September 23, 2005).

Freeman, A.M. III. 1993. *The Measurement of Environmental and Resource Values: Theory and Methods*. Resources for the Future, Washington, DC, 516 p.

Gibert, J. 1994. Groundwater ecology from the perspective of environmental sustainability. In Gibert, J., Danielopol, D.L., and Stanford, J. (eds.), *Groundwater Ecology*. Academic Press, San Diego, CA, pp. 3–13.

Gillig, D., McCarl, B.A., Jones, L.L., Boadu, F. 2004. Economic efficiency and cost implications of habitat conservation: An example in the context of the Edwards Aquifer region. *Water Resources Research*. doi: 10.1029/2003WR002749. URL: https://agupubs.onlinelibrary.wiley.com/doi/full/10.1029/2003WR002749 (accessed October 21, 2019).

Glennon, R. 2002. *Water Follies: Groundwater Pumping and the Fate of America's Fresh Waters*. Island Press, Washington, DC, 314 p.

Global Development Research Center (GDRC). 2005. *The Concepts of Water Footprint and Virtual Water*. URL: http://www.gdrc.org/uem/footprints/water-footprint.html (accessed October 22, 2005).

Gonzalez, A.S. 1989. Basic economic concepts applied to groundwater management. In *Groundwater Economics: Selected Papers from a United Nations Symposium, Barcelona, Spain*. Elsevier Science Publishers B.V., Amsterdam, the Netherlands, pp. 3–22.

Hanemann, M. 2005. *The Value of Water*. University of California, Berkeley CA. URL: http://are.berkeley.edu/courses/EEP162/spring06/Readings/valuewater.pdf (accessed November 10, 2006), 35 p.

Hardin, G. 1968. The tragedy of the commons. *Science*, 162, 1243–1248.

Hersko, T.. 2017. State's first water market will allow farmers to buy, sell groundwater. *VC Star*, July 17, 2017. URL: https://www.vcstar.com/story/money/business/2017/07/17/states-first-water-market-allow-farmers-buy-sell-groundwater/448023001/ (accessed October 11, 2019).

Higgott, R.A. 2005. *Globalization: The Benefits and The Threats*. Asia-Europe Foundation. URL: http://www.asef.org/asef-uni/3_infohub/word/Globalization%20(Benefits%20and%20Threats)%20-%20Richard%20Higgott.doc (accessed October 22, 2005).

Hirshleifer, J. 1976. *Price Theory and Applications*. Prentice-Hall Inc, Englewood Cliffs, NJ, 506 p.

Incisive Interactive Marketing LLC (IIM). 2005. *Stats, Advertising; Click ZNetwork.* URL: http://www.clickz.com/stats/sectors/advertising/article.php/181901 (accessed September 24, 2005).

IndiaMart. 2005. *Manufacturers Directory, Agricultural Equipment, Irrigation Equipment & Systems, Drip Irrigation Systems.* URL: http://catalogs.indiamart.com/products/drip-irrigation-systems.html (accessed September 24, 2005).

International Water Management Institute (IWMI). 1998. *Groundwater Depletion: The Hidden Threat to Food Security?* URL: www.iwmi.cgiar.org/Press/brief2.htm (accessed September 23, 2005).

Job, C. 1997. A summary of state wellhead protection programs. *Ground Water Monitoring and Remediation*, 17 (2), 61–63.

Lancaster, K. 1966. A new approach to consumer theory. *Journal of Political Economy*, 74, 132–157.

Maler, K.G. 1974. *Environmental Economics: A Theoretical Enquiry.* Johns Hopkins University Press for Resources for the Future, Baltimore, MD, 278 p.

Microsoft Corporation (MSC). 2005. *"Advertising," Microsoft® Encarta® Online Encyclopedia.* URL: http://encarta.msn.com (accessed September 24, 2005).

mnmeconomics. Undated. The demand for public goods. URL: https://mnmeconomics.wordpress.com/2011/10/22/the-demand-for-public-goods/ (accessed June 24, 2020).

National Research Council (NRC). 1994. *Alternatives for Ground Water Cleanup.* National Academies Press, Washington, DC. URL: https://www.nap.edu/read/2311/chapter/4 (accessed October 15, 2019).

National Research Council (NRC). 1997. *Valuing Ground Water, Economic Concepts and Approaches.* National Academy Press, Washington, DC, 189 p.

Natural Resources Defense Council (NRDC). 1999. *Bottled Water: Pure Water or Pure Hype.* URL: http://www.nrdc.org/water/drinking/bw/bwinx.asp (accessed September 24, 2005).

New York Times Company (NYTC). 2005. Online advertising to boom, despite Dot.Com gloom, spending to grow to $16.5 billion by 2005. In About Inc. URL: http://retailindustry.about.com/library/bl/bl_jup0816.htm (accessed September 24, 2005).

Nicholson, W. 1979. *Intermediate Microeconomics and Its Application.* The Dryden Press, Hinsdale, Illinois, 614 p.

Notenboom, J. and van Gestel, K. 1992. Assessment of toxicological effects of pesticides on groundwater organisms. In *Proceedings of the First International Conference on Ground Water Ecology, Tampa, FL*, Stanford, J.A. and Simons, J.J. (eds.). American Water Works Association, Bethesda, MD, pp. 311–317.

Pennsylvania State University Extension. 2006. *A Primer on Water Quality Credit Trading in the Mid-Atlantic Region.* By Charles Abdalla. URL: https://extension.psu.edu/a-primer-on-water-quality-credit-trading-in-the-mid-atlantic-region (accessed October 24, 2019).

Petrosian, R. 2019. *Groundwater as a Commodity in Texas.* URL: https://aipg-tx.org/groundwater/(accessed October 24, 2019)

Organization for Economic Cooperation and Development (OECD). 2005. *Glossary of Statistical Terms.* URL: http://stats.oecd.org/glossary/index.htm (accessed November 12, 2006).

Rain Bird Corporation. 2005. *The Intelligent Use of Water™ Advertising Campaign.* URL: http://www.rainbird.com/iuow/resources/ads.htm (accessed September 24, 2005).

Reel, M. 2005. Skepticism prevails at trade talks. *Washington Post.* November 6, 14, 15 p.

Samuelson, P.A. 1964. *Economics: An Introductory Analysis.* McGraw-Hill Book Company, New York, 838 p.

Schiffler, M. 1998. *The Economics of Groundwater Management in Arid Countries; Theory, International Experience and a Case Study of Jordan.* Frank Cass Publishers, London, U.K., 394 p.

Skurray, J., Pandit, R., Pannell, D.. 2013. Institutional impediments to groundwater trading: the case of the Gnangara groundwater system of Western Australia. *Journal of Environmental Planning and Management*, 56, pp. 1046–1072.

The Trust for Public Land (TTPL). 2003. *Valuing a Water Right. Chapter 10 in Water Acquisition Handbook.* URL: https://www.tpl.org/sites/default/files/cloud.tpl.org/pubs/water-handbook/ca-waterhandbook-chapter10.pdf (accessed April 24, 2020).

Tufts University. 2004. *New Consumption Guidelines Issued for Water, Sodium, and Potassium.* Health and Nutrition Newsletter. URL: http://healthletter.tufts.edu/issues/2004-04/consumption_guidelines.html (accessed September 22, 2005).

U.S. Census Bureau (USCB). 2005. Total midyear population for the World: 1950–2050. *International Database.* URL: http://www.census.gov/ipc/www/worldpop.html (accessed October 22, 2005).

U.S. Environmental Protection Agency (USEPA). 2013. *Nonpoint Source Program and Grants Guidelines for States and Territories.* URL: https://19january2017snapshot.epa.gov/sites/production/files/2015-10/documents/319-guidelines-fy14.pdf (accessed October 23, 2019).

U.S. Environmental Protection Agency (USEPA). 2016. *Frequently Asked Questions About Water Quality Trading*. URL: https://www.epa.gov/npdes/frequently-asked-questions-about-water-quality-trading#overview (accessed October 23, 2019).

U.S. Environmental Protection Agency (USEPA). 2019. Water quality trading under the national pollutant discharge elimination system program. *Federal Register*, 84, 49293–49297. URL: https://www.federalregister.gov/documents/2019/09/19/2019-20324/water-quality-trading-under-the-national-pollutant-discharge-elimination-system-program (accessed October 23, 2019).

UNESCO. 2004. *Water Footprint*. URL: http://www.waterfootprint.org/ (accessed October 22, 2005); GDRC, *The Concepts of Water Footprint and Virtual Water, 2005*. URL: http://www.gdrc.org/uem/footprints/water-footprint.html (accessed October 22, 2005)

United Nations (UN). 1992. *United Nations Conference on Environment & Development; Agenda 21*. Rio de Janerio, Brazil, 3 to 14 June 1992. URL: https://sustainabledevelopment.un.org/content/documents/Agenda21.pdf (accessed October 13, 2019).

United Nations Education, Scientific and Cultural Organization (UNESCO). 2004. *Water Footprint*. URL: http://www.waterfootprint.org/ (accessed October 22, 2005).

Weiner, S.D. 2006. *Wilfredo Pareto*. URL: http://sdweiner.home.texas.net/d/archive/theory-method/WILFREDO_PARETO.htm (accessed September 2, 2006).

Wheeler S.A., Schoengold, K., Bjornlund, H. 2016. Lessons to Be Learned from Groundwater Trading in Australia and the United States. In A.J. Jakeman, O. Barreteau, R.J.K. Hunt, J.D. Rinaudo, A. Ross (eds) *Integrated Groundwater Management*. Springer, Cham.

Wheeler, S.A., Schoengold, K., Bjornlund, H., Wheeler, S.A. 2017. Lessons to be learned from groundwater trading in Australia and the United States. In *Integrated Water Management*. pp. 493–517.

Young, R.A. 2005. *Determining the Economic Value of Water: Concepts and Methods*. Resources for the Future Press, Washington, DC, 357 p.

Zilberman, D. and Schoengold, K. 2013. The use of pricing and markets for water allocation. *Canadian Water Resources Journal*, January 23, 2013. pp. 47–54. URL: https://doi.org/10.4296/cwrj300147(accessed October 11, 2019).

6 Groundwater Access and Supply

Wells, Aquifer Recharge and Natural Infrastructure

ACCESS IS FUNDAMENTAL TO USE

Access to groundwater is the first condition to bring it into human use and value. Since water is essential to all life, water and access to it have an inherent value. The cost of creating properly installed access to groundwater (i.e., a well) affects the groundwater producer's supply cost for water. Access to groundwater may determine habitability of an area: consider the oasis in the desert, the Biblical Jacob's well or the high-capacity High Plains irrigation well. Gaining admission to the land overlying the groundwater of interest is a basic access requirement for groundwater. The factors that influence access to groundwater and its quality affect its supply cost and ultimately its price to consumers. Before obtaining access to groundwater for whatever purpose, it is best to consider available information about the subsurface before conducting field exploration for groundwater and installing one or many wells. Hydrogeologic conditions of the subsurface, water demand implications for well size and well type will also factor into well development costs. Access to groundwater may also provide the potential for supply augmentation as well as disposal, using the subsurface for water supply and as a sink for treated and untreated wastes. Furthermore, wells may be used to produce other substances of commercial value in addition to water. Ecological aspects of groundwater access should receive further examination. Groundwater protection is fundamental to reliable, safe groundwater supply.

A TYPICAL WELL

Wells typically provide access to groundwater. A well is defined as an artificial excavation into the ground, usually a dug, bored or drilled hole or tunnel, that penetrates a water-yielding underground zone and allows water to flow or to be pumped to the land surface through a pipe, referred to as well casing (ITFM, 1997; TGF, 2020). A typical well has features common to most wells (Driscoll, 1986; USGS, 1997, pp. 18–32):

Annular space	The space between the drill hole and the casing
Casing	Usually iron, stainless steel or PVC (polyvinyl chloride) pipe through which the groundwater is pumped to the surface.
Screen Gravel pack	A slotted section of casing that allows water into the well.
	Gravel in the annular space around the screened interval to allow groundwater to move easily to reach the screen.
Backfill	Sometimes native materials drilled up and put back into the annular space above the gravel pack.
Surface seal	Cement is used to seal the upper annular space above the backfill to the ground surface and create a pad around the casing at the wellhead to keep surface runoff from traveling down along the casing and potentially contaminating the well water.

(*Continued*)

Annular seal	Often, cement fills the annular space above the gravel pack or backfill. Some states require the first 3 m of annular space below the ground surface to be grouted with cement to keep surface runoff from carrying contaminants down along the casing to groundwater and screen.
Grout	Another term for cement used to fill the annular space.
Protective well stack casing	A sleeve of steel casing, larger in diameter than the well casing, that is set in concrete at the wellhead and rises above the top of the well casing to protect it
Packer	Cement or other impermeable material that is placed between the gravel pack and the backfill to isolate the gravel around the screen to prevent water (and contaminant) movement from above and below the screen. A packer may be used in conjunction with backfill.
Pitless adaptor	Device below ground surface and frost line in a well that allows the sanitary transmittal of water from the well through a line to a building.
Pump	Typically, a motorized device that pulls water from the aquifer and pushes it up the well. Some pumps may be at the wellhead, in which case, they would pull water up the casing. Pumps can also be hand and foot powered.
Apron	At ground surface, a concrete pad or apron is installed around the casing to keep surface runoff away from the well casing.
Well/pump house	A building or shelter built around and over the well or pump to keep the wellhead from being accidentally damaged and allow protected access.

Each of the items above affects the cost of an installed well because they require labor and materials. Depending on how they are installed and what materials used, these items will also significantly affect the cost of operating and maintaining the well. The features for simple monitoring and municipal wells are described in Figure 1.6.1. A domestic household well will have similar features to a municipal well.

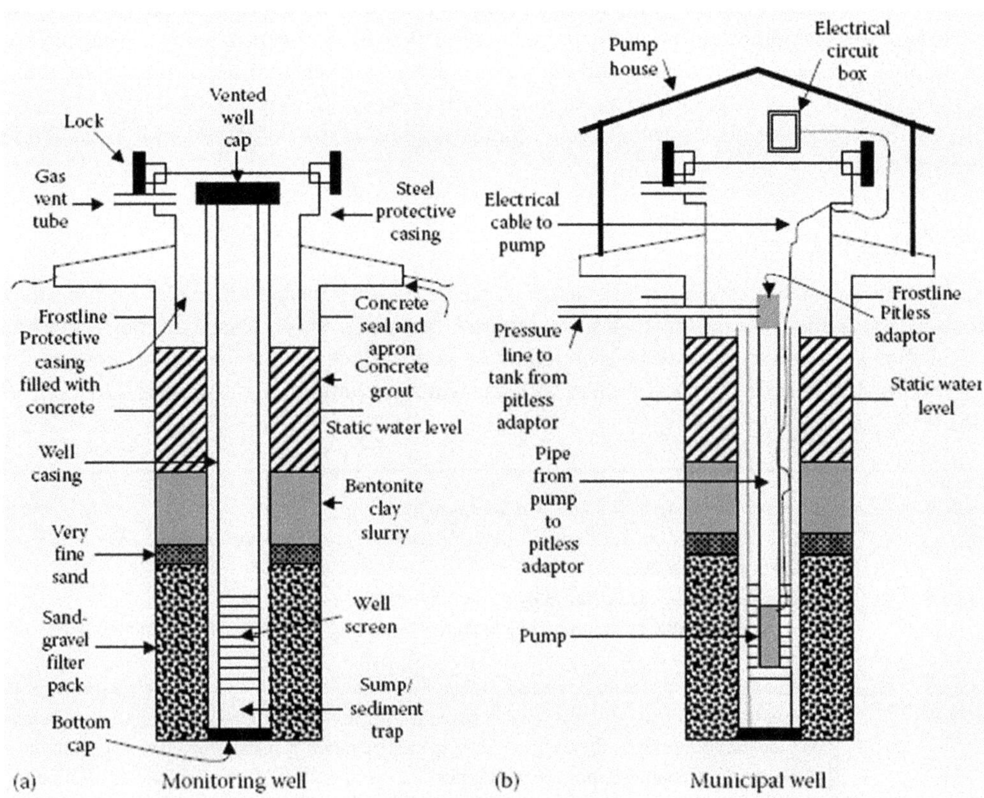

FIGURE 1.6.1 Typical well design in unconsolidated aquifers. Generalized monitoring and municipal well diagrams (unconsolidated geological setting).

GROUNDWATER RESOURCE INVESTIGATION

Elements of an Investigation

A groundwater resource investigation assesses the volume and quality of the resource that may be produced physically and economically. The investigation of the groundwater resource for water supply is more expensive than that of surface waters, while developing the access to groundwater is less capital intensive than for surface water (Foster, 1989, p. 53). The major steps in exploration and investigation of groundwater are well defined.

Investigation of groundwater occurrence and availability addresses the sources and discharge locations of the resource and the characteristics of the geologic matrix in which it is contained. The principal steps in an investigation program may include the following steps, also referred to as a feasibility study (Driscoll, 1986, pp. 150–204; RMC, 1990, pp. 34–35; NGWA, 2016a):

- Defining the boundaries of the area to be supported by the water supply or of remedial interest.
- Determining the objectives of the investigation.
- Conducting a preliminary study with available information (also referred to as a Phase I environmental assessment), including project constraints, such as water rights and laws; previous site use, ownership, possible contamination, existing water conditions and previous site investigation results; title search; endangered or threatened species identification.
- Establishing an initial hydrologic budget for the area.
- Implementing an investigation program, including testing existing wells for levels and production capability, geophysical surveys, water quality analysis, drilling additional wells to fill information needs, leading to drilling production or dedicated monitoring wells.
- If the site and subsurface will be used for aquifer storage and recovery for municipal water supply, analyze the compatibility of the influent water quality with the aquifer matrix.
- With the technical site and well information, conduct a life-cycle economic cost/benefit analysis of the proposed project. Life-cycle cost/benefit analysis will be covered in Volume II, Chapter 5.

Groundwater investigation has incorporated the technologies of aerial and satellite remote sensing. An expanded paradigm of groundwater exploration has emerged with modern technologies, referred to as "megawatershed exploration," focusing on groundwater occurrence in deep fractured consolidated rock (Bisson and Lehr, 2004). Subsurface heterogeneity and complexity further add to the investigatory challenge and cost. Other factors affecting the cost of groundwater resource evaluation for large-scale development include acquiring sufficient information on aquifer recharge and storage and on natural and anthropogenic constituents affecting quality. Satellite imagery from the Gravity Recovery and Climate Experiment (GRACE) can be used to conduct gravity measurements over time and calculate changes in groundwater storage. These remote measurements can be combined with monitoring well data to obtain an understanding of large areal groundwater occurrence and quality (Gholizadeh, 2020).

Foster (1989, p. 55) lists the following points as significant in planning and implementing an investigatory program:

1. Small-scale development factors related to individual rural and residential wells:
   a. Location and depth of groundwater producing zones
   b. Variabilities of groundwater quality
   c. Water table (piezometric level)
2. Large-scale development factors, in addition to the factors above, affecting large urban, irrigation and industrial water needs:
   a. Water demand
   b. Subsurface storage properties
   c. Recharge rates

d. Relationships with surface waters (streams, lakes and wetlands)
e. Intrusion of brackish or saline waters

To address these factors, the field investigation would comprise installing and examining aquifer properties with an exploratory well and may also include addition of monitoring wells. At this stage, geophysical logging and packer tests would be performed to evaluate aquifer heterogeneity. As a well is installed, the well driller will keep a written log of the geologic material brought to the ground surface for reference and usually reported to local or state offices responsible for authorizing or overseeing wells. If the investigation involves aquifer storage and recovery, testing the water quality of source and receiving waters and of the aquifer matrix for geochemical compatibility is essential. Aquifer pumping tests to determine hydraulic properties and well specific capacity will characterize groundwater recharge and recovery potential. These results will provide the critical technical information for project design (NGWA, 2016a).

Analysis of Available Information

Possibilities for evaluating these factors are enhanced in locations where previous investigation has already occurred and information is available. A first step to obtain access to groundwater is to examine the available information (Foster, 1989). A variety of sources can provide the necessary information on the depth to groundwater, direction of flow, seasonal conditions, geology of the subsurface, previous land use, costs of drilling, water quality and potential sources of contaminants. Locally, the public health office may have a good idea of conditions relative to depth and quality. Both the state or provincial and national geological surveys of many countries have detailed information on most areas, which is often available on the Internet as well as in hardcopy report format.

In the evaluation stage, project managers can examine the results of important parameters, such as recharge and storage. While this stage may suggest more information is needed, use of modeling and performance of sensitivity analyses can guide the use of further field investment. Further major field investment may not improve the overall results (Foster, 1989, p. 61). Significant effort should be focused on aquifer response, especially in areas of complex hydrogeology. More complete descriptions of steps in the investigation and evaluation process are found in the U.S. Geological Survey (USGS, 1997) and other references (Driscoll, 1986; Brassington, 1988; Schwarz, 1989; RMC, 1990) and will not be detailed here. This resource evaluation will guide the development stage relative to location, number, spacing, size, depth of wells and groundwater production.

While individuals think that they may know exactly where they want a well for residential, industrial, municipal or monitoring purposes, checking available information may make the decision to invest in a well more efficient (i.e., save money) in the long run. For example, knowing the depth to groundwater and seasonal fluctuations of the water table will help determine how long the screen length should be or how deep to drill the well for a continuing source of supply during dry seasons. Knowing the direction of flow and that an adjacent area was used for other purposes (such as a waste or sludge disposal site) may indicate that the well should be placed more distantly from certain features of the site. Aerial photos combined with property title and environmental permit searches may identify important information on previous land use and potential contaminants on-site or nearby that would affect the quality of groundwater. The cost of obtaining this range of information is small in comparison with the cost of problems that could lie ahead once drilling is underway or the well installed and the water testing over time showing a deterioration of water quality.

Part of the initial site assessment and preliminary plan development in establishing the direction and rate of flow and aquifer recovery from production need to be established from information on water table levels combined with detailed hydrogeologic data about the aquifer of interest and other pumping in the area. If existing wells cannot be used, then new monitoring wells may be installed to establish these needed data for further groundwater analyses. The groundwater project would enter the investigatory phase at this point.

Investigatory groundwater monitoring wells may be installed to satisfy initial concerns about the suitability of a site for more permanent production wells. Hydrogeologists will also consult available field data from state and national geological surveys, reported well logs and local drilling companies. On the basis of the data, a hydrogeologist may prepare a map of the area with water table contours to guide his or her drilling, well installation and other geophysical investigation.

Cost of Investigation

Costs of these assessment factors will influence investigation and development. Early groundwater investigation may be funded by government to provide the basis for future groundwater development, considering such information as a public good. In the private development case, economic return on groundwater use will be a constraint on expenditures for groundwater resource evaluation. Public sector projects will likely focus on least-cost assessment of options having similar results. In any case, success of the investigation is driven significantly by the associated expenditures.

Foster (1989) offers a framework to evaluate investigation costs. First, a comparison of subsurface investigation method cost to drilling savings is done. If *m* and *n* are two different methods with costs of C_m and C_n and their average cost of a successful borehole can be described as *d/P*, *d* being average drilling cost and *P* the probability of success, then the method selected has a cost differential less than the saving in drilling costs, defined as:

$$(C_n - C_m) < ([d_n / P_n] - [d_m / P_m]). \quad (1.6.1)$$

We would choose method *n* if this relationship held for that method's costs. P, the probability of success, will be strongly affected by subsurface factors as well as aquifer yield-drawdown relationships and the groundwater quality acceptability (Foster, 1989, p. 56).

Second, the maximum expenditure that would be warranted for groundwater investigation could be estimated by

$$C_{\max} = (d[1 - P_n] / P_n), \quad (1.6.2)$$

where

d is the drilling cost

P_n is the probability of success for method

nP_n will typically be less than 1.0 (or 100%) for any method, since success does not just consider groundwater occurrence, but also yield-drawdown and water quality, and may include other factors specific to a location of potential water use. Curves for Equation 1.6.2 relating $C_{\max}$ to P_n can be developed (Foster, 1989, p. 57).

While the probability of success cannot be known in advance, a minimum accepted probability of success could be calculated by

$$P_{\min} = (dP_n) / (d - C_n P_n), \quad (1.6.3)$$

where the symbols are as previously used.

Equation 1.6.3 gives the fraction of boreholes that meet the success factors to allow a particular method to be considered for use in field investigation.

Several approaches to improve the cost-effectiveness of groundwater investigations may be considered. Typically, groundwater development activities evolve in stages: investigation, evaluation and development. Investigation involves risk capital investment. At this stage, integrating available data to identify high-cost aspects of the project, considering alternatives and examining possible drilling sites based on hydrogeologic assessment can provide information supportive of flexible decision-making. Geophysical exploration may include, but not be limited to (along with relative cost indication) (1) electrical resistivity measuring the slope of the water table and the concentration of ions in groundwater through transmitting current through the ground and recording voltage loss (low cost), (2) seismic refraction (medium cost)/reflection (high cost measuring compression waves traveling through the subsurface and reflecting off different densities of rock, (3) gravimetric surveys, measuring differences in densities of rock, saturation and faults (high cost), (4) electromagnetic surveys, recording the time for electromagnetic waves to reach different densities and reflecting back (low cost), (5) ground penetrating radar (low cost) and (6) borehole methods recording electrical resistivity, gamma radiation, neutron energy loss, temperature changes, acoustic patterns and fluid flow movement of the stratigraphic sequence in a borehole (high cost) (Driscoll, 1986, pp. 168–202; Sanner and Abbas, undated). Remote sensing and gravity measurements by satellites also provide further techniques of identifying the presence and elevation of groundwater (O'Hanlon, 2003).

Options for dealing with high cost factors can be considered before major investment is made. One course of action is to address one area at a time, concentrating on geophysical exploration and drilling equipment and coordinating results through a chief field hydrogeologist who can make immediate decisions on well depth, completion and screening of appropriate intervals for production and water quality acceptability.

Documentation of costs for groundwater resource investigation is not extensively and readily available. Foster (1989, p. 59) found that in the United Kingdom costs per production borehole for low-yield rural water supply wells were $2000 ($3050 in 2007 $US) compared with $50,000 ($76,260 in 2007 $US) for high-yield irrigation or municipal water wells. For high-yield wells, Foster (1989, p.59) reported costs of $100–$1000 ($153–$1525 in 2007 $US) per km^2 and $100–$250 ($153–$381) per m^3/day for groundwater production. Schwarz (1989, p. 74) identified that in Spain investigation wells cost one-tenth that of production wells. For a two-year exploration of a 10,000 km^2 area, a breakdown of total groundwater investigation costs of $600,000 (1989 $US) was experts, 1/2; field labor, 1/6; and equipment, 1/3. For 1989 in comparison, he found that rural water supply studies in Chad and Mali cost $5 per km^2 and $25 per km^2, respectively (1989 $US).

Recent experience in well siting investigation indicates (VIS, 2019):

- For a municipal well, doing a detailed Phase 1 study (analyzing geology and well logs, potential quantity, groundwater quality, potential contaminant sources, land ownership, utility location and economic data), costs may be between $10,000-$25,000.
- An irrigation well siting investigation, assuming water of known quality and no source protection problems, may cost around $6,000.
- For a managed aquifer recharge well, site investigation costs may be $40,000 to over $200,000, depending on site complexity and project objectives.

Megawatershed exploration for groundwater is an approach developed based on investigation of groundwater occurrence in fault and fracture zones in bedrock that transcends watershed boundaries and topographic features. Estimates by the principal investigators of this method suggest that accessible groundwater resources may be 10–100 times larger than current calculations for many water-stressed parts of the world (EarthWater Global, 2009). The approach combines investigation using currently available geological data with satellite-based remote sensing of faults and fractures in deep bedrock and geographic information systems to analyze fracture width, density and connectedness and relate those characteristics to fluid flow (Bisson and Lehr, 2004, p. 136). Use of this approach has led to developing high-volume groundwater supplies (5,000–75,000 m^3/day depending on location) in Trinidad, Tobago, Sudan, Somalia and New Hampshire. Long-term effects on water tables and on the quality and habitat of connected coastal receiving waters have not been documented.

COSTS AND BENEFITS OF GROUNDWATER INVESTIGATION

Consideration of costs and benefits of a resource investigation guide the determination of proceeding on an investment in groundwater supply. Practically, a groundwater development program, in particular, in an area with no other water source alternative and with scarce economic resources, should proceed by evaluating the benefits to society or the communities served and the costs of developing the groundwater resource. Chapters 9, 12 and 13 will go into more detail on comparing costs and benefits. Box 1.6.1, Cost-Benefit Model for Groundwater Investigation, provides an initial view of cost–benefit analysis applied to groundwater field investigations.

BOX 1.6.1 COST-BENEFIT MODEL FOR GROUNDWATER INVESTIGATION

A simplified approach to evaluate costs and benefits of groundwater resource investigation (Schwarz, 1989) considers success and failure of multiple attempts (subsurface explorations) to find groundwater, drawing on Bayesian analysis and applying the resulting probabilities. The outcomes are: (1) finding water (success), with probability P_q and (2) not finding water (failure), with probability $1–P_q$. Many factors related to local conditions affect success, such as achieving a particular continuous pumping rate, location within a specified distance of the point of use and finding water within a predetermined depth. Failure is then the opposite of these success factors.

The model description is:

Probability of water found to provide quantity q (a priori estimate) = P_q
Probability of water not found in sufficient quantity = $1–P_q$
Cost of subsurface exploration providing sufficient information = C_I
Probability of subsurface exploration producing required information establishing groundwater availability (a priori estimate) = P_I
Probability of failing to obtain sufficient information from subsurface exploration = $1–P_I$
Cost of groundwater development of quantity $q = C_q$
Benefits (discounted to net present value) of groundwater supply of quantity $q = B_q$

Incorporation of probabilities for the quantity desired and the success of investigation in the model addresses the notion of planning for "expected" success and benefits.

The decision maker may choose among three groundwater development policies after evaluating the available information:

1. Assume successful investigation, develop resource
2. Attempt to develop resource without investigation, relying solely on available information
3. Do not develop resource

Evaluation of these policies considers expected "net benefits":

4. Investigation followed by development: $(Bq–Cq)Pq – (CI/PI)$
5. No investigation, but proceed with development: $Bq\ Pq – Cq$
6. No investigation and no development: 0

The decision rules for development proceeding use the net benefits framework:

Investigation followed by development is substantiated based on two circumstances applying to the situation:

1. When $(B_q–C_q)P_q–(C_I/P_I) > 0$, i.e., net benefits must be greater than zero, or alternatively rearranging for $(B_q–C_q)P_q\ P_I > C_I$, i.e., development benefits must exceed investigation costs

(Continued)

BOX 1.6.1 (Continued)

2. When $(B_q–C_q)P_q–(C_I/P_I) > B_q\ P_q–C_q$, i.e., net benefits from investigation are greater than net benefits from development with no prior investigation

Development without investigation is warranted when: $B_q\ P_q–C_q > 0$, i.e., net benefits are greater than zero, or restated, $B_q\ P_q > C_q$, i.e., benefits are greater than costs.

The ranking process for selecting groundwater development projects may evaluate:

1. C_I/P_I *using cost-effectiveness to choose* among investigation approaches for the same project
2. $(B_q-C_q)P_q–(C_I/P_I)$ applying the net benefits criterion to select among investigation approaches for several development projects
3. Depending on water need of a specified quantity, an alternative approach could consider ranking by P_q, balancing investigation costs with reduced variance of obtaining the quantity desired.

Source: Adapted and summarized from Schwarz, J. 1989. *Groundwater Economics: Selected Papers from a United Nations Symposium Held in Barcelona, Spain,* Elsevier Science Publishers B.V., Amsterdam, the Netherlands. Pages 67–81.

Groundwater resource investigation for water supply has benefits that have some elements that have been quantified in the available information (EarthWaterGlobal, 2009):

- Capital cost: low, compared with alternatives of dams and reservoirs and desalination.
- Treatment requirements: low, assuming high-quality water not needing treatment.
- Operating costs: low, assuming principal cost is for operating and maintaining wells and pumps.
- Environmental effects: confined mainly to the subsurface and groundwater fracture flow.
- Land requirements: areas around wellheads, approximately 0.04 hectare/well.
- Water supply reliability: high, assuming continued natural flow sources.
- Time to initial production: 6 months from beginning of project compared with 10 years for dams and 1–5 years for desalination projects.
- Contaminant risk: low, since the sources are deep.

LAND OR PROPERTY ENTRY

Fundamentally, access to groundwater means obtaining entry to the land in the vicinity of the aquifer of interest and, if the property does not have a free flowing spring, having the right to install a well and produce groundwater (NGWA, 2018). The conditions for land to be entered for groundwater production involve:

1. Ownership or permission of the owner or trustee to use the land.
2. Ensuring that the property rights allow subsurface intrusion and groundwater production.
3. Permission of the governmental jurisdiction(s) in the area to install a well and produce groundwater.
4. Location and distance from underground features, such as pipelines and cables or distance from drains, septic tanks and cesspools.

5. Overland access and egress for drilling equipment to enter and leave the property.
6. The capacity of the land to accommodate well-drilling equipment, including sufficient land area and overhead space to enter and position a drill rig.

Costs of land or property entry, equipment positioning, permission for well installation and groundwater production must be considered in the cost of groundwater access. Otherwise, groundwater access may not be possible. These land and property costs will vary by location and must be evaluated on a site-by-site basis.

FACTORS AFFECTING WELL LOCATION

Many factors affect well location. As will be discussed here, well location has both a horizontal (or areal) and vertical (or depth) component. Some of the major factors include (adapted from NGWA, 2018)

1. Property ownership and its access to others
2. Water rights
3. Sources or pathways of contamination
4. Setback distances from contaminant sources or underground features
5. Geology
6. Elevation, slope and drainage of land
7. Depth to desired water-bearing zone
8. Location of other wells
9. Subsurface structures (avoidance of)
10. Overhead access for drill rig
11. Natural water quality
12. Size of well and flow rate
13. Accessibility for well cleaning, treatment, repair, testing, inspection and related activities

Property ownership will be a major determining factor as to the location of wells, either for water supply or for monitoring. Unless access to the wellhead can be controlled, a certain amount of liability is associated with a well. Therefore, any owner agreement to have a well on his/her property is significant. Liability can take several forms: if not properly marked or fenced, the wellhead may not be highly visible and be a physical hazard; if the wellhead casing or surface apron is damaged, contaminants can potentially enter the aquifer affecting the well's water quality and possibly that of other wells; subsurface engineered structures need to be located prior to construction, but even with this precaution taken, well installation can cut or penetrate such structures, disrupting other services and potentially creating a dangerous situation. These situations create additional cost for well installation, especially if not taken into account prior to decisions of location and installation. Setback distances from other features should be taken into account so as not create future costs because of potential contamination or interferences. Table 1.6.1 highlights key setback distance factors for domestic wells.

Volume of water to be pumped also affects both the vertical location of the well completion and the location across the area being considered for the well. Pumping by its nature draws water from the zone beyond the well and if pumping is excessive, it can affect production at the surrounding wells on neighboring properties. When large volumes of water are needed, it may be necessary to install larger capacity wells in deeper aquifers. This, of course, raises the cost of production. However, one larger, deeper, higher capacity well may be preferable to several shallower wells. Where water tables are dropping at high rates due to water use, deeper high-production wells must be used. The long-term problem with this approach in some locations is that such pumping may draw water from the overlying shallower aquifers, thus impacting the water available in all the useable aquifers. Impermeable geologic strata (also called aquicludes) are never totally impermeable, allowing water to move down to lower aquifers, albeit, at a slower rate, which can be accelerated by deep, high-volume pumping.

TABLE 1.6.1
Setback Distances for Locating Domestic Wells
Minimum distances from the wellhead should be used in locating a well, unless state or local codes or regulations require greater setback distances.

Feature	Setback (meters; feet)
Cesspool (receiving raw sewage)	60.96 m; 200 ft
Pit, privy, filter bed	15.24 m; 50 ft
Septic tank, tile sewer, foundation drain	15.24 m; 50 ft
Iron sewer with approved mechanical joints	3.05 m; 10 ft
Pumphouse floor drain	0.61m; 2 ft
Property boundary	1.52 m; 5 ft
Outer boundary of any road	6.10 m; 20 ft
Landfill, garbage dump	60.96 m; 200 ft

Source: Adapted from: National Ground Water Association. 2018. Planning for A Water Well. http://www.wellowner.org/basics/planning-for-a-water-well/

Factors influencing well design and installation include (USGS, 1997, p. 20)

- Nature of the subsurface materials that overlie and comprise the aquifer (ranging from unconsolidated sand to fractured bedrock with dissolution channels)
- Well casing and screen material depending on subsurface conditions
- Screen length and type
- Diameter of casing and screen (or open borehole)
- Depth to static water level
- Depth to the top of the aquifer of interest
- Depth to the zone in the aquifer to be monitored or produced
- Expected duration of use

Local or jurisdictional requirements will also affect the materials and other factors of the well. These requirements may vary from state to state and include well termination at or above ground surface, depth of grout, grout mix specifications, protective casing use, type of casing material, casing dimensions, well capping type, discharge control, disinfection, correction of damage, yield tests and well sample tests, as well as abandonment procedures (Job and Gabanski, 1987). Table 1.6.2 demonstrates the differences among U.S. states in well construction requirements. Notably, some states have requirements for well termination 107 cm (42 inches) above the highest known flood elevation and other states do not reference any elevation. Depth of grout in the annular space ranges from no minimum to 9.14 m (30 feet). Cap requirements to safeguard direct access to the well casing vary from none to welded or locking caps. Thus, in some states, a well may cost significantly more than in adjacent states.

Additional factors for locating high-volume irrigation, industrial or urban water supply wells may include (Niñerola, 1989, p. 84)

- Energy costs for pumping water to the ground surface
- Distance from the wellhead or field to the location of use
- Distance for extending energy supply to the site
- Ease of access for machinery and materials

TABLE 1.6.2
Selected Well Construction Requirements in 12 North Central U.S. States

State[a]/ Well Type Affected	Location Requirement	Termination Requirement[h]	Minimum Grouting Depth[h]	Grout Mix Specifications	Casing Type Specifications	Locking Cap Requirements	Disinfection Requirements
Colorado/ all	—	30.48 cm above GS[b]	3.05–6. lm	—	—	—	Yes
Illinois/all	Yes	60.96 cm above HFE[c]	-	Yes	PVC[d] or steel	—	—
Indiana/ PWS[e]	—	—	—	—	—	—	—
Iowa/all	Yes/45.72 m from contaminated Source	45.72 cm above GS 106.68 cm above HFE	6.1m	Yes	Steel	No (welded or threaded steel cap)	Yes
Michigan/ all	Yes/Health Officer approval in certain cases[f]	30.48 cm above GS	7.62 m	Yes	PVC; no steel for monitoring wells	Yes	Yes
Minnesota/ Mon[g]	Yes/in contaminated Sources	30.48 cm above GS 60.96 cm above HFE	9.14 m	Yes	PVC or steel	Yes	Alternative methods
Nebraska/ Mon	Depends on geology	30.48 cm above GS	3.05 m	Yes	PVC or steel	—	Yes
North Dakota/ all	Yes/away from contaminated Sources	30.48 cm above GS	6.1m	Yes	PVC or steel	NO (welded plate)	Yes
Ohio/All	Yes/away from contaminated Sources	20.32 cm above GS	Aquifer to GS	—	PVC or steel	—	Yes
South Dakota/ NonPWS	Yes/away from contaminated Sources	30.48 cm above GS 60.96 cm above HFE	6.1m	Yes	PVC or steel	—	Yes
Wisconsin/ Mon	Yes/for groundwater monitoring	30.48 cm above GS	—	Yes	PVC or steel	Yes	—
Wyoming/ all	Yes/away from pollution sources	60.96 cm above HFE 30.48 cm above GS	3.05 m	Yes	PVC or steel	No (temporary seal)	Yes

Source: Modified from Job, C. and Gabanski, G., Monitoring wells need consistent regulation, *Superfund '87, Proceedings of the 8th National Conference,* Washington, DC, The Hazardous Materials Control Research Institute, Silver Spring, MD, November 1987 (Reprinted in *Waste Age,* 1988, 164–170).

[a] State requirements drawn from the following sources:

Colorado: Rules and regulations governing construction of wells and the installation of pumping equipment.

Illinois: Water well construction code, Chapter III 1/2.

Indiana: Indiana code 91–11, 13.

Iowa: Iowa Water Supply Facilities Design Standard, Chapter 3.

(*Continued*)

TABLE 1.6.2 (*Continued*)
Selected Well Construction Requirements in 12 North Central U.S. States

Michigan: Michigan Act 368, P.A. 1978, Part 127 and Rules; Act 315, P.A. 1969.

Minnesota: Minnesota Rules Chapter 4725.

North Dakota: North Dakota Century Code, Chapters 43–35 and 33–18.

Nebraska: Guidelines for Design and Construction of Water-Quality Monitoring Wells, NRS Chapter 12.

Ohio: ORC-3745-9, ORC-3745-41, ORC 3701-28.

South Dakota: South Dakota Code Chapter 74:02404.

Wisconsin: Guidelines for Monitoring Well Installation; Wisconsin Administrative Code Chapter NR 110, Chapter NR 214.

Wyoming: Part III—Water Well Minimum Construction Standards; Chapters VIII, IX, X, Water Quality Rules and Regulations.

[b] GS means "ground surface."

[c] HFE means "highest-known flood elevation."

[d] PVC means "polyvinyl chloride."

[e] PWS means "public water supply."

[f] Health Officer Approval needed near polluted sources; permission needed for flood areas.

[g] Mon means "monitoring well only."

[h] 15.24cm = 6in, 30.48cm = 12in, 60.96m = 24in; 3.05m = 10ft, 6.1m = 20ft, 9.14m = 30ft

These factors may also affect long-term operation and maintenance of the groundwater production site and its wells.

Job and Gabanski (1987) show that the differences in state well requirements, in particular, for monitoring wells can have a significant result on costs, which can vary from state to state. They found that for two wells in the same hydrogeologic conditions (glacial till of interbedded clay and sand with some boulders) and depth (15.24 m; 50 ft), differences in requirements might cause one well to be twice as expensive as another well for the same purpose (monitoring). Most of the cost difference is in the type of casing used, steel or polyvinyl chloride (PVC) plastic. If the less expensive PVC casing is used with the more protective package at the wellhead, the cost difference is about 25%. These differences are described in Table 1.6.3. More stringent requirements add nearly 30% more to costs, in this example. Additional requirements would increase costs further.

More recent published monitoring well installation cost information for California's Central Valley indicates the following per unit cost results for a range of depths without a description of the geology (Choy, 2016). Factors affecting costs cited are geology, casing type, well depth and borehole diameter.

Well Depth	Cost per Unit
30.48m (100ft) or less	\$230/m (\$70/ft) to \$328/m (\$100/ft)
30.48m-152.4m (100ft-500ft)	\$492/m (\$150/ft) to \$656/m (\$200/ft)
Over 152.4m (500ft)	\$820/m (\$250/ft) to \$1,148/m (\$350/ft)

Not included in the costs above are land acquisition, permitting and other expenses.

Comparing the costs per unit for the recent results above for California to the results of Table 1.6.3 for the north central United States for depths of less than 30.48m (100ft) with both in 2018\$US, the

TABLE 1.6.3
Comparative Costs for Monitoring Wells under Different Regulation Requirements

Well 1[a]	Cost[b] (1987 $US)	Well 2[a]	Cost[b] (1987 $US)
Drilling and casing		*Drilling and casing*	
Depth: 15.24 m (50 ft) Drilling with truck rig and 8.26 cm (3.25 in) hollow stem auger Soil sampling conducted Stainless steel screen (5.08 cm × 3.05 m; 2 in X 10 ft) and casing		Depth: 15.24 m (50 ft) Drilling with truck rig and 8.26 cm (3.25 in) hollow stem auger Soil sampling conducted PVC[c] screen (5.08 cm × 3.05 m; 2 in X 10 ft) and casing, flush grade	
Subtotals	$1400–$1550		$750–$900
Pack/grout/seal		*Pack/grout/seal*	
Sand and gravel pack Cement grout from Seal to GS[d] Bentonite seal		Sand and gravel pack Cement grout 3.05 m (10 ft) to GS[d] Native backfill up to 3.05 m (10 ft) from GS[d]	
Subtotals	$300		$250
Well protection	$250	*Well protection*	$10
Protective casing Three protective posts Locking cap		Threaded steel cap	
Subtotals			
Totals	$1950–$2100		$1010–$1160
$133/m or $41/ft		$104/m or $32/ft	

Source: Job, C. and Gabanski, G., Monitoring wells need consistent regulation, *Superfund '87, Proceedings of the 8th National Conference,* Washington, DC, The Hazardous Materials Control Research Institute, Silver Spring, MD, November 1987 (Reprinted in *Waste Age,* 1988, –170). Results reflect costs for north central United States.

[a] Assumes glacial till of interbedded clay and sand with some boulders.

[b] Costs do not include mobilization or utility clearance costs and will vary for each state.

[c] PVC means "polyvinyl chloride" plastic.

[d] GS means "ground surface."

costs are somewhat similar: the California cost range is $243–$347/m ($74–$106/ft) while the north central US cost range is $232/m–$297/m ($71–$91/ft).

WELL PERMITS

Installation requirements for wells vary from state to state, as noted, or country to country. Some local jurisdictions even have well installation requirements. These requirements are to protect the well owner and to protect against contamination of the aquifer. The types of well installation requirements are

1. Well installation standards
2. Well installer certification
3. Well permit

Well Installation Standards

Some jurisdictions have requirements on the depth of the screen below the ground level or the water table, depth of grout, type of grout, surface (wellhead) completion, type of casing, setbacks and other requirements, based on the type of well (Job and Gabanski, 1987).

Well Installer Certification

Other jurisdictions may have requirements concerning the knowledge, training and experience of well installers to be met before a well installer is allowed to operate a well installation business in the state. Often, knowledge of specific well standards, such as those set by voluntary standards organizations as ASTM, is required to be tested (Job and Gabanski, 1987).

Well Permit

In some states (e.g., Ohio), a property owner must apply for a well permit. Depending on the water volume to be produced and the water use, the state will decide whether a well permit be issued. A key factor is whether pumping would interfere with other groundwater uses nearby. Domestic well permit application requirement examples and costs are provided for selected states and localities in Box 1.6.2.

BOX 1.6.2 DOMESTIC WELL PERMIT APPLICATION REQUIREMENTS AND COSTS FOR SELECTED STATES

Colorado

State Water Well Permit Requirements & Regulations

Applicant name, address and contact; full legal description and location of the property including (Sec,township,range + .1/4-1/4) and / or G.P.S coordinates Northing & Easting using Nad 83 Datum; well use, pumping rate, depth, amount withdrawn, aquifer completion, water rights; copy of a warranty deed describing the property; driller license.

Water Well Permit fee = $100

Maryland

Frederick County Health Department Well Application

Applicant name, street address and contact; subdivision; Section or Block; Lot; acreage or sq ft; Tax Map, Parcel, Tax ID No.; type of water supply now; well to furnish water to Home, Farm, Public or Private Water Co, Industrial/Commercial; Test Well; Heat Pump

Frederick Co.Maryland water well permit fee = $160

New Mexico

Office of the State Engineer

Applicant name, address and contact information; NM State Plane or Lat/Long; County; quarter or half, section, township, range; Land Grant Name (if applicable); Lot no. Block no., Unit/Track, Subdivision; Hydrographic Survey; Land owner; purpose of use; well driller and license no.; approximate depth; outside diameter of well casing.

New Water Well Permit Fee = $125

Anaheim, California

Anaheim Public Utilities

Well Location address; Site plan (including roads, sewers, surface water, and possible sources of contamination within 200 feet of well site); well name; well type; boring diameter;

(*Continued*)

BOX 1.6.2 (Continued)

casing diameter; depth; screen intervals; start date; vertical diagram detailing depths, sizes, and thicknesses of well casing, annular seal, and screens; pertinent geological features.

New Water Well Permit Fee = $221

Olmstead County, Minnesota

Olmstead Count Planning Department

Property owner and address; contractor and address, well location (subdivision, lot, block, township, range section, quarter), well address, list of properties to be served by well; well use; minimum isolation distances to sources of contamination, buildings, underground utilities, streams and water bodies, areas subject to flooding; casing material; depth of geologic formations penetrated, total depth, static water level; borehole diameter; number of dwellings served; well appurtenances to be installed

New Water Well Permit Fee = $639

Monitoring Well Permit Fee = $439

Sources: Anaheim (California) Public Utilities. 2019. Well/Boring Permit. URL: Anaheim.net/DocumentCenter/View/4066/Well-Permit-Application (Accessed May 5, 2020).

Colorado Division of Water Resources. 2020. General Purpose Water Well Permit Application. URL: dwrpermitsonline@state.co.us and signnow.com/jsfiller-desk17/?projectid+451201009#053938f7fafee1d1622a12264caa41a2 (Accessed May 5, 2020).

Frederick County, Maryland, Health Department. 2020. Frederick County Health Department Well Application. URL: health.frederickcountymd.gov/DocumentCenter/View/3001/WELL-PERMIT-TEMPLATE?bidid= (Accessed May 5, 2020).

New Mexico Office of the State Engineer. 2017. Application for Permit to Use Underground Waters. URL: ose.state.nm.us/WR/Forms/WR-01%20Application%20for%20Permit%20to%20Use%20Use%20Underground%20Waters%20(72-12-1.1.%2072-12-1.2.%20or%2072-12-1.3)_2017-06-30_final.pdf (Accessed May 5, 2020).

Olmstead Count Planning Department. 2020. New Well Construction Permit Application. URL: co.olmstead.mn.us/planning/applicationsfees/BldgWellSeptic/Documents/Well%20Permits/New%20Well%20Construction%20Permit%20Application.pdf (Accessed May 5, 2020).

All of these jurisdictional requirements are for the protection of the well users or owners. While they add cost, the states having the standards have determined the standards to be of value in protecting the health and welfare of its citizens from inappropriate well installation practices and in minimizing damages to adjacent groundwater users.

WELL DRILLING AND INSTALLATION

The use of the well and conditions of the area and subsurface should guide construction of a well through drilling and installation. Well construction is detailed in other texts (Driscoll, 1986; USGS, 1997). Unconsolidated sand and sand and gravel typically provide easier drilling conditions and are probably the least costly. Box 1.6.3 describes briefly different well drilling methods.

BOX 1.6.3 DRILLING METHODS

Dug wells. Constructed by using simple hand earth-removal tools (shovels, trowels, etc.), usually to several meters in depth, with a brick, stone or concrete wall for casing and open at the bottom.

Drilled wells. Require a motorized drill rig and downhole equipment to loosen and remove unconsolidated material or cuttings from consolidated geologic formations. Several techniques may be used, the principal ones being

- *Cable tool.* This method uses gravity to drive a drill bit on the end of a cable hanging from the tower of a vehicle mounted drill rig into the subsurface. Drill cuttings are removed by pumping water in the borehole to become slurry with the cuttings and then pumping the slurry out of the hole. Wells in unconsolidated material have reached depths of 305 m.
- *Rotary drilling.* This technique also uses a drill rig and turns a drill bit into the subsurface to loosen the geologic material. Either water or air is pumped into the borehole to combine with the cuttings and be pumped out. Variations on this technique have achieved depths in excess of 610 m.
- *Air hammer.* A drill pipe of special alloy steel is driven in successive rapid strikes against the rock with the cuttings brought to the surface by the air used for the hammer strokes.
- *Jet percussion.* A high-velocity water stream is used in combination with the percussion drilling to remove cuttings and keep the drill bit clean so it is striking uncut rock. Depths ranging from 61 m to 305 m have been reached with this technique.
- *Hydraulic percussion.* This technique is similar to jet percussion except that a check valve allows the water-cuttings mixture to be captured and drawn up the casing with each stroke.
- *Jetting washbore.* A high-velocity water stream is jetted into a casing to remove unconsolidated earth materials.
- *Earth augur.* The vehicle-mounted augur turns and bores in the subsurface and brings the loosened material to the surface. Depths of 76.2 m for a bucket augur, 36 m for a solid-stem augur and 91.5 m for a hollow-stem angur have been achieved with this method.
- *Other variations.* Variations of the above drilling methods are described in detail by Driscoll (1986).

Driven wells. This technique can only be used in unconsolidated geologic material that do not contain rocks. Hollow metal casing with a well point can be driven by hand to depths of 9.1 m or machine to 15.2 m or more.

Sources: Driscoll, F.G., *Groundwater and Wells*, Johnson Filtration Systems, Inc., St. Paul, MN, 1986, 268–339.

Niñerola, S., Groundwater Economics: Selected Papers from a United Nations Symposium Held in Barcelona, Spain, Elsevier Science Publishers B.V., Amsterdam, the Netherlands, 1989, 90–96.

Photographs in this box provided through the courtesy of Richard Laton and the National Ground Water Association and reproduced with permission.

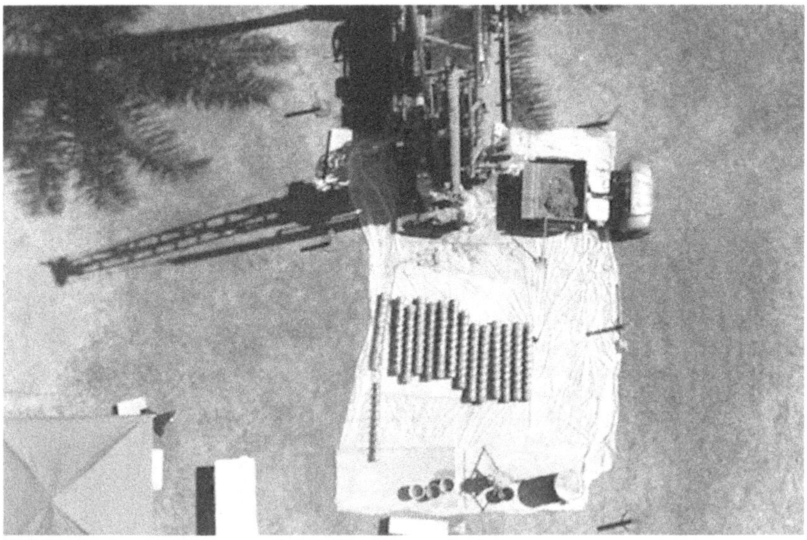

Hollow stem drilling rig

Mud rotary drilling rig

Drill bits

Air rotary drilling rig

More drill bits

TABLE 1.6.4
Comparison of Average Monitoring Well Installation Costs by Well Type and Hydrogeologic Setting in the United States

Installed Well Costs per Monitoring Well

	Well Depth		Hydrogeologic Setting (1995 $US)		
Well Type	**(m)**	**Well Casing**	**Sand**	**Till**	**Bedrock**
1	15.24	PVC	1,577	2,005	2,771
2	30.48	PVC	3,170	4,253	4,646
3	30.48	Steel	4,964	6,015	6,279
4	60.96	Steel	12,757	14,865	13,448

Unit (per m) Costs for Monitoring Wells

	Well Depth		Hydrogeologic Setting (1995 $US)		
Well Type	**(m)**	**Well Casing**	**Sand**	**Till**	**Bedrock**
1	15.24	PVC	31.54	40.10	55.42
2	30.48	PVC	31.70	42.53	46.46
3	30.48	Steel	49.64	60.15	62.79
4	60.96	Steel	63.78	74.33	67.24

Monitoring wells of different types in different hydrogeologic settings used in monitoring well construction are presented in Table 1.6.4. The table clearly shows that hydrogeologic settings of sand are the least expensive to install a well in, of till are intermediate in cost and of bedrock are the most expensive. Also, steel-cased wells are more expensive than PVC-cased wells. Deeper wells tend to be steel cased.

Table 1.6.4 also shows that for a particular hydrogeologic setting, stainless steel casing is more expensive to install than PVC casing by 57% for a 15.2 m (50 ft) well and 41% for 30.5 m (100 ft) well. Generally, costs increase with depth and more complicated hydrogeologic setting. This increase is because of the increased cost of stainless steel, and the time and labor it takes to add more drilling flights and generally complete repetitive tasks over longer distances in the borehole with each meter of depth.

Installation costs for some production wells are provided in Table 1.6.5. Deeper wells and more complex hydrogeologic settings have higher costs. Bedrock presents the greatest difficulty in drilling and typically is most expensive for well installation. Care must be taken in well installation that conditions are not created that could cause aquifer contamination (USGS, 1997, p. 47).

Additionally, Navarro (1989, p. 50) evaluated costs of irrigation wells in Spain as a percentage of total production cost, exclusive of externalities costs, and found the following cost distribution:

- Drilling: 9.7%
- Pump installation: 15.4%
- Operation and maintenance: 41.1%
- Water distribution: 29.6%.

WELL SIZE

High Capacity Wells

Well size depends on its use. A large capacity water supply well may be 30 or more centimeters in diameter and made of iron casing. Its depth may vary from 30 to 460 m (100 to 1,500 ft) or more. Large irrigation wells may also be of this size. The more water that needs to be produced over a given time, the larger the diameter of the casing must be.

TABLE 1.6.5
Drilling and Installation Costs for Production Wells

Various (anonymous by request) water well installers in the United States provided information in 1989 and 1995 on the costs of installing municipal wells. Locations relying on aquifers in deeper bedrock formations have higher costs, as would be expected.

Municipal Water Supply Wells
Capacity and Price of Average Well Installed from Examples in the United States
($US for Year Indicated)

	August 1989		September 1995	
Location	**m³/Day**	**Price**	**m³/Day**	**Price**
Upper Midwest	2,840	$80,000	2,840	$93,000
Central Midwest	2,840	$75,000	2,840	$100,000
Gulf Coast	4,920	$175,000	4,920	$500,000
Southwest	6,435	$150,000	8,328–16,277	$400,000–$500,000

Price of 5678 m³/Day Well Installed in the United States
($US for Year Indicated)

Location	August 1989	September 1995
Upper Midwest	$125,000	$145,000
Central Midwest	$150,000	$200,000
Gulf Coast	$200,000	$600,000[a]
Southwest	$140,000	—[b]

Source: Various industry sources.

[a] Typically deeper wells in excess of 305 m.

[b] Drilling larger, deeper wells at that time than previously installed.

Monitoring Wells and Piezometers

A monitoring well used to check water level or quality may be 4 or more cm (1.5 inches) in diameter and may be made of iron, stainless steel or PVC, depending on water quality testing requirements. A monitoring well used to determine water level and test physical characteristics of the aquifer is called a piezometer. Piezometers can also be installed underwater in streambanks or shorelines to test the groundwater discharge or recharge relationships and even to collect water samples for water quality tests. In-stream piezometers can be quite small. Mini-piezometers having an internal diameter of 10 mm (0.394 inches) or smaller may be installed to depths of less than two meters (6.56 feet) in a stream bed. Larger in-stream piezometers (diameter of 15–25 mm or 0.591–0.984 inches; may also be larger) are also used and installed to several meters in unconsolidated stream sediments. The cost of a larger in-stream piezometer system (40 heads and base pipes and 150 extension pipes) and installation equipment, including generator, is approximately $2,000 (Burk and Cook, 2015).

PRIVATE DOMESTIC WELLS

Most private domestic wells are typically installed to draw water from the first aquifer and are deep enough to provide water during drier seasons when recharge is less and the water table may fall (be lower). Approximately, 34 million people in the United States are served by private domestic wells (NGWA, 2016b).

Capital Cost and Related Factors

Cost for installation of a private domestic well with a pump to bring the groundwater to the subsurface can average from $49 to $98 per meter ($15 to $30 per foot). The depth of residential wells in the United States may range from 9 to 122 meters (30 to 400 feet) (Zimmerman, 2015; CHI, 2019). The price per meter or foot may also include well permit fees and well seal, cap and apron (Brenner, 2019). A well inspection fee may also be charged in some localities. An easily installed well may cost from $2,000 to $10,000, depending on depth and geology. Beyond the well installation, waterlines and electricity to the well need to be installed as well as a control panel and water treatment system, if water quality needs to be addressed, which can add $2,000 to $8,000 to the cost of delivering water to the house (Zimmerman, 2015; CHI, 2019). Private wells are less costly to install, generally. This installation cost depends on the location and type of geology. Sandy subsurface environments are more easily drilled. However, in rocky subsurfaces, such as glacial till with boulders, or in bedrock, well installation is more difficult and costly. Shallow bedrock wells may often not be reliable. Residences locating in areas of shallow bedrock aquifers may have other serious problems such as wet basements. The location of wells and residential activities should be suited to the hydrologic conditions. Otherwise, additional costs not anticipated in the original investment will be incurred. These costs may include needing to truck water in, inability to use basements because construction has changed underground flow causing wet basements and wet yards that cannot be used during certain times of the year. The point here is that completing a domestic well may not mean the location is suitable for residential use.

A well in a low-yield formation that produces 568 L/day (150 gallons/day) may be adequate to sustain a family of four. However, this rate of production is probably at the margin, given the water use of today's water using appliances and bathroom fixtures in the United States.

If water quality needs improvement, residential units that treat at either point of entry (POE) to the home or point of use (POU) can cost from $248 to $1,542, depending on technology and treatment needs (see Chapter 7) (USEPA, 2007).

Operation and Maintenance

Operating a domestic well each day has costs for pumping water up the well. At an electricity price of $0.11/kWh (US EIA, 2019), the cost to pump 568 liters per day (150 gallons/day), a minimal amount of water for a family of four in the United States in 2019, up a 9-meter (30-foot) deep well is $0.0031 per day. This same volume pumped up a 122-meter (400–foot) would cost $0.041 per day (See Box 1.6.5 for calculations for residential groundwater use of 1,136 liters/day or 300 gallons/day.) Treating the water may done by a variety of means if the water quality needs improvement. Water treatment will be covered in Chapter 7. Depending on the residential technology, operating costs may have an average daily cost of $0.41.

Maintaining the well can be aided by having an annual well inspection. Routine well inspection provides for checking of proper operation and that conditions exist for safe water supply. A well inspection involves examining the wellhead and the area around the well for: damage to well casing and apron; appropriate setback distances from buildings and potential microbial and chemical contamination sources; adequate equipment ventilation; secure electrical connections; and necessary elevation to reduce floodwater effects. An inspection of a private domestic water well by a qualified water well professional may range from $80 to $325 depending on the location and conditions of the well (CWT, 2018; Job, 2017).

MONITORING WELLS

Monitoring well costs vary by hydrogeologic setting and well type. Table 1.6.4 summarizes more information from eight well drilling companies across the United States. Based on the figures in

Table 1.6.4, steel casing installed costs about 35 to 55% more than PVC casing installed, depending on the hydrogeologic setting, with sand settings less than bedrock settings.

To monitor several groundwater zones, multilevel monitoring wells may be constructed rather than installing a monitoring well for each zone (referred to as cluster type installation). For example, a multilevel monitoring well installed to a depth of 50 feet with four monitoring zones may cost from $1650 to $7300, depending on the geologic conditions and monitoring requirements. The multilevel approach uses a single well to sample many zones, thereby having less disturbance of the subsurface (Cooper, 2019).

LARGE WATER SUPPLY WELL

In part because of competitive pricing within the well drilling industry, well drilling and installation costs are not well documented in the literature. Niñerola (1989) provides estimated unit (per meter) costs for municipal and irrigation wells in Spain in 1987 for different types of installation: percussion drilling and reverse rotary drilling. Notably, deeper wells and larger casing show diseconomies of scale for well installation costs with both greater depth and larger casing per meter and diameter, respectively. Niñerola found reverse rotary drilling unit costs to be 71%–90% that of percussion drilling depending on depth and casing diameter.

MUNICIPAL WELLS

Water supply well costs vary across a country, based on the factors of well location (noted above), the volume of water needed and regional labor and equipment rates. Table 1.6.5 gives a perspective of how variable these well costs may be for public water supply wells in the United States. From the information provided, it appears that for the regions compared, municipal water well prices generally rose faster than the national inflation rate. While these are not a statistically valid result, the data do show some interesting trends. The locations in the eastern half of the United States relied on shallower municipal wells, while the more western locations contracted for deeper and higher capacity wells. This circumstance also correlates with increased precipitation and increased potential for recharge in the more humid east. Larger volumes of water pumped from deeper locations require more energy, thus increasing pumping costs as a part of operation and maintenance costs. When larger volumes of higher quality water are consistently demanded, but are more remote (from deeper aquifers), the costs of water availability are higher to meet those requirements. Costs for a 5678 m^3/day (1.5 million gallons/day) municipal well ranged from $145,000 to $600,000 (1995 $US) depending on the hydrogeologic setting.

Glotfelty (2017) examined well installation and operation and maintenance costs for 70 high capacity municipal wells in central Arizona drilled from 1993 to 2011. The analysis examined three types of well screen performance and differences in capital and O&M costs by screen type with normalized costs for a 45.7 cm, 365.8 meter (18-inch diameter, 1200-foot) well. The three screen types were: low-carbon steel (lower capital cost), high strength low alloy steel (intermediate capital cost) and stainless steel (highest capital cost). The nature of the steel affected the timing and amount of scale build up on the screen and thus water flow (i.e., reduced groundwater flow through the well screen from scale clogging the screen slots), screen cleaning costs, well and pump replacement costs and energy costs over a 75-year life-cycle horizon. While the stainless steel screen had the highest capital cost, it had the lowest life-cycle cost (this will be covered in more detail in a later chapter) because of reduced scale accumulation. The capital costs (in 2011$US) of the LCS screened well were $512,300, for HSLA screened well were $567,000 and for SS screened well were $776,400, in 2019$US, $576,400, $638,000 and $873,600, respectively.

In implementing the results of an investigation of groundwater supply using the megawatershed method in Trinidad and Tobago, Bisson worked with the water authority to develop 75,708 m^3 per day supply for the two islands at a capital cost of less than $40 million (compared to a dam and reservoir proposal estimated to cost $60 million) with operating costs of $0.09 per m^3 (Bisson, 2004).

Agricultural Irrigation Wells

Agricultural irrigation wells are typically high-capacity (high pumping rate) wells. These wells draw a substantial amount of water for short periods during the year and growing season, usually over 3 months. Factors in sizing and locating a well include (Birt, 2013):

- Estimating the volume of irrigation water needed from each well – Some states may place a limit of the quantity of groundwater to be pumped over a specific time period.
- Knowledge of the aquifer – Well logs for nearby wells will be useful in understanding the local geology and depth to water table; a professional well driller or geologist working in the area can provide important information from experience and state well log records.
- Drilling a test hole – The test hole will provide information on soil and geologic formation characteristics, water table depth and water samples to check water quality for the crop and for well cleaning maintenance as the well becomes plugged from mineralization.
- Obtaining a permit to drill an irrigation well – The state may require information on well bore diameter, geologic formations being drilled, well casing size and material, and elevation of the well screen, latitude-longitude coordinates of the well and property boundary delineation.
- Proximity to neighboring wells – Because of large volume of groundwater withdrawals in a season and over years, irrigation pumping should not interfere with nearby water well users, whether agricultural, residential or commercial.

A significant benefit from agricultural irrigation is increased crop productivity with approximately 50% of crop income derived from 17% of the cultivated land that is irrigated (IA, 2019). An associated benefit is reliability of farm production when precipitation is variable and drought occurs. A related benefit is farm income maintenance through periods of irregular weather patterns and drought. A portion of irrigation water not consumed in evapotranspiration may return to the aquifer benefiting future groundwater use.

The cost of irrigation wells in the US Midwest may range from $50,000 to $75,000 (2013$US; $54,000 to $81,000 in 2018$US) for a 30–90 meter (100–300 foot) well with pump. Wells in bedrock may be 150–460 meters (500–1,500 feet) deep with costs of over $100,000 ($108,000 in 2018$US). While stainless steel screens may be more expensive, they may have longer useful life (similar to municipal well screen application) (Birt, 2013).

INJECTION WELLS AND HYDRAULIC FRACTURING

Injection Wells

Injection wells emplace fluids below ground surface in porous geologic formations. Fluids may be oil and gas production brines or municipal wastewater and are injected into deep wells to protect overlying fresh groundwater sources separated by impermeable geologic strata. Often, injection of fluids is done under pressure. The geologic formations used may be deep sandstone or limestone, or a shallow soil zone (USEPA, 2016). Materials for injection well casing are steel alloys or fiberglass with the outer casing set in a chemically resistant grout (Clark et al., 2018). Injection wells have a range of applications. Typically, injection wells are used for disposal of water, wastewater and liquid waste into the subsurface. Figure 1.6.2 depicts an injection well. Uses of deep injection wells have expanded to include landfill leachate, contaminated groundwater, reverse osmosis concentrate and fish farm wastewater (Clark et al., 2018). For wastewater disposal, these wells are regulated in the United States under the Safe Drinking Water Act, Section 1414, which prohibits injection of waste into or above an aquifer used, or potentially used, for drinking water purposes and must meet federal and state treatment standards. In the European Union, the Water Framework Directive 2000/60/EC, Article 11(3) (j) prohibits injection of regulated pollutants into groundwater, controls the injection of waters containing substances from oil, gas and mining extraction, and construction activities and authorizes reinjection of waters for geothermal production. The governmental programs designed

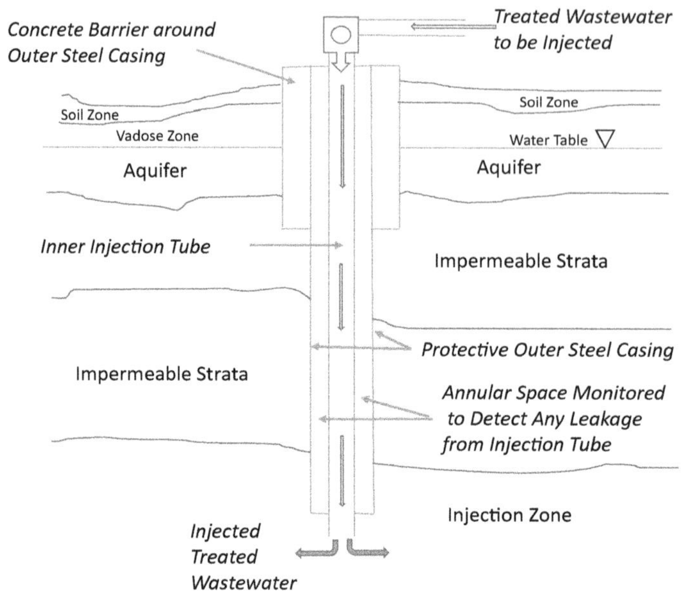

FIGURE 1.6.2 Underground injection well.

to protect drinking water from injected wastes are known as "underground injection control" programs. These programs also issue permits for deep injection of wastes below drinking water aquifers that have an impermeable zone below them.

Deep injection wells must be specifically designed, often with double casing if the injection is under pressure – outer casing to prevent release of wastewater if inner injection tube leaks, and tested for integrity so that no waste would be released from cracks or seams in the casing to drinking water aquifers through which the injection well may pass. Chapter 7 describes the Underground Injection Control Program in the United States in more detail. This program controls the largest volume of waste disposed into land or water in the United States. A similar program exists in the European Union.

Oil and gas brines produced in oil and gas production, a large portion of wastewater injected nationally, need to be disposed, usually through injection back into geologic formations. These brine disposal wells are 1,070 to 3,050 meters (3,500 to 10,000 feet) deep (Clark et al., 2018). Over 50,000 regulated oil and gas injection and disposal wells have been drilled in the state of Texas alone with 70% being active. Of these wells in Texas, approximately 20% are used for brine disposal (PBPA, 2015). Wastewater may also be disposed using deep injection wells. These wells in Florida range in depth from 305 to 1,370 meters (1,000 to 4,500 feet). Deep well injection capital cost may range from \$4 million - \$6 million with annual operating costs of \$100,00 to \$150,000 (Clark et al., 2018).

HYDRAULIC FRACTURING INJECTION WELLS

Injection wells used in hydraulic fracturing of rock containing oil and gas can enhance recovery of oil and gas, utilize large quantities of water to do so. Production of methane gas from extensive deep geologic formations in which it exists in microscopic pores may be stimulated by fracturing those strata, allowing the gas to be released along the fractures. About 95% of new wells drilled in the United States as of 2016 are hydraulically fractured (API, 2019). The fracturing takes place when wells are drilled into these formations and liquids are injected under high pressure causing the strata

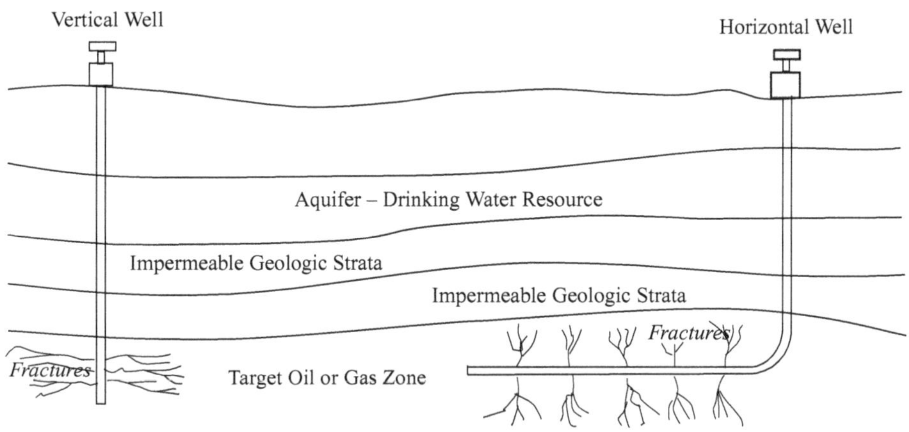

FIGURE 1.6.3 Generalized hydraulically fractured oil or gas well orientations.

to fracture. Wells are then used to collect and produce the methane to sell commercially. Hydraulic fracturing of strata is now typically done by drilling horizontally at a prescribed depth based on the geologic formation. Hydraulic fracturing wells have been drilled at lengths of over 12,200 meters (40,000 feet) but typical distances are about 2,290 meters (7,500 feet) (Larson, 2016). Spacing of wells over the geologic formation determines the number of wells drilled. Spacing between wells in large plot (as large as 405 hectares or 1,000 acres or more) can be 150 to 460 meters (500 to 1,500 feet) (Shahkarami et al., 2016). Several wells can be drilled on each site or pad (Eshkalak et al., 2015). Figure 1.6.3 portrays general orientations for vertically and horizontally drilled hydraulically fractured oil or gas injection wells.

Cost components of hydraulic fracturing include (Eshkalak et al., 2015):

- Lease acquisition and royalty cost – a significant capital cost, with average leasing of $3,500 or more per acre and 15% royalty
- Site preparation and permitting fees – if 10 pads with 5 wells each, estimated costs of $400,000/pad
- Drilling and completion costs – drilling horizontal well, casing, cementing and hydraulic fracturing stimulation may average $0.5 million per 305 meters (1000 feet).
- Operating costs – labor, maintenance, materials and administration costs average $1 per 28.3 cubic meters (1,000 standard cubic feet) of natural gas
- State and federal corporate income taxes – depending on state and federal tax rates

Costs to install and complete a hydraulic fracturing injection well depend on the location, length and geologic formation, among other factors, and run from $3–7 million (Eshkalak et al., 2015; USEIA, 2016; Shahkarami et al., 2016). Operating costs for performing a hydraulic fracturing stage or event are approximately $300,000 (Eshkalak et al., 2015) with operating costs per well per month of $2,000 (Satter and Iqbal, 2016). Techniques and tools for hydraulic fracturing have improved and regulations for well design and operations have been established to protect groundwater. Figure 1.6.4 gives a generalized timeline of activities over which hydraulically fractured oil or gas production wells incur costs. Table 1.6.6 provides a further perspective on hydraulic fracturing injection well costs. Approximately 35% of hydraulic fracturing costs are for the well casing, cement, pumps and equipment with another 40% for rig and drilling fluids, proppant and completion fluids and flow back.

Environmental concerns are amount of water used in the hydraulic fracturing process (AGI, 2019); the fluids used must be monitored to ensure that they do not escape and contaminate groundwater

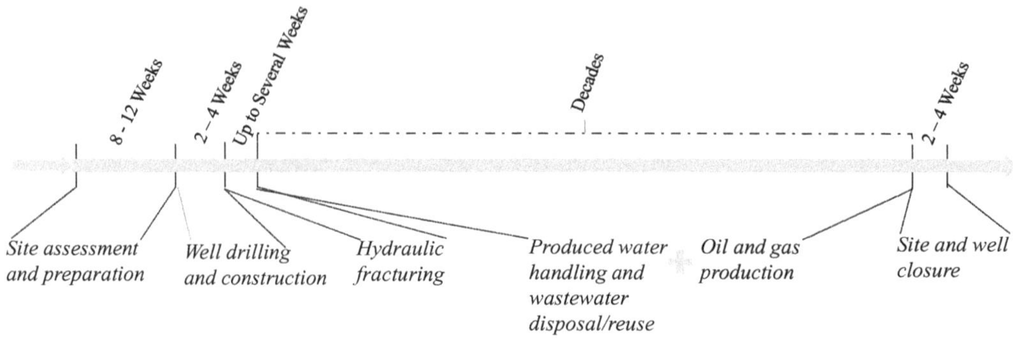

FIGURE 1.6.4 General timeline of activities for a hydraulically fractured oil or gas production well

Source: U.S. Environmental Protection Agency. 2016. Hydraulic Fracturing for Oil and Gas: Impacts from the Hydraulic Fracturing Water Cycle on Drinking Water Resources in the United States. EPA-600-R-16-236ES.

TABLE 1.6.6
Hydraulic Fracturing Injection Well Costs (in 2015$US)

Cost Estimates for Hydraulic Fracturing of Natural Gas Formation:	
Capital Cost/Well	$4.9-8.3 million
Completion Cost/Well	$2.9-5.6 million
Breakdown of approximately 75% of total costs:	
Rig and drilling fluids	15%
Casing and cement	11%
Frac Pumps, Equipment	24%
Proppant	14%
Completion fluids, flow back	12%

Rig and drilling fluids costs - overall drilling activity and dependent on well depth.
Casing and cement costs - casing design required by local well conditions and materials.
Fracturing Pumps and equipment costs - pressurizing equipment and power.
Proppant costs - amount and rates for particular type of material used as proppant for fractures.
Completion fluids, flow back costs - sourcing and disposal of water and other materials used in hydraulic fracturing and other treatments dependent on geology and play location.

Source: US Energy Information Agency. 2016. Trends in U.S. Oil and Natural Gas Upstream Costs. URL: https://www.eia.gov/analysis/studies/drilling/pdf/upstream.pdf

that is currently used or could be used in the future as a source of drinking water (Blend, 2002), and the earthquakes induced by the process. Hydraulic fracturing of an oil and gas well can use hundreds of thousands of cubic meters (millions of gallons) of water to start the rock fracture operation. This water demand is particularly problematic in arid regions that have other significant water demands for municipal and irrigation uses. Advanced hydraulic fracturing technology can inject brackish and saline water, including oil and gas field brines (referred to as "produced water") that are extracted with hydrocarbon production. The amount of produced water available varies by region of the country and may be very small in relation to hydraulic fracturing demand. In other areas, such as Texas, extensive underground brackish water deposits are used in the fracturing process (AGI, 2019).

Typical constituents of hydraulic fracturing fluids include: water, sand, resins, ceramics, gels, acids, biocides, potassium chloride, peroxydisulfates and corrosion inhibitor (Satter and Iqbal, 2016). The oil and gas industry has developed standards and practices to protect groundwater, including practices for: (1) attaining well integrity and fracture containment, (2) managing environmental aspects during well planning, construction and execution, (3) environmental protection for onshore oil and gas production operation and leases and (4) isolating potential flow zones during well construction (API, 2019).

The injection of hydraulic fracturing fluids into the subsurface can induce earthquakes that can affect people and infrastructure. These earthquakes are caused by the fluid injection near pre-existing faults and through the disposal of wastewater from the fracturing process by underground injection well. States most affected are California, Texas, Oklahoma and Ohio. A 5.7 Richter scale magnitude earthquake in Oklahoma affected homes and other buildings near the injection area (Earthworks, 2017).

WELL PUMPS

Types and Operation

A range of pump types can be applied to bring groundwater to the ground surface. These include for residential use, submersible (down in the well) and non-submersible (above-ground) pumps with power sources being electric, diesel, solar, animal-driven and hand and foot pumps. Clearly, a spectrum of costs is associated with such pumps. Relative to electric-driven pumps, above-ground jet pumps have a life expectancy, depending on the amount of sediment in the water being pumped, of 5–15 years and for submersible pumps, 9–15 years (Worst, 2019). Box 1.6.4 provides a range of prices for different pump types.

The cost of operating a pump to produce groundwater must involve well depth and volume of water withdrawn. Properly operating pumps should have efficiencies of 70% or more. Poorly operating pumps may only have efficiencies of 40% or less and can consume much more energy in operation. The weight of the water that the pump must bring to the surface has to be multiplied by the distance up the well casing that it has to move, the pumping rate (cubic meters per minute), the efficiency and the power or fuel cost to determine the energy cost. This cost can be substantial and often factors in decisions about whether a well can be economically operated. Box 1.6.5 provides a hypothetical example of calculations for groundwater pumping energy costs.

BOX 1.6.4 COST OF GROUNDWATER PUMPS

Many combinations of pump and motor capability are available. Some examples of costs are

Residential application

1 Submersible (3/4- 5 horse power): $230-$1200
2 Non-Submersible Jet (0.5-2 horse power): Shallow, $100-$500; Deep, $675-$745
o Solar-powered: $2,000-$4,000
o Hand-powered: $150+

Municipal application

1 1.60 Horse-power, 1.893 m^3/min (500 gpm) submersible pump, motor, controls: $15,000+ (US$2019)
2 Horse-power, 1.893 m^3/min (500 gpm), submersible pump, motor, controls: $34,000+ (US$2019)

Source: A variety of industry sources.

BOX 1.6.5 ENERGY COSTS TO PUMP GROUNDWATER

What is the cost for the average family in the United States to pump groundwater for household use?

Assumptions:

Average Household Water Use: 300 gallons/day (1,136 liters/day = 1,136 kilograms/day)

Pumping Rate: 300 gallons/day (1,136 Liters/day)

Water Lift Efficiencies:

Electric pump efficiency: 70%

Drive Train efficiency: 95%

Electric motor efficiency: 90%

Well Depth: 100 feet (30.48 meters)

1 kilogram – meter (force) = 0.000002724 kWh (electricity)

U.S. Electricity Price (avg.): $0.13/kWh

Calculations:

Force needed to lift water up well each day:

1,136 kilograms × 30.48 meters = 34,625.28 kg – m

Convert to electric power needed:

34,625.28 kg – m × (0.000002724 kWh/kg – m) = 0.094319 kWh

Account for water lift efficiencies:

$$0.094319\,\text{kWh per day}/(0.7\times0.95\times0.9)=0.1576\,\text{kWh}/\text{day}$$

Cost to pump 300 gallons/day (1,136 liters/day) up a Household Well of depth 100 feet (30.48 meters):

$$0.1576\,\text{kWh}/\text{day}\times\$0.13/\text{kWh}=\$0.02/\text{day}$$

Since the depth of the well through which the groundwater would be pumped enters multiplicatively into the series of equations in this calculation, the cost of pumping from different depths with other assumptions remaining the same can be considered here:

Well Depth ft/m	Pumping Cost per Day
50/15.24	$0.010
100/30.48	$0.020
150/45.72	$0.031
200/60.96	$0.041
300/91.44	$0.061
400/121.92	$0.082

Calculation for a Municipal Well:

Assumptions:

Well Depth: 100 feet (30.48 m)

Pumping rate: 1,000 gallons/min (3.785 m^3/min)

Municipal Well Pumping: 24 h

Water Lift Efficiencies: same as above

(Continued)

BOX 1.6.5 (Continued)

U.S. Commercial Electric Price: $0.11/kWh

Cost to pump 1.44 million gallons/day (5,450 cubic meters/day) up a Municipal Well of depth 100 ft (30.48 m):

646.493 kWh/day × $0.11/kWh = $71.11/day

Note: for calculation using diesel oil powered equipment: 1 liter diesel oil = 10.0 kWh

Sources: U.S. Environmental Protection Agency. 2018. How we use water. URL: epa.gov/watersense/how-we-use-water (accessed May 6, 2020).

National Centre for Engineering in Agriculture. 2015. Fundamentals of energy use in water pumping. URL: cottoninfo,com.au/sites/default/files/documents/Fundamentals%20EnergyFS_A_3a.pdf (Accessed May 6, 2020).

US Energy Information Administration. 2019. Electric Power Monthly; Average Price of Electricity to Ultimate Customers. URL: https://www.eia.gov/electricity/monthly/epm_table_grapher.php?t=epmt_5_3 [average price across all sectors = $0.106/kWh]; US Energy Information Administration. 2019. Petroleum and Other Liquids; Gasoline and Diesel Fuel Update. URL: https://www.eia.gov/petroleum/gasdiesel/

The University of California Cooperative Extension, Tulare County, Energy and cost required to lift or pressurize water. Publication No. IG6–96, 2005, URL: http://cetulare.ucdavis.edu/pubgrape/ig696.htm (accessed February 24, 2005).

BOX 1.6.6 COST OF SOLAR-POWERED PUMPS IN GROUNDWATER APPLICATIONS

Solar-powered pumps have decreased in cost. Industry sources indicate that the cost ranges for example applications as follows (2019$US):

- 10 m^3/day, pump, controller, solar panels, 30 m well: $1,400
- 1.9 m^3/day, pump, controller, solar panels,120 m well: $2,400

Solar-powered pump use example: Livestock Watering Using Solar-Powered Pump

Livestock watering in Wyoming (USA) is successfully provided with water using solar-powered pumps in wells. For example, in Kaycee, Wyoming, yearly average solar insolation provides the equivalent of 5.26 kWh/m^2/day. Costs for a solar-powered system to pump 20.8 m^3/day (5,500 gpd) in summer and 15.1 m^3/day (4,000gpd) in winter include (2013$US):

- Submersible pump, motor, controller with pumping rate of 0.038 m^3 per minute (10gpm) and 50.3-meter (165-foot) lift - $1,800
- 1,500-watt solar array - $4,500

(To convert US$ in 2013 to US$ in 2019, multiply by 1.09. This conversion does not take into account technology advances that may affect prices.)

Source: Geiger, Milton. 2013. Solar-Powered Livestock Water Pumping: Technical and Economic Feasibility. University of Wyoming, School of Energy Resources. URL: http://www.uwyo.edu/uwexpstn/centers/sheridan/_files/2013-field-days-pdfs/solar-stock-water.pdf

Solar-Powered Pumps

Box 1.6.4 provides information on solar pump costs in livestock water application, but the technology can be applied in many other circumstances requiring water pumping. This technology may have broader application for small water suppliers as more attention is placed on reducing energy costs and the "carbon footprint" of energy generated by fossil fuel.

Solar power pumping has a number of factors that contribute to interest in its application: system reliability, 25-year warranties for wind and hail damage, ability to locate equipment where needed, stable costs and reduced labor cost to maintain. The technology can be applied to existing wells or new well installations and can pump from wells of 1,000 feet or more in depth but with increasing cost compared to shallower wells. In a livestock watering use, the solar panels can be mobilized to reach pastures in rotational grazing. Time of year use considerations include that more solar panels may be needed for pumping in winter, but water needs may be lower at those times (Geiger, 2013). Life-cycle costs of a solar powered system as a function of hydraulic load may be from 20 to 50% that of diesel pump system costs (EmCon, 2006).

Wind-Powered Pumps

With continued concern about energy prices and fossil-fuel production with carbon footprints, wind power well pumps are receiving interest. Windmills are one of the oldest means of pumping water. In the United States, power from windmills provided many farms with groundwater since the mid-1800s. Use of wind power is being advanced by research in the United States by the Department of Energy (USDOE, 2004a, 2004b). Applications are varied, including individual systems, municipal water supply, groundwater remediation (USEPA, 2004; Girma et al., 2015) and irrigation (Vick et al., 2000; Rehman and Sahin, 2012). Feasibility depends on wind area classification (wind reliability), depth to water and amount of water to be pumped, availability of transmission lines, cost of land, use of land and cost of alternative power sources. Table 1.6.7 gives costs for a wind-electric pump system at two locations in Texas, showing economies of scale.

OTHER WELL TYPES

Other types of wells serve a range of purposes.

Dug Wells

Hand-dug wells are usually older wells, which may range from a pit a meter deep to deeper wells of several tens of meters with stonewalls in rural settings. Dug wells are excavated using hand tools and are the most common technique to provide water from a shallow aquifer in rural regions of developing countries. These wells are usually for rural domestic use. The water from them is typically untreated. While the storage reservoir of a dug well may be accessible by bucket, often, especially for deeper dug wells, a pipe with an intake (open end) is installed in the well with a pump at the ground surface which may be operated manually or with a motor. Large dug wells may have brick, stone or concrete for the inside (wall) of the excavation. Before use, dug wells should be disinfected and protected for runoff or objects that could fall into the well. A concrete apron around the wellhead will also provide protection from runoff and aid in access to the well. Dug wells are typically of low cost and represent simple technology to produce groundwater as needed. The performance of a dug well depends on the geological conditions, but usually are high given the shallow nature of the well (Bruni, 2019).

TABLE 1.6.7
Cost of Wind-Powered Groundwater Pumping Systems in Texas and Ethiopia

In Texas, 2000

Windmill Power Output	Location	Pumping Depth	System Specifications	Cost
1.5 kW	Wheeler County, Texas	40 m	Wind turbine 18.5 m Rohn 25G tower poly pipe in well 1.1 kW 3-phase 230V submersible motor 0.75 kW 15-stage centrifugal pump	$8,000
10 kW	Lubbock County, Texas	45 m	Wind turbine 40 m Rohn 45G tower 60 m of galvanized pipe 5.8 kW 3-phase 230V submersible motor 3.8 kW 15-stage centrifugal pump	$35,000

Ethiopia, 2015

Windmill Water Output	Location	Pumping Depth	System Specifications	Cost
10 m^3/day	Siyadberand Wayu, Ethiopia	73 m	Wind turbine 16 m tower poly pipe in well pump system - single acting piston pump, submersible pump diameter 115 mm reference area 24.87 m^2 rotor diameter 25 mm hydraulic power 337 W windmill life 20 years pump life 10 years	$2,329

Source: Vick, B.D. et al., Wind-powered drip irrigation systems for fruit trees, *Presented to 2000 ASAE Annual International Meeting*, Milwaukee, WI, July 9–12, 2000, 13, URL: http://www.cprl.ars.usda.gov/Drip%20Irrigation/vick_asae2000.pdf (accessed April 24, 2008).

Source: Girma, M.; Molina, M.; and Assefa, A. 2015. Feasibility study of a wind powered water pumping system for rural Ethiopia. AIMS Energy, 3 (4): 851-868. URL: http://www.aimspress.com/fileOther/PDF/energy/energy-03-00851.pdf

Bamboo Tube Wells

Bamboo tube wells are used in locations in which bamboo is readily accessible and groundwater is shallow. These wells have been widely used in south Asia. Usually hand or foot pumps are used to produce groundwater. Bamboo tubewells are addressed in Chapter 7.

Geothermal Heat Pump Wells

A quality of groundwater that is typically forgotten is its temperature, which groundwater heat pumps may use to provide heating and cooling to residences and commercial establishments. Because groundwater is usually fairly constant in temperature – about 12.5°C – efficient heat pumps have been developed to exchange heat in the subsurface with groundwater. Geothermal energy is considered renewable. The principal concern with groundwater heat pump use was that substances from the exchange process could be released to groundwater. Manufacturers have been cognizant of this circumstance and have been careful to properly design these systems to prevent such release. Most state or local regulatory agencies require that groundwater used in heat pumps be injected back into the aquifer from which it was withdrawn (Rafferty, 2003).

Typically, geothermal heat pump wells are drilled to depths of 45.7 to 121.9 meters (150 to 400 feet) and may be as deep as 182.9 meters (600 feet). To heat and cool a home, the relation of home size to well depth is about 46.5 m^2 (500 square feet) of floor space requires 60.96 meters (200 feet) of well depth. If a home or building space measures or exceeds 139.4 m^2 (1,500 square feet), then more than one well may be needed. In most cases, the geothermal well and pipe installation outside the home will cost approximately $5,000. This cost does not include the heat pump and ductwork (GPC, 2019). A 232.3 m^2 (2,500 square foot) home with a 60,000 BTU demand for both heating and for cooling may cost $20,000 to $25,000, about two times the cost of conventional heating, cooling and hot water systems, but are expected to reduce utility cost from 40 to 60%. In the United States, a tax credit applies continuing into 2019 (Energyhomes.org, 2019).

Brine Production Wells

While brine and brackish waters may be a by-product of oil and gas production and often reinjected for disposal in the subsurface, underground brines may be produced by wells for raw materials in manufacturing chemical products. As noted in Chapter 3, brine production provided the basis for the beginning of the chemical industry. The location of the industry in its early stages was dictated by the location of brine production wells (Brandt, 1997). Notably, brines in some locations around the world contain lithium in concentrations of commercial value. Extraction of lithium carbonate from brines is more economical than from the ore, spodumene, with fewer environmental effects (Deberitz, 2006). Lithium has significant application in battery production.

Dewatering Wells

In the case of construction in areas with water tables determined to be "high" relative to the subterranean activity, access to groundwater may be for the purpose of dewatering an aquifer – temporarily or indefinitely drawing down the water table – during construction to have a dry work site below the ground surface (Powers, 1992). Depending on the area and volume of water involved, numbers and capacities of wells would need to be determined appropriately. Construction site dewatering may only affect a limited area for a small local project, or, in cases such as mining, could lower the water table in an extensive subsurface zone. For example, a gold mining operation in Nevada (western United States) lowered the water table 503 m to remove the ore, which has in turn reduced the baseflow of the Humboldt River 32.2 km away (Glennon, 2002, pp. 175–176). Use of available information about the groundwater at a site or adjacent sites and for existing nearby wells aid in the planning and design of a dewatering project (Powers, 1992, pp. 190–202) and will help estimate costs. Since groundwater would be removed from the site, it would need to be disposed off-site, typically in a sewer or stream and not usually back to groundwater. Usually, the local jurisdiction or state requires a permit to discharge this groundwater to sewers or streams, even though it may be clean and uncontaminated. The cost of planning, well installation and production and disposal of the groundwater from dewatering may be a significant access cost, depending on the volume of groundwater to be disposed. Factors affecting dewatering costs are presented in Table 1.6.8. Dewatering costs can be highly variable based on volume, depths and tunneling requirements.

The benefit of dewatering a site is the ability to work in a dry space below the ground surface. This condition has resulted in many important commercial structures being constructed in cities and towns around the world, generating economic returns. Dewatering of a site may also cause unintended, but potentially predictable, effects that may damage the use of the overlying and adjacent land and water, including

- Ground settlement/land subsidence
- Reduced yield of other wells on adjacent lands
- Reduced baseflow of streams in the vicinity

TABLE 1.6.8
Dewatering Cost Factors

Construction Site Cost Estimating Factors (Exclusive of Off-Site or External Side Effects and Costs)

Basic Data	Wage rates Total Cost/Unit Time of Construction Equipment, Including Drill Rigs Cost of Fuel or Power	
Mobilization	Dewatering equipment	Well screen and casing Well pump Wellhead and fittings Discharge lines
	Jetting and developing equipment	Pumps High-pressure hoses Holepunchers and casings Surge blocks Air and water jet pipes
	Sumping equipment	Pumps and controls Extension cables Hoses Discharge lines
	Standby equipment	Generators Switch gear Fuel tanks Automatic startup devices
	Utilities (all) Housing for generators and control equipment Electrical substation	Transformers High line Installation charge
	Engineering expense	System Design Field testing On-site engineer
Installation and removal	Crew cost per day Equipment cost per day	
Operation and maintenance	Operating and maintenance labor and supervision Fuel or power Maintenance material Chemicals or removal of encrustration Equipment repair and major overhaul	
Other costs	Labor and energy escalation rates for long-term operations Payroll taxes and insurance State and local taxes on materials and gross revenue Contractor margin for overhead and profit Contingency allowance for unknown geological or hydrological circumstances	

Source: Powers, J.P., *Construction Dewatering: New Methods and Applications*, John Wiley & Sons, Inc., New York, 1992, 433–440.

- Water quality effects, such as salt water intrusion or accelerated migration of contaminant plumes
- Timber structures below the original water table being attacked by aerobic organisms as the water table declines
- Wetlands ecology being disturbed by reduced water supply as the water table drops
- Trees and vegetation with roots in the water table may dehydrate

(Powers, 1992, pp. 65–71; Glennon, 2002, pp. 175–176)

These adverse effects may create costs to the overlying and adjacent landowners. The magnitude of these costs will be determined by the purpose and size of the project and the volume, rate and duration of the groundwater withdrawal.

DEEP GEOTHERMAL ENERGY WELLS

Deep and superheated geothermal energy production involves accessing the extreme hot water and steam from deep below the ground surface in excess of a temperature of 300°C. Total dissolved solid concentrations may be as high as 250,000ppm, substantially higher than seawater. Drilling procedures are similar for reservoirs. However, once steam conditions are encountered, high-pressure air drilling methods are used (California DOGGR, 2003; Benjaminsen, 2015). Deep geothermal resources exist in 37 of the United States, with high temperature resources in the western states and moderate and low temperature resources on the East and West coasts (Abel and Walker, undated). Typical well installation parameters and cost are given in Box 1.6.7. Costs to install injection wells at geothermal power sites may be approximately 20% more than the installation costs for production wells because of alternative drilling methods that must be used (Rafferty, 2007, p. 2). Well materials, usually steel casing and cement in the annular space around the casing, must be able to survive extreme temperatures and corrosive conditions. Wells may be 30%–50% of the cost of the entire geothermal production unit (Benjaminsen, 2015). Other products may be derived from geothermal water such as sulfuric acid, silica (used in concrete) and zinc, which reduce waste disposal costs and enhance revenue from these energy projects. Other costs include well and wastewater discharge permits. Additional special disposal may apply based on the chemical content of the water. Access to geothermal waters for use in energy production reduces environmental costs, which would have been incurred by other sources: geothermal production emits very small amounts of carbon dioxide and sulfur dioxide and none of the nitrogen oxides (U.S. Department of Energy, 2003). In the United States, initial cost of field development and power plant is approximately $2,500 per kilowatt of installed capacity with operating and maintenance costs of $0.01 to $0.03 per kWh (USDOE, 2019).

BOX 1.6.7 GEOTHERMAL ENERGY WELLS AND COST

The following parameters generally describe conditions affecting geothermal well access to superheated water and steam, including well cost, in the United States.

Well depth: 91.4–3658 m
Water/Steam temperatures: 180°C–315°C
(180°C is considered minimal for direct electric power generation).
Pressure in well: Fluid: 220 bars
Well diameter: Near surface: 30.48–33.02 cm
At depth: 15.24–30.48 cm
Well life: 30 years
Well cost: $700,000–$3,000,000

Sources: U.S. Department of Energy, *Geothermal Energy Program: What Is Geothermal Energy?* 2003, URL: http://www.eere.energy.gov/geothermal/geoimpacts.html (accessed September 6, 2003); International Solar Energy Society. 2010. Drilling 10,000 m deep geothermal wells. http://www.renewableenergyfocus.com/view/12469/drilling-10-000-m-deep-geothermal-wells/)

IN SITU AND ON-SITE GROUNDWATER TREATMENT WELLS

When groundwater is contaminated by human activities, access to groundwater may be fundamental in remediating the resource. Access for remediation typically involves many of the steps and costs previously identified (USEPA, 1989), including

1. Access to the property
2. Permission from the property owner or trustee and the relevant governmental jurisdictions to install wells and produce the groundwater
3. Permission of the owner/trustee and jurisdictions to treat the groundwater on-site and release the treated groundwater from the property
4. In cases where the groundwater is to be treated in situ (in the subsurface environment without being produced), permission must be given by the owner/trustee and governmental agency having jurisdiction to apply chemical or microbiological substances on the ground or into the subsurface to treat contaminated groundwater.

The third and fourth steps are particularly problematic in many situations. Some states and federal agencies require application for a permit to discharge the treated groundwater to a sewer, if in a sewered area, or a stream or other body of water, including an aquifer. Special application for an underground injection permit may be needed if the treated water is to be injected back into the ground or the aquifer, assuming that the jurisdictional authority allows that practice, which it may not. In situ treatment usually requires further permission. The cost of identifying the information needed to satisfy these permit requirements, preparing the permits and providing explanation to the regulatory authorities with jurisdiction over these steps must be factored into the cost of access to groundwater for remedial purposes.

Remediation involving wells may include (Shevah and Waldman, 1995; Reddy, 2008):

- Dual vacuum extraction wells – fluid and vapor removal using or not using downhole pumps.
- Air stripping – pumping water from wells to a tower filled with packing material down through which the contaminated water flows as air is blown into the packing material with effluent water discharged to a recharge trench or well or to a surface discharge point or sewer.
- In situ vapor stripping – air is injected down a tube in the well pushing contaminated water up the well casing with volatile organic chemicals (VOCs) are collected at the top of the well while the partially treated water is returned to the vadose zone to reinfiltrate.
- Air sparging wells – air is forced into the contaminated aquifer, VOCs partition to vapor phase and migrate upward to the vadose zone to be captured by vacuum extraction.
- In-situ flushing wells – one well (or set of wells) is used for injection of reactive agents with the reaction occurring in the aquifer and a second well (or set of wells) extracts the groundwater with the reactive agents and contaminants which is then treated with recovery of the reactive agent that can be reused and the contaminants disposed.
- Monitoring wells - used in combination with these technologies above to sample groundwater and observe progress of treatment.
- Remedial costs are covered in Chapter 7, Groundwater Quality Treatment and Waste Disposal.

SPRINGS

Springs allow groundwater to flow to the ground surface and be easily accessed for human use and other purposes. Notably, the Potomac River North Branch starts as a spring along the Maryland-West Virginia border (Gude, 1984, p. 4). As noted previously, springs may maintain a unique habitat for aquatic life. Some springs are used for water supply directly. In the early days of the European settlement of the United States, some springs were diverted to provide waterpower to turn grist mill wheels. Springs are also used to provide drinking water. In these cases, a springhouse is often built to protect the spring and collect water in a cistern, which is then connected to a distribution line. Jurisdictions may regulate the use of spring water just as they regulate groundwater brought to the ground surface. The design and construction of spring catchments will be specific to each location because of geologic complexity but can be cost-effective for water supply (Meuli and Wehrle, 2001).

GROUNDWATER PROTECTION

Protecting groundwater quantity and quality is essential to water supply for communities relying on groundwater for their principal water source. Using supply side steps, quantity protection may include ensuring that aquifer recharge zones have limited modification through local zoning measures, low-impact development, artificial aquifer recharge and water recycling and reuse policies. On the demand side, conservation may be an important step in water supply management.

For groundwater quality protection, managing actual or potential sources of contamination around wellheads and in recharge areas will be important. This approach is also referred to as wellhead protection (USEPA, 1996). Examination of the wellhead and recharge areas for hydrogeologic vulnerability and ground surface susceptibility to a range of contamination types is fundamental to protecting groundwater. Sources of contamination include industrial activities using chemical and microbial substances, gasoline stations, agricultural activities and residential area maintenance. Contaminants may comprise volatile organic compounds, synthetic organic compounds, nitrates, radionuclides and microorganisms (Frederick County, 2019).

Several studies have shown that the ratio of cost of groundwater protection compared to its benefits of avoiding groundwater contamination may be on the order of 1 : 5 to 1 : 200, depending on community size and prospective groundwater contamination sources (USEPA, 1996; Williams and Fenske, 2004). Other supply-related benefits associated with wellhead protection include (Job, 1996):

- Reduced operating costs due to process changes
- Lower industrial water treatment costs
- Improved chemical management that reduces emergency response costs
- Attractive economic environment for community development

AQUIFER RECHARGE

Natural Groundwater Recharge

Natural groundwater recharge is a hydrologic process in which percolation of water (rainwater or snowmelt) from the ground surface through the soil and vadose (unsaturated) zones reaches the water table or transmission of water through a stream bed (stream leakage) adds to water contained in an aquifer (Davis and DeWiest, 1970, p. 39; NRC, 1994, pp. 14, 282; NRC, 1997; Winter et al., 1998; Llamas and Custodio, 2003). Natural recharge generally increases the volume of groundwater in storage which raises the water table. The rate of recharge can be affected by alterations in the land surface (Dunne and Leopold, 1978, p. 194). Aquifer recharge adds to water supply for the range of groundwater uses.

Managed Aquifer Recharge

Managed aquifer recharge, or MAR, is the deliberate or intentional placement of water by engineered infiltration or well in the subsurface environment to increase the amount of water stored in aquifers for later use or environmental improvement. Managed aquifer recharge has in the past been referred to as "artificial recharge"; however, this terminology carries with it recharge that may not be intentional and in some locations perhaps even undesirable, including uncontrolled recharge from storm events carrying contamination that will not receive treatment. Managed aquifer recharge (MAR) increasingly is being relied on for recycling and reusing treated wastewater and stormwater as well as storing excess surface water, particularly in regions where water is scarce. MAR provides sustainable groundwater resources and associated infrastructure for the well-being of communities and application in industry, including incorporating the hydrologic cycle and natural processes in groundwater management through (NGWA, 2016c):

- Storing water underground to reduce evaporative loss and protect groundwater quality
- Facilitating recharge by use of green infrastructure and native-vegetated landscapes
- Maintaining groundwater levels to avoid nonrecoverable land subsidence
- Preserving groundwater levels, flow and quality for ecosystem biodiversity

The seven components of a managed aquifer recharge project are (AGNWC, 2009):

- Water capture zone – an area for collecting water (such as stormwater or other residual water) for recharge
- Pretreatment – treatment of water to be used for recharge to improve its quality, such as detention (to allow solids to settle) or filtration (to remove solids), including achieving compatible quality with the geologic matrix and characteristics of groundwater in the aquifer
- Recharge – replenishing groundwater by infiltration basin or injection well
- Subsurface storage – natural underground zone of an aquifer to hold water before eventual treatment and/or distribution
- Recovery – replenished groundwater withdrawn by well to be treated before use
- Post treatment – after recovery, additional water treatment for intended use
- End use – consumer application of water for intended purposes, such as drinking, industrial processing or irrigation

Purpose and uses of water from managed aquifer projects include (AGNWC, 2009):

- Water supply diversification, augmentation and security
- Responding to groundwater demand
- Protection from aquifer salination and saltwater intrusion
- Reducing saline concentrations in aquifers
- Mitigation of stormwater damage
- Enhancing community recreation, aesthetic, land value and ecological diversity
- Coastal water quality improvement through interception, treatment and storage of urban runoff and discharges
- Supporting volume and quality of streamflow for ecosystem restoration

Aquifer storage zones may exist at depths ranging from 75 to 900 m (ASRF, 2003). The piezometric surface of these storage zones can be 10 m over the ground surface to 300 m or more underground (ASRF, 2003). Ambient water quality native to the storage zones varies widely from fresh (potable with no contaminant removal) to brackish with as much as 5000 mg/L of total dissolved solids (ASRF, 2003).

The International Groundwater Resources Assessment Centre has documented information on 262 MAR projects in the United States in 28 states. Uses of the water identified for these projects include domestic, industrial and agricultural water supply, ecological purposes and water quality management. Main hydrologic objectives specified are maximizing natural storage and aquifer

TABLE 1.6.9
Summary of Managed Aquifer Recharge Projects in the United States Reported to IGRAC

- 262 Managed Aquifer Recharge projects recorded for the United States as of November 7, 2018
- 28 States identified with MAR facilities/projects: AK, AZ, AR, CA, CO, FL, HI, ID, IN, IA, KS, KY, MO, NE, NV, NJ, NM, NY, NC, ND, OH, OR, SC, TX, UT, VA, WA, WI
- Distribution of MAR projects among top 5 states:
  - California – 87 projects
  - Florida – 29 projects
  - Oregon – 26 projects
  - New Jersey – 21 projects
  - Arizona – 19 projects
- Oldest MAR project recorded in portal: 1902 – Tennet Pond Well Field, Old Bridge, New Jersey
- Methods reported
  - Well injection - 141
  - Infiltration basin – 102
  - Induced Bank Filtration – 9
  - In-channel Modification – 7
  - Rainwater - 2
  - Excess Irrigation – 1
- Water source for recharge
  - River water – 96
  - Reclaimed wastewater – 38
  - Lake water – 27
  - Groundwater - 26
  - Tap water – 22
  - Stormwater – 16
  - Brackish water – 1
  - No reported source – 36
- Water Use
  - Domestic – 46
  - Industrial – 18
  - Agriculture – 16
  - Ecological – 5
  - Water Quality management – 1
  - Research – 1
  - Remainder not reporting use – 175
- Main Hydrologic Objective
  - Maximize Natural Storage – 64
  - Aquifer Management – 60
  - Other Benefits – 25
  - Ecological Benefits – 5
  - No reported objective – 107

Source: IGRAC. 2018. Managed Aquifer Recharge Portal. https://apps.geodan.nl/igrac/ggis-viewer/viewer/globalmar/public/default

management. Worldwide, nearly 1,200 MAR projects in over 50 countries are recorded for Europe, Asia, Africa, North America, South America and Oceania (IGRAC, 2018). Table 1.6.9 provides summary information of the MAR projects in the United States reported to IGRAC. MAR projects are distributed around the country with most being in the western United States. The main means of recharge are injection well and infiltration basin. The principal water sources are river water and treated wastewater with significant uses being domestic, industrial and agricultural supply.

Aquifer Storage and Recovery

Aquifer storage and recovery (ASR) is a type of MAR and is defined as "the storage of water in a suitable aquifer through a well during times when water is available, and recovery of the water from the same well during times when it is needed" (Pyne, 1995). ASR may utilize injection wells to place treated water into the subsurface for storage and use at a later time. The same well used for injection is also used to recover the stored water. Also, this technology may employ recharge areas, often artificially created wetlands or ponds (Dillon, 2002).

Aquifer Storage, Transfer and Recovery

Aquifer storage, transfer and recovery (ASTR) is also a type of MAR and uses the subsurface for storage by injecting water that is typically treated before injection as well as that may use the subsurface for physical-chemical degradation and treatment in the aquifer using one set of wells and then producing the stored water at a different set of wells further away from the point of injection. ASTR has been found to produce water of potable quality but long-term assessment should be done on a site-specific basis (Page et al., 2010). This technology may also employ recharge or infiltration basins.

Figure 1.6.5 depicts ASR and ASTR systems basic operation.

General Types of Recharge

The general types of ground water recharge are (Llamas and Custodio, 2003; AGNWC, 2009):

- Natural: Rainfall, streamflow or snowmelt that is introduced to groundwater
- Artificial: Water placed by intentional or unintentional human action on or in the ground
- Managed: Intentional infiltration of water to subsurface storage by human action through infiltration basins or injection well including potential pre- and post-treatment for groundwater quality control and for intended end use
- Enhanced: The rate of infiltration of precipitation or runoff into the ground is augmented by human action
- Induced: Aquifer pumping near a surface water body draws water into the aquifer
- Rejected: When precipitation or runoff at the ground surface cannot infiltrate because the infiltration capacity of the soil zone has been reached or the water table is at the ground surface

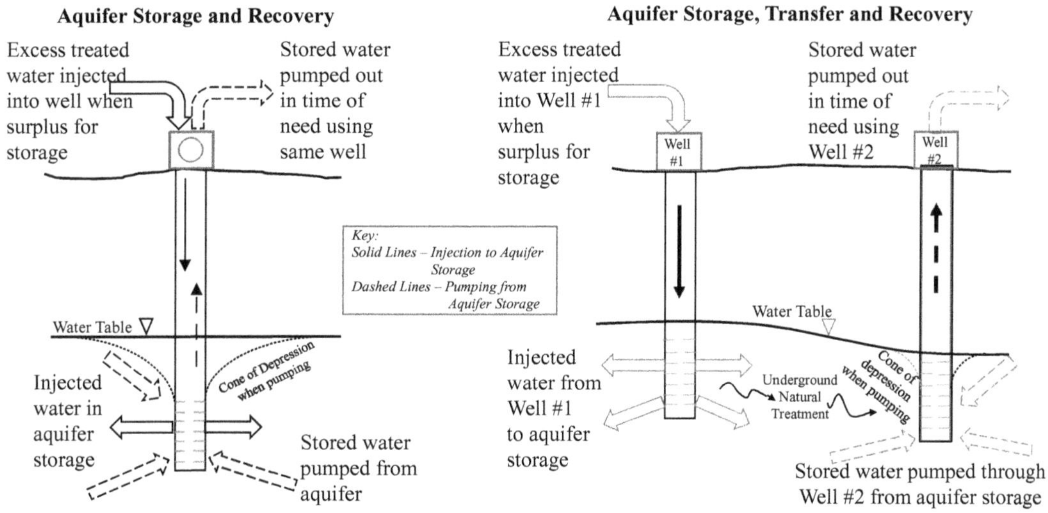

FIGURE 1.6.5 Basic operation of ASR and ASTR systems.

Recharge Area

Natural aquifer recharge areas are land surfaces and stream channels within a region where the prevailing geologic conditions allow infiltration rates which contribute to the replenishment of groundwater and may also create a potential for contamination of groundwater resources (Sumner, 2018). Recharge areas for managed aquifer recharge are often engineered infiltration basins covering hundreds to thousands of square feet or more. For managed aquifer recharge by injection well, the recharge area may be considered to be the immediate zone around the well, such as the concrete apron at the well casing, although natural recharge to the affected aquifer may take place over an extended area significantly beyond the well casing and apron. Ecological issues have also been raised that may cause the artificial recharge manager to incur costs for ensuring recharge of reclaimed waters. These include clogging of recharge areas and scrapping the bottom of recharge ponds (de los Cobos, 2002; Olsthoorn and Mosch, 2002).

Recharge sources may include (Dunne and Leopold, 1978; NRC, 1994; USEPA, 2018a):

- Rainfall
- Snowmelt
- Septic system leachate from adequately operating system
- Drinking water from a public water treatment system
- Treated municipal wastewater
- Treated and untreated (depending on location's contaminant sources) stormwater
- Treated irrigation return flow
- Untreated groundwater and surface water
- Reclaimed or recycled water

Quality of Recharge Sources

The quality of recharge can be influenced by chemical substances and microorganisms released on the ground surface above the uppermost aquifer that are taken up by precipitation and carried into the ground (Dunne and Leopold, 1978, p. 194; NRC, 1993, pp. 14, 18) or by the placement of water with chemical and/or microbial substances below the ground surface by human action (NRC, 1993, pp. 14, 18; NRC, 1994, pp. 14, 282). Wastewaters that may be qualitatively appropriate sources for groundwater recharge are municipal wastewater, stormwater runoff and irrigation return flows that are treated to control concentrations of chemical and microbial substances (NRC, 1994, p. 35–96). Industrial wastewaters and runoff may be acceptable but would need site-specific evaluation for its potential use as a recharge source (NRC, 1994, p. 35).

Hydrogeological Factors

Hydrogeological factors important in siting a MAR project include (Casanova et al., 2016):

- Unsaturated zone must allow water to infiltrate and reach the aquifer
- Aquifer must have storage capacity for infiltrated water
- Preferred sites have rather low diffusivity: relatively low permeability and high storage capacity
- Formations with interstitial porosity (such as sandy and sandstone formations)
- Formations with both interstitial and fracture porosity (such as chalk)
- Need certainty of recharge water quality compatibility with reactive potential of the aquifer matrix and particularly in the unsaturated zone

- Site maintenance provides groundwater protection from pollutant discharges
- Wellhead protection area should include entire well or well-field catchment area to preclude non-point source pollution

(Casanova et al., 2016; Parker, 2018).

Managed Aquifer Recharge Technology

Infiltration Methods

Infiltration basins (ponds, tanks) to improve aquifer levels and recharge water quality through attenuation in the unsaturated zone including:

- infiltration ponds include reservoirs and facilities that manage stormwater/runoff
- treated wastewater allowed to infiltrate
- reservoir pavements
- recharge pits
- drainage trenches
- vegetated ditches
- mounds systems
- sand filters
- septic drain fields
- dry wells
- sand or check dams
- distributed projects
- flood managed aquifer recharge
- low impact development, including porous pavement

The technology allows water to pass through the unsaturated zone prior to entering the aquifer. Soil contact potentially degrades pathogens and harmful inorganic and organic substances. Infiltration rates can take account of basin characteristics: high rates if water quantity is an objective; low rate if objective includes water quality improvement of infiltrated water.

Geo-purification in the unsaturated zone may improve if plants grow in the watershed reducing erosion and clogging and may have bacteria that biodegrade some inorganic and organic pollutants. Colder water with higher viscosity infiltrates more slowly, decreasing notably in winter. Water with less treatment may have more organic matter that promotes bacterial growth, reducing porosity and inducing clogging through biofilm growth in the subsurface. To avert or slow clogging, water pre-treatment prior to infiltration may employ settling ponds or sand filters at sites before reaching infiltration points. To change the recharge water chemistry, inorganic compounds can be added at the infiltration basins. Alternating wet and dry times of operation allow decompaction and maintenance of the basin floor. A critical factor for using polluted or impaired water for infiltration is not to design infiltration features, such as drains, that would avoid percolation of the water through the soil zone thereby avoiding or minimizing any natural degradation of contaminants (NRC, 1994).

Direct Injection Method

Injection wells used for Aquifer Storage and Recovery (ASR) and Aquifer Storage Transfer and Recovery (ASTR) to recharge injection wells used for Aquifer Storage and Recovery (ASR) and Aquifer Storage Transfer and Recovery (ASTR) to recharge confined (or semi-confined) aquifers and/or to create hydraulic barriers.

Filtration Methods

Riverbank filtration uses pumping wells near the riverbank that lower the water table with a hydraulic head difference between the river and the groundwater, thereby inducing recharge to the aquifer. Dune filtration through water infiltration from ponds is constructed in dunes with water extracted by wells or from ponds at lower elevation.

Large Area Source Water Quality Protection Methods

Under the Safe Drinking Water Act, community water systems can designate wellhead protection areas around public water supply wells and manage potential contaminant sources in those areas to reduce the probability of contamination to their groundwater supplies. Also under the Safe Drinking Water Act, communities relying on groundwater (including communities on private wells) as their sole or principal source of water supply can petition the US Environmental Protection Agency for designation as a Sole Source Aquifer Area that results in all federally assisted projects being reviewed that would be constructed or in operation in that area to ensure that the project or activity will not contaminate their aquifer.

Implementation Steps in MAR Project Siting

Five necessary steps for siting a MAR project include (Casanova et al., 2016):

1. preliminary feasibility assessment of recharge system on selected site using existing data or modeling
2. design of the recharge system
3. detailed verification study
4. pilot or experimental system at a scale to conduct preliminary tests
5. extrapolation to an operational scale

To support preliminary tests of the selected site, monitoring of water quality is recommended including the physical-chemical parameters: water saturation, water pressure, temperature, conductivity, redox potential, pH of the infiltration water.

Additional testing should include sampling and analysis of gas and water in the unsaturated zone, sampling of the soil for mineralogical and microbiological analyses and permeability testing of soil and sub-soil on the recharge site.

Benefits of Managed Aquifer Recharge

Since MAR projects often involve some level of construction, they have been examined for benefits compared with costs. In certain situations, the benefits far outweigh costs in the decision process employed. Several benefit and cost comparisons are presented in Box 1.6.8 and Table 1.6.10. As a potential response to drought, variation in water supply, and climate change, benefits of aquifer storage may include:

Direct Benefits

(Reichard et al., 2004; CSIRO, 2019; Sharma 2020)

- Water supply security/maximize water storage (long-term and seasonal)
- Use of the transmission and treatment capacities of the subsurface environment
- Natural physical-chemical-biological degradation of contaminants
- No evaporative loss
- Low-cost, low-energy water source with reduced costs for pumped storage
- Reduced greenhouse gas emissions

TABLE 1.6.10
Some Benefit-Cost Comparisons Reported for Aquifer Storage and Recovery Projects

Country	Benefit/Cost	Reported By (in Source Below)
The Netherlands	Avoid expensive treatment plants and reduce land use intensity in "absorptive" (recharge) areas	Hoogendoorn
The United States	Two seepage trenches 6 m deep, 45 m long (infiltration rate 2.2–2.7 ML/d), cost $15,000 each trench, four spreading basins (infiltration rate 0.6 m/day), cost $30,000 each basin five recharge wells 36–8 m completed, 450 mm diameter, PVC casing, 2.5 mm slot screen, cost $130,000 per well	Legg and Sagstad
India	Percolation tanks with 10–25 mm/d infiltration rate with catchment area of 10–50 sq. km, cost $61,000–$84,000 US 2002, Benefit–cost ratio ranges from 1.3 to 2.0	Limaye
India	Cavity tube well, benefits $3000–$4000 US 2002, benefit-cost ratio = 25	Malik et al.
Australia	Large-scale stormwater treatment/groundwater recharge/ water reuse scheme, cost $9 million AUS 2002, payback period 10 years or more	Chaudhary and Pitman

Source: Dillon, PJ., (ed.) Management of Aquifer Recharge for Sustainability, A.A. Balkema Publishers, Lisse, the Netherlands, 2002, 567.

- Lower costs for permitting, construction and operation
- Replenishment of depleted aquifers/restore groundwater levels/reduce water table decline
- Location where water needed
- No loss of land for farming or habitat for water storage
- Restores groundwater-dependent ecosystems, including river flows
- Management of subsidence and saltwater intrusion

Indirect Benefits

- Buffer and existence values of maintaining aquifers, particularly to offset drought
- Avoided costs of alternative water supply during drought
- Avoided impacts to communities during water shortages
- Conjunctive management of ground and surface waters

Cost-Effectiveness

Different sources have considered the cost-effectiveness of managed aquifer recharge. Halihan (2016) cites surface water storage costs of $1 Billion/km^3 of water stored in reservoirs as compared to aquifer storage costs of $100,000/km^3. California researchers examining alternatives to address that state's water needs found cost ranges by source technology as SWIE (2014):

- Groundwater recharge $90-$1,100 per acre-foot
- Reservoir expansion $1,700-$2,700 per acre-foot
- Seawater desalination $1,900-over $3,000 per acre foot

A major factor in cost-effectiveness of MAR projects over land-based reservoirs is the cost of land to provide sufficient storage (AGNWC, 2009).

BOX 1.6.8 EXAMPLES OF MANAGED AQUIFER RECHARGE FOR MUNICIPAL SUPPLY AND ENVIRONMENTAL SUSTAINABILITY

Manage Saltwater Intrusion and Aquifer Quality of Potomac Aquifer, Suffolk, Virginia

- Issue: saltwater intrusion because of extensive use of Potomac Aquifer for groundwater supply and sea level rise
- May 2018, Hampton Roads Sanitary District, Virginia began operation of advanced treatment of wastewater and producing up to one million gallons per day of drinking water quality and injecting the water into the Potomac Aquifer – creating a freshwater barrier and protecting groundwater quality
- Cost of treatment and injection facilities $25 million with goal of producing more than 100 million gallons recharging the Potomac Aquifer daily by 2030 for eastern Virginia with current population of 1.7 million people

Improve Stream Flow Near Madison, Wisconsin

- Issue: Cogeneration facility at University of Wisconsin draws on Yahara River during low flow and affects aquatic life sustainability
- Recharges aquifer with water from Lake Mendota to be stored and used to pump into Yahara River when low flow occurs during dry weather and cogeneration facility uses river water
- Joint project of Madison Gas and Electric and University of Wisconsin-Madison
- Negligible effects on Lake Mendota levels
- Groundwater recharge is from stormwater runoff collection and infiltration at the Odana Hills Golf Course infiltration pond which is filtered, pumped to higher ground and infiltrated through soil
- The recharge project was completed in May 2006 with benefits of
  - Help replenish aquifer supplying Lake Wingra and nearby springs, re-establishing natural system existing previously
  - Reduce excess sediments and nutrients that flow through storm sewers to Lake Wingra

Drought Reserve in Bernillo County, New Mexico

- Issue: Store water for drought years
- Developed by Albuquerque Bernalillo County Water Utility Authority after successful demonstration and in operation in 2014
- Conjunctive surface water and groundwater management using up to 3,000 acre-feet of bank-filtered Rio Grande surface water migrating through the 500-foot-thick vadose zone profile beneath channel to recharge Santa Fe Group aquifer system for later use during droughts

Increase Municipal Water Supply in Riverside, California

- Issue: Sustain local water supply
- Infiltration recharge basin with 10-month construction to be completed in spring 2019
- Replenish aquifer initially with up to 1,800 acre-feet/year- supply for 3,600 homes

(*Continued*)

BOX 1.6.8 (Continued)

- Basin to recharge up to 3,500 acre-feet of stormwater, nonpotable water and potential recycled water in the future
- Water from recovery well to be treated by the Arlington Desalter (reverse osmosis).
- Total cost $4.54 million including design cost is $500,000 and annual estimated O&M cost $100,000 and design life of 30 years
- Projected cost of producing water from the Victoria Recharge Facility $385/AF ($1.18/1000gal)

Treatment at MAR Sites

Natural subsurface treatment and degradation have been a benefit identified for managing water at MAR sites. Dillion and others (AGNWC, 2009, p.32) calculated the equivalent cost of treatment that an MAR project site could potentially provide from the residence time of water in the subsurface environment. The subsurface provides several treatment functions continually, including filtration for colloids, disinfection (inactivation of viruses, protozoa and bacteria harmful to people) and degradation of some trace organics. Each groundwater is different and must be evaluated individually to determine its natural treatment capabilities. Table 1.6.11 provides a perspective on the cost equivalents of water treatment in the subsurface for some removal processes (In the table, original 2008 Australian dollar costs are presented and then adjusted to 2020$US for comparison purposes.) These costs represent avoided cost savings that MAR projects offer assuming that managed aquifer recharge occurs with the quality of the water being compatible for other biochemical constituents to be placed in the subsurface and not cause adverse water quality reactions to occur there.

Other contaminants not adequately treated naturally in the subsurface would require other pretreatment before being infiltrated or injected underground or post-treatment after being recovered from subsurface storage. Geochemical testing would need to occur prior to MAR project initiation to determine natural subsurface treatment capabilities.

TABLE 1.6.11
Costs of Treatment-Equivalency for Water in MAR Projects

	Flow and Costs for Treatment Technology Equivalencies					
Flow or Expenditure	**Biofiltration/Slow Sand Filtration**		**Rapid Sand Filtration**		**Chlorination**	
Flow (ML/d)	**30**		**30**		**10**	
	$Australian in 2008	**Inflated to $US in 2020**	**$Australian in 2008**	**Inflated to $US in 2020**	**$Australian in 2008**	**Inflated to $US in 2020**
Capital Expenditure ($/ML/d)	8.30	9.27	0.40	0.45	0.026	0.029
Operating Expenditure ($/kL)	0.01	0.01	0.01	0.01	0.01	0.01

Monetary Conversion: 2008$A to 2008$US = 0.955; 2008$US to 2020$US = 1.17Source: Abstracted and Modified from: Australian Government National Water Commission. 2009. Managed aquifer recharge: An Introduction. Waterlines Report Series No. 13. URL: https://www.readkong.com/page/managed-aquifer-recharge-an-introduction-peter-dillon-9032877?p=1 (Accessed May 28, 2020).

Maintenance of Aquifer Recharge Basins

Aquifer recharge basins require maintenance for continued performance. Infiltration performance may decline over time and should be addressed. Three events may occur contributing to this decline in performance: fine sediments settle and accumulate in the infiltration zone, infiltration slows and microbiological growth occurs that clogs the sediment with a slime. Excavating, or scraping, to remove this accumulated sediment and microbial growth may be done as scheduled maintenance, such as every five years (NYSDOT, 2015; Nagaoka, 2018).

Unanticipated Costs of Aquifer Recharge

Because water sources for managed aquifer recharge facilities may include chemically treated water or may have received treated water that has become mixed with natural water, the water in the ground may undergo chemical reaction or may react with the geological matrix and affect whether the groundwater can be used as drinking water. Some states may have regulations or other requirements regarding what water sources are used for aquifer recharge. The following conditions may constrain use of groundwater that has received recharge or injected waters and may create unanticipated costs for its potential use as drinking water (USEPA, 2018a):

- Pathogens from water that is not disinfected prior to recharge or injection, such as injection of raw water and treated effluent.
- Disinfected water with soluble organic carbon that is injected possibly allowing disinfection byproducts to form in the aquifer including trihalomethanes and haloacetic acids (regulated under the Safe Drinking Water Act).
- Differences in pH and reduction-oxidation potential between injected water and the aquifer mobilizing metals and radionuclides and potentially resulting in release of arsenic, iron, manganese or radionuclides (all regulated contaminants) from the rock to the groundwater.
- Insufficiently acidic recharge or injected water possibly causing carbonate precipitation in carbonate aquifers resulting in well clogging.
- Treated and untreated recharge waters contaminated with per- and polyfluoroalkyl substances (PFAS) that are very stable in the environment, have been identified as causing cancer and other serious health problems in very low concentrations and have been used around the world in firefighting, heat-resistance applications and stain-resistant cloth among other uses (EWG, 2020). Other emerging contaminants of concern may raise similar challenges for waters used for aquifer recharge.

Response to these conditions may add cost to use some recharge waters by contributing to the need for further treatment of water used to recharge groundwater or the groundwater itself before it can be used for drinking water purposes.

Additionally, sufficient attention should be given to site assessment, onsite and adjacent environmental conditions, permit requirements, and potential offsite effects (e.g., water table rise flooding basements and gravel pits, induced seismicity) in the planning of MAR projects to ensure that major adverse impacts can be mitigated and costs recognized. (Pyne et al., 2020) Anticipating onsite and offsite conditions based on thorough review of existing data early in project planning may minimize additional costs in designing and developing a MAR project.

NATURAL AND GREEN INFRASTRUCTURE

Natural Infrastructure

Valued Functions

Natural infrastructure comprises the innate ecosystem features and functions that perform essential services and have value for human and animal populations, used strategically to provide, protect and/or conserve resources for current and future purposes and sustain and enhance life (WRI, 2013;

Ozment et al., 2015). For groundwater considerations and without engineered characteristics, these features include but are not limited to: aquifers, springs, wetlands, soil and vadose zones, recharge areas and discharge zones of stream, lake and coastal channels and beds, and less permeable geologic strata of confined aquifers. Functions address groundwater and energy storage, flow and transmission; overburden support of aquifers; habitat for flora and fauna and microscopic creatures; stream flow maintenance; associated aquatic life maintenance; filtration, degradation and adsorption of contaminants; precipitation and floodwater adsorption; climate regulation; stream baseflow support for wildlife and navigation; and perhaps other functions not currently understood. These features and functions provide the basis for life in much of the world due to accessible underground water supply. In the world today, these features and functions are often complemented by engineered green infrastructure to extend their purpose and value.

Areas of natural infrastructure are often seen as having high near-term opportunity costs if not developed for residential, industrial or agricultural purposes. At the margin for example, wetlands may be drained or filled for residential, industrial or agriculture projects. Conversely, watersheds with more extensive forested area provide higher levels of groundwater recharge resulting in less stormwater runoff and lesser concentrations of nutrients and sediment in surface waters as compared to urban and agricultural areas (Ozment et al., 2015).

The subsurface environment can also provide natural degradation of contaminants. For example, stormwater make accumulate microorganisms and chemical pollutants as it travels across the ground surface. In some communities, stormwater may be collected and treated before released to local surface waters or may be infiltrated to ultimately reach groundwater. Depending on the depth and texture of the soil zone, microbiologic contaminants (pathogens) may be filtered or adsorbed in the soil zone with their movement constrained that die-off or inactivation occurs (Brumley et al., 2018). More research is needed on the degradation of contaminants entering the soil and upper vadose zones from stormwater infiltration into the subsurface and the resultant groundwater quality effects to maintain safe water supply.

Benefits of Functions

Storage benefits of natural infrastructure are significant. In addition to providing habitat for fish and wildlife-dependent species as well as berries, timber, rice, materials for medicines and water quality improvement (USEPA, 2018c), wetlands provide flood protection by storing floodwaters. A study of the cost-effectiveness of wetland restoration on the US Gulf Coast indicates that for every $1 expended on restoring wetlands would provide $3.50 of direct flood-reduction benefits (Reguero et al., 2018). The flood control functions of wetlands include: reducing the magnitude and duration of storm events, retention and dissipation of runoff, decrease in velocity of stormwater flow, above-ground and underground storage of water and sediment retention (MDE, 2019). Additionally, groundwater recharge and storage projects in the western United States have a cost range per 1,233 m^3 (1 acre-foot) of water of $90 to $1,100 (median cost of $390 in 2014) compared to expansion of surface water reservoirs with costs of $1,700 to $2,700 (Stanford University, undated).

Another fundamental function of aquifers, the transmission and distribution of groundwater, has been relied on for centuries to provide water to springs and hand-dug wells long ago and even today and continues to provide water supply for half the world's population through simple as well as high-capacity engineered wells. Groundwater transmission also supports streams with groundwater discharge maintaining baseflow for aquatic life and drinking water for downstream users. Studies done of four cities in North Carolina and six cities in Ohio of the cost of installing water transmission and distribution pipe document the alternative cost that aquifers provide naturally at no cost. Depending on the pipe size (10.2–30.5 cm; 4–12 inches) and geologic conditions, installation of water pipe costs from $5 to $46 per meter ($15 to $152 per foot) in the communities evaluated (BCC Research, 2016). A 16-kilometer (10-mile) water pipeline in rural North Carolina cost $309 per meter ($94 per foot) (DMP, 2015). These costs do not include costs of pumping water to move it from source to user. Annual maintenance costs may be 5% of capital costs (USBR, 2005). A large California city estimated that its 85 kilometers (53 miles) of transmission line had on average an estimated life of 70 to 90 years depending on installation date and that 50% will need to be rehabilitated, 25% will need sliplining (inserting an interior pipe wall) and 25% will need to be replaced (SSWD, 2011). Installing and maintaining water pipe are significant costs in supplying water through an engineered system that the natural infrastructure of aquifers provide at no cost, referred to as an avoided-cost benefit.

Green Infrastructure

Green infrastructure related to groundwater is a cost-effective, sustainable, engineered means to manage stormwater by using the subsurface capabilities to degrade contaminants and store water to increase groundwater supply and quality for the benefit of a community. Green infrastructure methods to improve a community's groundwater availability and quality include (USEPA, 2018b; MPCA, 2011):

- Managed aquifer recharge
- Underground infiltration
- Rain gardens/Bioretention basin/Biofiltration basin
- Bioswales
- Permeable pavements
- Constructed wetlands

Benefits for Green Infrastructure

Benefits of green infrastructure include (ConcreteNetwork, 2019; USEPA, 2018b; Terhell et al., 2015)

- Reduction in stormwater runoff volume sent to streams detention areas or causing flooding (avoided costs of loss of surface water flora and fauna and of damages from flooding)
- Increase storage of water in aquifer (value of water supply)
- Reduction of space requirements to collect and retain stormwater (avoided costs of rights-of-way, property acquisition and maintenance)
- Lessening demand for and capital and maintenance cost of storm sewers (avoided costs of sewer and drain construction and maintenance)
- Reduction of capital, operation and maintenance costs of stormwater treatment (avoided costs of treatment plant construction and operation)
- Treatment of pollutants through natural processes if adequate soil and vadose zone conditions (avoided costs of treatment plant and construction)

Monetary values for the benefit of increasing groundwater in local aquifer storage would be the price of water to a community's water supply customers.

Costs of Green Infrastructure

Costs for green infrastructure components are addressed in Table 1.6.12. More highly engineered components have greater costs to install and maintain. Green groundwater-related infrastructure has a longevity of up to 40 or 50 years. Annual maintenance costs range from 1 to 18% of capital costs depending on green infrastructure component with an unweighted average of 3.4 percent.

TABLE 1.6.12
Costs of Green Infrastructure Components (Other than MAR)

Component	Construction Cost		Maintenance Cost		Expected Life
(w/Cost Dates: Constr/Main)	Per m²	Per ft²	Per m²	Per ft²	(years)
Rain gardens (2007–2009/2005)	$55.44–172.77	$5.15–16.05	$3.34–6.57	$0.31–0.61	25–50
Bioswales (2005–2007/2005–2007)	$59.20–258.34	$5.50–24.00	$0.65–2.26	$0.06–0.21	20–50
Permeable Pavements					
Pavers (2016/2005-2008)	$53.82–107.64	$5.00–10.00	$1.08–2.48	$0.10-0.23	15–50
Porous Asphalt (2016/2003-2009)	$5.38-10.76	$.50-1.00	$0.97-2.48	$0.09-0.23	15-40
Porous Concrete (2016/2003-2009)	$21.53-69.97	$2.00-6.50	$0.97-2.48	$0.09-0.23	20-40
Gravel (2007-2009/2000-2006)	$18.51-64.59	$1.72-6.00	$0.22-0.54	$0.02-0.05	20–40

Sources: Green Values. 2013. Green Values; National Stormwater Management Calculator. URL: http://greenvalues.cnt.org/national/cost_detail.php; University of Maryland-Extension. 2016. Permeable Pavement Fact Sheet; Information for Howard County, Maryland Homeowners. URL: https://extension.umd.edu/sites/extension.umd.edu/files/_docs/programs/master-gardeners/Howardcounty/Baywise/PermeablePavingHowardCountyMasterGardeners10_5_11%20Final.pdf

BENEFITS OF PROPERLY INSTALLED AND MAINTAINED WELLS

While well permits and construction standards may add costs, the perspective of local governments is that they are for the protection of the well owner and adjacent groundwater users. Having well construction standards provides some assurance to the well owner that the well meets best science and engineering practice, as long as the state or local government keeps the standards current. A properly constructed well can operate for decades without mechanical or contaminant problems, assuming the original well placement has not been encroached on by land uses that could affect its water quality or use. Having a properly constructed well saves the owner from the costs of proper abandonment of an old well (often requiring pulling the casing out of the ground and carefully cementing the hole from the bottom up to the ground surface) and installing a new well, if a replacement well is required.

Properly constructed wells also eliminate pathways for contamination around the wellhead by grading the ground away from the well and installation of a concrete apron. Construction also should provide for grouting (cementing) the annular space around the well casing 1.5 m or more below the surface. This construction keeps surface water runoff away from the wellhead and from going down the well. The apron and the annular grouting also keeps contaminated runoff from moving to the casing and down along the casing to the water table. Should the well water become contaminated, the well user may suffer illness of short or long duration, depending on the contaminant. This illness is a cost to the well owner and may include medical expenses and lost income. Avoiding these costs is a benefit to the well owner. Knowing the construction of the well may give the owner some understanding of the potential risk for contamination and exposure to the contaminants.

Since groundwater moves in the subsurface, reducing the risk of contamination to the well's water supply may also reduce the risk of contaminating the groundwater used by a neighbor.

Unfortunately, neighbor has sued neighbor to address nearby contaminant sources that are a cost to the contaminating property owner for correction. While not all wells will be in the contaminant pathway, proper construction can help avoid such costs, thereby benefiting the well owner.

Well permitting is usually related to the density and production of wells. The well permit process considers existing adjacent groundwater production and the use and production of the proposed well. A well permit may not be granted if the area's groundwater production is not sustainable and would result in groundwater mining, lowering water tables, drying up neighboring wells and causing lost water production at neighboring wells, thereby affecting the use and value of the property. Proper management of the density of wells provides benefits to current and future groundwater users by ensuring sufficient water sources.

(Note that western states using prior appropriation groundwater doctrine do not consider adjacent users groundwater production, since it is already allocated by appropriation. That right allows the owner to produce a specific volume of water despite impacts on junior appropriators nearby. These effects could include groundwater mining, resulting in costs of installing deeper wells and more powerful pumps to lift the water further. These costs may be incurred by all users in the area, not just the large producers. Furthermore, the volume of water available in the future is likely reduced, even with constant or increased demand, since in many western locations, precipitation is low and evaporation rates exceed precipitation rates)

COSTS OF IMPROPERLY INSTALLED OR MAINTAINED WELLS

Wells not properly installed or poorly maintained may result in costs to users. Improper grouting of wells, no or cracked apron at the wellhead or no well house can allow contaminants down along the well casing and annular space and into the aquifer. Cracked well casing can also allow contaminants to enter the well and allow pumped water to escape before reaching ground level. Injection wells with cracked casing can release contaminants to groundwater. At least three injection well sites are locations of remedial actions in the United States (Paque, 2003). Such problems may pose health risks and economic costs to well owners and users and, because groundwater migrates, also could present costs to adjacent groundwater users.

ECOLOGICAL ASPECTS OF ACCESS TO GROUNDWATER AND THE SUBSURFACE

The ecological aspects of access to groundwater and the subsurface are emerging as a preeminent water issue with environmental or total systems implications for water management. Groundwater reinforces the concept that "everything is connected to everything else." Just as groundwater inherently responds to nature's physical, chemical and biological forces to achieve balance, human action is effecting continued changes influencing the balance in the subsurface and all that is connected with it. It has been estimated that the earth's largest amount of biomass exists under the ground surface. This is also the location of the largest volume of freshwater available for human use: the aquifers of the subsurface environment. The very pathways through intergranular space below ground are also routes for contaminating the resource. Since most groundwater nearer the ground surface moves, aquifers become a transport mechanism for what is released to them. Briefly, we will consider the ecological aspects of access to groundwater and the subsurface. This is a subject that needs more serious documentation through research and evaluation.

UNSATURATED ZONE AND SHALLOW AQUIFERS

Residual chemicals released intentionally or unintentionally to the ground or beneath it can move through the interstitial subsurface space and be degraded or adsorbed to some particles depending on their electrical charge, if they are adapted to do so or move through it. If they move through it, they can reach or access the first groundwater in the saturated zone to be transported to adjacent locations. In some cases, large volumes of chemicals have overwhelmed the subsurface environment and its ability to adsorb or degrade them, ultimately being received in high concentrations that contaminate the groundwater for future users. Past chemical waste disposal practices were based on the belief that this subsurface environment could carry away these wastes, not to be a problem any more. We now know that is not the case. The aquifers can carry these residuals to places that may kill vegetation and other wildlife, contaminate streams and befoul water supplies.

Additionally, since groundwater exists in the interstitial spaces of the subsurface, we might think of the subsurface as large holder (even as a subterranean "sponge") of water. This capability of the subsurface to allow water to percolate into it and be held evens out the unpredictable sequences of precipitation for ecological purposes and human use. This capability also moderates extreme precipitation events to lessen flood conditions, a benefit to wildlife and human beings.

Shallow aquifers are accessed by plants through their root systems. This groundwater access provides life-sustaining water to the plants and organisms associated with them. These plants also transpire water as vapor back to the atmosphere and maintain their part of the hydrologic cycle. Other shallow groundwater moves through fractures and even caverns in limestone and dolomite and provide habitat for larger forms of wildlife, including fish.

Rivers and Streams

Most rivers and streams are discharge points for groundwater. While rivers can also recharge groundwater, they act to release groundwater stored below ground from past precipitation for gradual release for various purposes that humans have imagined and actually implemented: drinking water supply, wildlife support, waste removal and dilution, navigation and transport, and recreation. Certain fish and other aquatic wildlife depend on the quantity and quality (specific natural chemical concentrations and temperature) of groundwater discharges to the streams (Glennon, 2002, pp. 4–8). Most rivers and even smaller streams flow when there is no precipitation in their watersheds, because groundwater is discharging to them through their streambeds during these times. Rivers can also be dried up from producing so much accessible groundwater that the water table is lowered below the riverbed, no longer able to discharge to it, as happened to the Santa Cruz River in southern Arizona (Glennon, 2002). Other uses of the river are eliminated and wildlife are destroyed.

Wetlands

Many wetlands are discharge points for near-surface groundwater. This relationship allows wildlife to access groundwater for their own existence. Since wetlands are highly productive zones, waters flowing to them with nutrients may be degraded and recycled through the biological actions of the wetlands flora and microscopic organisms. Wetlands, in turn, can provide important habitat for many fauna, including spawning zones for fish.

Access to groundwater is typically viewed from the human perspective. Access to the groundwater and the capabilities of the subsurface should also be considered in the larger context of the ecological system and its myriad of interrelationships. This connection of groundwater to "everything else" indicates that the cumulative actions, while seemingly isolated, in the aggregate affect the larger world in ways we may not consider. We all, humans and wildlife – even microorganisms we forget but depend on to degrade the residuals we want to be rid of – share access to groundwater in ways not always or typically contemplated. How we balance this access with each other can have profound effects on the resource, its future use, and interpersonal, interjurisdictional and international relations.

SUMMARY

Access to the groundwater resource is the key to obtaining an economic groundwater supply. Access is affected by a number of factors, including land entry rights, hydrogeologic setting and depth of the water table and well. The cost of installing a well affects the user's or producer's cost of supplying groundwater. Developing the approach to groundwater access starts with checking land entry requirements and then proceeds to exploration. Costs of well installation increase with depth and with complexity of the hydrogeologic setting. Costs of groundwater access vary by location, purpose and requirements for properly producing groundwater in a locality. Jurisdictional well installation requirements, while adding costs up front in the well installation process, are for the protection of the consumer and adjacent property owners who use groundwater. Well installation costs can vary from location to location because of differences in requirements, and sometimes costs in one locality may double those in another. Benefits of properly installed wells for groundwater access include avoided costs of reinstallation, potential illness and damage to neighboring well users, as well as maintenance of property use and value. Other types of wells include dug, injection, heat pump exchange, hydraulic fracturing, remedial, dewatering, geothermal and in situ treatment. Shallow wells that allow untreated wastewaters into the subsurface may impose costs on adjacent groundwater users if the subsurface is not able to degrade contaminants from nonpoint sources that flow into stormwater drains. Injection of any substances into the subsurface should meet regulatory requirements and objectives for the use of adjacent groundwater without costs to water users. Disposal of produced waters typically requires a discharge permit and associated costs. Improperly installed and maintained wells can serve as conduits of contamination to groundwater and result in health risks and economic costs to owners and nearby groundwater users. Managed aquifer recharge is gaining wide acceptance as communities around the world adopt this method to better manage water supplies and protect against drought. A wide range of techniques are now developed to enable storm and excess water to infiltrate through the soil zone for natural treatment to augment aquifers as water sources. Benefits of MAR include natural treatment, no evaporative loss, buffer for drought, no loss of farmland for reservoirs, restoring ecological systems and reducing subsidence risk. Wildlife also rely on groundwater directly for plant transpiration and through maintenance of streamflow for watering.

FOR FURTHER CONSIDERATION

1. What major aquifers supply the area you live in with water? Are the aquifers unconsolidated or bedrock, unconfined or confined? Which will cost more to have a well installed?
2. What different kinds of wells (private domestic, municipal, irrigation, injection, monitoring) have been installed in your county? How deep are these wells?
3. What is the well permit or authorization process in your state and/or county? Is there a fee or charge a well permit or authorization application?
4. Do you have artificial recharge facilities in your area? Contact the recharge facility office and find out what water they use to recharge the aquifer and what volume is recharged. What is the unit price of the water when it is sold to customers?
5. What would the cost of a well to supply your home with groundwater be? Check your water use and calculate the amount in cubic meters or gallons per minute of use over a day, considering that certain times of higher demand occur. Inquire with the state as to groundwater geology and water level, then contact a local well driller on the cost of a well including a pump.
6. Calculate the cost of diesel oil-powered pumping to produce water from a municipal well described in Box 1.6.5 Energy Costs to Pump Groundwater.

REFERENCES

Abel, F. and Walker, B. Undated. *Tapping Geothermal Energy, Heat from Within the Earth.* URL: https://naldc.nal.usda.gov/download/IND81006338/PDF

American Geosciences Institute (AGI). 2019. *Water Sources for Hydraulic Fracturing.* URL: https://www.americangeosciences.org/critical-issues/factsheet/pe/water-sources-hydraulic-fracturing

Aquifer Storage and Recovery Forum (ASRF). 2003. URL: http://www.asrforum.com (accessedJune 21, 2003).

Australian Government National Water Commission. 2009. Managed aquifer recharge: An Introduction. Waterlines Report Series No. 13. URL: https://www.readkong.com/page/managed-aquifer-recharge-an-introduction-peter-dillon-9032877?p=1 (accessed May 28, 2020).

BCC Research. 2016. *Special Research Study: Comparison of Water Main Pipe Installation Lengths and Costs in North and South Carolina: Raleigh, Charlotte, and Spartanburg/Greenville.* URL: https://www.uni-bell.org/portals/0/E-NEWS/MediaFiles/bcc_pipe_report_carolinas.pdf; 2016. Special Research Study: Comparison of Water Main Pipeline Installation Lengths and Costs in Ohio Columbus, in Delaware County, including the County as well as Delaware, Dublin, Westerville, and Powell [and Dayton]. URL: https://www.greenbuildingsolutions.org/wp-content/uploads/2016/11/BCC-Report-Pipe-Costs-OH-Cities-Feb-25-2016.pdf

Benjaminsen, C. 2015. Drilling the world's hottest geothermal well. In *ScienceNordic*, October 24, 2015. URL: http://sciencenordic.com/drilling-worlds-hottest-geothermal-well

Birt, N. 2013. *7 Questions to Ask Before Drilling an Irrigation Well.* URL: https://www.agweb.com/article/7_questions_to_ask_before_drilling_an_irrigation_well_naa_nate_birt/

Bisson, R.A. 2004. Revolutionary groundwater method reveals megawatersheds in Trinidad & Tobago. In *WaterWorld*, Water and Wastewater International. URL: http://www.pennnet.com/Articles/Article_Display.cfm?Section = ARTCL&ARTICLE_ID = 215858& VERS ION_NUM = 1&p = 20 (accessed February 8, 2009).

Bisson, R.A. and Lehr, J.H. 2004. *Modern Groundwater Exploration.* John Wiley & Sons, Inc., Hoboken, NJ, 309 p.

Blend, J. 2002. *Potential Benefits and Costs to Montanans of Coalbed Methane Development in Montana (draft).* Montana Department of Environmental Quality. URL: http://www.deq.state.mt.us/CoalBedMethane/CBMBenefitCost.htm (accessed July 26, 2003).

Brandt, E.N. 1997. *Growth Company, Dow Chemical's First Century.* Michigan State University Press, East Lansing, MI 649 p.

Brassington, R. 1988. *Field Hydrogeology. Geological Society of London Professional Handbook Series.* Open University Press, Milton Keynes, UK

Brenner, L. 2019. *The Average Cost of Drilling a Water Well.* URL: https://www.hunker.com/13401521/the-average-cost-of-drilling-a-water-well

Brumley, J., Marks, C., Chau, A., Lowrance, R., Huang, J., Richardson, C., Acree, S., Ross, R., and Beak, D.. 2018. *The Influence of Green Infrastructure Practices on Groundwater Quality: The State of the Science.* U.S. Environmental Protection Agency, Washington, DC, EPA/600/R-18/227.

Bruni, M. 2019. *Dug Wells. Sustainable Sanitation and Water Management Toolbox - SSWM.info*, URL: https://sswm.info/sswm-university-course/module-4-sustainable-water-supply/further-resources-water-sources-hardware/dug-wells

Burk, L., and Cook, P.G. 2015. A Simple and Affordable System for Installing Shallow Drive Point Piezometers. *Groundwater Monitoring and Remediation*, April 29, 2015.

California Division of Oil, Gas and Geothermal Resources. 2003. *Geothermal Production Wells.* URL: http://www.consrv.ca.gov/DOG/geothermal/general_info/production_wells.htm (accessed August 30, 2003).

Casanova J., Devau N., and Pettenati M. 2016. Managed aquifer recharge: An overview of issues and options. In: Jakeman A.J., Barreteau O., Hunt R.J., Rinaudo J.D., Ross A. (eds). *Integrated Groundwater Management.* Springer, Cham.

Choy, J. 2016. *High Quality Groundwater Data Isn't Always Easy or Cheap, But It Is Necessary.* Stanford University; Water in the West. URL: http://waterinthewest.stanford.edu/news-events/news-insights/high-quality-groundwater-data-isn%E2%80%99t-always-easy-or-cheap-it-necessary

Clark, B., Markley, M., and Kundral, S. 2018. Wastewater deep injection wells for wastewater disposal industries tap a unique resource. *WasteAdvantage Magazine.* URL: https://wasteadvantagemag.com/wastewater-deep-injection-wells-for-wastewater-disposal-industries-tap-a-unique-resource/

Clean Water Testing LLC (CWT). 2018. *Well Water and Septic Inspections*. http://www.cleanwatertesting.com/local-water-testing-services/well-septic-inspections/

Commonwealth Scientific and Industrial Research Organisation (CSIRO). 2019. *Managed Aquifer Recharge*. URL: https://research.csiro.au/mar/

ConcreteNetwork.com. 2019. *Use Of Pervious Concrete Eliminates Over $260,000 In Construction Costs*. URL: https://www.concretenetwork.com/pervious/design-ideas/pervious-concrete-washington.html

Cooper, K. 2019. *Multilevel Monitoring Well Installations*. Solinst Canada Limited. URL: https://www.solinst.com/resources/papers/401k.php

Cost Helper, Inc (CHI). 2019. *How Much Does Well Drilling Cost?* URL: https://home.costhelper.com/well-drilling.html

Davis, S., and DeWiest, R. 1970. *Hydrogeology*. Wiley and Sons, New York. 463 p.

Deberitz, J. 2006. Lithium; production and application of a fascinating and versatile element. *Die Bibliothek der Wissenschaft*, vol. 2. 82 p.

de los Cobos, G. 2002. The aquifer recharge system of Geneva, Switzerland: A 20 year successful experience. In *Management of Aquifer Recharge for Sustainability*, P.J. Dillon (ed.). A.A. Balkema Publishers, Lisse, the Netherlands, pp. 49–52.

Dillon, P.J.. (ed.). 2002. *Management of Aquifer Recharge for Sustainability*. A.A. Balkema Publishers, Lisse, the Netherlands, 567 pp.

Driscoll, F.G. 1986. *Groundwater and Wells*. Johnson Filtration Systems, Inc., St. Paul, MN, 1089 p.

Dunne, T., and Leopold, L. 1978. *Water in Environmental Planning*. WH Freeman and Company, San Francisco. 818 p.

Earth Water Global. 2009. *Megawatershed Paradigm*. URL: http://www.earthwaterglobal.com/paradigm.htm (accessed February 8, 2009).

Earthworks. 2017. *Fracking-related Earthquakes*. https://earthworks.org/issues/fracking_earthquakes/

EmCon Consulting Group. 2006. *Feasibility Assessment for the Replacement of Diesel Water Pumps with Solar Water Pumps*. For Namibian Ministry of Mines and Energy. URL: http://pubdocs.worldbank.org/en/511871475091277205/1-Replacement-diesel-GEF-UNDP-Namibia-2006.pdf

Energyhomes.org. 2019. *What is a Geothermal Heat Pump?* URL: http://www.energyhomes.org/renewable-technology/howgeoworks.html

Eshkalak, M.O., Aybar, U., and Sepehmoori, K. 2015. On the feasibility of re-stimulation of shale wells. *Petroleum Science*, August 2015, Volume 12, Issue 3, pp 553–559. *URL*: https://link.springer.com/article/10.1007/s12182-015-0036-7

Foster, S.S.D. 1989. Economic considerations in groundwater resource evaluation. In *Groundwater Economics: Selected Papers from a United Nations Symposium, Barcelona, Spain*, Elsevier Science Publishers B.V., Amsterdam, the Netherlands, pp. 53–65.

Frederick County, Maryland. 2019. *Wellhead Protection*. URL: https://frederickcountymd.gov/1327/Wellhead-Protection

Geiger, M. 2013. *Solar-Powered Livestock Water Pumping: Technical and Economic Feasibility*. University of Wyoming, School of Energy Resources. URL: http://www.uwyo.edu/uwexpstn/centers/sheridan/_files/2013-field-days-pdfs/solar-stock-water.pdf

Gholizadeh, M.H., Melesse, A.M., Reddi, L. 2016. A comprehensive review on water quality parameters estimation using remote sensing techniques. *Sensors (Basel)*, 16(8):1298. URL: https://www.mdpi.com/1424-8220/16/8/1298/htm (accessed December 17, 2020).

Girma, M., Molina, M., and Assefa, A. 2015. Feasibility study of a wind powered water pumping system for rural Ethiopia. *AIMS Energy*, 3 (4), 851–868. URL: http://www.aimspress.com/fileOther/PDF/energy/energy-03-00851.pdf

Glennon, R. 2002. *Water Follies: Groundwater Pumping and the Fate of America's Fresh Waters*. Island Press. Washington, DC, 314 p.

Glotfelty, M.. 2017. Life-Cycle Economic Analysis of Water Wells. Water Well Journal, March 31, 2017 (with permission) URL: https://waterwelljournal.com/life-cycle-economic-analysis-of-water-wells/ (Accessed December 17, 2020); communication, 2016.

Gude, G. 1984. *Where the Potamac Begins: A History of the North Branch Valley*. Seven Locks Press, Cabin John, MD.

Halihan, T. 2016. *The Future of Water: Will Our Data Beat Our Instincts*. Published October 20, 2016, by the National Ground Water Association.

Intergovernmental Task Force on Monitoring Water Quality (ITFM). 1997. *Conceptual Frameworks for Ground Water Quality Monitoring*. 112 p. URL: http://acwi.gov/gwfocus.pdf (accessed August 4, 2009).

International Groundwater Resources Assessment Centre (IGRAC). 2018. *MAR Portal; Global Inventory of Managed Aquifer Recharge Schemes*. URL: https://www.un-igrac.org/special-project/mar-portal

Irrigation Association (IA). 2019. *Benefits of Irrigated Agriculture and Economic Benefits of Efficient Irrigation*. URL: https://www.irrigation.org/IA/Advocacy/Position-Statements/Benefits-of-Irrigated-Agriculture/IA/Advocacy/Benefits-of-Irrigated-Agriculture.aspx?hkey=145446d0-abf8-420c-842d-127429d299ee; https://www.irrigation.org/IA/Advocacy/Position-Statements/Economic-Benefits-of-Efficient-Irrigation/IA/Advocacy/Economic-Benefits-of-Efficient-Irrigation.aspx?hkey=bc90c13e-59b5-4169-a596-2ee321eafc81

Job, C. and Gabanski, G. 1987. Monitoring wells need consistent regulation. In *Superfund 87, Proceedings of the 8th National Conference*, The Hazardous Materials Control Research Institute, Washington, DC, Silver Spring, MD. November 1987 (Reprinted in *Waste Age*, August 1988. pp. 164–170).

Job, C. 1996. *Business Benefits of Wellhead Protection-Case Studies: Dayton*. US Environmental Protection Agency, Ohio; Xenia, Ohio; and Pekin, Illinois. EPA 813-B-95-004.

Job, C. 2017. Responding to flooded wells; A sustainability Issue. *Water Well Journal*, November 30, 2017. https://waterwelljournal.com/responding-flooded-wells-2/

Larson, D. 2016. Energy pipeline: Pushing the limits — Companies increasingly up their game in well length, productivity. *In Greeley Tribune*, July 21, 2016.

Llamas, R., and Custodio, E. 2003. *Intensive Use of Groundwater*. Swets and Zeitlinger B.V., The Netherlands. p. 11.

Maryland Department of the Environment (MDE). 2019. *About Wetlands*; Stormwater and Flood Control. URL: https://mde.maryland.gov/programs/Water/WetlandsandWaterways/AboutWetlands/Pages/stormwater.aspx

Minnesota Pollution Control Agency (MPCA). 2011. *Best Management Practices Construction Costs, Maintenance Costs, and Land Requirements*. URL: https://www.pca.state.mn.us/sites/default/files/p-gen3-13x.pdf

Nagaoka, G. 2018. Personal communication regarding aquifer recharge pond maintenance for the Santa Clara Valley Water District, California, USA.

National Ground Water Association (NGWA). 2016a. *Best Suggested Practices for Aquifer Storage and Recovery*. BSP26-ASR. URL: https://my.ngwa.org/NC__Product?id=a185000000BYugHAAT (Accessed December 17, 2020).

National Ground Water Association (NGWA). 2016b. *Groundwater Use in the United States of America*. URL: https://www.ngwa.org/docs/default-source/default-document-library/groundwater/usa-groundwater-use-fact-sheet.pdf?sfvrsn=5c7a0db8_4 (Accessed December 17, 2020).

National Ground Water Association (NGWA). 2016c. Sustainable Groundwater Resources; NGWA Position Paper.

National Ground Water Association (NGWA). 2018. *Planning for a Water Well*. URL: http://www.wellowner.org/basics/planning-for-a-water-well/.

National Research Council (NRC). 1993. *Ground Water Vulnerability Assessment*. National Academy Press. 204 p.

National Research Council (NRC). 1994. *Ground Water Recharge Using Waters of Impaired Quality*. National Academy Press. 283 p. URL: https://www.nap.edu/download/4780 (Accessed December 17, 2020).

National Research Council (NRC). 1997. *Valuing Ground Water*. National Academy Press. 189 p. URL: https://www.nap.edu/download/5498 (Accessed December 17, 2020).

Navarro, A. 1989. Economics of groundwater works. In *Groundwater Economics: Selected Papers from a United Nations Symposium, Barcelona, Spain*. Elsevier Science Publishers B.V., Amsterdam, the Netherlands, pp. 25–51.

New York State Department of Transportation (NYSDOT). 2015. *Design, Construction, and Maintenance of Recharge Basins*. 130 p. URL: https://www.dot.ny.gov/divisions/engineering/technical-services/technical-services-repository/GDP-8b.pdf

Niñerola, S. 1989. Economic aspects of groundwater exploitation. In *Groundwater Economics: Selected Papers from a United Nations Symposium, Barcelona, Spain*. Elsevier Science Publishers B.V., Amsterdam, the Netherlands, pp. 83–97.

O'Hanlon, L. 2003. Grace map shows earth is very lumpy. *Science in News*, July 23, 2003. URL: http://www.csr.utexas.edu/GRACE/publications/press/03–07–29-ABConline.pdf (accessed February 13, 2009).

Olsthoorn, T.N. and Mosch, M.J.M. 2002. Fifty years artificial recharge in the Amsterdam dune area. In *Management of Aquifer Recharge for Sustainability*, P.J. Dillon (ed.). A.A. Balkema Publishers, Lisse, the Netherlands, pp. 29–33.

Ozment, S., DiFrancesco, K. and Gartner, T. 2015. Natural Infrastructure in the Nexus 2015. 36 p. URL: http://www.iwa-network.org/wp-content/uploads/2016/06/Natural-Infrastructure-in-the-Nexus_Final-Dialogue-Synthesis-Paper-2015.pdf

Page, D., Dillion, P., Vanderzalm, J., Toze, S., Sidhu, J, Barry, K., Levett, K., Kremer, S., and Regel, R. 2010. Risk Assessment of Aquifer Storage Transfer and Recovery with Urban Stormwater for Producing Water of a Potable Quality. Technical Reports : Groundwater Quality. URL: https://pdfs.semanticscholar.org/099a/6a83f8c756eb3d1ac6778d1073a4082dd9c0.pdf

Paque, M. 2003. *Personal communication*. In *Ground Water Protection Council Conference*, Alexandria, VA, March 2003.

Parker, T.K. 2018. *Managed Groundwater Storage; Managed Aquifer Recharge in California*. American Geosciences Institute Webinar, July 18, 2018.

Permian Basin Petroleum Association (PBPA). 2015. Well Disposed. *The Permian Basin Petroleum Association Magazine*. URL: https://pboilandgasmagazine.com/well-disposed/

Powers, J.P. 1992. *Construction Dewatering: New Methods and Applications*. John Wiley & Sons, Inc., New York. 492 p.

Pyne, R.D.G. 1995. *Groundwater Recharge and Wells*. Lewis Publishers, Boca Raton, FL, 376 p.

Pyne, D., Burchard, G., and Musick, M. 2020. Personal communication with author to National Ground Water Association Managed Aquifer Recharge Work Group. December 17, 2020.

Rafferty, K. 2003. Geoheat center: Klamath Falls, Oregon. *Bulletin*, 18 (2). URL: http://geoheat.oit.edu/bulletin/bull18–2/art40.htm (accessed July 26, 2003).

Rafferty, K. 2007. *A Spread Sheet for Geothermal Direct Use Cost Evaluation*. URL: http://geoheat.oit.edu/software/cost.pdf (accessed April 13, 2007).

Reddy, K.R. 2008. Physical and chemical groundwater remediation technologies. In *Overexploitation and Contamination of Shared Groundwater Resources*, C.J.G. Darnault (ed.). Springer Science+Business Media B.V. pp. 257–274.

Reguero, B.G., Beck, M.W., Bresch, D.N., Calil, J., and Meliane, I. 2018. Comparing the cost effectiveness of nature-based and coastal adaptation: A case study from the Gulf Coast of the United States. *PLoS ONE* 13(4): e0192132. doi:10.1371/journal.pone.0192132

Rehman, S., and Sahin, A.Z. 2012. Wind power utilization for water pumping using small wind turbines in Saudi Arabia: A techno-economical review. *Renewable and Sustainable Energy Reviews* 16(7):4470–4478. https://www.researchgate.net/publication/257548419_Wind_power_utilization_for_water_pumping_using_small_wind_turbines_in_Saudi_Arabia_A_techno-economical_review

Reichard, E.G., Raucher, R.S., and Nishikawa, T. 2004. Economic valuation of aquifer storage projects. *In Presented to the American Geophysical Union*, Fall Meeting 2004. URL: http://adsabs.harvard.edu/abs/2004AGUFM.H33D0505R (accessed February 21, 2009).

Roscoe Moss Company (RMC). 1990. *Handbook of Ground Water Development*. Wiley-IEEE Press, Hoboken, NJ, 493 p.

Sanner, B. and Abbas M.A. Undated. How can geophysical exploration help to determine GSHP ground properties? URL: http://intraweb.stockton.edu/eyos/energy_studies/content/docs/proceedings/SANNE1.PDF (accessed February 13, 2009).

Satter, A. and Iqbal, G.M. 2016. *Hydraulic Fracturing Design*. Science Direct. URL: https://www.sciencedirect.com/topics/engineering/hydraulic-fracturing-design

Schwarz, J. 1989. Economic aspects of groundwater exploration and assessment. In *Groundwater Economics: Selected Papers from a United Nations Symposium, Barcelona, Spain*. Elsevier Science Publishers B.V., Amsterdam, the Netherlands, pp. 67–81.

Shahkarami, A., Wang, G., and Belyadi, H. 2016. *Horizontal Well Spacing and Hydraulic Fracturing Design Optimization: A Case Study on Utica-Point Pleasant Shale Play. Unconventional Resources Technology Conference*, San Antonio, Texas, USA, 1–3 August 2016. URL: https://www.francis.edu/uploadedFiles/Content/Academics/School_of_Sciences/Engineering/Engineering_Faculty_Page/Alireza_Shahkarami_Documents/2016%20URTeC%20Paper-2459851.pdf

Sharma, S.K. 2020. *Strategies and Policies for Artificial Recharge of Ground Water*. URL: https://www.indiawaterportal.org/sites/indiawaterportal.org/files/S.K.Sharma_GWE.pdf (accessed April 15, 2020).

Shevah, Y. and Waldman, M. 1995. In-situ and on-site treatment of groundwater. *Pure &App/Chern*, 67, 8/9, 1549–1561. URL: http://citeseerx.ist.psu.edu/viewdoc/download?doi=10.1.1.531.4597&rep=rep1&type=pdf

Sumner, Washington (State). 2018. Sumner Municipal Code Chapter 16.48, Aquifer Recharge Area, passed June 18, 2018. URL: http://www.codepublishing.com/WA/Sumner/html/Sumner16/Sumner1648.html

Terhell, S., Cai, K., Chiu, D., and Murphy, J. 2015. *Cost and Benefit Analysis of Permeable Pavements in Water Sustainability*. University of California-Davis, Water Management. URL: http://watermanagement.ucdavis.edu/files/5414/3891/2393/A03_Terhell_Cai_Chiu_Murphy_ESM121_FinalReport.pdf

The Groundwater Foundation. 2020. *What is Groundwater?* URL: https://www.groundwater.org/get-informed/basics/whatis.html (accessed December 17, 2020).

U.S. Bureau of Reclamation (USBR) 2005. *Park City & Snyderville Basin Water Supply Study*; Appendix-Cost Estimates. URL: https://www.usbr.gov/uc/provo/rm/snyderville/pdfs/appdx.pdf

U.S. Department of Energy (USDOE). 2003. *Geothermal Energy Program: What Is Geothermal Energy?* URL: http://www.eere.energy.gov/geothermal/geoimpacts.html (accessed September 6, 2003).

U.S. Department of Energy (USDOE). 2004a. *Wind Power: Today and Tomorrow* 36 p. URL: http://www.nrel.gov/docs/fy04osti/34915.pdf (accessed April 24, 2008).

U.S. Department of Energy (USDOE). 2004b. *Wind Turbine Powers Ground-Water Circulation Well*. URL: http://www.clu-in.org/products/newsltrs/tnandt/view.cfm?issue=0904.cfm#1 (accessed April 24, 2008).

U.S. Department of Energy (USDOE) 2019. *Geothermal FAQs*. URL: https://www.energy.gov/eere/geothermal/geothermal-faqs

U.S. Environmental Protection Agency (USEPA). 1989. *A Guide on Remedial Actions for Contaminated Ground Water*. Solid Waste and Emergency Response Directive 9283.1-2FS. April 1989.

U.S. Environmental Protection Agency (USEPA). 1996. Benefits and Costs of Prevention: Case Studies of Community Wellhead Protection-Volume 1. EPA-813-B-95-005.

U.S. Environmental Protection Agency (USEPA), Technology Innovation Program. 2004. Wind turbine powers ground-water circulation well. In *Technology News Trends* (September 2004). URL: http://www.clu-in.org/products/newsltrs/tnandt/view.cfm?issue=0904.cfm (accessed October 3, 2009).

U.S Environmental Protection Agency (USEPA). 2007. *Cost Evaluation Of Point-Of-Use and Point-Of-Entry Treatment Units For Small Systems: Cost Estimating Tool And User Guide*. EPA 815-B-07-001.

U.S. Environmental Protection Agency (USEPA). 2016. *General Information About Injection Wells*. URL: https://www.epa.gov/uic/general-information-about-injection-wells

U.S. Environmental Protection Agency (USEPA). 2018a. *Aquifer Recharge and Aquifer Storage and Recovery*. URL: https://www.epa.gov/uic/aquifer-recharge-and-aquifer-storage-and-recovery (accessed April 15, 2020).

U.S. Environmental Protection Agency (USEPA). 2018b. What is Green Infrastructure? URL: https://www.epa.gov/green-infrastructure/what-green-infrastructure

U.S. Environmental Protection Agency (USEPA). 2018c. *Why are Wetlands Important?* URL: https://www.epa.gov/wetlands/why-are-wetlands-important

U.S. Environmental Protection Agency. 2018. *How We Use Water*. URL: https://www.epa.gov/watersense/how-we-use-water (accessed December 17, 2020).

U.S. Geological Survey (USGS). 1997. Guidelines and Standard Procedures for Studies of Ground-Water Quality: Selection and Installation of Wells, and Supporting Documentation. Water-Resources Investigations Report 96–4233.

Vick, B.D., Clark, R.N., and Evett, S.R. 2000. *Wind-Powered Drip Irrigation Systems for Fruit Trees. Presented to 2000 ASAE Annual International Meeting*, Milwaukee, WI, July 9–12, 13 pp. URL: http://www.cprl.ars.usda.gov/Drip%20Irrigation/vick_asae2000.pdf (accessed April 24, 2008).

Various Industry Sources (VIS). 2019. Communication with various industry consultants, not to be cited.

Williams, M.B. and Fenske, B.A. 2004. *Demonstrating Benefits of Wellhead Protections Progams*. AWWA Research Foundation.

Winter, T.C., Harvey, J.W., Franke, O.L., and Alley, W.M. 1998. *Groundwater and Surface Water: A Single Resource*. US Geological Survey Circular 1139.

World Resources Institute (WRI). 2013. *Natural Infrastructure; Investing in Forested Landscapes for Source Water Protection in the United States*. Edited by Todd Gartner, James Mulligan, Rowan Schmidt, and John Gunn. 140 p. URL: https://www.wri.org/sites/default/files/wri13_report_4c_naturalinfrastructure_v2.pdf

Worst, R.C. 2019. *How long should my pump last?* URL: www.rcworst.com/blog/How-Long-Should-My-Pump-Last

Zimmerman, M. 2015. *How Much Does a Residential Water Well Cost?* URL: http://www.zdrillerteam.com/residential-water-cost/

OTHER REFERENCES CONSULTED

American Petroleum Institute (API). 2019. *Hydraulic Fracturing: Unlocking America's Natural Gas Resources*. URL: https://hydraulicfracturing.com/#/?section=energy-and-opportunity

British Columbia Ministry of Agriculture and Food. 1994. *Livestock Watering Factsheet: Livestock Water Pumping Using Solar Energy*. Order No. 590.306–4. Agdex: 778 (June). URL: http://www.agf.gov.bc.ca/resmgmt/publist/500series/590306–4.pdf (accessed February 24, 2005).

Davis-Martin-Powell Engineers and Surveyors (DMP). 2015. *Yadkin County Water System Improvements, East Bend Water Transmission Main WIF-1853*, Engineering Report and Environmental Information Document. 108 p.

El Sheikh, R.A. and Hamdan, S.M. 2002. Artificial recharge of groundwater in Palestine: A new technique to overcome water deficit. In *Management of Aquifer Recharge for Sustainability*, P.J. Dillon (ed.). A.A. Balkema Publishers, Lisse, the Netherlands, pp. 413–417.

Environmental Working Group (EWG) 2020. *UPDATE: Thousands of Industrial Facilities Likely Discharging Toxic 'Forever Chemicals' Into Air and Water*. URL: https://www.ewg.org/news-and-analysis/2020/04/updated-thousands-industrial-facilities-likely-discharging-toxic-forever (accessed April 15, 2020).

Geothermal Pros and Cons (GPC). 2019. *Introduction to Geothermal Drilling*. URL: https://geothermalprosandcons.net/introduction-to-geothermal-drilling/

Legg, C. and Sagstad, S. 2002. Optimization and use of various recharge techniques for reclaimed wastewater at a sensitive site in Glendale, Arizona. In *Management of Aquifer Recharge for Sustainability*, P.J. Dillon (ed.). A.A. Balkema Publishers. Lisse, the Netherlands, pp. 333–338.

Meuli, C., and Wehrle, K. 2001. Spring Catchment. Series of Manuals on Drinking Water Supply, Volume 4. Swiss Centre for Development Cooperation in Technology and Management. URL: http://www.clean-water-for-laymen.com/support-files/skatspringcatchments.pdf

Miller, T. 2003. Aquifer storage and recovery in Wisconsin: Case studies—the good, the bad, and the ugly. *In Presented at the Wisconsin Ground Water Association Conference*, Geneva, WI, April 9, 2003. URL: http://www.wgwa.org/conference/papers/miller.htm (accessed June 21, 2003).

Sacramento Suburban Water District (SSWD). 2011. *Water Transmission Main Asset Management Plan*. URL: http://www.sswd.org/home/showdocument?id=1002

Stanford Woods Institute for the Environment (SWIE). 2014. *Understanding California's Groundwater*. URL: http://waterinthewest.stanford.edu/groundwater/recharge/ (accessed May 7, 2020).

The University of California Cooperative Extension, Tulare County. 2005. Energy and cost required to lift or pressurize water. Publication No. IG6-96, 2005. URL: http://cetulare.ucdavis.edu/pubgrape/ig696. htm (accessed February 24, 2005).

Toolbase Services (TBS). 2008. *Low Impact Development (LID) Practices for Storm Water Management*. URL: http://www.toolbase.org/Technology-Inventory/Sitework/low-impact-development (accessed April 24, 2008).

Tsuchihashi, R., Asano, T., and Sakaji, R.H. 2002. Health aspects of groundwater recharge with reclaimed water. In *Management of Aquifer Recharge for Sustainability*, P.J. Dillion (ed.). A.A. Balkema Publishers, Lisse, the Netherlands, pp. 11–20.

U.S. Energy Information Agency. 2016. *Trends in U.S. Oil and Natural Gas Upstream Costs*. URL: https://www.eia.gov/analysis/studies/drilling/pdf/upstream.pdf

U.S. Energy Information Administration (US EIA). 2019. *Electric Power Monthly; Average Price of Electricity to Ultimate Customers*. URL: https://www.eia.gov/electricity/monthly/epm_table_grapher.php?t=epmt_5_3 [average price for residential customers = $0.1289/kWh]

U.S. Environmental Protection Agency (USEPA), Office of Water. 1997. *Ambient Ground Water Quality Monitoring Cost Analysis*. EPA 816-R-97-013.

7 Groundwater Quality Treatment and Waste Disposal

MANY POTENTIAL CONTAMINANT SOURCES

Groundwater containing contaminants may need treatment before use to meet its expected quality. Or, for continued use, groundwater may need special protection from wastes disposed in the subsurface "sink." Treatment is a "last resort" if other groundwater protection barriers are unsuccessful. Furthermore, groundwater may have naturally occurring constituents at high concentrations that minimize its use for certain purposes, such as drinking or industrial processes and irrigation. It may also have anthropogenic contaminants that make it unhealthful. These contaminants may reach the subsurface environment through leaching and percolating to groundwater, or may have been consciously and actively introduced underground, because the current economics of waste disposal indicate that this is the least-expensive disposal repository. In any case, groundwater may need to be treated before use, in particular, if no alternative sources of water with the expected or preferred amount and quality were available at the same or reasonable cost.

The major sources of groundwater quality problems in the United States (USEPA, 1996; USEPA, 2019a) and the European Union (EEA, 1999; indicated by*) include:

- Underground storage tanks
- Pesticide applications* and storage
- Septic tanks
- Fertilizer applications* and storage
- Landfills*
- Surface impoundments
- Aboveground storage tanks
- Land application
- Animal feedlots and waste disposal systems
- Shallow injection wells and floor drains
- Mining and mine drainage
- Road salting
- Urban runoff
- Transportation of materials
- Pipelines and sewer lines
- Saltwater intrusion*
- Waste tailings
- Irrigation practices*
- Deep injection wells
- Abandoned production and disposal wells
- Naturally-occurring radionuclides
- Floodwater in locations of water wells

Contaminants emanating from this range of sources may include: petroleum compounds, nitrate, chloride, metals, organic chemicals and pesticides, inorganic chemicals and pesticides, bacteria, viruses, protozoa and radionuclides.

The cost of treatment in economic terms translates into "abatement costs" for contaminants that can be compared with the damages from contaminants affecting the water quality, to determine the efficient result for a community or nation to obtain the desired quality of groundwater. Abatement costs may include not only the costs of purchasing, installing and operating a particular technology at a treatment plant but also purchase and transaction costs of permits for land and construction, chemical use for treatment and waste disposal from removal of the contaminants. The cost of underground disposal should include the damages to the subterranean ecosystem or to the future users of this subsurface zone.

First, groundwater treatment for immediate drinking water use is examined. Second, the best practices in the watershed are considered as important alternatives to water treatment for some contaminants. Subsequently, groundwater remediation for future use is discussed. Finally, an examination of groundwater and the subsurface as a "waste sink" is presented, which suggests increasing pressure and competition for the subterranean environment.

SUBSURFACE CONTAMINANT SINK AND NATURAL CAPITAL TREATMENT

While a range of treatment for groundwater contamination may be undertaken, consideration should be given to the inherent nature of the subsurface environment and its influence on groundwater quality. The subsurface portion of the ecosystem is complex. The soil zone may be thin or thick, of different textures and chemical compositions and even non-existent in rocky ridges or outcrops. Strata, the geologic layers, are similarly thin or thick, even tens to hundreds of feet in depth, of different compositions with angular folds and thrusts as well as fractures. As water infiltrates and percolates through the subsurface, the soluble components of the subsurface add to the chemistry of the water, as it becomes more mineralized. Geochemical reactions and microbial chemical processes also change groundwater quality. Some minerals may be healthful and others not, so treatment may be needed.

As precipitation picks up different constituents on the ground surface, these chemicals and microorganisms may be carried into the subsurface affecting groundwater quality. The complexity of the subsurface environment affects how these chemicals and microorganisms move through it. The U.S. Environmental Protection Agency (1985) has characterized 48 different generic hydrogeologic settings in which groundwater may exist. Each setting has its own complexities at a particular location. Each location has its own vulnerability to chemical and microbial contaminants conditioned on the subsurface hydrogeological factors present (USEPA, 1993a). Depending on the characteristics of the chemicals and microorganisms interacting with the underground geologic matrix, the chemicals and microbes in the subsurface pore water may travel very slowly or somewhat more quickly, measured as millimeters per decade or meters (feet), even kilometers (miles) in karstic conditions, per day. Chemicals and microbes may be filtered, adsorbed, changed in chemical reaction and thus constrained in movement. Microorganisms can die off, if the subsurface matrix is fine-grained, appropriately reactive or otherwise restraining their movement. If waste chemicals or product residuals have been in the ground for many years because the subsurface was used as a disposal sink for them, they may have migrated a great distance beyond their disposal site and contaminated an area as large as 2,023 km^2 (780 square miles) (USDA, 1997), with the potential to impact many water supplies.

The natural capital of the subsurface portion of the ecosystem can function to provide some treatment for chemical and pathogens. The opposite is also operable. The locations that are more vulnerable to contamination, such as sand and gravel and fractured bedrock aquifers, may be highly productive groundwater sources because their groundwater conditions are more transmissive. While those same conditions can make groundwater both vulnerable and productive, adequate distances from contaminant sources (e.g., poorly maintained septic systems) to water wells can provide natural treatment and protection services to water sources. "Adequate distances" refers "setbacks" of wells from potential contaminant sources (see Volume 1, Chapter 6) that are widely recognized and incorporated in local and state water codes (see Volume 1, Chapter 4). These setbacks provide the time for natural treatment to occur, reducing the contamination potential. Subsurface conditions should be examined carefully before placing chemical and microbial contaminants on or under the ground and in proximity to current or potential water sources. Depending on subsurface conditions, the groundwater may be expensive to remediate for use once contaminated. Therefore, substantial hydrogeologic knowledge, investigation and planning is needed to decide how to utilize the subsurface environment and whether and how to treat changed groundwater quality to provide resilient continuing safe resource use.

TREATMENT FOR DRINKING WATER

Natural Capital Treatment Service

Groundwater in the United States has traditionally been considered the water source least susceptible to contamination by indicator bacteria or human pathogens (USGS, 2017). The service of the subsurface natural capital provides level of treatment for chemicals and microbes, as noted above. Unknowingly in ages past and knowingly in state water well codes across the nation, prescription of setbacks for placement of wells in relation to septic systems and sewer lines has served to protect 15-21 million households that self-supply themselves onsite through private water wells and wastewater treatment (USCB, 2008, LaFond, 2015). Properly located and maintained water wells and septic systems should provide households with many years of safe water and wastewater treatment service drawing on the protective conditions of the subsurface. A wellhead protection area around a public water well can provide a similar function through managing distances of potential contaminant sources from the well combined with the protective features of the subsurface environment. Hydrogeologic conditions must be thoroughly assessed to determine the level of natural treatment that can be expected. Box 1.7.1 gives a perspective on well water protection from viruses and other pathogens and the role of setbacks in providing safe well water, as well as costs for well and septic system maintenance.

BOX 1.7.1 PROTECTING WELL WATER FROM VIRUSES AND OTHER PATHOGENS

Viruses in Groundwater

The principal potential sources of household well water contamination by viruses and other pathogens is from onsite wastewater treatment (septic) systems and nearby wastewater sewer lines. Other sources of fecal contamination may include underground storage tanks, waste disposal systems, animal manure and compost, and sewage sludge applied to fields in agricultural areas (Krauss and Griebler, 2011) About 50 percent of waterborne disease outbreaks are related to contaminated groundwater. (Craun *et al.*, 2010) A substantial number of these outbreaks are at businesses and small water systems that may not disinfect their water or do not regularly monitor their groundwater supply. Outbreaks from household wells have received less documentation. The U.S. Environmental Protection Agency has already set standards for viruses and other pathogens in drinking water (USEPA, 2020c).

Viruses and other pathogens may be picked up and carried in stormwater runoff, although they are usually present at low levels (NRC, 1994) and generally the occurrence of pathogens in groundwater is less than in surface water. (Gundy *et al.*, 2009) Cracks in the well riser, casing, and apron at the wellhead and faulty sealing of the well cap can allow stormwater into the well or down along the well casing to the water table circumventing the soil and vadose zone that can naturally protect groundwater.

(*Continued*)

BOX 1.7.1 (Continued)

Viruses in human fecal waste present the major disease concern for well water. These viruses in human fecal matter have been associated with outbreaks. Underground, pathogens can be inactivated through filtration, adsorption and die-off. With filtration, the pathogen, although very small, may be too large to move through the soil and become trapped. Adsorption occurs when the microbe attaches to a soil particle and is restrained in movement. Die-off can occur when temperature, pH or other conditions destroy the organism. Thus, soil can be effective in removing virus and other pathogens. However, fractured rock and karst formations may facilitate virus migration underground (Gundy *et al.*, 2009).

The COVID-19 coronavirus is spread from contact with contaminated surfaces or airborne releases from another person. Research has detected the COVID-19 virus in the feces of some infected patients but its infectiousness from that potential source is not known. Transmission risk is considered to be low based on other coronavirus outbreaks. No fecal-oral transmission of COVID-19 has been documented at this time. Other coronaviruses have had rapid die off in wastewater, with 99.9 % inactivation in 2-3 days. Some studies indicate that coronaviruses may possibly survive in groundwater for several weeks (Gundy *et al.*, 2009) but based on knowledge to date, the risk to private wells is extremely low (NGWA, 2020).

Setbacks and Septic Systems

Virus inactivation is critical for protection of any water source, including groundwater and the wells producing the water. A key step in activation that is already recognized in local building codes across the nation is "setback" – a requirement that a water supply well be at least a certain distance from a septic system or sewer line to ensure that natural degradation of chemicals and die-off of harmful organisms occur in the subsurface if a release would occur that may endanger groundwater pumped by a well (see for example: MPCA, 2019; MCDPS, 2018; CDNR, 2005) The setback approach as a barrier to contamination of wells is similar to the concept of wellhead protection: keep potential sources of contamination away from wells. EPA expects a properly managed septic system to treat COVID-19 the same way it safely manages other viruses often found in wastewater (USEPA, 2020).

The setback distances take into account the soils and subsurface geology of an area or state to enable chemical degradation and virus or other microbe inactivation to occur. For example, in Minnesota, setback distances from water wells are 20 feet for sewer lines, 50 feet for septic tanks and leach fields. In other states, the minimum setback from a water well may be a range, such as 50 feet but may be 25 feet depending on hydrogeologic information for the site, as in Colorado. Additionally, onsite wastewater treatment (septic) systems treat virus (Heger, 2020), which will minimize that potential source of virus.

The next protective step is proper maintenance of the well and septic system. Stormwater may carry viruses and other pathogens. Inspection of water wells should identify cracks in the well casing, riser, and apron at the wellhead through which floodwater could enter the well or carry contaminants down along the casing to the water table. Wells submerged in floodwater may be more vulnerable to contamination from viruses (CDC, 2015).

Treatment Systems

Groundwater in the United States has traditionally been considered the water source least susceptible to contamination by indicator bacteria or human pathogens. (USGS, 2017) After use of setbacks for wellhead protection and well and septic systems maintenance, water treatment is an optional barrier, if determined to be needed after well water testing. Chlorination, ozonation, distillation, ultraviolet (UV) radiation, and reverse osmosis remove virus. (NRC,

(*Continued*)

BOX 1.7.1 (Continued)

1994; WQA, 2016) Distillation, ultraviolet treatment and reverse osmosis technologies are cost-effective virus removal processes available at a household level as point-of-entry/point of use equipment (WQA, 2016; CDC, 2014).

Cost of Testing, Inspection and Maintenance

Before purchasing a treatment system, a well owner who is concerned about possible virus or other pathogenic contamination of their household well should have the well tested. The cost of well water sample testing for fecal indicator organisms is about $35 (if other more comprehensive testing is done, $25 each for heterotrophic plate count and adenosine triphosphate analyses). To plan ahead, well inspection may be considered. Costs for inspection of a private domestic water well conducted by a qualified water well professional may range from $80-$325 depending on the location and conditions of the well. (Job, 2017; CWT, 2018; CCHS, 2014) Ideally, well inspections would be done annually to avoid or minimize the possibility of conditions that may contribute to unsafe well water. If well disinfection/chlorination is indicated, this procedure can be performed by a qualified water well professional with costs ranging from $500 to $1,000 depending on the size and depth of the well and other well characteristics (Beitsch, 2018; pers. comm, 2019), and may be sufficient to address microbial contamination of the well.

If water testing indicates that the septic system may need maintenance or replacement, a qualified onsite wastewater treatment professional should be contacted. Septic system maintenance may cost from $250 to $375, depending on location and site conditions, and should be done at least every three years. (Heidenry, 2018; CHI, 2020) Replacement of a conventional household gravity septic system may cost $8,000 to $15,000, or more for larger houses. (Boone, 2014; Hurd, 2019).

Source: Adapted and expanded from: National Ground Water Association. 2020. Groundwater, Wells and Coronavirus. URL: https://www.ngwa.org/publications-and-news/covid-19-news-and-resources/groundwater-wells-and-coronavirus (Accessed April 22, 2020).

The useful life of a properly located and maintained water well may be up to 50 years depending on well casing material and water quality (RMC, undated) with pumps being replaced every 8–15 years depending on pump type (LoPresti, undated). The life span of a well-designed and maintained septic system ranges from about 20 to 30 years (BAC, 2019).

Household Treatment

Treatment of water to facilitate its use as household or community drinking water is the subject of much research and industry. Testing of the household water supply to determine whether contaminants of concern occur in the water source or pipes is a useful first step (NSE, 2019). Groundwater treatment can be done at the individual household level for highly mineralized water, referred to as "hard" water. Other processes can remove a range of contaminants at both the household and the community levels through a central water treatment plant. Household treatment of groundwater for drinking can range from water softening of highly mineralized water at the "point of entry" (POE) to a house, to more sophisticated "point of use" (POU) devices at faucets or on countertops for inorganic, organic and radiologic constituents. In the United States, approximately 42 million people are served by individual household (domestic) water systems (USGS, 2018). Box 1.7.2 provides information on a range of household water treatment options. Table 1.7.1 matches the potential contaminants with the treatment options, while Table 1.7.2 presents the costs for some treatment types. Importantly, the most economically efficient choice from among the options should be based on the objective being addressed: for example, choosing a faucet-mounted device for water used for drinking and cooking does not address any concerns about the dermal or inhalation exposure to the water. The costs can range from $25 to $12,000 (2002$US) depending on the contaminant(s) of concern and the type of exposure that is to be controlled by the treatment option. The options that provide the most comprehensive response to a range of potential contaminants and address all major exposure pathways may be more expensive: installing a new well or connecting to a public water supply.

BOX 1.7.2 HOUSEHOLD GROUNDWATER TREATMENT OPTIONS

Individual well-owner water treatment options (based on treatment objectives) (Water System Council, 2002)

1. Disinfection: treatment of the well to remove bacteria
2. Point of entry: treatment located where water is supplied to water pipes within a home
3. Point of use: treatment device under a sink or on a countertop to filter contaminants from water used for drinking or cooking
4. Multiple treatments: a combination of treatments to remove contaminants for all household purposes, often near the water storage tank

Specific types of treatment (Water System Council, 2002)

Activated carbon filter
A highly porous, absorbent material, usually made from coal or wood, is used to filter contaminants, such as excess chlorine, and to reduce soluble materials, such as organic chemicals and radon.

Activated carbon block filter
Activated carbon is molded into a cartridge filter with a much greater absorption capacity and speed than a granular carbon filter. Specialized media may be added to target specific contaminants.

Chlorination
Chlorination is added to water to destroy unhealthy bacteria and control microorganisms and to remove dissolved iron, manganese and hydrogen sulfide. Shock chlorination of a private well uses concentrations of chlorine that are 100–400 times the amount found in municipal water supplies. The highly chlorinated water is held in the pipes for 12–24 h before it is flushed out and the system is ready for use again. (Also see *Disinfection.*)

Coagulation
Chemicals neutralize the electrical charges of fine particles (contaminants) in water, making it easier to remove the particles by settling, skimming, draining or filtering.

Disinfection
Chemicals such as chlorine, iodine, ozone or hydrogen peroxide are used to destroy disease-producing bacteria without eliminating all the microorganisms. The treatment may also

(*Continued*)

BOX 1.7.2 (Continued)

involve steps such as distillation, microfiltration, ultrafiltration, boiling or the use of ultraviolet light. (Also see *Chlorination.*)

Distillation

Organic and inorganic contaminants are separated from water through a combination of evaporation, cooling and condensation.

Electrodialysis

An electric current is used to remove ions (an atom or group of atoms) from water through a semipermeable membrane (which allows select molecules to pass and blocks others).

Water softening to remove calcium and magnesium (Hardness)

The hardness of groundwater is measured by the amount calcium carbonate in it, which is reported as parts per million (ppm). Water is typically considered as hard when it has more than 120 ppm of calcium carbonate (Driscoll, 1986, p. 829).

Water softening can be accomplished by either lime softening or ion exchange.

Lime softening – applies lime to precipitate calcium carbonate and/or magnesium hydroxide from water, which after settling, is filtered out.

Ion exchange – can remove calcium and magnesium as well as ions of iron, manganese, strontium and other inorganic constituents by passing water through a resin medium to be exchanged for sodium ions. The medium needs to be regenerated regularly and can last for many years (Driscoll, 1986, pp. 802–803).

Oxidizing filter

A type of filter that changes the balanced state of the dissolved molecules, making them insoluble and, therefore, filterable.

Oxidizing chemical injection

Agents such as oxygen, ozone, chlorine or peroxide are used to attract electrons, and thus, they can be removed from water.

Reverse osmosis

Pressure is used to force the water molecules through a semipermeable membrane (it allows select molecules to pass and blocks others). The pressure forces the molecules to flow in the reverse direction, moving from a concentrated solution to a dilute solution, thus, diluting the molecules in water. To make these devices effective, water may need to be pretreated with chlorine or oxidation.

More comprehensive descriptions of these and other treatment technologies are available from the following sources:

Water System Council, *Drinking Water Treatments* (Factsheet) (May 2002), URL:http://www.watersystemscouncil.org/upload/wellcare/in serttreat.pdf (accessed March 9, 2003).

Water Systems Council, (Factsheet) (2019) URL: https://www.watersystemscouncil.org/download/wellcare_information_sheets/well_water_testing_&_treatment_information_sheets/DrinkingWaterTreatmentsandCostsFINAL.pdf (Accessed August 24, 2019)

Driscoll, F.G., *Groundwater and Wells,* Johnson Screens, St. Paul, MN, (1986) 1089.

U.S. Environmental Protection Agency (USEPA), http://epa.gov/ogwdw/safewater/standard (accessed April 5, 2003).

TABLE 1.7.1
Household Groundwater Treatment Options for Selected Contaminants

Contaminant	Treatment Options
Sand, silt, rust	Point of Entry: Cartridge filter, distillation
Algae	Point of Entry: Chlorination, ultraviolet light, distillation; Point of Use: Reverse osmosis
Bacteria	Disinfection: Chlorination, ultraviolet light, ozone treatments, reverse osmosis
Viruses	Point of Entry: Ultraviolet light, distillation
Arsenic	Point of Entry: Activated Alumina, Anion Exchange, Distillation, Oxidation/Filtration; Point of use: Reverse osmosis, distillation or cartridge-type removal devices
Calcium	Point of Entry: Ion exchange, distillation; Point of Use: Reverse osmosis
Chromium	Point of Entry: Coagulation, ion exchange or lime softening; Point of Use: Reverse osmosis
Iron	Point of Entry: Shock chlorination, ion exchange, activated carbon filtration or oxidation/filtration
Nitrate	Point of Entry: Ion exchange, electrodialysis or distillation; Point of Use: Reverse osmosis
Lead	Point of Entry: Activated Alumina, Activated Carbon, Ion-exchange; Point of Use: Reverse osmosis
Radium	Point of Entry: Cation exchange, distillation or electrodialysis; Point of Use: Reverse osmosis
Radon	Point of Entry: Aeration devices or granular activated carbon (GAC) filters
Sulfur and manganese	Point of Entry: Sulfur – distillation, reverse osmosis or ion exchange; manganese and hydrogen sulfide – shock chlorination, water heater modification, activated carbon filter, oxidizing filter or oxidizing chemical injection
TCE (trichloro ethylene)	Point of Entry: Packed tower aeration (GAC filter with reverse osmosis, distillation)
Benzene and other petroleum solvents	Point of Entry: GAC filters, distillation; Point of Use: Reverse osmosis
Pesticides	Point of Entry: GAC filters, distillation; Point of Use: Reverse osmosis

Sources: Driscoll, F.G., *Groundwater and Wells*, Johnson Screens, St. Paul, MN (1986), 827.Water Systems Council, *Well Water Treatment Options and Costs* (information sheet) (October 2009), https://www.watersystemscouncil.org/download/wellcare_information_sheets/well_water_testing_&_treatment_information_sheets/DrinkingWaterTreatmentsandCostsFINAL.pdf (accessed August 24, 2019).

Central Water Supply Treatment

Central water supply treatment providing water to a community of known and accepted quality has a long history. Wells and springs serving many families and entire communities had existed in ancient times. Documented treatment of water dates back to Sanskrit and Egyptian times of 2000–1400 BC (Baker, 1949). In the United States, in 2019, 94.4 million people (30% of all people served by community water systems) received their groundwater supply from over 38,900 central water systems (USEPA, 2019b), and the quality of the water from these systems in the United States has been regulated under the Safe Drinking Water Act standards described in Volume 1, Chapter 8. Significantly, many European countries have very large populations relying on central water systems supplied by groundwater: in Austria, Denmark, France, Germany, Hungary, Luxembourg, Netherlands and Switzerland, with 99%, 99%, 62%, 72%, 96%, 69%, 68% and 83% of the centrally supplied population using groundwater sources, respectively. These centrally supplied systems typically treat the groundwater to ensure that it meets the national and local standards for drinking water quality.

Removal of contaminants can be accomplished through the use of different treatment processes applied to the larger volumes in municipal systems. In the United States, central water systems serve communities ranging in size from 25 to millions of people. In many very small central water systems, historically, groundwater has been considered clean enough that it is often untreated after leaving the wellhead. With the more frequent incidences of microorganisms contaminating groundwater, treatment may be needed more often. Larger groundwater systems typically provide some level of treatment depending on the contaminants needing to be removed and their concentration.

TABLE 1.7.2
Household Treatment Costs

Treatment System	Estimated Costs[a] (2019$US, Unless Otherwise Noted)
Water softening	$500–$10,000 (average $3,000)[1]
Activated carbon filtration	Faucet-mounted: $30–$60[2]
	Under the sink: $130–$550[3]
	Whole house: $900–$1200[4]
Distillation	Countertop: $90–$240[5]
	Automatic: $1400–$2500[6]
Reverse osmosis	Single tap: $200–$400[7] Reverse osmosis replacement filters $40 per year
Bottled water	$21-$29 per month for a family of 3-5 persons[8]
New well replacement	30-120 meters (100-400 feet) depth, pump, water lines to house, control panel and treatment: $4,000-$20,000[9]
Public water system	$12,000 or more per household hookup depending on the distance to water main, along with monthly water use payments

Sources: ImproveNet.com. Water Softener Price Guide URL: https://www.improvenet.com/r/costs-and-prices/water-softeners (Accessed August 24, 2019).

Faucetguide.net. 2019. 6 Best Faucet Water Filters. URL: https://www.faucetguide.net/best-faucet-water-filter-reviews/ (Accessed August 24, 2019).

Waterfiltercomparisons.com. 2019. Under Counter Water Filters. URL: https://www.faucetguide.net/best-faucet-water-filter-reviews/ (Accessed August 24, 2019)

Amazon.com. 2019. Whole House Carbon Filter. URL: https://www.amazon.com/whole-house-carbon-filter/s?k=whole+house+carbon+filter (Accessed August 24, 2019).

Amazon.com. 2019 Water Distiller Countertop. URL: https://www.amazon.com/s?k=water+distiller+countertop&i=tools&ref=nb_sb_noss_1 (Accessed August 24, 2019).

H_2OLabs. 2019. Durastill Automatic Water Distillers. URL: http://www.h2olabs.com/p-181-8-gallonday-automatic-4-gallon-reserve.aspx (Accessed August 24, 2019); Aquanui. 2019. Home Water Distiller. https://myaquanui.com/ URL: (August 24, 2019).

U.S. Water Systems. 2019. Under Sink Reverse Osmosis. URL: https://www.uswatersystems.com/drinking-water/reverse-osmosis-systems/under-sink-reverse-osmosis (Accessed August 24. 2019); Advanced Purification Engineering Corporation. 2019. Reverse Osmosis Water System. URL: https://www.freedrinkingwater.com/products/ (Accessed August 24, 2019).

Cost How Much.com. 2019. How Much Does A Water Delivery Service Cost. URL: http://www.costhowmuch.com/f/water-delivery-service.htm (Accessed August 24, 2019).

Cost Helper, Inc. (CHI). 2019. How Much Does Well Drilling Cost? URL: https://home.costhelper.com/well-drilling.html (Accessed August 24, 2019) Zimmerman, Mike. 2015. How Much Does a Residential Water Well Cost? URL: http://www.zdrillerteam.com/residential-water-cost/ (Accessed August 24, 2019).

Water Systems Council, Well Water Treatment Options and Costs (information sheet) (October 2009), https://www.watersystemscouncil.org/download/wellcare_information_sheets/well_water_testing_&_treatment_information_sheets/DrinkingWaterTreatmentsandCostsFINAL.pdf (accessed August 24, 2019).

[a] Estimated costs are one-time capital costs and do not include operation and maintenance costs, unless otherwise noted.

Table 1.7.3 provides an overview of the treatment steps of a central water treatment system. It also gives the costs of several treatment plants.

Treatment costs for central water systems may vary based on the volume of water and the contaminants to be treated. Table 1.7.4 displays the range of costs for capital and operation and maintenance costs for a range of specific treatment technologies at the central water treatment plants.

Economies of scale can be clearly observed from the results reported in Tables 1.7.3 and 1.7.4 for different technologies. For example, for the Chlorination technology, the capital costs for the 38 m^3/day plant are $3857/1000 m^3, while the capital costs for a 37,854m^3/day plant are $1849/1000$m^3$.

TABLE 1.7.3
Central Water Treatment Plants and Costs

Principal Water Purification Stages at Central Water Treatment Plants:

1. Primary treatment – Pumping from aquifers or rivers, and screening for larger objects (stones, paper, large organic matter)
2. Secondary treatment – Removal of fine solids and most of the contaminants through coagulation, flocculation, filtration and membranes
3. Tertiary treatment – Polishing, pH adjustment, carbon treatment to remove taste and odors, disinfection and temporary storage sufficient for disinfectants to be effective.

Selected Water Treatment Plant Costs:

Municipality	Treatment Identified	Plant Size (m^3/Day)	Cost (for Year Listed)	Year	Cost per m^3/Day
Breckenridge Colorado	Conventional	18,930	$30,000,000	2015	n.a.
Tampa Bay, Florida	Reverse osmosis	12,870	US$8,100,000	2007	US$629
Seward, Nebraska	Reverse osmosis	15,140	US$4,300,000	2004	US$284
Longmont, Colorado	Secondary	113,560	US$40,200,000	2006	US$354
Clearwater, Florida	—	227,125	US$79,000,000	2000	US$347
Salt Lake City, Utah	Conventional	264,980	US$135,000,000	2003	US$509

Likewise, the scale is significant for the operation and maintenance costs of these technologies: for the 38 m^3/day plant, the cost is $\$31.70/m^3$ and for the 37,854 m^3/day plant, the cost is $\$0.07/m^3$. While calculating the abatement costs, the economies of scale must be considered when relating the cost estimates to the population affected by contaminants.

Table 1.7.4 also shows that some technologies may be more expensive than others to treat the same contaminant. Coagulation-assisted microfiltration is the most expensive technology at all design flows. On the other hand, enhanced lime softening is the least expensive at smaller design flows, whereas chlorination is the least expensive at larger flows. This circumstance points out the reason that it is important not to use only one technology, but rather let the economics of the situation indicate, in part, the appropriate technology, with everything else being equal.

Definitions

Chlorination: Chlorine is added to water to remove dissolved arsenic, as well as other inorganic contaminants such as iron, manganese and hydrogen sulfide, and to destroy unhealthy bacteria and control microorganisms.

Enhanced coagulation/filtration: A treatment process that changes the physical or chemical properties of the colloidal or suspended solids and enhances agglomeration that subsequently to allow the solids to settle out or be filtered out. This process is enhanced by increasing coagulant dosage, reducing pH or both.

Coagulation-assisted microfiltration: It is a coagulation treatment process, except that microfiltration is used rather than a conventional gravity filter. Removal can be adjusted by changing the pH.

Enhanced lime softening: This is the typical lime softening process (applying lime to the water), except for adjusting the pH to greater than 10.5 by increasing the lime dosage and possibly increasing the soda ash dosage.

Activated alumina: This is a physical/chemical process in which ions in the water are sorbed to the oxidized activated alumina in a bed.

TABLE 1.7.4
Costs of Different Treatment Technologies at Central Water Treatment Plants

The U.S. Environmental Protection Agency carried out an extensive review of costs associated with a range of central treatment plants that could provide arsenic removal. The costs were derived from industry sources and modeling. While these treatments could also remove other contaminants, the costs presented do not consider treatment of other contaminants and do not include land purchase costs. However, the costs include housing the treatment plant, waste disposal and permit. The results of this examination are presented as follows:

Costs Based on Water Treatment Plant Flow (Plant Size) in Cubic Meters/Day (m^3/day)[a] with Estimated Costs in Thousands (1000s) of 1995$US

Selected	38		380		3785		37854	
Technology[a]	**Capital[b]**	**O&M[c]**	**Capital[b]**	**O&M[c]**	**Capital[b]**	**O&M[c]**	**Capital[b]**	**O&M[c]**
Chlorination	$14.6	$1.2	$14.6	$1.6	$14.6	$5.8	$70.0	$27.6
Enhanced coagulation/ filtration	$7.2	$0.3	$8.6	$1.0	$18.1	$14.0	$864.7	$157.4
Coagulation- assisted microfiltration	$140.0	$22.0	$460.0	$33.0	$2,100.0	$66.0	$11,400.0	$200.0
Enhanced lime softening	$7.9	$0.6	$12.9	$3.4	$33.1	$30.2	$1,235.0	$275.6
Activated alumina	$15.4	$6.0	$61.7	$23.0	$430.5	$200.5	$4,213.1	$1,917.1
Anion exchange	$23.0	$5.8	$68.6	$12.1	$349.6	$52.2	$3,228.4	$390.0
Greensand filtration	$12.0	$8.0	$90.0	$13.3	$590.0	$66.3	$4,000.0	$596.6

[a] The relationship between the treatment plant flow and the approximate population served is: 38 m^3/day: 100 persons; 380 m^3/day: 1000 persons; 3785 m^3/day: 10,000 persons; 37,854 m^3/day: 100,000 persons.

[b] Capital costs are one-time treatment plant costs estimated for design flow and technology specified.

[c] O&M are operation and maintenance costs on annual basis estimated for average flow.

Anion exchange: A physical/chemical process in which contaminant ions in the water are exchanged for ions from a solid resin media.

Greensand filtration: Glauconite sand is coated with manganese oxides, and exchange ions with the contaminant ions, as water passes through the sand.

Source: U.S. Environmental Protection Agency (USEPA), Technologies and Costs for Removal of Arsenic from Drinking Water, EPA-815-R-00-010/December 1999, URL: www.epa.gov/ogwdw/ars/treatments_and_costs.pdf (accessed April 5, 2003).

Desalination

As noted in Volume 1, Chapter 3, groundwater brines, in the past, have had economic value in the production of chemicals. As climate change may lead to less precipitation in some areas resulting in less groundwater recharge, other potential water sources may be utilized, such as brackish and saline groundwater, in water-short areas (Walha et al., 2007). For example, the Texas (USA) Water Development Board identified 3.3 trillion cubic meters of brackish groundwater in Texas that could be desalinated (TWDB, 2008). As many as 60% of the community water systems in the United States overlie saline and brackish groundwaters. Typically, brackish waters contain 1000–10,000 ppm of total dissolved solids, with saline waters having greater than 10,000 ppm of total dissolved solids. As treatment technology, such as reverse osmosis or nanofiltration, has demonstrated lower costs for the production of high-quality water by removing drinking water contaminants effectively, brackish groundwater is now attractive as an alternative water source.

Of note from a groundwater perspective is the method of obtaining seawater for desalination. An analysis in 2012 indicated that construction of intake and outfall structures may be a significant portion of the desalination project cost, even up to a third of the cost. Subsurface options may include beach wells (low cost), horizontal direction drilled wells (medium cost), radial wells (medium cost) and seabed infiltration gallery (high cost) and would likely offer less pretreatment requirements. Submerged, surface or co-located intakes may have lower construction costs but higher pretreatment costs which would continue in operation and maintenance (WRA, 2012).

Desalination of seawater and saline and brackish waters in deep geological formations has been considered as the "backstop" technology for water supply (Frederick, 1995), except in places that must actively seek other water sources. The technologies for desalination include reverse osmosis, mechanical vapor compression, distillation, nanofiltration and electrodialysis reversal. The cost ranges for these technologies are from \$0.45 to \$6.56/m^3 for seawater and \$0.25 to \$0.70/m^3 for brackish water from underground sources (Younos, 2005; NRC, 2008). Since the 1960s to 2005, seawater desalination costs have fallen by 60% (Frederick, 1995; Younos, 2005). This cost, when compared with the average drinking water supply costs of water systems to consumers in the United States in 2004 was US\$0.53/m^3 (USEPA, 2004a), and in 2005, in Germany, it was \$0.68/m^3; in the United Kingdom, \$0.46/m^3; and in France, \$0.42/m^3 (UKTI, 2007). In 2015, the world's largest desalination plant in Sorek, Israel, was pricing its water at \$0.58/m^3 (Talbot, 2015). In 2015, the average household cost of public water supply in the United States was \$1.36/m^3, assuming 0.76 m^3 per household per day (200 gallons/day) for a family of four (Statista, 2019). Similar average household water costs in 2013 for three coastal cities of Los Angeles, Houston and New York City and Tucson, a groundwater-supplied system, were \$1.35, \$1.33, \$1.19 and \$0.96, respectively (Walton, 2013). At that time, it appears that desalination had opportunity to support other population centers needing water as a cost-efficient alternative.

Over the period 1996 to 2016, power costs for desalination have been reduced by 80%. For seawater desalination by reverse osmosis, capital and power costs are 37% and 44%, respectively, while for brackish water, they are 54% and 11%, respectively (see Table 1.7.5). Clearly, power costs are a major factor in removal of a greater concentration of dissolved solids in seawater (Younos, 2005). Desalination is discussed further under "Waste Releases to Groundwater and the Subsurface Environment" in conjunction with waste concentrate disposal from treatment of brackish and saline groundwater.

Large capacity plants do exhibit economies of scale (Younos, 2005). Capital costs for seawater desalination plants using reverse osmosis in early planning stages in 2003 in California ranged from \$70 million for a 45,425 m^3/day plant proposed for Los Angeles, to \$272 million for a 189,270

TABLE 1.7.5
Central Treatment Plant and Production Costs for Desalination by Reverse Osmosis

	Brackish Water (US\$ per m^3)	Seawater (US\$ per m^3)
Production costs		
Average	\$0.34	\$1.16
Range	\$0.18–\$0.70	\$0.45–\$6.56
Cost factors	(Percent of costs)	(Percent of costs)
Fixed costs	54	37
Electric power	11	44
Other O&M	35	19

Source: Younos, T., J. Contemp. Water Res. Edu., 132, 39, 2005.

m^3/day plant proposed for San Diego. Thus, the economies of scale can be given as \$1541/m^3/day capacity at the smaller Los Angeles plant versus \$1437/m^3/day capacity at the larger San Diego plant. The San Diego Carlsbad desalination plant was producing drinking water from seawater for \$1.85/m^3 in 2017 (CDPP, 2017). San Diego's average water consumer price in 2017 for a family of 4 using 0.76 m^3 per household per day (200 gallons per day) was \$2.74/m^3 (Walton, 2018).

Subsurface Disposal and Release of Drinking Water Treatment Residuals

Drinking waters often contain naturally occurring constituents, which, in high concentrations, may be harmful to human health, as noted in Volume1, Chapter 8. Removal of these high concentrations results in wastes that must be disposed. Constituents that may be treated include: arsenic, barium, cadmium, chromium, lead, mercury, silver, radium-226 (Ra-226), radium-228 (Ra-228), gross alpha particle activity, beta particle and photon radioactivity and uranium. For solid wastes, disposal options include solid-waste landfill, hazardous-waste landfill or radioactive-waste landfill, depending on the concentration of the constituents and whether they are radioactive. For liquid wastes, disposal options are discharge to waters (which requires a governmental permit), release to a waste treatment plant (which requires a permit) and injection into deep geologic strata (also requiring a permit) (USEPA, 2006a, 2006b, 2006c). Thus, residual disposal is an additional cost apart from the treatment of drinking water.

Watershed Management

Watershed or river basin management has emerged as an integrative approach to improve water quality with regard to ground and surface waters, their interaction and managing contaminant sources affecting both, within the catchment. As groundwater and surface water are hydrologically connected in a watershed and contaminants can flow from one resource to the other, it is important to examine the dynamic relationships among these factors. Nonpoint sources of contaminants, sources not originating from one defined point, and/or spread across large areas, should be considered for their variable inputs of contaminants. Furthermore, whether one source is more cost-effectively controlled should be carefully evaluated before major expense is incurred in controlling any particular source of contaminants in a watershed. This approach will require modeling of the water flow and contaminant movement in the watershed. Likewise, the most effective management of point sources needs to be considered in the same way, as well as between the point and nonpoint sources. Research is needed to begin to better understand the tradeoffs among the contaminant control strategies in watersheds and the most economically efficient approaches. Sometimes, the laws that recognize only a portion of the water resource, either groundwater or surface water, obstruct a comprehensive evaluation and efficiently targeted water quality improvement actions from occurring.

WASTE RELEASES TO GROUNDWATER AND THE SUBSURFACE ENVIRONMENT

Groundwater and its subsurface environment receive many types of waste releases, potentially affecting their quality at a particular site or, if left unaddressed for a long time, even under a large area. As noted previously, billions of pounds of agricultural chemicals and pesticides are applied to the ground each year, some of which percolate through the subsurface to groundwater. The unused portion of these chemicals is a residual waste. Past physical, chemical and biological wastes that were considered properly disposed in the ground continue to be the sources of contamination to groundwater around the world (Sampat, 2000). Deep injection of wastes is expected to increase, partly in response to concerns about global warming: disposal of both carbon dioxide (CO_2) emissions and desalination wastes. There are numerous small private and local wastewater systems that treat and discharge wastewater to the subsurface and groundwater. These demands, along with the ever increasing demands for groundwater, especially in areas of drought, present a significant pressure and competition for groundwater and the subsurface – at times in conflict, where the same groundwater may be used for human and animal water supply as well as being a sink for wastes.

Onsite Wastewater Disposal

Many small and individual water systems serve populations whose domestic waste is treated by onsite or near-site septic systems. Box 1.7.3 describes an onsite wastewater disposal (septic) system in general terms. A typical residential septic system may release 1893 liters per day of sewage, on average, to the subsurface if it is working properly (Lucas, 2007). In the United States as of 2000, approximately 21 million homes have septic systems (LaFond, 2015), of which as many as up to 60% may not work properly, discharging untreated domestic wastewater to the subsurface and groundwater (Lucas, 2007) that is potentially used as a drinking water source.

Properly maintained onsite septic systems may be financially beneficial to homeowners in terms of saving money. Typical maintenance includes a regular inspection and pumping. These maintenance costs are less than the cost of replacing a failed system. Table 1.7.6 describes the life-cycle costs of a septic system.

BOX 1.7.3 ONSITE WASTEWATER DISPOSAL (SEPTIC) SYSTEM DESCRIPTION

A typical septic system has four main components: a pipe from the home, a septic tank, a drainfield and the soil. Microbes in the soil digest or remove most of the contaminants from the wastewater before it eventually reaches groundwater.

Pipe From the Home
All household wastewater exits through a pipe to the septic tank.

Septic Tank
The septic tank is a buried, watertight container typically made of concrete, fiberglass or polyethylene. It holds the wastewater long enough to allow solids to settle out (forming sludge) and oil and grease to float to the surface (as scum). It also allows partial decomposition of the solid materials. Compartments and a T-shaped outlet in the septic tank prevent the sludge and scum from leaving the tank and traveling into the drainfield area. Screens are also recommended to prevent the solids from entering the drainfield. Newer tanks generally have risers with lids at the ground surface to allow easy location, inspection and pumping of the tank.

Drainfield
The wastewater exits the septic tank and is discharged into the drainfield for further treatment by the soil. The partially treated wastewater is pushed along into the drainfield for further treatment every time new wastewater enters the tank. If the drainfield is overloaded with too much liquid, it will flood, causing sewage to flow to the ground surface or create backups in plumbing fixtures and prevent treatment of all wastewater. A reserve drainfield, required by many states, is an area suitable for a new drainfield system if the current drainfield fails. This area should be treated with same care as the septic system.

Soil
Septic tank wastewater flows to the drainfield, where it percolates into the soil, which provides final treatment by removing harmful bacteria, viruses and nutrients. Suitable soil is necessary for successful wastewater treatment.

Sources: Abstracted from U.S. Environmental Protection Agency (USEPA), Homeowner's Guide to Septic Systems-Short Version, 2003c, URL: http://www.nesc.wva.edu/NSFC/pdf/homeowner_guide_long.pdf (accessed February 11, 2007).

TABLE 1.7.6
Septic System Costs

The price of a septic system depends on house size (number of bedrooms), soil suitability, local labor and material costs. The price may range from $3000 to $75,000. The main costs of a septic system are (1) installation cost, (2) maintenance and repair costs and (3) replacement costs if the system fails.

Installation Costs

System Type	Cost
Standard, gravity-fed tank and trench system	$3,000–$20,000 (depending on the location) National average (2019): $5,808 Typical Range (2019): $3,095-$9,189
Cluster system (each home with septic tank share common drainfield)	$5,000–$8,000
Engineered systems (mounds, sand/peat filters, aerobic systems and constructed wetlands)	$6,000–$15,000
Evapotranspiration system	$44,000
Maintenance Costs	
Standard, gravity-fed tank and trench systems with 1–3 years of regular inspection and pumping	$75–$150
Typical annual costs of drainfield or mound system	$30–$500 if replacing pumps in mound systems
Annual costs of constructed wetlands or sand and peat filters	$50–$1,700, depending on the discharge method and monitoring requirements
Annual costs for multiple-household systems	$200–$1500 per household
Lifetime costs over 20 years:	
Individual home systems	
Trenches and mounds	$6,300–$13,000
Alternative treatment systems	$13,500–$32,000
Multi-household systems	
Trench or mound systems	$18,500–$25,000
Alternative treatment systems	$18,000–$44,500

Sources: Abstracted from Laundry Alternative Inc., (LAI) (2005a), Septic System Price, URL: http://www.laundry-alternative.com/septic_system_price.htm (accessed February 11, 2007)

Abstracted from Stony Brook-Millstone Watershed Association (SBMWA), Funding for Septic Systems, 2004, URL: http://www.thewatershed.org/images/uploads/Funding%20brochure.pdf (accessed February 11, 2007).

Home Advisor. 2019. How Much Does a Septic Tank Cost? URL: https://www.homeadvisor.com/cost/plumbing/install-a-septic-tank/ (Accessed August 24, 2019).

Underground Injection

The underground injection control (UIC) program in the United States was briefly presented in Volume 1, Chapter 3, with the mention that the program is designed to place hazardous wastes into deep geologic zones below the underground sources of drinking water. In both the United States and the European Union, disposal of hazardous substances into groundwater is prohibited. Box 1.7.4 describes the UIC program in the United States.

Nearly one million shallow injection wells in the form of Class V injection wells exist in the United States alone. Some UIC wells commonly used and permitted by regulation are septic tanks for domestic wastewater and cisterns to collect runoff from roofs. The UIC wells that are not permitted are floor drains in service station bays that allow wastewater containing oils and other chemical products or wastes to enter the subsurface and intersect the water table below, and dry cleaners, as well as septic systems used by smaller commercial operators that dispose chemicals from their production or service processes. In these industries requiring chemical

BOX 1.7.4 THE UNDERGROUND INJECTION CONTROL PROGRAM IN THE UNITED STATES

The UIC Program is very significant in managing chemical releases and waste disposal in the United States. In 2001, of the 1.27 billion kilograms (2.8 billion pounds) of chemical releases (including waste disposal) reported by the industry sources to the Toxic Release Inventory (required under Section 313 of the Emergency Planning and Community Right-to-Know Act - EPCRA), 3.5% were disposed by underground injection through wells into the subsurface. The top chemical releases via underground injection were: nitrate compounds, ammonia, methanol, manganese compounds, zinc compounds, lead compounds, hydrogen fluoride, chromium compounds, arsenic compounds and barium compounds.

In 2017, underground injection accounted for 4% of the 1.76 billion kilograms (3.88 billion pounds) of chemical releases. Excluding mining chemical and waste releases, the major land disposal releases including underground injection in 2017 were for barium and barium compounds, manganese and manganese compounds, and zinc and zinc compounds.

The UIC Program works with State and local governments to regulate injection wells to prevent them from contaminating drinking water resources. The EPA defines the five classes of wells according to the type of waste that they inject and where the waste is injected.

Class I wells are technologically sophisticated wells that inject large volumes of hazardous and nonhazardous wastes into deep, isolated rock formations that are separated from the lowermost underground source of drinking water (USDW) by many layers of impermeable clay and rock.

Class II wells inject fluids associated with oil and natural gas production. Most of the injected fluid is brine that is produced when oil and gas are extracted from the earth (about 10 barrels for every barrel of oil).

Class III wells inject super-hot steam, water or other fluids into mineral formations, which is then pumped to the surface and extracted. Generally, the fluid is treated and reinjected into the same formation. More than 50% of the salt and 80% of the uranium extraction in the United States is produced in this way.

Class IV wells inject hazardous or radioactive wastes into or above the underground sources of drinking water. These wells are banned under the UIC program because they directly threaten the quality of underground sources of drinking water.

Class V wells use injection practices that are not included in the other classes. Some Class V wells are technologically advanced wastewater disposal systems used by industry, but most are "low-tech" holes in the ground. Generally, they are shallow and depend on gravity to drain or "inject" liquid waste into the ground above or into underground sources of drinking water. Their simple construction provides little or no protection against possible groundwater contamination, and hence, it is important to control what goes into them.

Class VI wells are wells used for injection of carbon dioxide (CO_2) into underground subsurface rock formations for long-term storage, or geologic sequestration. UIC regulations mandate the consideration of a variety of measures to assure that injection activities will not endanger underground sources of drinking water (USDWs).

Sources: USEPA 2001 TRI Public Data Release Materials, URL: https://epa.gov/toxics-release-inventory-tri-program/2001-tri-public-data-release-materials (Accessed August 25, 2019); 2017 TRI National Analysis, URL: https://epa.gov/trinationalanalysis/report-sections-tri-national-analysis (Accessed August 25, 2019).

USEPA, Underground Injection Control Well Classes, URL: https://epa.gov/uic/underground-injection-control-well-classes (Accessed August 25, 2019).

disposal, the chemicals should be treated, recycled or returned to the manufacturer, if such arrangements can be made. While this would add cost, it protects the community water supply. However, stormwater drains exist extensively for the control of overland flow from storm events and may drain into cisterns (rather than sewer lines) without treatment. These waters, while recharging groundwater, may have contaminants that may not expeditiously be broken down or degraded in the subsurface and could reach nearby water-supply wells. With greater population densities, there is an increasing need to dispose of stormwater and wastewaters associated with increased use of chemicals in all kinds of products. These waters accumulating nonpoint sources of pollution can have a detrimental effect on groundwater. Users of groundwater near these drains may have to incur the cost of treatment which may be an additional cost that they did not previously have.

The economic significance of the UIC program is that the governments give the right of discharge to waste producers through access to deep geologic formations for the largest volume of controlled waste disposal. The technical expectation is that the disposal permanently remains in the formations below drinking water aquifers. This disposal is modeled and no major incidents of migration of wastes to the aquifers used as public water sources have occurred in the United States. Only the waste concentrations are monitored as they are injected; no monitoring of the formations above the injection zone takes place, except that many water-supply wells exist in shallower formations and the water from them is monitored regularly under the Safe Drinking Water Act and State drinking water laws. Box 1.7.4 compares the distribution of major toxic substances disposal in the United States between the years 1987 and 2001, including underground injection, as well as other land disposal methods that may use the subsurface as an intentional or unintentional disposal path and sink. Based on receiving waste site classification used, it is unclear to what extent underground injection as a method for disposal of toxic chemicals in the United States has changed over time, as indicated in Table 1.7.7. Additionally, nontoxic wastes and small unreportable quantities of toxic wastes that are disposed by underground injection are not accounted for. The U.S. EPA reported in 2001 that underground injection wells are used to dispose over 34 million cubic meters of hazardous waste every year, and over 7.6 million cubic meters of brine from oil and gas operations every day in the United States (USEPA, 2003d). In 2012, oil and gas brine, also referred to as "produced water," amounted to 9.2 million m^3/day. These waters can be treated for other uses (PWS, 2017).

Subsurface Releases and Modifications Associated with Global Climate Change

Residual releases to the subsurface and groundwater are projected to be associated with responses to brackish and saline groundwater treatment in areas with no other lower-cost alternative water source and to global climate change. These releases include injection and sequestration of CO_2 in deep geologic strata and desalination wastes that may result from the need to find alternative water sources because of saltwater intrusion into coastal aquifers, owing to sea-level rise as well as in arid inland locations requiring the use of alternative brackish or saline groundwaters as water supplies. Significant assessments and technology research are being carried out to address these issues.

Carbon Dioxide

Current global emissions of CO_2 are approximately from 22.6 (IPCC, 2005, p. 241) to 25 Gt (Greenblatt, 2006) (1 gigaton = 1 billion metric tons or 1 trillion kilograms), some of which may be capable of being stored deep underground, depending on the location of the source relative to geologic formations. Globally, 35% of CO_2 emissions are from electricity and heat production, and 21% each from manufacturing/construction and transportation (IPCC, 2005, p. 83). The benefits of addressing CO_2 emissions may include inhibiting global rise in temperatures and associated losses

TABLE 1.7.7
Distribution of Toxic Substances Released and Reported in the United States for the Years 2001 and 2017

	2001		2017	
Total releases	2.8 million metric tons		1.76 million metric tons	
Number of chemicals	650		595	
Percent distribution by receiving media or process	Other onsite land releases	54%	On-site Land Releases (incl. landfill and underground injection)	70%
	Air emissions	27.3%	On-site Air emissions	15%
	Transfers offsite to disposal	9.4%	Off-site disposal or other releases	10%
	Surface water discharge	3.6%	On-site Surface Water discharge	5%
	UIC Class I wells	3.1%		
	UIC Class II-V wells	0.4%		
	Onsite RCRA Subtitle C landfills	2.2%		
		100%		100%

These results indicate a 17% increase in land disposal of toxic chemicals in the United States over the period from 2001 to 2017. While the changed accounting makes it unclear whether underground injection changed significantly, with an overall increase in land disposal, monitoring of groundwater quality may need to be enhanced to ensure the safety of groundwaters used for drinking water, animal watering and food production.

Sources: U.S. Environmental Protection Agency (USEPA), Toxic Release Inventory Program 2001. URL: http://www.epa.gov/tr:/

U.S. Environmental Protection Agency (USEPA), TRI National Analysis 2018. URL: https://www.epa.gov/trinationalanalysis/comparing-industry-sectors Accessed August 24, 2019).

in biodiversity and sea-level rise affecting wetlands and coastal lands (IPCC, 2002). Major costs for the capture and deep storage of CO_2 emissions – a "greenhouse" gas – range widely. The capital costs of capture and compression for 907,200 metric tons of CO_2 in Algeria in 2005 were $100 million (USDOE, 2005a). "Using present technology, estimates of sequestration costs are in the range of $110–$330/metric ton of carbon emissions avoided" (USDOE, 2005b). Candidate geologic zones for CO_2 storage and their estimated potential capacities are given in Table 1.7.8.

TABLE 1.7.8
Candidate Geologic Zones for CO_2 Storage and Potential Capacities

	U.S. (Gt)	Global (Gt)
Depleted oil and gas fields	98	675–900
Unminable coal seams	7	15–200
Deep (800+ meter) saline aquifers	500	1,000–10,000
Total	605	1,700–11,000

Source: Greenblatt, J., Driving a wedge through global warming: Existing technologies to bridge the gap between business-as-usual and stable global emissions, in Annual Meeting, Committee on Environmental Improvement, American Chemical Society, San Francisco, CA, September 9, 2006, URL: http://membership.acs.org/c/cei/Greenblatt.pdf (accessed March 23, 2007).

The economic damages due to underground injection of CO_2 have been estimated at $0.002/kilowatt-hour for coal-fired power production (Smekens and van der Zwaan, 2004, p. 7), responsible for 25% of the CO_2 volume (equivalent to 2.27 billion metric tons/year) produced in the United States (NRDC, 2007). Other specific environmental effects in the subsurface environment from underground injection and storage of CO_2 may include (Smekens and van der Zwaan 2004, pp. 4–6):

- Acidification of groundwater, potentially affecting groundwater supplies of drinking water if the CO_2 migrates from the deep storage location, and also having potential effects on aquatic life habitat
- Modification of the hydrodynamic properties of underground geological layers, affecting the water extraction potential of certain sources
- Loss of integrity of deep geologic strata
- Dissolution of deep geologic strata
- Well "blowouts," which could be catastrophic for human and animal life in the vicinity. Slow or rapid release of CO_2 from geologic formations, thus reintroducing CO_2 to the atmosphere.

The costs of deep underground injection of wastes potentially affecting water supply and other ecological purposes related to the economy should be examined in the future. The costs of disposal do not equate to value of the subsurface environment: if only costs of disposal are considered, then the value of the subsurface for its disposal capacity and other uses/purposes displaced is considered as zero, which may not be equitable either in the near-term or long-term to current or future users of this inner-earth zone. Migration of CO_2 across the boundaries is also a factor in evaluating the effects of underground storage (Rubin, 2005). Measurement, monitoring and verification at an injection site are important for ensuring that environmental factors are addressed, including the steps of site characterization, simulation, modeling, injection array design and placement, baseline monitoring, operational monitoring during injection and array monitoring during and after injection. The extent of potential CO_2 storage is presented in Table 1.7.8. The cost estimates for CO_2 capture and storage (CSS) are presented in Table 1.7.9. Enhanced oil recovery (EOR) or gas recovery (EGR) using injected CO_2 resulting from electrical production may be balanced against a portion of the cost of capture and storage.

These costs incurred for the indicated investments alone will not ensure by themselves that no leakage or other damage would occur. Potential leakage may be significant in the long term[1] (Rubin, 2005) and needs more research. Box 1.7.5 provides a perspective of the types of monitoring that may be necessary to verify successful subsurface disposal of CO_2. Additionally, these costs do not include any risk premium for deleterious effects that might be projected.

- Alternatives to subsurface disposal of CO_2 include (USEPA, 2006a, 2007a)
- Promoting more forest growth
- Increasing the carbon content in the soils and development of humus
- Injection into oceans
- Stimulating photosynthesis of oceans through addition of iron particles

These alternatives may have ecosystem effects that need further research to examine their potential success, cost and external impacts.

Sea-Level Rise and Saltwater Intrusion

With sea-level rise resulting from global climate change, groundwater sources of water supply will be affected by saltwater intrusion. Three-fourths of the population of the United States live within 200 km of the coast (USGS, 2007). Saline water intrusion is also a problem in the coastal zone of Europe (EEA, 2003), and along Asian and African coasts where cities pump significant quantities of groundwater (Howard and Gelo, 2003). A cost of saltwater intrusion includes the closure of near-coast wells and the drilling of new wells, importation of water from wells or water sources more distant from the coast (EEA, 2003, p. 20) or desalination of groundwater turned brackish from saltwater intrusion, such as at Cape May, New Jersey. The cost of well drilling and installation is examined in Volume 1, Chapter 6.

TABLE 1.7.9
Cost Estimates for CO_2 Capture and Storage (CCS)

- Groundwater-related costs for a large-scale demonstration project using low-cost CO_2 for the capture of 90,718–5,443,108 metric tons/year (Friedman, 2005)

Activity	Cost Range	Frequency
Detailed predrill assessment	\$2–\$4 million	Once
Injection wells (1–2) and monitoring wells (3–8)	\$3–\$8 million	Once
Monitoring (multiple methods)	\$2.2–\$6.4 million	Annually
Analysis and modeling	\$5–\$7 million	Annually
Post-injection sampling/recompletion	\$3–\$8 million	Once

- CO_2 capture, storage and monitoring costs (400–800 MW coal power plant net with capture 300–700 MW) (IPCC, 2005)

Capture with current technology:	2002\$US 21–32/metric ton CO_2
Storage depending on geologic conditions:	2002\$US 0.5–7.3/metric ton CO_2
Monitoring depending on geologic conditions:	2002\$US 0.1–0.3/metric ton CO_2

- Estimated CCS cost for new power plants using current technology (IPCC, 2005; Rubin, 2005; USDOE, 2005c)

Power Plant System	Natural Gas Combined Cycle Plant (2002\$US)	Pulverized Coal Plant (2002\$US)	Integrated Gasification Combined Cycle Plant (2002\$US)
Reference plant production cost[a] (without capture) – IPCC	0.03–0.05	0.04–0.05	0.04–0.06
Added cost of CCS with geological storage – Rubin	0.01–0.03	0.02–0.05	0.01–0.03
Added cost of CCS with EOR storage – Rubin	0.01–0.02	0.01–0.03	0.00–0.01
USDOE power plant including geologic storage	0.06–0.09		

- Cost of CO_2 Avoided

	(2002\$US/ton)	(2002\$US/ton)	(2002\$US/ton)
Reference plant[a] with CCS (geological storage)	\$40–\$90	\$30–\$70	\$15–\$55
Reference plant[a] with CCS (EOR storage)	\$20–\$70	\$10–\$45	\$(-5)-\$30
USDOE power plant including geologic storage		\$50–\$200	

Sources: Rubin, E.S., IPCC special report on carbon dioxide capture and storage, U.S. Climate Change Science Program Workshop, Washington, DC, November 14, 2005, URL: www.climatescience.gov/workshop2005/presentations/breakout_2ARubin.pdf (accessed March 23, 2007).United States Department of Energy (USDOE), The Cost of Carbon Dioxide Capture and Storage in Geologic Formations, 2005c, URL: www.netl.doe.gov (accessed March 23, 2007).

[a] Reference power plant assumptions: 400–800 MW without capture, 300–700 MW net power output with capture, 90% efficiency of CO_2 recovery.

Additionally, a projected 20% of coastal wetlands may be lost by 2080 due to sea-level rise (IPCC, 2002).

Inland Brackish and Saline Water Intrusion

Inland areas where intensive pumping of fresh groundwater has led to brackish and saline water intrusion (e.g., EPWU, 2007) may pursue alternative water supplies. Inland southwestern United States communities have experienced similar brackish water intrusion to their freshwater aquifers from pumping for water supply (Sheng and Devere, 2005). El Paso, Texas has been producing

BOX 1.7.5 MEASUREMENT, MONITORING AND VERIFICATION OF SUBSURFACE CO_2 DISPOSAL

The following techniques have been identified for use in monitoring the subsurface disposal of CO_2:

CO_2 fate and transport models

- Reservoir models (target formation to vadose zone)
- Geochemical models
- Geomechanical models
- CO_2 equation of state at reservoir conditions

Plume tracking

- Surface to borehole seismic monitoring
- Micro-seismic monitoring
- Cross-well tomography
- Reservoir pressure monitoring
- Observation wells/fluid sampling

CO_2 leak detection

- Vadose zone soil/water sampling
- Air sample/gas chromospectrometry
- Infrared-based CO_2 in air detectors
- Vegetation growth rates
- CO_2 tracers, natural and introduced
- Subsurface monitoring wells

Land surface deformation

- Satellite and air plane-based monitoring

Mitigation

- De-pressure target formation

Sources: Benson, S.M. and Myer, L., Presentation to the IPCC Workshop for Carbon Capture and Storage, November 18–21, 2002, Regina, AB, Canada, URL: http://arch.rivm.nl/env/int/ipcc/docs/css2002/ccs02-10.pdf (accessed April 6, 2007), pp. 3–8.

United States Department of Energy (USDOE) (2006) Carbon Sequestration Technology Roadmap and Program Plan, URL: http://www.netl.doe.gov/publications/carbon_seq/2006%20Sequestration%20Roadmap%20 FINAL.pdf (accessed April 6), p. 27.

brackish groundwater for treatment to meet drinking water standards and fulfill its water supply demand. Arid regions may receive less precipitation as a result of climate change and thus less recharge (IPCC, 2008), indicating a need to pursue alternatives for water supply, such as inland brackish water sources.

Desalination Wastes

Coastal areas already using groundwater to an extent that is depleting their historically used aquifers and causing saltwater intrusion and future subterranean saltwater advances from sea-level rise may already have turned to desalination of seawater, removing salts through thermal or membrane technologies. The concentration of dissolved salts in ocean water is approximately 35,000 parts per million (ppm), whereas typical freshwater from wells used for drinking is 1000ppm or less (USGS, 2003). In addition, the brackish and saline waters from interior continental locations may be used as water sources in the arid regions where those waters may be the only ones available and must be desalinated before use by people. The cost of water from desalination with reverse osmosis and other technologies was discussed earlier and averages \$0.34/m^3 for brackish water and \$1.16/m^3 for seawater (Younos, 2005; USGS, 2003). The discharged wastewater from desalination plants includes (CCC, 1993):

- High salt concentrations in the range of 46,000–80,000 ppm
- Higher temperatures than ocean waters (about 5E F increase at the point of discharge)
- Turbidity levels above ocean water
- Oxygen levels below those of ocean waters for some technologies
- Chemicals from pretreatment of the feedwater which may include biocides, sulfur dioxide, coagulants (e.g., ferric chloride), CO_2, polyelectrolytes, antiscalants (e.g., polyacrylic acid), sodium bisulfite, antifoam agents and polymers
- Chemicals used in flushing the pipelines and cleaning the membranes in plants using reverse osmosis technology (which may include sodium compounds, hydrochloric acid, citric acid, alkalines, polyphosphate, biocides, copper sulfate and acrolein)
- Chemicals used to preserve membranes of reverse osmosis technology (e.g., propylene glycol, glycerine or sodium bisulfite)
- Organic chemicals and metals contained in feedwater that become concentrated in the desalination process
- Metals that are picked up by the brine in contact with plant components and pipelines.

At coastal locations, these wastewaters are usually discharged back into the ocean. Higher salinity in ocean waters that receive discharge from desalination plants affect the embryo development of certain aquatic life in those coastal areas (CCC, 1993); however, appropriately sized and placed diffusers used in conjunction with this discharge can minimize the marine impact (USNOAA, 2005). However, disposal of desalination wastes through injection into deep geological strata is an alternative (TWDB, 2006). The capital costs for a 37,854 m^3/day (10 million gallons/day) deep injection desalination waste disposal facility may range from \$3.3 to \$10.8 million, with operating costs of \$300,000–\$400,000 per year (Foldager, 2003, p. 48). Deep injection of oilfield brine as a waste product of oil production has already been carried out (Manning and Thompson, 1995, p. 114). A more recent analysis (USEPA, 2006d) indicated that the cost of drilling and constructing a drinking water treatment (reverse osmosis) residuals injection well with a 40–50 year useful life may be \$500,000 to \$1.25 million, depending on specific requirements.

Remediation of Chemical Residuals in Groundwater

Past chemical and biological disposal practices (most of them prior to the late 1970s) often assumed that such wastes could be applied on the land or buried underground and be held at that location without causing any harm or costs to other people or property. The advent of extensive groundwater monitoring throughout the United States in the 1970s revealed groundwater contamination in places never imagined such as Love Canal, New York; Bridgeport, New Jersey; and Riverside, California. Scientists began to realize that contaminants are not bound in the subsurface, but migrate and can be carried by groundwater movement. At that time, over 30,000 sites of potential uncontrolled hazardous waste disposal were identified in the United States through activities under the Comprehensive Environmental Response, Compensation and Liability Act (CERCLA), also known as "Superfund." Other active waste disposal sites that involved contaminated groundwater were identified under the

TABLE 1.7.10
Examples of Groundwater Contaminated Areas

Location	Surface Area Underlain with Contaminated Groundwater	Contaminant Source	Contaminants
New Brighton/ Arden Hills, MN	65 km^2	Munitions production	Volatile organic compounds (VOCs), semi-VOCs, metals, polychlorinated biphenyls (PCBs), cyanide, pesticides and explosives
San Gabriel Valley, El Monte, CA	78 km^2	Wide range of industrial activities	Tetrachloroethene (PCE), trichloroethene (TCE) and other VOCs
Shelton, Central Platte Valley, NE	2023 km^2	Farming (90% area cultivated, 85% irrigated)	Nitrate–nitrogen levels greater than the U.S. Maximum Contaminant Level of 10 parts per million
Hanford Site, WA	225 km^2 reaching Columbia River	Plutonium production	Nitrate, chromium, carbon tetrachloride, uranium, technetium-99, tritium, strontium-90, iodine-129
Bitterfield, Germany	26 km^2	Open pit lignite mining activities and related chemical industries	Benzene, chlorobenzene, 1,2-dichlorobenzene, 1,4-dichlorobenzene, trichloroethylene (TCE), *cis*-1,2-dichloroethylene (DCE) and *trans*-1,2-DCE sulfate and chloride
Yejsk, Russia	53 km^2	Military reservation petrochemical spills and leaks	Hydrocarbons
Dundigal Town, India	7 km^2	Leather tanning	Tannery waste

Sources: United States Environmental Protection Agency (USEPA), EPA/ROD/R09-00/114 EPA Superfund Record of Decision: San Gabriel Valley (Area 1) EPA ID: CAD980677355 OU 05, El Monte, CA09/29/2000, 2000.

United States Environmental Protection Agency (USEPA), NPL Fact Sheets for Minnesota: New Brighton/Arden Hills, 2004a, URL: http://www.epa.gov/R5Super/npl/minnesota/MN7213820908.htm (accessed May 19, 2006).

Mondal, N.C. and Singh, V.S., Proceedings of the Asia Pacific Association of Hydrology and Water Resources (APHW) Second Conference, Singapore, July 5–8, 2004.

United States Department of Agriculture (USDA), Water Quality Information Center, Nebraska Management System Evaluation Area, Shelton Site, 1997, URL: http://www.nal.usda.gov/wqic/wgwq/neb1.shtml (accessed May 19, 2006). Washington (State) Department of Ecology (WDE), Cleaning Up Hanford's Groundwater, Ecology Publication Number 02-05-015, 2005.

Walkow, F. et al., Abstracts; Workshop on the SAFIRA Project, Bitterfield, Germany, 78, November 17–18, 1999, URL: http://www.ufz.de/data/ufzbericht23-00523.pdf (accessed May 19, 2006).

Schafer, K.W., International Experience and Expertise in Registration Investigation, Assessment, and Clean-Up of Contaminated Military Sites, 1996, URL: http://www.umweltbundesamt.de/altlast/web1/berichte/mooreeng/ dmeng_t.htm (accessed May 19, 2006).

Resource Conservation and Recovery Act (RCRA). In addition, other abandoned waste sites have been identified around the world. For example, refer to Table 1.7.10, Examples of Groundwater Contaminated Areas, and Sampat (2000).

Though many of the sites are not considered as locations of extensive contamination, more than 85% of the nationally listed sites in the United States had contamination that affected groundwater quality and, thereby, its use for drinking water (USEPA, 1991). By the end of 2000, the National Priorities List (NPL) had 1450 sites for cleanup. Dense nonaqueous phase liquids (DNAPLs), chemicals that sink in water, have been a continuing source of contamination at approximately 60% of these sites (USEPA, 1993b). Perhaps, as many as 8–10 sites involve injection of wastes through shallow wells into the subsurface and contaminated groundwater (Paque, 2003). The most commonly found contaminants in groundwater at NPL sites are volatile organic and inorganic chemicals. Table 1.7.11 lists the most commonly found contaminants at NPL sites with contaminated groundwater.

The costs of remediating groundwater contamination are considerable. In 2003, the average cost of Superfund site cleanup was estimated to be about $10 million (USEPA, 2003b). More recently in 2016, the average cost of remediating an uncontrolled hazardous waste site is estimated to be $17 million for less complex sites and $195 million for "mega sites" of greater complexity (Probst, 2017) with higher costs due to more complex sites. Some of these costs include:

- Sampling the contaminated materials and the surrounding soil, air and water
- Installing security measures
- Disposing contaminated containers and debris
- Excavating and disposing contaminated soil, wastes and debris
- Pumping contaminated liquids from overflowing lagoons
- Draining or skimming off contaminants
- Controlling detonation or explosions of ordnance
- Applying alternative treatment technologies
- Restoring the site
- Relocating threatened individuals or finding them temporary shelter
- Providing alternate water supplies
- Installing decontamination devices

In situations where the contamination of aquifers is extensive or deep, cleanup of groundwater may not be economically or technically possible. However, where remediation is practical, it can occur in the subsurface, called "in-situ" cleanup, or at the ground surface nearby, referred to as "ex-situ" remediation. In-situ technologies include bioremediation, chemical treatment, permeable reactive barriers, phytoremediation, physical extraction and barriers or capping. Ex-situ remediation comprises pumping groundwater to the ground surface and treating it, referred to as "pump and treat." These ex-situ technologies include advanced oxidation processes, chemical treatment, bioremediation, physical separation, air stripping and thermal treatment (FRTR, 2009). Box 1.7.6 provides a brief description of the technologies. Table 1.7.11 reviews the costs of different technologies.

As Table 1.7.12 indicates, the cost to treat contaminated groundwater is highly variable, depending on the contaminant and the technology. The cost and effectiveness of groundwater remediation are also affected by many other factors. Natural subsurface characteristics include the geology of the location and depth of the water table. More permeable subsurface conditions may facilitate pumping groundwater or subsurface treatments at lower costs. Greater depth to reach water will increase the pumping costs. The contaminant factors are obviously important. The contaminant type and concentration will determine the technologies that can be applied most effectively. Furthermore, the remediation goal will also influence technology selection. Often groundwater is used as a drinking water source, and hence, the goals may be driven by the maximum contaminant levels set for that use.

If the institutions (governments, water suppliers or public) decide that drinking water is no longer a goal for the resource, then alternative target levels may be set. The volume of groundwater contaminated and the time within which the remediation is expected to occur will determine the extent of technology deployment: large volumes and short timeframes will probably require greater inputs to be applied, increasing the cost of remediation. Technology to treat contaminated groundwater will continue to evolve, affecting costs and treatment methods used.

Of the 1,343 contaminated sites on the Superfund National Priorities List in the United States, 80% involve groundwater (USEPA, 1996a) and 414 sites have been deleted from the list as they

BOX 1.7.6 BRIEF DESCRIPTION OF REMEDIAL TREATMENT TECHNOLOGIES

Ex-situ Technologies

Advanced Oxidation Processes – Combine oxygen from the oxidizing agents with the contaminants and transform it to different substances and may use ultraviolet radiation and strong oxidizing agents (hydrogen peroxide or ozone) to improve efficiency.

Precipitation/Coagulation/Flocculation – Chemicals added to transform contaminants to solids to be precipitated, coagulated or flocculated and removed.

Separation (distillation, filtration/ultrafiltration/ microfiltration, freeze crystallization, membrane pervaporation and reverse osmosis) – Separation techniques concentrate contaminated wastewater through physical and chemical means.

Ion Exchange – Ion exchange removes ions from the aqueous phase by exchange with counter ions on the exchange medium.

Bioremediation – Uses microorganisms in bioreactors or constructed wetlands to destroy or stabilize contaminants and may add nutrients, oxygen or other chemicals to enhance the action of the organisms.

Air Stripping – Once pumped to the ground surface, the contaminated groundwater flows through a media that allows the contaminants to volatilize into the air.

Thermal Treatment – Heat is applied to the contaminated groundwater to destroy or remove the contaminants.

Granular Activated Carbon – Ground water is pumped through a series of canisters or columns containing activated carbon to which dissolved organic contaminants adsorb. Periodic replacement or regeneration of saturated carbon is required.

Sprinkler Irrigation – A relatively simple treatment technology that sprays contaminated wastewater into the air to volatilize VOCs and remove them from contaminated wastewater

In-situ Technologies

Bioremediation – Treating contaminated groundwater in place underground by introducing microorganisms as well as chemicals that may stimulate those organisms to degrade or stabilize the contaminants.

Bioslurping – Combines the two remedial approaches of bioventing and vacuum-enhanced free-product recovery. Bioventing stimulates the aerobic bioremediation of hydrocarbon-contaminated soils. Vacuum-enhanced free-product recovery extracts LNAPLs from the capillary fringe and the water table.

Chemical Treatment – Injecting chemicals that treat the contaminants in groundwater in place.

Dual Phase Extraction--A high vacuum system is applied to simultaneously remove various combinations of contaminated ground water, separate-phase petroleum product and hydrocarbon vapor from the subsurface.

Hydrofracturing – Injection of pressurized water through wells cracks low permeability and over-consolidated sediments. Cracks are filled with porous media that serve as substrates for bioremediation or to improve pumping efficiency.

In-well Air Stripping-- Air injected into a double screened well lifts the water in the well and forcing it out the upper screen allowing the dissolved phase VOCs to transfer to the vapor phase and then drawn off for treatment.

Thermal Extraction--Steam is forced into an aquifer through injection wells to vaporize volatile and semivolatile contaminants. Vaporized components rise to the unsaturated zone where they are removed by vacuum extraction and then treated.

Passive/Reactive Treatment Walls – Creation of a zone in the aquifer for contaminated groundwater to move through, which will treat the water using other chemical, physical or microbiological technologies.

Phytoremediation – Plants roots draw the contaminants out of the groundwater, which may serve as a food source or may be concentrated in their cell structure and then harvested and disposed as a hazardous waste.

Containment – In cases where other approaches are too costly or infeasible, a barrier to contain the contaminants in place may be constructed, using a range of materials depending on the type of containment, such as grout, plastic, ice or steel, which can be placed over (capping), under or around the contaminated zone.

Sources: Federal Remedial Technologies Roundtable (FRTR), Remediation Technologies Screening Matrix and Reference Guide, Version 4.0, 2007, URL: https://frtr.gov/scrntools.htm (accessed August 23, 2019).

Driscoll, F.G., Groundwater and Wells, Johnson Screens, St. Paul, MN, 1986, 1089.

TABLE 1.7.11
Most Frequently Found Contaminants in Groundwater at National Priority List (Superfund) Sites in the United States

Organic Contaminants		Inorganic Contaminants
Trichloroethylene	Vinyl chloride	Lead
Tetrachloroethylene (TCE) or (perchloroethene, "PCE")	Xylene	Chromium
	Ethylbenzene	Arsenic
Chloroform, benzene	Carbon tetrachloride	Cadmium
Toluene	Phenol	Mercury
1,1,1-Trichloroethane	Methylene chloride	Copper
Polychlorinated biphenyls (PCB)	1,2-Dichloroethane Pentachlorophenol (PCP)	Zinc
1,2-Trans-dichloroethylene	Chlorobenzene	Nickel
1,1-Dichloroethane	Benzo (a) pyrene	Cyanide
1,1-Dichloroethene		Barium

Source: U.S. Environmental Protection Agency (USEPA), Common Contaminants Found at Superfund Sites, 2003a, URL: http://www.epa.gov/superfund/resources/chemicals.htm (accessed April 5, 2003).

TABLE 1.7.12
Costs of Groundwater Treatment

Technology	Contaminants Treated	Approximate	Unit Cost
Ex-situ	Advanced oxidation processes	VOCs, SVOCs, DNAPLs, LNAPLs, pesticides, pathogens, metals, creosote, TPH, MTBE	\$0.03–\$3.00 per cubic meter[a]
Ex-situ	Precipitation/Coagulation/ Flocculation Ion Exchange	Metals, radionuclides Metals, Radionuclides	\$5-\$11 per cubic meter[a] \$0.08 to \$0.21 per cubic meter[a]
Ex-situ	Sprinkler Irrigation	VOCs	Not available for Superfund contaminants
Ex-situ	Separation (distillation, filtration/ ultrafiltration/ microfiltration, freeze crystallization, membrane pervaporation and reverse osmosis)	VOCs, SVOCs, pesticides and suspended particles	\$0.36 to \$1.20 per cubic meter[a]; some proprietary technologies may cost more
Ex-situ	Air stripping	VOCs, SVOCs, LNAPLs, TPH, MTBE	\$0.11–\$0.55 per cubic meter[a]
Ex-situ	Granular Activated Carbon Bioreactor Constructed Wetlands Containment/Deep Well Injection	VOCs, SVOCs, pesticides, PFOS, some metals SVOCs, fuel hydrocarbons and any biodegradable organic material Organic matter, nutrients, suspended sediments, trace metals and other toxic materials VOCs, SVOCs, fuels, explosives and pesticides	\$0.32-\$1.70 per cubic meter for 400 m^3/day average flow \$6-\$44 per cubic meter[a] \$0.36/cubic meter (amortized over 10 years); \$0.12/cubic meter (amortized over 30 years) Not available for Superfund wastes

(*Continued*)

TABLE 1.7.12 (*Continued*)
Costs of Groundwater Treatment

Technology	Contaminants Treated	Approximate	Unit Cost
In-situ	Enhanced Bioremediation (oxygen enhancement with air sparging, nitrate enhancement)	Nonhalogenated VOCs, nonhalogenated SVOCs and fuels	$10 to $20 per cubic meter[b] (oxygen enhancement); $40 to $60 per cubic meter[b] (nitrate enhancement)
	Natural Attenuation	VOCs, SVOCs, fuel hydrocarbons, pesticides, some metals	Monitoring and modeling costs
In-situ	Chemical treatment	VOCs, SVOCs, PCBs, DNAPLs, LNAPLs, pesticides, pathogens, TPH, MTBE, metals, radionuclides	Variable with technology, in-situ chemical oxidation through recirculation (ISCOR) costs approximately $77 per cubic meter[b]
In-situ	Permeable reactive barriers	VOCs, SVOCs, DNAPLs, pesticides, pathogens, metals, TPH, BTEX, radionuclides, nitrate, arsenic, PCBs, creosote	Installation costs $59–$370 per cubic meter[a]
In-situ	Phytoremediation (enhanced rhizosphere biodegradation, hydraulic control, phyto-degradation, phyto-volatilization)	VOCs, SVOCs, PCBs, NDNAPL, pesticides	$5-$18 per square meter[b]
	Air Sparging	VOCs and fuels	$24-$84 per cubic meter[b]
	Bioslurping (bioventing and vacuum-enhanced free-product recovery)	Fuels (free product)	$15 per liter[b]
	Horizontal directional wells	Wide range of contaminant groups	$60 to $250 per meter - hydraulic bi-directional thrust drilling; $330 per meter – sonic drilling (conventional PVC/HDPE well installation $164 per meter)
	Dual Phase Extraction	VOCs and fuels	$30-$72 per cubic meter[b]
	Thermal Treatment (forced hot water/steam into aquifer, then vacuum extraction)	VOCs, SVOCs and fuels	$65-$390 per cubic meter[b]
	Hydrofracturing	Wide range of contaminant groups	$1,000-$1,500 per fracture
	In-Well Air Stripping	Halogenated VOCs, SVOCs and fuels	Not available
In-situ	Passive/Reactive Treatment Walls	VOCs, SVOCs, inorganics	Generally installation costs range from $1,267-$2,580 per cubic meter of treatment wall; $0.10-$0.27 per cubic meter of groundwater treated[b]
In-situ	Physical Barriers/Containment	Wide range of contaminant groups	$540 to $750 per square meter[c]

Source: Federal Remedial Technologies Roundtable (FRTR), Remediation Technologies Screening Matrix and Reference Guide, Version 4.0, 2007, URL: https://frtr.gov/scrntools.htm (accessed August 23, 2019).

[a] Of influent media.

[b] Of subsurface media.

[c] Volumetric costs not available.

have been determined to have met remedial action objectives (USEPA, 2018a). In 2019, 25 sites were deleted or proposed to be deleted of which 80% involved groundwater contamination. Of these groundwater contaminated sites, 80% drew on institutional controls as part of the resolution of remedial objectives indicating that no further response is needed to protect human health and the environment. Institutional controls are non-engineered approaches to minimize human exposure to contamination or protect the integrity of the remedy. Typically, institutional controls limit land and/or water use to protect human health, such as zoning restrictions on use of groundwater or installation of drinking water wells. One-third of the groundwater sites in 2019 had institutional controls on groundwater use of access beyond the Superfund site and all groundwater sites had restrictions on use of groundwater on the site (FR, 2019). The implication of the institutional controls is that groundwater at the site and in some cases offsite cannot be used indefinitely, potentially constraining land and water development and use in or around those communities affected. Many previously remediated and deleted sites have benefited from redevelopment which may also have restrictions on types of use (USEPA, 2018b).

A technology not listed earlier is that of replacing the contaminated groundwater resource with another water source, either ground or surface water, and even bottled or mobile tank-delivered water. The costs for installing a new well for a single home or a community have been addressed in Volume 1, Chapter 6. In addition, the costs of delivering water from an alternative source should also be considered. Delivery costs include those of installing a transmission line and then maintaining it. In 2000, estimates of water distribution pipe costs in the United States ranged from US$80,000/km for small communities serving 3,300–10,000 people, to US$1.1million/km for large urban communities serving more than 500,000 people (USEPA, 2002, p. 65).

REMEDIATED SITES AND REDEVELOPMENT

Of the 1,910 sites on the 2020 U.S. Superfund list, 1,214 had all or a part of the site in a "completed construction" status, including sites that have institutional controls limiting future development. Of the sites that have completed construction and are remediated or in institutional control, 602 have been redeveloped and in use. These sites are a significant success, as they contribute productively to the local economies. Since 2011, the redeveloped sites, also often referred to as "brownfields," cumulatively have nearly 10,000 new businesses with over 200,000 jobs and have generated over $58 billion in sales revenue (see Table 1.7.13). Additionally, these sites provide local sales and property tax revenue to the incorporated areas in which they are located.

TABLE 1.7.13
Estimated Benefits of Superfund Site Redevelopment, 2011–2019

Year	Sites in Reuse with Economic Data	Number of Businesses	Annual Sales[a]	Jobs	Annual Employment Income[a]
2019	602	9,188	$58.3 billion	208,468	$14.4 billion

[a] Adjusted to 2019 USD using the Consumer Price Index (CUUR0000SA0, not seasonally adjusted, U.S. city annual average).

Source: U.S. Environmental Protection Agency (USEPA). 2020a. Construction Completions at NPL Sites - by State. URL: https://www.epa.gov/superfund/construction-completions-npl-sites-state; NPL Site Totals by Status and Milestone. URL: https://www.epa.gov/superfund/npl-site-totals-status-and-milestone (Accessed April 18, 2020).

WATER RECLAMATION AND REUSE

Many communities and countries around the world have reached or are expected to reach the limits of their water resources (USEPA, 2004a, 2004b). This circumstance was also demonstrated in the California drought of 2014–2016, with communities unable to supply water to their residents (Benson, 2016; Grossi, 2017). Increasing demand in population centers contributes to greater demand on all other economic sectors and their derived demand for water: domestic, commercial, industrial and agricultural. In 2004, an estimated 6.4 million m^3 of water was reused in the United States. Likewise, wastewater reclamation and reuse projects are active in the European Union focusing on applying treated wastewater for irrigated agriculture (EU, 2009). Water reuse often begins with reclamation, defined as the treatment of wastewater to meet particular water quality criteria for further use in another beneficial application. Thus, reuse is the actual deployment of treated wastewater for beneficial uses, such as irrigation or industrial cooling (Asano et al., 2007). Water reclamation in areas of limited water provides a "new" or "alternate" water source. Water reuse is often cost-effective when (USEPA, 2004a, 2004b):

- Considering obtaining the needed water supplies from distant sources
- Ground or surface water supplies are being depleted or have limited availability
- Treating a raw water supply source of lower quality
- Wastewater must be treated to stricter discharge standards

Health concerns are addressed with the reuse, such as ensuring that the application for the intended use satisfies the customer requirements, and the prevention of (1) improper system operation, (2) cross-connections with potable water lines and (3) improper use of nonpotable water.

Table 1.7.14 highlights the water reuse applications that are found to be beneficial as well as the factors affecting groundwater recharge applications.

TABLE 1.7.14
Some Beneficial Water Reuse Applications

Factor	Recharge Basins	Vadose Zone Injection Wells	Direct Injection Wells
Aquifer type Pretreatment requirements	Unconfined Low technology	Unconfined Removal of solids	Unconfined or confined High technology
Estimated major capital costs (2004US$)	Land, engineered basin $4.5 million+ and distribution system		$500,000–$1,500,000/well
Capacity	100–20,000 m^3/hectare-day	1000–3000 m^3/d/well	2000–6000 m^3/d/well
Maintenance requirements	Drying and scraping	Drying and disinfection	Disinfection and flow reversal
Estimated life cycle	>100 years	5–20 years	25–50 years
Soil/aquifer treatment	Vadose zone and saturated zone	Vadose zone and saturated zone	Saturated zone

Sources: U.S. Environmental Protection Agency (USEPA), Drinking Water Costs & Federal Funding, EPA 816-F-04–038, 2004a, URL: http://epa.gov/ogwdw/sdwa/30th/factsheets/pdfs/fs_30ann_dwsrf_web.pdf (accessed August 16, 2007).U.S. Environmental Protection Agency (USEPA), Guidelines for Water Reuse, EPA/625/R-04/108 450p, 2004b, URL: http://www.epa.gov/ord/NRMRL/pubs/625r04108/625r04108.pdf (accessed January 7, 2009).

An equation for reclaimed water service, which includes the cost factors for providing reclaimed water (USEPA, 2004a, 2004b), can be given as:

$$\text{Total reclaimed water service} = [\text{reclaimed water treatment} - \text{treatment to permitted disposal standards}] + \text{additional transmission} + \text{additional distribution} + \text{additional storage} \quad (1.7.1)$$

Many examples of costs of delivering reclaimed wastewater exist (USEPA, 2004b). One example is San Marcos, Texas, in which the analysis found that the cost of:

- Treated and delivered reclaimed water was \$0.07–\$0.14/m^3
- Alternative water supplies in the region was \$0.11–\$0.24/m^3
- The reclaimed water charge established by the city was \$0.18/m^3

The charges for the use of reclaimed water have a wide range. During 2001–2002, for selected communities in six states in the United States (Texas, Colorado, Nevada, Florida, Illinois and Hawaii), the range of reclaimed water charges for metered usage was \$0.03–\$2.69/m3 (USEPA, 2004b).

The U.S. Environmental Protection Agency found the following benefits calculated as financial credits of reclaimed water projects in California (USEPA, 2004b):

- Water supply, \$0.24–\$0.89/m^3
- Water supply reliability, \$0.08–\$0.11/m^3
- Effluent disposal, \$0.16–\$1.62/m^3
- Downstream watershed, \$0.32–\$0.65/m^3
- Energy conservation, \$0.00–\$0.19/m^3

The 2017 Texas State Water Plan projected water reuse strategies in the state will produce 1.365 billion m^3 (1.107 million acre-feet) per year by 2070, about 13% of new water supplies. If implemented, direct potable reuse will provide 107 million m^3/year (87,000 acre-feet/year) by 2070, indirect reuse will provide 800.5 million m^3/year (649,000 acre-feet/year) and direct reuse will provide 457.6 million m^3/year (371,000 acre-feet/year) (TWDB, 2019). The cities of Big Spring (partly groundwater-supplied) and Witchta Falls, Texas, USA, have implemented direct potable reuse of wastewater after extensive research and testing. The Big Spring plant in operation in 2013 used microfiltration, reverse osmosis and ultraviolet disinfection to treat 7,750 m^3/day (2 million gallons/day) of wastewater effluent in its treatment system costing \$14 million (Martin, 2014).

Internationally, reclaimed water use is significant in countries with arid climates that rely heavily on groundwater. In a survey of 23 northern African, Middle Eastern and central Asian countries for the years 1990–1991, water reclamation was found to range from 0.002% to 15%. The three countries with the greatest use of reclaimed water at that time were: Israel, 10%; Cyprus, 11%; and Kuwait, 15% (USEPA, 2004b). In 2009, because of aquifer depletion and limits on water resources, Australia, Singapore and Spain also reuse significant amounts of water.

ECONOMIC CONSIDERATIONS FOR GROUNDWATER IN TREATMENT AND DISPOSAL

From an economic standpoint, alternative technologies must be evaluated for the same objective or goal. Treating groundwater for drinking water purposes only as compared to groundwater for all uses, including drinking water, are vastly different objectives. The risk of contamination from residuals and the disposal of contaminants in the subsurface as a waste sink should be considered in a cost evaluation. An economic approach requires alternatives to be methodically evaluated. Different goals may have different, or, conversely, similar alternatives. Some technologies remove more contaminants than others. The economic efficiency evaluation requires that we determine the alternative that provides groundwater of an accepted quality at least cost. Thus, we must consider the initial costs for installing the treatment or disposal technique and the expenses of operating, maintaining and replacing the equipment or plant in the future from a life-cycle perspective. These costs can be subsequently incorporated into the price of groundwater delivered to the consumer who can decide how much water to purchase and use. If a community uses a large volume of water, its leaders have to decide the most efficient way to access, treat and deliver that water: many small wells and treatment plants, or larger wells with a large treatment plant. The technology costs mentioned earlier suggest economies of scale, with smaller units having higher costs per unit of water treated and delivered.

With regard to global warming and its effects, a consensus of scientists and nations indicates that this phenomenon is underway and that an immediate response is necessary (IPCC, 2001). Relative to related groundwater effects, the potential capability of the subsurface to store CO_2 is perceived to be a significant benefit for all humankind, and major research is being carried out to successfully implement this (USDOE, 2007b). The locations of storage will need to be evaluated on a case-by-case basis to determine the costs into the future and the scale of deep geologic storage of this greenhouse gas if this option is widely used. The data and modeling suggest that rising temperatures are related to increased severity of droughts as a result of elevated evaporation and water consumption (ADEWR, 2003), and that the land area affected by serious drought may be extensive (USNCAR, 2005), which have implications for reduced groundwater recharge and lower water tables, implying less groundwater for supply. In 2001, the Intergovernmental Panel on Climate Change concluded that over the twentieth century were statistically significant and that the ocean levels, as measured by tidal gage data, had increased to 0.1–0.2 m – and are expected to rise as much as 0.9 m by 2100 (IPCC, 2001). This fact indicates that the coastal aquifers may already have been affected, and issues related to the use of near-coastal wells and activities in the coastal areas may emerge as costs to be mitigated, including the use of deep geologic storage of CO_2. These activities, such as deep well injection of CO_2 emissions and management of groundwater resources in coastal, arid and drought-stricken areas, suggest a significant need for investment in the subsurface and groundwater protection and supply, as well as for research and practice of using these subterranean resources sustainably, while balancing the benefits with costs.

With regard to deep geologic disposal, a nation or jurisdiction holds the subsurface rights in trust for its people, to ensure adequate consideration of the subterranean environment in decision making at all levels, including public and private entities. The possibility of initial and ongoing charges for use of this public trust space, such as a disposal tax or fee based on volume to cover administrative and local groundwater protection costs of underground wastewater injection regulation or septic system evaluation and placement, for example, may reflect the economic value of its capacity and the other uses and purposes to which it is currently or could be engaged, including ecosystem support. Without such charges, the value of this space is equated to be zero, promoting its exploitation until some adverse consequence is observed. Waiting until the occurrence of adverse effects may have deleterious results that may not be obvious at the time.

From an environmental perspective, if a homeowner, community or corporation must treat its groundwater because of a contaminating activity degrading the quality of its water supply, the expense of treatment is a "damage cost." However, who would pay this cost and what are the key environmental economic issues? The costs may be greater than simply treating the groundwater. In situations where people have become ill, the costs may be greater. Furthermore, if the groundwater quality (or watertable level) changes, then for what purpose can a property owner use his or her land? These changes are also considered as costs with a different set of benefits. An efficient resolution is to compare the abatement costs of the contaminating entity with the damage costs of the recipients of the unwanted contamination. These considerations will be addressed subsequently. Taking a watershed or river basin approach to allow a holistic view of actions, effects, benefits and damages for water-related results will enable an evaluation of alternatives with a more balanced potential for acceptable outcomes.

SUMMARY

Groundwater users may have a preference for high-quality water. If groundwater is used for drinking and other purposes resulting in human contact, then the consumers may expect that the quality meets the public health standards for drinking water. If the quality does not meet the users' requirements, then treatment may be necessary. For people using individual wells, a range of treatment options are available, including POE to the household (such as a water softening unit) that treats water for all water uses in a house, or POU that treats only water for that tap. The costs vary according to the technology used and the volume of water needing treatment. Most groundwater users are supplied water through a central water supplier that can provide treatment for an entire community and spread the costs of that treatment over many more units of groundwater. If groundwater becomes contaminated with wastes that are not properly disposed, remediation of groundwater can be carried out in places underground or above the ground, if pumped to the surface. The treatment costs may vary based on the contaminant, extent of contamination, the subsurface environment and the remedial goals. Water treatment and desalination produce wastes, which must also be managed and disposed if they cannot be recycled. Brackish groundwater desalination and water reclamation and reuse are significant in regions with limitations on groundwater availability.

Concerns about global climate change also present challenges to groundwater use, with associated responses that involve disposal of "greenhouse" gases and desalination waste in deep geologic strata, suggesting a greater competition for use of the subsurface environment. Efficiency as an economic goal indicates that we should evaluate alternatives of technology and scale and select the least-cost approach that provides the quality of groundwater which the community desires. Environmental and public health effects of poor quality groundwater may increase the overall cost of its contamination beyond treatment alone. A watershed approach to waste treatment and disposal affecting groundwater may provide an opportunity to consider the costs and benefits in a holistic setting and a balance among people's needs and resource uses. To promote the efficient use of the public trust subsurface environment, particularly for disposal, its value should not be equated to zero. Public charges for subsurface use, such as fees to cover administrative and resource protection costs of an underground injection or septic system oversight program, might facilitate its best use, especially where significant risk is associated with the disposal.

Remedial costs for contaminated sites are significant. Technology has advanced to facilitate remediation. Thirty percent of Superfund sites have been redeveloped from 2011–2019 with substantial numbers of businesses, jobs and sales, and providing local tax revenue.

FOR FURTHER CONSIDERATION

1. What water treatment and wastewater disposal methods are used in your area? Contact the local sanitarian and discuss water treatment and disposal methods. What do they cost to install (check online)?
2. Ask the local sanitarian what the most prevalently used private individual household water treatment and wastewater disposal systems are used in your area. Contact a water treatment and wastewater disposal contractor and ask about cost.

3. Is there a Superfund hazardous waste site with contaminated soil and groundwater near you? Research online the objectives, methods and costs for remediating the site. What costs have been incurred to date for which remedial actions? What are the annual operation and maintenance costs of the site? How long is it projected to remediate the site? Does the remedial approach include institutional controls on land use? If so, what resources are being unused because of continuing contamination? Consider the costs of this loss of use and quantify those costs.

NOTE

1 Rubin (2005) estimated the potential leakage as up to 10% in 100 years and up to 40% in 500 years.

REFERENCES

Alley, W. M. and Alley, R.. 2017. Pathogens. In *High and Dry: Meeting the Challenges of the World's Growing Dependence on Groundwater*, 195–202. New Haven: Yale University Press.

Alley, W.M. and Job, C.. 2020. Groundwater, Coronavirus and Wells. *National Ground Water Association; Publications and News*. URL: https://www.ngwa.org/publications-and-news/covid-19-news-and-resources/groundwater-wells-and-coronavirus (Accessed December 18, 2020).

American Water Works Association (AWWA). 2004. *Water Treatment Plant Design*, 4th edn. McGraw-Hill Professional, New York, 896 p.

Asano, T., Leverenz, H.L., Tsuchihashi, R., and Tchobanoglous, G. 2007. *Water Reuse: Issues, Technology and Applications*. McGraw-Hill Professional, New York, 1570 p.

Australian Department of Environment and Water Resources (ADEWR). 2003. *Climate Change—An Australian Guide to the Science and Potential Impacts*. P. Barrie (ed.), URL: http://www.greenhouse.gov.au/science/guide/index.html (accessed April 22, 2007).

Baker, M.N. 1949. *The Quest for Pure Water*. The American Water Works Association, Inc., New York, 527 pp.

Beitsch, Rebecca. 2018. Few wells after major flooding from hurricanes. The Salem News, December 17, 2018. [Quoting Merritt Partridge, President, Partridge Well Drilling Co., Inc., Jacksonville Florida.] URL: https://www.salemnews.com/news/national_news/few-wells-after-major-flooding-from-hurricanes/article_d0798100-dbdc-5adb-af98-1938c0944657.html (Accessed April 21, 2020).

Benson, S.M. and Myer, L. 2002. *Monitoring to ensure safe and effective geologic sequestration of carbon dioxide. Presentation to the IPCC Workshop for Carbon Capture and Storage*, Regina, AB, Canada, November 18–21, 2002, 15 pp. URL: http://arch.rivm.nl/env/int/ipcc/docs/css2002/ccs02-10.pdf (accessed April 6, 2007).

Benson, Thor. 2016. How a California town completely ran out of water. In *Business Insider*, Feb. 21, 2016. URL: https://www.businessinsider.com/california-town-completely-ran-out-of-water-2016-2 (Accessed August 22, 2019).

Black Veatch Holding Company (B& V). 2006. *Colorado Water Treatment Plant Earns Design-Build Award*. URL: http://www.bv.com/wcm/press_release/10232006_5201.aspx (accessed February 11, 2007).

Boone, Mary. 2014. *How Much Do Septic Systems Cost to Install?* URL: https://www.zillow.com/blog/cost-to-install-septic-system-150191/ (Accessed April 22, 2020).

Building Advisor.com (BCA). 2019. Lifespan of a Septic System? URL: buildingadvisor.com/how-long-can-septic-system-last/ (accessed May 9, 2020).

California Coastal Commission (CCC). 1993. *Seawater Desalination in California*. URL: http://www.coastal.ca.gov/desalrpt/dtitle.html (accessed August 27, 2009).

California Energy Commission (CEC). 2003. *Unit Cost of Desalination*. URL: http://www.owue.water.ca.gov/recycle/desal/Docs/UnitCostDesalination.pdf (accessed January 19, 2009).

Carlsbad Desalination Plant Project (CDPP) 2017. Desal 101. URL: https://www.carlsbaddesal.com/desal-101.html (accessed August 24, 2019).

Centers for Disease Control and Prevention (CDC). 2014. *A Guide to Drinking Water Treatment Technologies for Household Use*. URL: https://www.cdc.gov/healthywater/drinking/home-water-treatment/household_water_treatment.html (Accessed April 23, 2020).

Centers for Disease Control and Prevention (CDC). 2015. *Overview of Water-related Diseases and Contaminants in Private Wells*. URL: https://www.cdc.gov/healthywater/drinking/private/wells/diseases.html (Accessed April 23, 2020).

Clean Water Testing LLC (CWT). 2018. *Well Water and Septic Inspections*. http://www.cleanwatertesting.com/local-water-testing-services/well-septic-inspections/ (Accessed April 21, 2020).

Clearwater Public Utilities, Florida (CPUFL). 2007. Reservoir No. 1 Project. URL: http://www.dear-water-fl.com/gov/depts/pwa/public_utils/projects/reservoir_1/FAQwtp.asp (accessed February 11, 2007).

Colorado Department of Natural Resources (CDNR), Division of Water Resources. 2005. *Rules and Regulations For Water Well Construction, Pump Installation, Cistern Installation, And Monitoring And Observation Hole/Well Construction*. URL: https://www.sos.state.co.us/CCR/GenerateRulePdf.do?ruleVersionId=4344 (accessed Aril 16, 2020).

Contra Costa Health Services (CCHS). 2014. *Small Water System Fee Schedule*. URL: https://cchealth.org/eh/small-water/pdf/Fees.pdf (accessed April 21, 2020).

CostHelper, Inc. (CHI). 2020. *Cost of Septic Tank Cleaning - Estimates and Prices Paid*. URL: https://home.costhelper.com/septic-tank-cleaning.html (accessed April 22, 2020).

Craun, G.F., Brunkard, J.M., Yoder, J.S., Roberts, V.A., Carpenter, J., Wade, T., Calderon, R.L., Roberts, J.M. Beach, M.J., and Roy, S.L.. 2010. Causes of outbreaks associated with drinking water in the United States from 1971 to 2006. *Clinical Microbiology Reviews* 23 (3): 507–528.

Driscoll, F.G. 1986. *Groundwater and Wells*. Johnson Screens, St. Paul, MN, 1089 p.

El Paso Water Utilities (EPWU). 2007. *Past and Present Water Supplies*. URL: http://www.epwu.org/water/water_resources.html (accessed February 16, 2009).

European Environment Agency (EEA). 1999. *Groundwater Quality and Quantity in Europe*. Copenhagen, Denmark, 123 p.

European Environment Agency(EEA). 2003. *Indicator Fact Sheet (WQ03b) Saltwater intrusion*. URL: http://ims.eionet.europa.eu/Specific_media/water/indicators/WQ03b%2C2003.1001/Fact_sheet_16_Saltwater_Inrusion_final.pdf (accessed August 27, 2009).

European Union (EU). 2009. *Water Re-use*. URL: http://www.waterreuse.eu/p_frames_page.html (accessed February 8, 2009).

Federal Register (FR). 2019. Various notices searchable in archives of federalregister.gov with keywords "superfund site deletion 2019".

Federal Remedial Technologies Roundtable (FRTR). 2007. *Remediation Technologies Screening Matrix and Reference Guide*, Version 4.0. URL: https://frtr.gov/scrntools.htm (accessed August 23, 2019).

Federal Remediation Technology Roundtable (FRTR0). 2009. *Glossary*. URL: http://www.frtr.gov/glossary/aterms.htm (accessed January 7, 2009).

Foldager, R.A. 2003. Economics of desalination concentrate disposal methods in inland regions: deep-well injection, evaporation ponds, and salinity gradient solar ponds. Thesis, New Mexico State University. Las Cruces, NM, URL: http://wrri.nmsu.edu/research/rfp/studentgrants03/reports/foldager.pdf (accessed August 27, 2009).

Frederick, K.D. 1995. America's water supply: Status and prospects for the future. *Consequences*, 1 (1), (Spring 1995). URL: http://www.gcrio.org/CONSEQUENCES/spring95/Water.html (accessed January 18, 2009).

Friedman, S.J. 2005. *Geologic Carbon Sequestration Science and Technology*. URL: www.nelson.wisc.edu/outreawch/energy2006/powerpoint/friedman.ppt#2 (accessed February 9, 2007).

General Accountabzzility Office (GAO). 2010. Superfund: EPA's Estimated Costs to Remediate Existing Sites Exceed Current Funding Levels, and More Sites Are Expected to be Added to the National Priorities List. (URL: https://gao.gov/assets/310/304124.pdf (Accessed August 25, 2019).

Greenblatt, J. 2006. *Driving a wedge through global warming: Existing technologies to bridge the gap between business-as-usual and stable global emissions. Annual Meeting, Committee on Environmental Improvement*, American Chemical Society, San Francisco, CA, September 9, 2006. URL: http://membership.acs.org/c/cei/Greenblatt.pdf (accessed March 23, 2007).

Grossi, M. 2017. The California Drought Isn't Over, It Just Went Underground. In Water Deeply, July 5, 2019. URL: https://www.newsdeeply.com/water/articles/2017/07/05/the-california-drought-isnt-over-it-just-went-underground (accessed August 22, 2019).

Gundy, P. M., Gerba, C. P. and Pepper, I. L. 2009. Survival of coronaviruses in water and wastewater. *Food Environmental Virology* 1: 10–14.

Heger, Sarah. 2020. Are Septic System Professionals at a Greater Risk for COVID-19? In On-Site Installer. (March 17, 2020) URL: https://www.onsiteinstaller.com/online_exclusives/2020/03/are-septic-system-professionals-at-a-greater-risk-for-covid-19 (Accessed April 21, 2020).

Heidenry, Margaret. 2018. *Septic System Maintenance Matters: Costs, Upkeep, and Items You Should Never Flush*. URL: https://www.realtor.com/advice/home-improvement/septic-system-maintenance/ (Accessed April 22, 2020).

Howard, K.W.F. and Gelo, K.K.. 2003. Intensive groundwater use in urban areas: The case of megacities. pp. 35–58. In *Intensive Use of Groundwater: Challenges and Opportunities*. A.A. Balkema Publishers. Lisse, the Netherlands, 478 pp.

Hurd, Gordon. 2019. *How Much Do Septic Tanks Cost? The Stomach-Churning Price Revealed*. URL: https://www.realtor.com/advice/home-improvement/how-much-does-a-septic-system-cost/ (Accessed April 22, 2020).

Intergovernmental Panel on Climate Change (IPCC). 2001. *Climate Change 2001:* Synthesis Report. A Contribution of Working Groups I, II, and III to the Third Assessment Report of the Intergovernmental Panel on Climate Change, Watson, R.T. and the core writing team (eds.). Cambridge University Press, Cambridge, U.K., and New York, 398 pp.

Intergovernmental Panel on Climate Change (IPCC). 2002. *Climate Change and Biodiversity: IPCC Technical Paper V*. URL: http://www.ipcc.ch/pub/tpbiodiv.pdf (accessed April 6, 2007).

Intergovernmental Panel on Climate Change (IPCC). 2005. *Carbon Dioxide Capture and Storage*. URL: http://www.ipcc.ch/activity/srccs/SRCCS_Contents_Foreword_Preface.pdf (accessed March 23, 2007).

Intergovernmental Panel on Climate Change (IPCC). 2008. *Climate Change and Water: IPCC Technical Paper VI*. URL: http://www.ipcc.ch/ipccreports/tp-climate-change-water.htm (accessed February 16, 2009).

Job, Charles. 2017. Responding to Flooded Wells; A sustainability Issue. Water Well Journal, November 30, 2017. URL: https://waterwelljournal.com/responding-flooded-wells-2/ (Accessed April 21, 2020).

Krauss, Steffen and Griebler, Christian. 2011. Pathogenic Microorganisms and Viruses in Groundwater. In Acatech Materialien-NR. 6. URL: https://www.researchgate.net/publication/285864145_Pathogenic_microorganisms_and_viruses_in_groundwater (Accessed April 21, 2020).

LaFond, K. 2015. Infographic: America's Septic Systems. In Circle of Blue, October 16, 2015. URL: https://www.circleofblue.org/2015/world/infographic-americas-septic-systems/ (accessed August 24, 2019)

Laundry Alternative Inc. (LAI). 2005a. *Septic System Price*. URL: http://www.laundry-alternative.com/septic_system_price.htm (accessed February 11, 2007).

Laundry Alternative Inc. (LAI). 2005b. *Septic System Price*. URL: http://www.laundry-alternative.com/septic_system_price.htm (accessed January 19, 2009).

Livingston, E.C. 2004. *Considerations for brackish groundwater use as a water supply for small water systems and municipalities. Presented to the New Mexico Brackish Groundwater Aquifer Assessment Workshop*, Albuquerque, New Mexico, January 15, 2004.

LoPresti, N. Undated. How Long Does A Water Well Pump Last? Here is the Average Life Span. yourh@jome.com/well-water-pump-lifespan/

Lucas, L. 2007. *Surface Septic Systems Failing in Illinois—Dumping Raw Sewage*. The Legal Record's News Bulletin. URL: http://www.legalrecord.net/surface-septic-systems-in-illinois.html (accessed February 10, 2007).

Manning, F.S. and Thompson, R.E. 1995. *Oilfield Processing of Petroleum: Crude Oil*. Pennwell Books, Tulsa, OK 400 p.

Martin, L. 2014. Texas Leads The Way With First Direct Potable Reuse Facilities In U.S. In Water Online, September 16, 2014. URL: https://www.wateronline.com/doc/texas-leads-the-way-with-first-direct-potable-reuse-facilities-in-u-s-0001 (accessed August 22, 2019).

Minnesota Pollution Control Agency (MPCA). 2019. *Subsurface sewage treatment systems well setbacks; The distances required between water supply wells and septic systems*. https://www.pca.state.mn.us/sites/default/files/wq-wwists4-36.pdf (accessed April 3, 2020).

Modesto, California. 2003. *Cost Estimates for Treatment Technologies*. Web site URL: www.modestogov.com/omd/01_ccr/pdf/phg_cost_treat.pdf (accessed April 5, 2003).

Montgomery County (Maryland) Department of Permitting Services (MCDPS). 2018. *Guidelines - Well Permit Plan*. https://www.montgomerycountymd.gov/DPS/Resources/Files/Land_Development/GuidelinesWellPermitPlan.pdf (accessed April 3, 2020).

Mondal, N.C. and Singh, V.S. 2004. Proceedings of the Asia Pacific association of hydrology and water resources (APHW) second conference, Singapore, July5–8, 2004

National Ground Water Association. 2003. Consumer price index annual adjustments for household well drilling costs, January 1992 to January 2001 (unpublished).

National Ground Water Association (NGWA). 2020. National Ground Water Association: Low Risk of COVID-19 Contaminating Private Water Wells. In *2020 Press Releases*, April 22, 2020. URL: https://www.ngwa.

org/publications-and-news/Newsroom/2020-press-releases/national-ground-water-association-low-risk-of-covid-19-contaminating-private-water-wells (Accessed April 23, 2020).

National Public Radio (NPR). 2007. *Census Bureau Adopts GPS to Find American Homes.* URL: npr.org/templates/story/story.php?storyId=5590541 (accessed February 10, 2007).

National Research Council. 1994. *Ground Water Recharge Using Waters of Impaired Quality.* URL: https://www.nap.edu/catalog/4780/ground-water-recharge-using-waters-of-impaired-quality (accessed April 21, 2020).

National Research Council (NRC). 2008. *Desalination: A National Perspective.* National Academies Press, Washington, DC. 298 p.

Natural Resources Defense Council (NRDC). 2007. *Global Warming Basics.* URL: www.nrdc.org/global-Warming/f101.asp#7 (accessed March 23, 2007).

National Sanitation Foundation(NSE). 2019. *Home Water Treatment System Selection.* URL: http://www.nsf.org/consumer-resources/water-quality/water-filters-testing-treatment/home-water-treatment-system-selection (accessed August 24, 2019).

Olasson Associates. 2004. *Seward Water Treatment Plant.* URL: http://www.oaconsulting.com/project_indi.asp?ID=107 (accessed February 11, 2007).

Organization of American States (OAS). 1997. *Source Book of Alternative Technologies for Freshwater Augmentation in Latin America and the Caribbean.* Web site URL: www.oas.org/usde/publication/unit/oea59e/ch35.htm (accessed April 5, 2003).

Paque, M. 2003. Personal communication from Michael Paque, Executive Director, Ground Water Protection Council, March 24, 2003.

Probst, K. 2017. *Superfund 2017: Cleanup Accomplishments and the Challenges Ahead. Prepared for the American Consulting Engineers Council.* URL: https://acec.org/default/assets/File/Superfund_2017_FINAL.pdf (accessed August 25, 2019).

Produced Water Society (PWS). 2017. *Produced Water 101.* URL: https: producedwatersociety.com/produced-water-101/ (accessed August 28, 2019).

Robillard, P.D., Sharpe, W.E., and Swistock, B.R. 2003. *Water Softening Factsheet 141.* Pennsylvania State University, Cooperative Extension Service. URL: http://www.age.psu.edu/extension/factsheets/F%20141.pdf (accessed March 8, 2003).

Roscoe Moss Company (RMC). Undated. Case Study; Increased Well Efficiency, Extended Lifetime and Reduced Maintenance through Selection of Stainless Steel Casing and Well Screen – Sun City and SunCity West, Arizona. Technical Memorandum 004-2 URL: roscoemoss.com/wp-content/uploads/techmemos/TechMemo004-2_Increased_Efficiency_of_Steel_Well_Casing_and.pdf (accessed May 9, 2020).

Rubin, E.S. 2005. *IPCC Special Report on Carbon Dioxide Capture and Storage. U.S. Climate Change Science Program Workshop,* Washington DC, November 14, 2005. URL: www.climatescience.gov/workshop2005/presentations/breakout_2ARubin.pdf (accessed March 23, 2007).

Salt Lake City (Utah) Department of Public Utilities (SLC). 2003. *Water District Breaks Ground for New Water Treatment Plant.* URL: http://www.ci.slc.ut.us/Utilities/NewsEvents/news2003/news6132003.htm (accessed February 11, 2007).

Sampat, P. 2000. *Deep Trouble: The Hidden Threat of Groundwater Pollution. Worldwatch Paper 154.* Worldwatch Institute. Washington DC, 55 p.

Schafer, K.W. 1996. *International Experience and Expertise in Registration Investigation, Assessment, and Clean-Up of Contaminated Military Sites.* URL: http://www.umweltbundesamt.de/altlast/web1/berichte/mooreeng/dmeng_t.htm (accessed May 19, 2006).

Sheng, Z. and Devere, J. 2005. Understanding and managing the stressed Mexico-USA transboundary Hueco bolson aquifer in the El Paso del Norte region as a complex system, *Hydrogeology Journal,* 13 (5–6), 813–825.

Smekens, K. and Van der Zwaan, B. 2004. *Environmental externalities of geological carbon sequestration effects on energy scenarios. FEEM Working Paper No. 58.04.* URL: http://ssrn.com/absract=541123.

Statista.com. 2019. *Average monthly residential cost of water in the U.S. from 2010 to 2018.* URL: https://www.statista.com/statistics/720418/average-monthly-cost-of-water-in-the-us/ (Accessed August 24, 2019).

Stony Brook-Millstone Watershed Association (SBMWA). 2004. *Funding for Septic Systems.* URL: http://www.thewatershed.org/images/uploads/Funding%20brochure.pdf (accessed February 11, 2007).

Talbot, David. 2015. Megascale Desalination. In MIT Technology Review, February 18, 2015. URL: (accessed August 24, 2019).

Texas Water Development Board. 2006. *Please Pass the Salt: Using Oil Fields for the Disposal of Concentrate from Desalination Plants; Report 366.* URL: http://www.twdb.state.tx.us/publications/reports/GroundWaterReports/GWReports/Report366.pdf (accessed August 27, 2009).

Texas Water Development Board (TWDB). 2008. *Brackish Groundwater Desalination One-Pager.* URL: http://www.twdb.state.tx.us/iwt/desal/facts/onepagebrack.html (accessed September 26, 2008).

Texas Water Development Board (TWDB). 2019. Water Reuse. URL: http://www.twdb.texas.gov/publications/shells/WaterReuse.pdf (accessed August 22, 2019).

United Kingdom Trade and Investment (UKTI). 2007. *Utility Costs in the UK: Information Sheet.* URL: http://www.ukinvest.gov.uk/10231/en_GB/0.pdf (accessed August 16, 2007).

United States Department of Agriculture (USDA). 1997. *Water Quality Information Center, Nebraska Management System Evaluation Area, Shelton Site.* URL: http://www.nal.usda.gov/wqic/wgwq/neb1.shtml (accessed May 19, 2006).

U.S. Census Bureau (USCB). 2008. Current Housing Reports, Series H150/07, American Housing Survey for the United States: 2007. U.S. Government Printing Office, Washington, DC.

U.S. Department of Energy (USDOE). 2005a. *Carbon Sequestration Newsletter.* March 2005. (Citing: Wall Street Journal (WSJ), February 4, 2005, http://online.wsj.com/public/us) URL: http://www.netl.doe.gov/sequestration (accessed February 4, 2007).

U.S. Department of Energy (USDOE). 2005b. *Carbon Sequestration R&D Overview.* November 6, 2005. URL: http://www.fossil.energy.gov/sequestration/overview.html (accessed February 4, 2007).

U.S. Department of Energy (USDOE). 2005c. *The Cost of Carbon Dioxide Capture and Storage in Geologic Formations.* URL: www.netl.doe.gov (accessed March 23, 2007).

U.S. Department of Energy (USDOE). 2006. *Carbon Sequestration Technology Roadmap and Program Plan*, 40 pp. URL: http://www.netl.doe.gov/publications/carbon_seq/2006%20Sequestration%20Roadmap%20FINAL.pdf (accessed April 6, 2007).

U.S. Department of Energy (USDOE). 2007a. *Carbon Sequestration: Ocean Sequestration.* URL: http://www.fossil.energy.gov/programs/sequestration/ocean/ (accessed April 6, 2007).

U.S. Department of Energy (USDOE). 2007b. *Carbon Sequestration: CO2 Storage.* URL: http://www.netl.doe.gov/technologies/carbon_seq/core_rd/storage.html (accessed April 22, 2007).

U.S. Department of Commerce (USDOC). 2003. America's Homes, Sweet Homes; Press Release CB03–182. URL: http://www.census.gov/Press-Release/www/releases/archives/census_2000/001543.html (accessed February 10, 2007).

U.S. Environmental Protection Agency (USEPA). 1985. DRASTIC: A Standardized System for Evaluating Ground Water Pollution Potential Using Hydrogeologic Settings. EPA/600/2-85/018. Washington, DC, 163 p.

U.S. Environmental Protection Agency (USEPA). 1990. *Toxics in the Community.* EPA 560/4-90-017. Washington, DC, 364 p.

U.S. Environmental Protection Agency (USEPA). 1991. *NPL Characterization Project.* EPA/540/8-91/069.

U.S. Environmental Protection Agency (USEPA). 1993a. Ground Water Resource Assessment. Office of Water. EPA 813-R-93-003. Washington, DC, 232 p.

U.S. Environmental Protection Agency (USEPA). 1993b. *Evaluation of the Likelihood of DNAPL Presence at NPL Sites, National Results.* EPA 540-R-93-073.

U.S. Environmental Protection Agency (USEPA). 1996a. National Water Quality Inventory; 1994 Report to Congress; Ground Water Chapters. EPA-813-R-96-001.

U.S. Environmental Protection Agency (USEPA). 1996. Ground Water Cleanup at Superfund Sites. EPA540-K-96-008. URL: https://semspub.epa.gov/work/HQ/157421.pdf (accessed August 25, 2019).

U.S. Environmental Protection Agency (USEPA). 1999a. *Office of Water, Office of Ground Water and Drinking Water Underground Injection Control.* URL: WWW.epa.gov/ogwdw (accessed October 9, 1999).

U.S. Environmental Protection Agency (USEPA). 1999b. *Technologies and Costs for Removal of Arsenic from Drinking Water.* EPA-815-R-00-010/December 1999. URL: www.epa.gov/ogwdw/ars/treatments_and_costs.pdf (accessed April 5, 2003).

U.S. Environmental Protection Agency (USEPA). 2020a. *Construction Completions at NPL Sites - by State.* URL: https://www.epa.gov/superfund/construction-completions-npl-sites-state; NPL Site Totals by Status and Milestone. (accessed April 18, 2020).

U.S. Environmental Protection Agency (USEPA). 2020b. *NPL Site Totals by Status and Milestone.* URL: https://www.epa.gov/superfund/npl-site-totals-status-and-milestone (Accessed April 18, 2020).

U.S. Environmental Protection Agency (USEPA). 2020c. *National Primary Drinking Water Regulation.* URL: https://www.epa.gov/ground-water-and-drinking-water/national-primary-drinking-water-regulations (Accessed April 21, 2020).

U.S. Environmental Protection Agency. 2020. *Will my septic system treat COVID-19?* URL: https://www.epa.gov/coronavirus/will-my-septic-system-treat-covid-19 (Accessed April 22, 2020).

U.S. Environmental Protection Agency (USEPA). 2002. *Community Water System Survey 2000; Detailed Tables and Survey Methodology*. EPA-815-R-02-005B. Washington, DC, 194 pp. URL: http://www.epa.gov/safewater/consumer/pdf/cwss_2000_volume_ii.pdf (accessed February 11, 2007).

U.S. Environmental Protection Agency (USEPA). 2003a. *Common Contaminants Found at Superfund Sites*. Web site URL: http://www.epa.gov/superfund/resources/chemicals.htm (accessed April 5, 2003)

U.S. Environmental Protection Agency (USEPA). 2003b. *The Facts Speak For Themselves: A Fundamentally Different Superfund Program*. Web site URL: http://www.epa.gov/superfund/whatissf/sf_fact4.pdf (accessed April 5, 2003).

U.S. Environmental Protection Agency (USEPA). 2003c. *Homeowner's Guide to Septic Systems—Short Version*. EPA 832-B-02-006. Washington, DC, 15 pp. URL: http://www.nesc.wvu.edu/NSFC/pdf/homeowner_guide_long.pdf (accessed February 11, 2007).

U.S. Environmental Protection Agency (USEPA). 2003d. *Underground Injection Control Program*. URL: www.epa.gov/OGWDW/uic.html (accessed August 23, 2003).

U.S. Environmental Protection Agency (USEPA). 2004a. *Drinking Water Costs & Federal Funding*. EPA 816-F-04-038. URL: http://epa.gov/ogwdw/sdwa/30th/factsheets/pdfs/fs_30ann_dwsrf_web.pdf (accessed August 16, 2007).

U.S. Environmental Protection Agency (USEPA) 2004b. *Guidelines for Water Reuse*. EPA/625/R-04/108 450 pp. URL: http://www.epa.gov/ord/NRMRL/pubs/625r04108/625r04108.pdf (accessed January 7, 2009).

U.S. Environmental Protection Agency (USEPA). 2006a. *Factoids, Office of Ground Water and Drinking Water*. URL: http://www.epa.gov/safewater/data/pdfs/statistics_data_factoids_2005. pdf (accessed February 10, 2007).

U.S. Environmental Protection Agency (USEPA). 2006b. *Carbon Sequestration in Agriculture and Forestry*. URL: http://www.epa.gov/sequestration/practices.html (accessed April 6, 2007).

U.S. Environmental Protection Agency (USEPA). 2006c. *A System's Guide to the Identification and Disposal of Hazardous and Non-Hazardous Water Treatment Plant Residuals*. EPA 816-F-06-011. URL: http://www.epa.gov/safewater/arsenic/pdfs/guide_arsenic_disposalhazardous-nonhazardous.pdf (accessed April 6, 2007).

U.S. Environmental Protection Agency (USEPA). 2006d. *Drinking Water Treatment Residual Injection Wells; Technical Recommendations*. URL: https://epa.gov/sites/production/files/2015-08/documents/dwtr_final_report_01-19-07.pdf (Accessed August 25, 2019).

U.S. Environmental Protection Agency (USEPA) 2018a. *Superfund: National Priorities List (NPL)*. URL: https://epa.gov/superfund/superfund-national-priorities-list-npl (accessed August 25, 2019).

U.S. Environmental Protection Agency (USEPA). 2018b. *Superfund Site Reuse/Redevelopment*. URL: https://epa.gov/superfund/superfund-sitwe-reuseredevelopment (accessed August 25, 2019).

U.S. Environmental Protection Agency (USEPA). 2019a. *Potential Well Water Contaminants and Their Impacts*. URL: https://www.epa.gov/privatewells/potential-well-water-contaminants-and-their-impacts (accessed August 24, 2019).

U.S. Environmental Protection Agency (USEPA). 2019b. *Safe Drinking Water Information System, Oracle BI Interactive Dashboard, Government Performance and Results Act (GPRA) Tool, GPRA Inventory*. URL: https://obipublic.epa.gov/analytics/saw/dll?PortalPages (accessed August 28, 2019).

U.S. Geological Survey (USGS). 1998. *Estimated Use of Water in the United States in 1995. Circular* 1200. 71 pp.

U.S. Geological Survey (USGS). 2003 *Desalination of Ground Water-Earth Science Perspectives*. Fact Sheet 075-03. URL: http://pubs.usgs.gov/fs/fs075-03/pdf/AlleyFS.pdf (accessed August 27, 2009).

U.S. Geological Survey (USGS). 2007. *Processes Influencing the Transport and Fate of Contaminated Sediments in the Coastal Ocean—Boston Harbor and Massachusetts Bay*, Bothner, M.H., and Butman, Bradford (eds.) U.S. Geological Survey Circular 1302, 89 p.

U.S. Geological Survey (USGS). 2008. *Summary of Selected U.S. Geological Survey Data on Domestic Well Water Quality for the Centers for Disease Control's National Environmental Public Health Tracking Program*. Scientific Investigations Report 2007– 5213. Prepared by Roy C. Bartholomay, Janet M. Carter, Sharon L. Qi, Paul J. Squillace, and Gary L. Rowe. Reston, VA, 57 p. URL: http://pubs.usgs.gov/sir/2007/5213/ (accessed March 6, 2009).

US Geological Survey (USGS), Michigan Water Science Center. 2017. *Bacteria and Their Effects on Ground-Water Quality*. https://mi.water.usgs.gov/h2oqual/GWBactHOWeb.html

U.S. Geological Survey (USGS). 2018. Estimated Use of Water in the United States in 2015. Circular 1441. URL: pubs.usgs.gov/circ/1441/circ1441.pdf (accessed August 24, 2019)

U.S. National Oceanic and Atmospheric Administration (USNOAA). 2005. *Monterey Bay National Marine Sanctuary: Resource Management Issues: Desalination*. URL: http://montereybay.noaa.gov/resourcepro/resmanissues/desalination.html (accessed August 27, 2009).

U.S. National Center for Atmospheric Research (USNCAR). 2005. Drought's Growing Reach: NCAR Study Points to Global Warming as Key Factor. News Release of January 10, 2005. URL: http://www.ucar.edu/news/releases/2005/drought_research.shtml (accessed April 23, 2007).

Van Dueren, J., Lloyd, T., Chetry, S., Lious, R., and Peck, J. 2002. *Treatment technology profiles. Section 4 in* Remediation Technologies Screening Matrix and Reference Guide, Version 4.0. URL: http://www.frtr.gov/matrix2/section1/toc.html (accessed February 18, 2003).

Voutchkov, N. 2016. *Desalination – Past, Present and Future. International Water Association*. URL: https://iwa-network.org/desalination-past-present-future/ (accessed August 24, 2019).

Walha, K., Ben Amar, R., Firdaous, L., Quemeneur, F., and Jaouen, P. 2007. Brackish groundwater treatment by nanofiltration, reverse osmosis and electrodialysis in Tunisia: Performance and cost comparison. *Desalination*, 207 (1–3), 95–106.

Walton, B. 2013. The Price of Water 2013: Up Nearly 7 Percent in Last Year in 30 Major U.S. Cities; 25 Percent Rise Since 2010. In Circle of Blue, June 5, 2013. URL: The Price of Water 2013: Up Nearly 7 Percent in Last Year in 30 Major U.S. Cities; 25 Percent Rise Since 2010 (accessed August 24, 2019).

Walton, B. 2018. Price of Water 2018: Utilities Revise Household Water Rate Formulas. In Circle of Blue, May 30, 2018. URL: https://www.circleofblue.org/2018/water-management/pricing/price-of-water-2018/ (accessed August 24, 2019).

Walkow, F. et al., 1999. Abstracts; Workshop on the SAFIRA Project, Bitterfield, Germany, 78, November 17–18, 1999, URL: http://www.ufz.de/data/ufzbericht23-00523.pdf (accessed May 19, 2006).

Water Industry News (WIN). 2000. *Tampa Bay water board approves contract for regional surface water treatment plant to US Filter/CDM*. URL: http://waterindustry.org/New%20Projects/usfilter-1.htm (accessed February 11, 2007).

Water Quality Association (WQA). 2016. Cost Benefits of Point-Of-Use Devices In Reduction of Health Risks From Drinking Water. Unpublished. Authors: Marc Verhougstraete, Kelly Reynolds, Akrum Tamimi, and Charles Gerba.

WateReuse Association (WRA). 2012. *Seawater Desalination Costs*; White Paper. URL: https://watereuse.org/wp-content/uploads/2015/10/WateReuse_Desal_Cost_White_Paper.pdf (accessed August 24, 2019).

Water Systems Council (WSC). 2019. *Drinking Water Treatments and Costs (Factsheet) 2019*. URL: https://www.watersystemscouncil.org/download/wellcare_information_sheets/well_water_testing_&_treatment_information_sheets/DrinkingWaterTreatmentsandCostsFINAL.pdf (accessed August 24, 2019)

Younos, T. 2005. The Economics of Desalination, *Journal of Contemporary Water Research and Education*, 132, 39–45.

8 Groundwater Consumption for Health and Food

We can consume groundwater in at least two significant ways – directly through drinking it and indirectly through food produced by using it as an input. Drinking groundwater is obvious if you have a well on your property or used by your community as a water supply. Less obvious to most people, except the farmers who monitor water needed and used for crops, is the consumption of groundwater through food. Irrigated agriculture is the largest use of groundwater in many countries. Groundwater may also be the primary water supply for animals consumed by humans. Groundwater may also become "virtual water" source for locations of water scarcity that import food grown in other regions. Virtual water may also be used for manufacturing other products requiring significant amounts of water for production. First, we will explore groundwater as a drinking water supply for human and animal health and then consider irrigated agriculture for food supplies that require it. The factors and outputs reviewed comprise important costs and benefits of producing and using groundwater for these purposes.

PART 1: GROUNDWATER AND HEALTH

If we consume groundwater, we must be concerned about its consequences for our health and that of other living organisms, as well as in commercial use. Water makes up 60% of the human body's composition and affects bodily hydration, conversion of food to energy, nutrient transfer and waste elimination (Schrecengost, 1998). In many locations, groundwater may be the only source of water supply and, therefore, critical to be maintained (McCabe et al., 1997). As noted previously, all commercial non-drinking water-related (agricultural, industrial and mining) applications account for about 78% of all fresh groundwater uses in the United States (USGS, 2018) and range from approximately 10% to 80% of groundwater uses in the European Union (EU) (EEA, 1999). The remainder is for drinking water.

While essential to the health of all living organisms, groundwater naturally contains minerals and microorganisms. Groundwater may also have chemical, biological and radiological constituents that were derived from human activities. In low concentrations, all of these substances may be harmless and, in some cases, even healthful. In other circumstances, at high concentrations, these constituents may present health threats. In 1980, an estimated 25,000 people around the world died everyday from drinking contaminated water (Danielopol et al., 2003, p. 114). An estimated 0.75–5.9 million illnesses result each year from consuming contaminated groundwater in the United States with 1400–9400 deaths as a consequence (Danielopol et al., 2003, p. 114). In both cases of naturally occurring and human-caused contaminants, governments establish health-based standards to protect people from illness and mortality. Typically, the establishment of these standards attempts to balance the cost of removing the contaminants with the health risk-reduction benefits of mandating the standard. Human health effects may be evaluated in economic terms. Economists usually assess policies having health effects through cost-benefit analysis and evaluate specific health interventions by applying cost-effectiveness analysis (Dickie, 2002). The first section of this chapter briefly highlights the health effects associated with consuming groundwater that may have a range of chemical, radiological and microbiological constituents for further economic examination and presents a production function for drinking water that addresses these substances in water for human consumption.

POSITIVE HEALTH EFFECTS OF NATURAL GROUNDWATER CONSTITUENTS

While myths exist about the curative abilities of springs, one prominent nineteenth-century European doctor classified the continent's mineral springs by chemical content and ailment treatment (Chapelle, 2000). The earliest chemical analyses of groundwaters in the United States occurred prior to 1820 and had several aspects in common: (1) they were done at springs or wells drilled into springs, (2) a commercial interest was associated with the groundwater source, (3) sample test results were given as "hypothetical compounds" and (4) the analyses were completed by medical doctors working as scientists (Davis, 2007). Spring water with purported healing benefits have previously attracted people in the United States, such as the hot springs of Virginia, USA (Figure 1.8.1). Spring houses with cisterns for water collection with no treatment also provided water for farms and communities (Figure 1.8.2). In the twentieth and twenty-first centuries, interest has focused on the negative health effects of contaminants – both natural and human-caused – in groundwater. As indicated in Chapter 2, many of these naturally occurring constituents are metals. Groundwater may also contain naturally occurring organic chemicals and microorganisms (Gibert et al., 1994). The concentrations of many of the natural chemical constituents found in groundwater are low and present minimal constraints on the use of groundwater (Driscoll, 1986, p. 98; note limitations below). Table 1.8.1 shows the ranges of dissolved solids (or minerals) in groundwater, as measured as total dissolved solids (TDS) in parts per million (ppm) and also referred to as "salinity."

Groundwater may contain trace amounts of chemicals that are important for human physiology. Because groundwater can dissolve chemicals in soil and rock, in part due to the length of time it is in contact with those materials, it can have a range of chemicals in solution. The longer

FIGURE 1.8.1 Healing spring, Hot Springs, Virginia.

FIGURE 1.8.2 Spring house along the Tanana River, Alaska.

TABLE 1.8.1
Simple Groundwater Classification Based on Total Dissolved Solids

Category	Total Dissolved Solids (mg/L)	Potential Uses
Freshwater	0–1,000	Drinking water, irrigation, industrial processes, cattle watering
Brackish water	1,000–10,000	Cattle watering (at low range of TDS)
Saline water	10,000–100,000	Raw material source of some chemicals
Brine water	Greater than 100,000	Raw material source of some chemicals

Source: Adapted from Driscoll, F.G., *Groundwater and Wells,* Johnson Screens, St. Paul, MN, 1986, 1089, p. 75. With permission

the groundwater is in the subsurface, the more mineralized it could become (Alley, 1993, p. 131). Chapelle (2000, citing Linn, 1893) notes the historical healthful aspects of the following chemicals known to be present in springs:

- Sulfur (dissolved hydrogen sulfide) – treatment of chronic rheumatism, gout, tuberculosis, paralysis and chronic bronchitis
- High-iron – treatment of anemia
- Iodine – treatment of goiter (thyroid gland enlargement)

Additionally, some groundwaters may have natural fluoride that helps prevent cavities in teeth. Groundwaters that are highly mineralized from calcium and magnesium are often referred to as "hard water," leaving pipes and taps encrusted with a hard residue. Calcium is an essential mineral for bones and muscle function. Table 1.8.2 lists the dissolved inorganic constituents that may be found in groundwater and indicates those which are considered essential minerals in human beings. In an economics context, the value of these dissolved minerals found in groundwater and

TABLE 1.8.2
Dissolved Inorganic Constituents in Groundwater and Essential Minerals

Dissolved Inorganic Constituent in Groundwater	Essential Mineral for Biological Processes	Principal Health Significance
Major Constituents (greater than 5 mg/L)		
Bicarbonate		
Calcium	Yes	Maintains bone health, blood pressure, reduce colon cancer and manage weight
Chloride	Yes	Maintains electrolyte balance with sodium or potassium
Magnesium	Yes	May have role in reducing chronic disease and hypertension
Silicon		Present in skin, fingernails, bones, connective tissues
Sodium	Yes	Helps regulate water balance and the distribution of fluids
Sulfate		
Minor Constituents (0.01–10.0 mg/L)		
Boron	Yes	Important for growth and health; reduces calcium loss from bones
Carbonate		
Fluoride		Tooth decay reduction
Iron	Yes	Main carrier of oxygen to all cells
Nitrate		
Potassium	Yes	Regulates water balance within the body
Strontium		Can replace calcium
Trace Constituents (less than 0.1 mg/L)		
Aluminum		
Antimony		
Arsenic		
Barium		
Beryllium		
Bismuth		
Bromide		
Cadmium		
Cerium		
Cesium		
Chromium	Yes	With insulin, removes glucose from blood; involved in fat metabolism
Cobalt	Yes	Part of Vitamin B12; red blood and other cell function
Copper	Yes	Important for growth, connective tissues and iron use
Gallium		
Germanium	Yes	Antioxidant and oxygen regulator
Gold		
Indium		
Iodide	Yes	Thyroid gland development and function controlling metabolism
Lanthanum		
Lead		
Lithium		Plays a key role in treating clinical depression
Manganese	Yes	Important for bone structure and metabolism
Molybdenum	Yes	Aids in final stages of making urine
Nickel	Yes	Factor in hormone, lipid and membrane metabolism
Niobium		

(Continued)

TABLE 1.8.2 (*Continued*)
Dissolved Inorganic Constituents in Groundwater and Essential Minerals

Dissolved Inorganic Constituent in Groundwater	Essential Mineral for Biological Processes	Principal Health Significance
Phosphate		
Platinum		
Radium		
Rubidium		
Ruthenium		
Scandium		
Selenium	Yes	Maximizes antioxidant enzyme
Silver	Yes	May be used as an external bactericide and disinfectant
Thallium		
Thorium		
Tin	Yes	Deficiency may be associated with hair loss, anorexia and acne
Titanium		
Tungsten		
Uranium		
Vanadium		
Ytterbium		
Yttrium		
Zinc	Yes	Important for growth and development and may improve immune system function in elderly
Zirconium		

Sources: Driscoll, F.G., *Groundwater and Wells*, Johnson Screens, St. Paul, MN, 1986, 1089.

Davis, S.N. and DeWiest, R.J.M., *Hydrogeology*, John Wiley & Sons, Inc., New York, 1966, 463.

Ryan-Harshman, M. and Aldoori, W., *Can. Fam. Physician*, 51 (5), 673–675, 2005, URL: http://www.pubmedcentral.nih.gov/articlerender.fcgi?artid=1472933 (accessed November 30, 2008).

RuralTech Services, Organic Facts, 2006, URL: http://www.organicfacts.net/health-benefits/minerals/ (accessed November 30, 2008).

Lenntech, B.V., Periodic Chart of Elements, 2008, URL: http://www.lenntech.com/Periodic-chart.htm (accessed November 30, 2008).

the groundwater in which they occur could be evaluated in a number of ways, including least-cost alternative means to deliver these minerals to people needing them, such as vitamin supplements, and to provide water from nearby sources if groundwater was in short supply or considered no longer locally available.

NEGATIVE HEALTH EFFECTS OF NATURAL AND ANTHROPOGENIC SUBSTANCES

Chemicals in groundwater may also have negative health effects. These chemicals can be natural occurring, meaning that they have been dissolved from soil and rock by groundwater and are carried by it and occur at high concentrations. They may also be human-caused or anthropogenic. Negative health effects (illness and death) can occur sooner in the case of contaminants with acute human responses, such as certain microorganisms, or later for those with a latency period before manifesting responses (Cropper and Sussman, 1990) from lower concentration long-term doses of inorganic or organic chemicals.

Naturally Occurring Contaminants

The soil and rock through which water in the subsurface flows consist of natural-occurring chemicals and compounds. Water dissolves these natural constituents and carries them as it flows in the ground and through aquifers. Chemicals occurring naturally in groundwater at high concentrations may be considered contaminants that constrain the use of groundwater. Some contaminants that occur naturally may also have human-caused sources. The United States Environmental Protection Agency (USEPA), the EU and the World Health Organization (WHO) have set health-based maximum concentration levels for contaminants in drinking water, which are often applied to groundwater when it is a source of drinking water. High concentrations of these naturally occurring contaminants may have adverse health effects. Table 1.8.3 provides the drinking water standards for these contaminants in the United States and potential health effects. Typically, the USEPA uses a risk range of 10^{-4} to 10^{-6} to establish these standards, as well as considers the capability and cost of technology to remove the contaminants. Box 1.8.1 summarizes the results of the process in the United States for setting a standard for arsenic.

Microorganisms also live in groundwater and in sufficiently high concentrations can cause illness and even death. Some organisms may occur naturally in groundwater (Table 1.8.7). Wildlife excrement can be one source for such organisms with a second source being domestic animal fecal matter. Environmental and health organizations have set levels of the organisms that can be typically tolerated in drinking water without serious health consequences. Table 1.8.3 provides one set of standards for these microorganisms and the associated potential health effects.

The average annual household costs categorized by system size are shown in Table 1.8.5. The average annual household costs for the homes served by the approximately 2,387 CWSs that require treatment are expected to be approximately \$32 per year. The disparity in household costs between system sizes is due to economies of scale. Larger systems are able to spread the costs they incur over a larger customer base.

The estimated average annual costs for CWSs categorized by system size, which exceed the final MCL of 10μg/L and are required to treat, are shown in Table 1.8.6.

Human-Caused or Anthropogenic Source Contaminants

Many contaminants of concern in groundwater are human caused or anthropogenic due to chemical, radiological, or biological production and use. These contaminants exist because we have manufactured and used chemicals in human activities from household maintenance and production processes to agricultural pesticide control and fertilization. Some chemicals have degraded into other chemicals. We have concentrated natural-occurring elements (such as uranium and its degradation products) through our uses or treatment of water. We have disposed of chemical, radiological and biological wastes. The U.S. Environmental Protection Agency reported that of the 86,073 chemicals it has on its chemical inventory, 42,037 chemicals were reported as actively used as of 2018 (USEPA, 2019b; Wiley Rein, 2019). All of these activities may result in these constituents entering the subsurface by direct placement (e.g., landfilling, well injection or storm drain) or by being carried through the soil and rock to the groundwater table by percolation and infiltration of water that does not runoff the ground surface. Some of these human-caused contaminants may be harmful to the health of people and animals. Clearly, pesticides are specifically produced to kill agricultural, house and garden pests but are designed to be used in small quantities that will not be injurious to human beings if indirectly consumed in low concentrations. Environmental protection agencies have set standards for such contaminants in drinking water and these have been applied to remediation of groundwater where it is used for water supply, such as at abandoned hazardous waste and chemical spill sites. Table 1.8.8 provides the drinking water standards used in the United States for these anthropogenic contaminants.

TABLE 1.8.3
Drinking Water Standards for Contaminants That May Be Naturally Occurring in Groundwater

Under the Safe Drinking Water Act (SDWA), the U.S. Environmental Protection sets health-based "maximum contaminant levels" (MCLs) and "treatment technique" standards for contaminants in water used for drinking supplied by public water systems. The standards below are current as of 2008 for inorganic and radionuclide constituents that may be in groundwater. Note that not all of them will be in all groundwaters. The implementing regulations require that the water be treated or otherwise meet the standards before being delivered to consumers. In setting the standards, the law (SDWA) requires the agency to balance the costs of implementing the standards with the health risk-reduction benefits, which may be denominated in dollars or illness and death averted. The EU and the WHO set similar standards, which may be different for some contaminants.

Inorganic Chemicals That Are Non-Radiologic

Inorganic Chemicals	MCLG[a] (mg/L)[b]	MCL or TT[a] (mg/L)[b]	Potential Health Effects from Ingestion of Water	Sources of Contaminants in Drinking Water
Antimony	0.006	0.006	Increase in blood cholesterol; decrease in blood sugar	Discharge from petroleum refineries; fire retardants; ceramics; electronics; solder
Arsenic	0[c]	0.010 as of 01/23/06	Skin damage or problems with circulatory systems, and may increase the risk of getting cancer	Erosion of natural deposits; runoff from orchards, runoff from glass and electronics production wastes
Asbestos (fiber >10 μm)	7 million fibers per liter	7 MFL	Increased risk of developing benign intestinal polyps	Decay of asbestos cement in water mains; erosion of natural deposits
Barium	2	2	Increase in blood pressure	Discharge of drilling wastes; discharge from metal refineries; erosion of natural deposits
Beryllium	0.004	0.004	Intestinal lesions	Discharge from metal refineries and coal-burning factories; discharge from electrical, aerospace and defense industries
Cadmium	0.005	0.005	Kidney damage	Corrosion of galvanized pipes; erosion of natural deposits; discharge from metal refineries; runoff from waste batteries and paints
Chromium (total)	0.1	0.1	Allergic dermatitis	Discharge from steel and pulp mills; erosion of natural deposits
Copper	1.3	TT[d] action level = 1.3	Short-term exposure: gastrointestinal distress Long-term exposure: liver or kidney damage People with Wilson's disease should consult their personal doctor if the amount of copper in their water exceeds the action level	Corrosion of household plumbing systems; erosion of natural deposits

(Continued)

TABLE 1.8.3 (*Continued*)
Drinking Water Standards for Contaminants That May Be Naturally Occurring in Groundwater

Inorganic Chemicals	MCLG[a] (mg/L)[b]	MCL or TT[a] (mg/L)[b]	Potential Health Effects from Ingestion of Water	Sources of Contaminants in Drinking Water
Cyanide (as free cyanide)	0.2	0.2	Nerve damage or thyroid problems	Discharge from steel/ metal factories; discharge from plastic and fertilizer factories
Fluoride	4.0	4.0	Bone disease (pain and tenderness of the bones); Children may get mottled teeth	Water additive which promotes strong teeth; erosion of natural deposits; discharge from fertilizer and aluminum factories
Lead	0	TT[d] action level = 0.015	Infants and children: Delays in physical or mental development; children could show slight deficits in attention span and learning abilities Adults: Kidney problems; high blood pressure	Corrosion of household plumbing systems; erosion of natural deposits
Mercury (inorganic)	0.002	0.002	Kidney damage	Erosion of natural deposits; discharge from refineries and factories; runoff from landfills and croplands
Nitrate (measured as nitrogen)	10	10	Infants below the age of 6 months who drink water containing nitrate in excess of the MCL could become seriously ill and, if untreated, may die. Symptoms include shortness of breath and blue-baby syndrome	Runoff from fertilizer use; leaching from septic tanks, sewage; erosion of natural deposits
Nitrite (measured as nitrogen)	1	1	Infants below the age of 6 months who drink water containing nitrite in excess of the MCL could become seriously ill and, if untreated, may die. Symptoms include shortness of breath and blue-baby syndrome	Runoff from fertilizer use; leaching from septic tanks, sewage; erosion of natural deposits
Selenium	0.05	0.05	Hair or fingernail loss; numbness in fingers or toes; circulatory problems	Discharge from petroleum refineries; erosion of natural deposits; discharge from mines
Thallium	0.0005	0.002	Hair loss; changes in blood; kidney, intestine, or liver problems	Leaching from ore-processing sites; discharge from electronics, glass and drug factories
Alpha particles	Non-Zero[c]	15 picocuries per liter (pCi/L)	Increased risk of cancer	Erosion of natural deposits of certain minerals that are radioactive and may emit a form of radiation known as alpha radiation

(*Continued*)

TABLE 1.8.3 (*Continued*)
Drinking Water Standards for Contaminants That May Be Naturally Occurring in Groundwater

Radiological Constituents

Radionuclides	MCLG[a] (mg/L)[b]	MCL or TT[a] (mg/L)[b]	Potential Health Effects from Ingestion of Water	Sources of Contaminants in Drinking Water
Beta particles and photon emitters	Non-Zero[c]	4 millirems per year	Increased risk of cancer	Decay of natural and man-made deposits of certain minerals that are radioactive and may emit forms of radiation known as photons and beta radiation
Radium 226 and radium 228 (combined)	Non-Zero[c]	5 pCi/L	Increased risk of cancer	Erosion of natural deposits
Uranium	Zero	30 mg/L as of 12/08/03	Increased risk of cancer, kidney toxicity	Erosion of natural deposits

Additionally, the USEPA has established standards for drinking water that address aesthetic qualities (taste, odor or appearance) for constituents that may naturally occur in drinking water but are not considered threats to human health.

Drinking Water Contaminant Secondary Standards

Aluminum: 0.05–0.2 mg/L	Fluoride: 2.0 mg/L	pH: 6.5–8.5
Color: 15 (color units)	Iron: 0.3 mg/L	Sulfate: 250 mg/L
Copper: 1.0 mg/L	Manganese: 0.05 mg/L	Total dissolved solids: 500 mg/L
Corrosivity: Noncorrosive	Odor: 3 threshold odor number	Zinc: 5 mg/L

Source: U.S. Environmental Protection Agency (USEPA). 2018. National Primary Drinking Water Regulations. URL: https://www.epa.gov/ground-water-and-drinking-water/national-primary-drinking-water-regulations (Accessed July 17, 2019).

[a] Maximum contaminant level (MCL): The highest level of a contaminant that is allowed in drinking water. MCLs are set as close to MCLGs as feasible using the best available treatment technology and taking cost into consideration. MCLs are enforceable standards. Maximum contaminant level goal (MCLG): The level of a contaminant in drinking water, below which there is no known or expected risk to health. MCLGs allow for a margin of safety and are nonenforceable public health goals; Treatment technique: A required process intended to reduce the level of a contaminant in drinking water.

[b] Units are in milligrams per liter (mg/L) unless otherwise noted. Milligrams per liter are equivalent to ppm.

[c] MCLGs were not established before the 1986 Amendments to the SDWA. Therefore, there is no MCLG for this contaminant.

[d] Lead and copper are regulated by a treatment technique that requires systems to control the corrosiveness of their water. If more than 10% of tap water samples exceed the action level, water systems must take additional steps. For copper, the action level is 1.3 mg/L, and for lead is 0.015 mg/L.

BOX 1.8.1 SETTING A HEALTH-BASED STANDARD FOR ARSENIC

Under Section 109 of the SDWA as amended in 1996, the United States Congress required the U.S. Environmental Protection Agency (EPA) to set a new standard for arsenic in drinking water by 2001. A 1999 report by the National Academy of Sciences concluded that the then existing standard of 50 parts per billion (ppb) did not achieve EPA's goal of protecting public health and should be lowered as soon as possible. While the standard EPA issued on January 22, 2001, applies to most public water systems (except those for transient use), those systems most affected are small groundwater-supplied systems. Arsenic is a human health concern, because it can contribute to bladder, skin and other cancers, as well as heart disease and diabetes. Arsenic is a naturally occurring element in rocks and soils and the water in contact with them releases arsenic to water supplies. When people either drink this water or eat animals that drink it or plants that take it up, they are exposed to arsenic. For most people in the United States, eating and drinking are the most common ways that expose people to arsenic, although it can also come from industrial sources. To protect human health, in January 2001, EPA revised the standard for arsenic in drinking water from 50ppb to 10ppb to be implemented by 2006, unless water systems obtain additional time to comply through an exemption.

Sources of Arsenic Contamination in Water

- Naturally occurs in rocks and soil, water, air and plants and animals.
- Released also by volcanic action, erosion of rocks, forest fires or through human actions.
- About 90% of industrial arsenic in the United States is currently used as a wood preservative.
- Also used in paints, dyes, metals, drugs, soaps and semiconductors.
- Also released by agricultural applications, mining and smelting.

Analysis of Arsenic Occurrence in Drinking Waters

In the late 1990s, the U.S. Geological Survey sampled 18,864 locations with wells, of which 2,262 were public water supply sources. For example, see Figure 1.8.3.

- Includes 1,528 counties with sufficient data included 76% of all large public water supply systems (serving more than 10,000 people) and 61% of all small public water supply systems (serving more than 1,000 and less than 10,000 people) in the United States.
- Summarized data associated arsenic concentrations in the groundwater resource with public water supply systems using groundwater.
- Exceeded targeted arsenic concentrations of 1, 2, 5, 10, 20 and 50 μg/L in the groundwater resource associated with 36%, 25%, 14%, 8%, 3% and 1%, respectively, of all public water supply systems accounted for in the analysis.
- Found higher levels of arsenic in groundwater sources more so than in surface water sources (i.e., lakes and rivers) of drinking water.
- Compared with the rest of the United States, western states have more systems with arsenic levels greater than 10ppb.
- Found parts of the Midwestern United States and New England have some systems whose current arsenic levels are greater than 10ppb, but more systems with arsenic levels that range from 2 to 10ppb.

While many systems may not have detected arsenic in their drinking water above 10ppb, there may be geographic "hot spots" with systems that may have higher levels of arsenic than the predicted occurrence for that area.

(Continued)

BOX 1.8.1 (Continued)

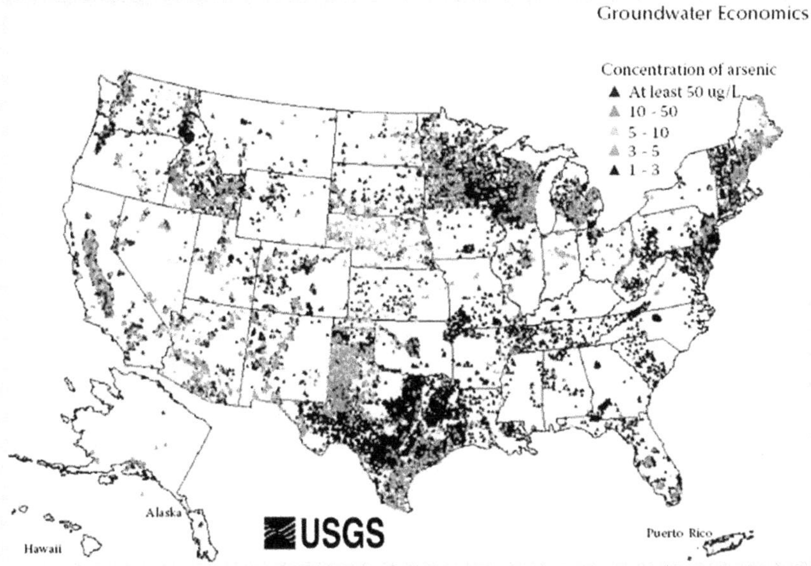

FIGURE 1.8.3 Map of concentrations of arsenic in the United States. (*Source:* U.S. Geological Survey).

Health Effects of Arsenic in Drinking Water

In most drinking water sources, the inorganic form of arsenic tends to be more predominant and can exert toxic effects after acute (short-term) or chronic (long-term) exposure. Although acute exposures to high doses of inorganic arsenic can cause adverse effects, such exposures do not occur from public water systems in compliance with the previous MCL of 50μg/L. The rule addresses the long-term, chronic effects of exposure to low concentrations of inorganic arsenic in drinking water, which studies indicate, include the following:

Cancerous effects: skin, bladder, lung, kidney, nasal passages, liver and prostate cancer; and *Non-cancerous effects:* cardiovascular, pulmonary, immunological, neurological and endocrine (e.g., diabetes) effects.

Regulatory Significance

EPA set the new arsenic standard for drinking water at 10ppb to protect consumers against the effects of long-term, chronic exposure to arsenic in drinking water. After careful consideration of the benefits and costs, EPA used its discretionary authority under 1412(b)(6) of SWDA and decided to set the drinking water standard for arsenic higher than the technically feasible level of 3μg/L, because EPA believed that the costs would not justify the benefits at this level. EPA believes that the final MCL of 10μg/L maximizes health risk reduction at a cost justified by the benefits.

Requirements of the Rule

All community water systems (CWS) (regularly serving 25 or more persons or 15 or more connections) and all nontransient noncommunity water systems (NTNCWS) (such as schools, restaurants, hospitals or factories that have their own water supply) that exceed the MCL of 10μg/L will be required to come into compliance 5 years after the publication of the final rule. Beginning with reports in 2002, all CWSs provide health information and arsenic concentrations in their annual consumer confidence report (CCR) for water that exceeds 1/2 the new MCL.

(Continued)

BOX 1.8.1 (Continued)

Of the 74,000 systems regulated by this MCL, approximately 3,000 systems will have to install treatment or take other steps to comply with this MCL. Water systems were required to meet this standard by January 2006.

People and Systems Affected by the Rule
About 3,000 (or 5.5%) of the nation's 54,000 CWSs and 1,100 (or 5.5%) of the 20,000 NTNCWSs (at the time of the rule promulgation) were to take measures to lower arsenic in their drinking water. Of the affected systems, 97% served less than 10,000 people.

Benefits of Setting the Standard at 10 ppb

Reducing arsenic from 50 to 10ppb will prevent

- 19–31 cases of bladder cancer per year; preventing 5–8 deaths from this cancer
- 19–25 cases of lung cancer, preventing 16–22 deaths from this cancer
- Numerous cases of other non-cancerous diseases such as diabetes and heart disease

Cost of the Regulation
USEPA estimated the total national annualized costs of treatment, monitoring, reporting, recordkeeping and administration for this rule to be approximately $181 million (using 1999 dollars at a 3% discount rate – Table 1.8.4). Most of the cost is due to the cost of installing and operating the treatment technologies needed to reduce arsenic in public water systems (both CWSs and NTNCWS). USEPA estimated the total treatment cost to be approximately $177 million per year. Annual monitoring and administrative costs were estimated to be about $2.7 million and States' costs were estimated to be approximately $1 million.

TABLE 1.8.4
Annual National System and State Compliance Costs (3% Discount Rate, 1999 $ In Millions)

System Costs	CWS	NTNCWS	Total
Treatment	$170	$7.0	$177
Monitoring/administrative	$1.80	$0.90	$2.70
State costs	$0.90	$0.10	$1.00
Total cost	$173	$8	$181

TABLE 1.8.5
Total Annual Costs ($, 1999) per Household for CWS

	Community Water System Size (by persons served)			
	25–500	501–3300	3.3k–10k	10k and above
Annual Household Costs	$327–$162	$71–$58	$38	$32–$0.86

(Continued)

BOX 1.8.1 (Continued)

TABLE 1.8.6
Average Annual Costs per CWS ($, 1999)

CWS System Size	Costs ($)
25–500	$6,494–$12,358
501–3,300	$22,100–$53,086
3,301–10,000	$111,646
10,001 and above	$531,584–$1,340,716

Sources: U.S. Environmental Protection Agency (USEPA), Arsenic Web site, 2001a, URL: http://www.epa.gov/safewater/arsenic.html (accessed December 2001).Job, C., *Ground Water Monitoring and Remediation*, Winter, 40–43, 2002

TABLE 1.8.7
Drinking Water Standards for Microbiological Contaminants That May Occur in Groundwater

Microorganisms	MCLG[a] (mg/L)[b]	MCL or TT[a](mg/L)[b]	Potential Health Effects from Ingestion of Water	Sources of Contaminants in Drinking Water
Cryptosporidium	Zero	TT[c]	Gastrointestinal illness (e.g., diarrhea, vomiting, cramps)	Human and fecal animal waste
Giardia lamblia	Zero	TT[c]	Gastrointestinal illness (e.g., diarrhea, vomiting, cramps)	Human and animal fecal waste
Heterotrophic plate count	N/A	TT[c]	HPC has no health effects; it is an analytic method used to measure the variety of bacteria that are common in water; the lower the concentration of bacteria in drinking water, the better maintained the water system	HPC measures a range of bacteria that are naturally present in the environment
Legionella	Zero	TT[c]	Legionnaire's disease, a type of pneumonia	Found naturally in water; multiplies in heating systems
Total coliforms (including fecal coliform and *Escherichia coli*)	Zero	5.0%[d]	Not a health threat in itself; it is used to indicate whether other potentially harmful bacteria may be present[e]	Coliforms are naturally present in the environment; as well as feces; fecal coliforms and *E. coli* only come from human and animal fecal waste
Turbidity	N/A	TT[c]	Turbidity is a measure of the cloudiness of water. It is used to indicate water quality and filtration effectiveness (e.g., whether disease-causing organisms are present). Higher turbidity levels are often associated with higher levels of disease-causing microorganisms such as viruses, parasites and some bacteria. These organisms can cause symptoms such as nausea, cramps, diarrhea and associated headaches	Soil runoff
Viruses (enteric)	Zero	TT[c]	Gastrointestinal illness (e.g., diarrhea, vomiting, cramps)	Human and animal fecal waste

(Continued)

TABLE 1.8.7 (*Continued*)
Drinking Water Standards for Microbiological Contaminants That May Occur in Groundwater

Source: U.S. Environmental Protection Agency (USEPA). 2018. National Primary Drinking Water Regulations. URL: https://www.epa.gov/ground-water-and-drinking-water/national-primary-drinking-water-regulations (Accessed July 17, 2019).

[a] *Maximum contaminant level (MCL):* The highest level of a contaminant that is allowed in drinking water. MCLs are set as close to MCLGs as feasible using the best available treatment technology and taking cost into consideration. MCLs are enforceable standards. *Maximum contaminant level goal (MCLG):* The level of a contaminant in drinking water, below which there is no known or expected risk to health. MCLGs allow for a margin of safety and are nonenforceable public health goals; *Treatment technique*: A required process intended to reduce the level of a contaminant in drinking water.

[b] Units are in milligrams per liter (mg/L) unless otherwise noted. Milligrams per liter are equivalent to ppm.

[c] EPA's surface water treatment rules require systems using surface water or groundwater under the direct influence *of* surface water to (1) disinfect their water and (2) filter their water or meet criteria for avoiding filtration so that the following contaminants are controlled at the following levels: *Cryptosporidium (as of 1/1/02 for systems serving >10,000 and 1/14/05 for systems serving <10,000) 99% removal.*

Giardia lamblia: 99.9% removal/inactivation.

Viruses: 99.99% removal/inactivation.

Legionella: No limit, but EPA believes that if *Giardia* and viruses are removed/inactivated, *Legionella* will also be controlled.

Turbidity: At no time can turbidity (cloudiness of water) go above 5 nephelolometric turbidity units (NTU); systems that filter must ensure that the turbidity go no higher than 1 NTU (0.5 NTU for conventional or direct filtration) in at least 95% of the daily samples in any month. As of January 1, 2002, turbidity may never exceed 1 NTU, and must not exceed 0.3 NTU in 95% of daily samples in any month.

HPC: No more than 500 bacterial colonies per milliliter.

Long-Term 1 Enhanced Surface Water Treatment (Effective Date: January 14, 2005); Surface water systems or (GWUDI) systems serving fewer than 10,000 people must comply with the applicable Long-Term 1 Enhanced Surface Water Treatment Rule provisions (e.g., turbidity standards, individual filter monitoring, *Cryptosporidium* removal requirements, updated watershed control requirements for unfiltered systems).

Filter Backwash Recycling: The Filter Backwash Recycling Rule requires systems that recycle to return specific recycle flows through all processes of the system's existing conventional or direct filtration system or at an alternate location approved by the state.

[d] More than 5.0% samples total coliform-positive in a month. (For water systems that collect fewer than 40 routine samples per month, no more than one sample can be total coliform-positive per month.) Every sample that has total coliform must be analyzed for either fecal coliforms or *E. coli.* If two consecutive TC-positive samples, and one is also positive for *E. coli* fecal coliforms, system has an acute MCL violation.

[e] Fecal coliform and *E. coli* are bacteria whose presence indicates that the water may be contaminated with human or animal wastes. Disease-causing microbes (pathogens) in these wastes can cause diarrhea, cramps, nausea, headaches or other symptoms. These pathogens may pose a special health risk for infants, young children and people with severely compromised immune systems.

Groundwater containing contaminants from whatever source can influence the quality and use of it for a range of health-related purposes. In the United States, SDWA requires that the USEPA consider the needs of sensitive subpopulations in setting any standards for drinking water contaminants (U.S. Congress, 1996). These sensitive groups may include children, pregnant women, the elderly and immunocompromised individuals. In many situations, groundwater is consumed directly or after treatment for drinking, or it may reach surface waters, then affecting their quality and use. Contaminated groundwater may impact the health of humans and animals when they drink or bathe in it. Furthermore, contaminated groundwater can affect food products, such as meat or vegetables, if it has been consumed or used in the growth or preparation of them (USEPA, 2003b). Since the 1980s, the U.S. Environmental Protection Agency received reports on more than 37,000 abandoned hazardous waste sites in the United States, and 80% of the most seriously contaminated sites involve groundwater contamination (USEPA, 1996a). Box 1.8.2 highlights the result of uncontrolled chemical use and disposal that resulted in contaminating a groundwater resource that in turn was used for drinking water supply and led to serious illness and death in one community.

TABLE 1.8.8
Drinking Water Standards for Anthropogenic Contaminants That May Occur in Groundwater

Organic Chemicals	MCLG[a] (mg/L)[b]	MCL or TT (mg/L)[b]	Potential Health Effects from Ingestion of Water	Sources of Contaminants in Drinking Water
Acrylamide	Zero	TT[c]	Nervous system or blood problems; increased risk of cancer	Added to water during sewage/ wastewater treatment
Alachlor	Zero	0.002	Eye, liver, kidney or spleen problems; anemia; increased risk of cancer	Runoff from herbicide used on row crops
Atrazine	0.003	0.003	Cardiovascular system or reproductive problems	Runoff from herbicide used on row crops
Benzene	Zero	0.005	Anemia; decrease in blood platelets; increased risk of cancer	Discharge from factories; leaching from gas storage tanks and landfills
Benzo(a)pyrene (PAHs)	Zero	0.0002	Reproductive difficulties; increased risk of cancer	Leaching from linings of water storage tanks and distribution lines
Carbofuran	0.04	0.04	Problems with blood, nervous system, or reproductive system	Leaching of soil fumigant used on rice and alfalfa
Carbon tetrachloride	Zero	0.005	Liver problems; increased risk of cancer	Discharge from chemical plants and other industrial activities
Chlordane	Zero	0.002	Liver or nervous system problems; increased risk of cancer	Residue of banned termiticide
Chlorobenzene	0.1	0.1	Liver or kidney problems	Discharge from chemical and agricultural chemical factories
2,4-D	0.07	0.07	Kidney, liver or adrenal gland problems	Runoff from herbicide used on row crops
Dalapon	0.2	0.2	Minor kidney changes	Runoff from herbicide used on rights of way
1,2-Dibromo-3-chloropropane (DBCP)	Zero	0.0002	Reproductive difficulties; increased risk of cancer	Runoff/leaching from soil fumigant used on soybeans, cotton, pineapples and orchards
o-Dichlorobenzene	0.6	0.6	Liver, kidney or circulatory system problems	Discharge from industrial chemical factories
p-Dichlorobenzene	0.075	0.075	Anemia; liver, kidney or spleen damage; changes in blood	Discharge from industrial chemical factories
1,2-Dichloroethane	Zero	0.005	Increased risk of cancer	Discharge from industrial chemical factories
1,1-Dichloroethylene	0.007	0.007	Liver problems	Discharge from industrial chemical factories
cis-1,2-Dichloroethylene	0.07	0.07	Liver problems	Discharge from industrial chemical factories
trans-1,2-Dichloroethylene	0.1	0.1	Liver problems	Discharge from industrial chemical factories
Dichloromethane	Zero	0.005	Liver problems; increased risk of cancer	Discharge from drug and chemical factories
1,2-Dichloropropane	Zero	0.005	Increased risk of cancer	Discharge from industrial chemical factories

(Continued)

TABLE 1.8.8 (*Continued*)
Drinking Water Standards for Anthropogenic Contaminants That May Occur in Groundwater

Organic Chemicals	MCLG[a] (mg/L)[b]	MCL or TT (mg/L)[b]	Potential Health Effects from Ingestion of Water	Sources of Contaminants in Drinking Water
Di(2-ethylhexyl) adipate	0.4	0.4	Weight loss, liver problems or possible reproductive difficulties	Discharge from chemical factories
Di(2-ethylhexyl) phthalate	Zero	0.006	Reproductive difficulties; liver problems; increased risk of cancer	Discharge from rubber and chemical factories
Dinoseb	0.007	0.007	Reproductive difficulties	Runoff from herbicide used on soybeans and vegetables
Dioxin (2,3,7,8-TCDD)	Zero	0.00000003	Reproductive difficulties; increased risk of cancer	Emissions from waste incineration and other combustion; discharge from chemical factories
Diquat	0.02	0.02	Cataracts	Runoff from herbicide use
Endothall	0.1	0.1	Stomach and intestinal problems	Runoff from herbicide use
Endrin	0.002	0.002	Liver problems	Residue of banned insecticide
Epichlorohydrin	Zero	TT[c]	Increased cancer risk, and over a long period of time, stomach problems	Discharge from industrial chemical factories; an impurity of some water treatment chemicals
Ethylbenzene	0.7	0.7	Liver or kidneys problems	Discharge from petroleum refineries
Ethylene dibromide	Zero	0.00005	Problems with liver, stomach, reproductive system or kidneys; increased risk of cancer	Discharge from petroleum refineries
Glyphosate	0.7	0.7	Kidney problems; reproductive difficulties	Runoff from herbicide use
Heptachlor	Zero	0.0004	Liver damage; increased risk of cancer	Residue of banned termiticide
Heptachlor epoxide	Zero	0.0002	Liver damage; increased risk of cancer	Breakdown of heptachlor
Hexachlorobenzene	Zero	0.001	Liver or kidney problems; reproductive difficulties; increased risk of cancer	Discharge from metal refineries and agricultural chemical factories
Hexachlorocyclopentadiene	0.05	0.05	Kidney or stomach problems	Discharge from chemical factories
Lindane	0.0002	0.0002	Liver or kidney problems	Runoff/leaching from insecticide used on cattle, lumber, gardens
Methoxychlor	0.04	0.04	Reproductive difficulties	Runoff/leaching from insecticide used on fruits, vegetables, alfalfa, livestock
Oxamyl (Vydate)	0.2	0.2	Slight nervous system effects	Runoff/leaching from insecticide used on apples, potatoes and tomatoes
Polychlorinated biphenyls (PCBs)	Zero	0.0005	Skin changes; thymus gland problems; immune deficiencies; reproductive or nervous system difficulties; increased risk of cancer	Runoff from landfills; discharge of waste chemicals

(*Continued*)

TABLE 1.8.8 (*Continued*)
Drinking Water Standards for Anthropogenic Contaminants That May Occur in Groundwater

Organic Chemicals	MCLG[a] (mg/L)[b]	MCL or TT (mg/L)[b]	Potential Health Effects from Ingestion of Water	Sources of Contaminants in Drinking Water
Pentachlorophenol	Zero	0.001	Liver or kidney problems; increased cancer risk	Discharge from wood preserving factories
Picloram	0.5	0.5	Liver problems	Herbicide runoff
Simazine	0.004	0.004	Problems with blood	Herbicide runoff
Styrene	0.1	0.1	Liver, kidney or circulatory system problems	Discharge from rubber and plastic factories; leaching from landfills
Tetrachloroethylene	Zero	0.005	Liver problems; increased risk of cancer	Discharge from factories and dry cleaners
Toluene	1	1	Nervous system, kidney or liver problems	Discharge from petroleum factories
Toxaphene	Zero	0.003	Kidney, liver or thyroid problems; increased risk of cancer	Runoff/leaching from insecticide used on cotton and cattle
2,4,5-TP (Silvex)	0.05	0.05	Liver problems	Residue of banned herbicide
1,2,4-Trichlorobenzene	0.07	0.07	Changes in adrenal glands	Discharge from textile finishing factories
1,1,1-Trichloroethane	0.20	0.2	Liver, nervous system or circulatory problems	Discharge from metal degreasing sites and other factories
1,1,2-Trichloroethane	0.003	0.005	Liver, kidney or immune system problems	Discharge from industrial chemical factories
Trichloroethylene	Zero	0.005	Liver problems; increased risk of cancer	Discharge from metal degreasing sites and other factories
Vinyl chloride	Zero	0.002	Increased risk of cancer	Leaching from PVC pipes; discharge from plastic factories
Xylenes (total)	10	10	Nervous system damage	Discharge from petroleum factories; discharge from chemical factories
Disinfection Byproducts	MCLG[a] (mg/L)[b]	MCL or TT1 (mg/L)[b]	Potential Health Effects from Ingestion of Water	Sources of Contaminants in Drinking Water
Bromate	Zero	0.010	Increased risk of cancer	Byproduct of drinking water disinfection
Chlorite	0.8	1.0	Anemia; infants and young children: nervous system effects	Byproduct of drinking water disinfection
Haloacetic acids (HAA5)	n/a[d]	0.060	Increased risk of cancer	Byproduct of drinking water disinfection
Total trihalomethanes (TTHMs)	None[e]	0.10	Liver, kidney or central nervous system problems; increased risk of cancer	Byproduct of drinking water disinfection
	n/a[d]	0.080		
Disinfectants	MRDL[a] (mg/L)[b]	MRDL[a] (mg/L)[b]	Potential Health Effects from Ingestion of Water	Sources of Contaminants in Drinking Water
Chloramines (as Cl_2)	MRDLG = 4[a]	MRDL = 4.0[a]	Eye/nose irritation; stomach discomfort, anemia	Water additive used to control microbes
Organic Chemicals	MCLG[a] (mg/L)[b]	MCL or TT (mg/L)[b]	Potential Health Effects from Ingestion of Water	Sources of Contaminants in Drinking Water

(*Continued*)

TABLE 1.8.8 (*Continued*)
Drinking Water Standards for Anthropogenic Contaminants That May Occur in Groundwater

Organic Chemicals	MCLG[a] (mg/L)[b]	MCL or TT (mg/L)[b]	Potential Health Effects from Ingestion of Water	Sources of Contaminants in Drinking Water
Chlorine (as Cl_2)	MRDLG = 4[a]	MRDL =4.0[a]	Eye/nose irritation; stomach discomfort	Water additive used to control microbes
Chlorine dioxide (as clo2)	MRDLG = 0.8[a]	MRDL = 0.8[a]	Anemia; infants and young children: nervous system effects	Water additive used to control microbes

Source: U.S. Environmental Protection Agency (USEPA). 2018. National Primary Drinking Water Regulations. URL: https://www.epa.gov/ground-water-and-drinking-water/national-primary-drinking-water-regulations (Accessed July 17, 2019).

[a] *Maximum contaminant level (MCL):* The highest level of a contaminant that is allowed in drinking water. MCLs are set as close to MCLGs as feasible using the best available treatment technology and taking cost into consideration. MCLs are enforceable standards. *Maximum contaminant level goal (MCLG):* The level of a contaminant in drinking water, below which there is no known or expected risk to health. MCLGs allow for a margin of safety and are nonenforceable public health goals. *Maximum residual disinfectant level (MRDL):* The highest level of a disinfectant allowed in drinking water. There is convincing evidence that addition of a disinfectant is necessary for control of microbial contaminants. *Maximum residual disinfectant level goal (MRDLG)*: The level of a drinking water disinfectant, below which there is no known or expected risk to health. MRDLGs do not reflect the benefits of the use of disinfectants to control microbial contaminants. *Treatment technique*: A required process intended to reduce the level of a contaminant in drinking water.

[b] Units are in milligrams per liter (mg/L) unless otherwise noted. Milligrams per liter are equivalent to ppm.

[c] Each water system must certify, in writing, to the state (using third-party or manufacturer's certification) that when acrylamide and epichlorohydrin are used in drinking water systems, the combination (or product) of dose and monomer level does not exceed the levels specified, as follows: (i) Acrylamide = 0.05% dosed at 1 mg/L (or equivalent) (ii) Epichlorohydrin = 0.01% dosed at 20 mg/L (or equivalent)

[d] Although there is no collective MCLG for this contaminant group, there are individual MCLGs for some of the individual contaminants: Trihalomethanes: bromodichloromethane (zero); bromoform (zero); dibromochloromethane (0.06 mg/L). Chloroform is regulated with this group but has no MCLG; Haloacetic acids: dichloroacetic acid (zero); trichloroacetic acid (0.3 mg/L). Monochloroacetic acid, bromoacetic acid and dibromoacetic acid are regulated with this group but have no MCLGs.

e MCLGs were not established before the 1986 Amendments to the SDWA. Therefore, there is no MCLG for this contaminant.

Valuing Health Risks

As we will see in subsequent discussions, a key part of economic analysis related to groundwater policies is the valuing in cost terms of human health risk. These costs are usually referred in such analyses as "damages." These costs are sometimes evaluated relative to the amount a group of people may be willing to pay to avoid the risk, usually for illness or death. As noted previously in this chapter, some negative health effects can occur very immediately, while others may take decades to manifest themselves (referred to as a latency period). One analysis indicated that, considering data on lifetime earnings potential and mortality rates, the amount "an individual is willing to pay today to reduce future risk equals the amount the same individual would be willing to pay in the future for a reduction in current risk, discounted back to the present" (Cropper and Sussman, 1990, p. 173). Comparing an individual of age 18 to one of age 60 in that context showed that the younger person who experienced a risk with a latency period had a discount factor that was one-twentieth that of the older person about to experience the effect of an immediate risk. In monetary terms, this implied a willingness by the younger person to pay $93,600 compared with $1.6 million by the elder person (both in 1985 US$). Willingness to pay and discounting will be addressed further Volume 2. If the risk associated with not exceeding a drinking water standard is reduced from six people to two people in one million, then four statistical lives are saved and a distribution of the population by age is used to calculate the value of risk reduction, as one approach in valuing health. Cropper's examination (1990b) suggested that the monetary value associated with willingness to pay to reduce

BOX 1.8.2 INDUSTRIAL CONTAMINATION OF A GROUNDWATER RESOURCE THREATENING A COMMUNITY'S WATER SUPPLY

In the early 1970s, townspeople in eastern Woburn (1990, pop. 35,943), Massachusetts, north of Boston, realized that they were experiencing significant levels of acute lymphocytic leukemia, especially among their children, and other diseases. In 1979, Woburn police investigated 184 barrels of industrial waste left on vacant land just west of the city's two wells in northeast Woburn. The barrels were removed before the contents were tested. However, the state environmental agency required the groundwater from the wells to be tested and reported high levels of trichloroethylene (TCE), 267ppb in one well and 183ppb in the other, as well as four other contaminants, including tetrachloroethylene (commonly referred to as "perc"). The USEPA had listed both of these contaminants as "probable" carcinogens at that time. Trichloroethylene is a solvent used for metal degreasing. Tetrachloroethylene is also a solvent used in industrial processes and dry cleaning. [*Note:* At that time, the USEPA had not yet issued a drinking water standard for either of these contaminants. The current MCLs for both of them have been set at 5ppb.] The wells were ordered closed. The wells had been installed to avert water shortages in the city and supply additional water, but most of the water served neighborhoods in southeastern Woburn. They were turned on and off numerous times after that discovery. Near the wells were the locations of past and active industrial operations. The Centers for Disease Control were eventually called in to evaluate the situation and found 12 families in east Woburn with members suffering from leukemia (some of whom had died), an incidence that was seven times the national occurrence. The trial that ensued documented uncontrolled chemical waste disposal by the industrial operations that contaminated the groundwater supplying the city's wells. The families eventually received an $8 million settlement, with $4.8 million going to trial costs and legal fees. The USEPA conducted a detailed investigation and determined in the early 1990s that its plan for remediating the soil and groundwater would require 50 years at an estimated cost of $69.4 million, making it the most expensive remediation in New England at that time.

Source: Harr, J., *A Civil Action,* Vintage Books, Random House, New York, 1996, 502.

Note: This reference documents the true story of a community, Woburn, Massachusetts, that dealt with the health issues from contaminated groundwater used for public water supply. The summary above was abstracted from that account reported by Jonathan Harr in his book A Civil Action.

the risk of immediate death may be the highest around an age of 40 years. A 1986 review of the "value-of-life" studies suggested a range of $2–$10 million for the value of a statistical life (Fisher et al., 1989). A 2003 review of the value of a statistical life used in U.S. federal policy development ranged from $3 million (1998$US) to $6.2 million (2000$US and based on a review of 26 studies), or, in 2018 dollars, a range of $4.6 million to 9.0 million (Viscusi and Aldy, 2003). Other methods to estimate the value of health-related considerations include evaluating the cost of illness and avoiding contaminated water (Levin and Harrington, 1995). Box 1.8.3 provides an example of estimating the value of a statistical life for policy decision-making.

Barriers Protecting Health

A series of natural and constructed barriers can provide significant health protection from consuming contaminated groundwater, but they are not absolute. Barriers are points that may be used to control the risk to health by minimizing the delivery of contaminants in water to the final user (Fewtrell and Bartram, 2001). The subsurface itself may provide a barrier for some contaminants.

BOX 1.8.3 EXAMPLE CALCULATION OF THE VALUE OF A STATISTICAL LIFE

"[V] alues [of a statistical life] are typically reported in units that match the aggregate dollar amount that a large group of people would be willing to pay for a reduction in their individual risks of dying in a year, such that we would expect one fewer death among the group during that year on average. This is best explained by way of an example. Suppose each person in a sample of 100,000 people were asked how much he or she would be willing to pay for a reduction in their individual risk of dying of 1 in 100,000, or 0.001%, over the next year. Since this reduction in risk would mean that we would expect one fewer death among the sample of 100,000 people over the next year on average, this is sometimes described as "one statistical life saved." Now suppose that the average response to this hypothetical question was $100. Then the total dollar amount that the group would be willing to pay to save one statistical life in a year would be $100 per person × 100,000 people, or $10 million. This is what is meant by the 'value of a statistical life.' Importantly, this is not an estimate of how much money any single individual or group would be willing to pay to prevent the certain death of any particular person."

Source: U.S. Environmental Protection Agency (USEPA). 2018. Mortality Risk Evaluation. URL: https://www.epa.gov/environmental-economics/mortality-risk-valuation (Accessed July 17, 2019).

Recognizing the characteristics of the contaminant to move through the subsurface or be impeded by the subterranean environment first may contribute significantly to health protection by minimizing risk of contamination to the groundwater resource. The value of the subsurface as a barrier may be evaluated as the least-cost alternative that would provide the same protection result or risk reduction. The value may also be determined by health costs avoided through preventing contamination, assuming that the barriers are fully effective, which may not be the case.

Wellhead Protection

Protecting the public health by reducing or eliminating the risk of contaminating the groundwater resource first through prevention is an essential barrier. The countries of the EU since the 1970s and the United States since 1986 have initiated programs referred to as wellhead protection to prevent contamination of groundwater used as a source of drinking water (also referred to as "source water protection" when surface water is included). Wellhead protection entails identifying potential sources of contamination around wells or their zones of recharge and managing them in ways that mitigate the risk of groundwater contamination.

The nature of the contaminants produced, used or disposed of, and in the case of microorganisms, living in, wellhead protection areas can affect groundwater quality and its potential to influence health. For example, contaminants characterized as dense nonaqueous phase liquids (DNAPLs) can sink deep into an aquifer, confounding their removal but leaving shallow wells less affected if the source of the contaminant is removed. Other contaminating substances, some with limited solubility in water, may float on top of the water table. An example is gasoline, which could be released from a leaking underground storage tank or from a spill. Such substances, if not removed from the subsurface, may imperil shallow wells used for water supply. Similarly, some microorganisms are longer-lived and can survive in the subsurface for extended time durations. Many microorganisms live in groundwater environments and some actually help degrade chemical contaminants (Gibert et al., 1994; NRC, 1994), but such activity may take years or decades.

The benefit of wellhead protection is the avoided cost of having to remediate contamination of groundwater after contamination occurs. Contamination of groundwater can cost millions of dollars to clean up. Analysis of 32 selected sites with groundwater contamination from volatile organic chemicals and metals showed average capital costs of $4.9 million for sites using pump-and-treat technology with $3.2 million in average annual operating costs over 6 years (USEPA, 2001a; in 2018$US, $6.96 million in capital costs and $4.54 million in operating costs). An examination of seven communities ranging in population from 700 to 60,000 across the United States that experienced contamination of their groundwater supply and then implemented wellhead protection found the potential savings from implementing wellhead protection significant (USEPA, 1996b). All seven wellfields were affected by volatile organic chemicals from a range of sources: dry cleaning fluids, uncontrolled waste disposal, leaking underground storage tanks and industrial residuals. Remediation costs ranged from $0.5 to $4 million (1994$US; $0.9 to $6.8 million in 2018$US). Secondary costs resulting from contamination were not evaluated, such as loss of fire protection or increased industrial water quality treatment costs. Wellhead protection costs were estimated to be $5,500 for the smallest town to $693,000 for a larger town ($9,300 to $1.2 million in 2018$US). For the seven towns and cities, the average per well contamination remedial cost was $933,000 ($1.6 million in 2018$US) and the average wellhead protection cost was $35,000 ($59,000 in 2018$US) per well. A separate study found a similar level of benefits and costs for wellhead protection (AwwaRF, 2004). The analysis assumed that the implementation of an effective wellhead protection program would result in recharge areas of the municipal wells free of contaminant sources and preclude contamination of a town's groundwater supply and wells.

Protective Aspects of the Subsurface

The prevalence of groundwater (nearly everywhere under the ground surface) and existence of natural processes that may possibly serve as a barrier to some contaminants are factors in groundwater's wide use for water supply (USGS, 2002). Groundwater by its location in the subsurface may have features that allow chemicals to degrade or microorganisms to die-off before they could harm one's use of the water. These subsurface natural capital features include (but may not be limited to)

- Depth – A groundwater table at significant depth may have an unsaturated zone above it that attenuates the flow of contaminants, providing sufficient time for degradation or die-off of some contaminants. Additionally, installing a well that has considerable depth past the groundwater table could provide a similar protection.
- Slow Movement – Some groundwater environments constrain water movement to be very slow so that if sufficient time passes between contaminant entry into the subsurface and the point of use, again degradation or die-off may occur. Not all groundwater environments or points of use provide this factor (see Chapter 2).
- Adsorption – Some soil and subsurface matrices that contain silts and clays provide electrical attraction of certain contaminant ions allowing them to be bound by the soil particles.
- Confinement – A confining strata situated above the aquifer may act as an impervious zone (but not perfectly impervious) to retain contaminants and keep them from moving into the aquifer (Jorgenson et al., 1998; USGS, 2002).

The regulated deep underground injection of liquid wastes relies on confinement and confining strata below aquifers used for water supply to keep contaminants from migrating upward into groundwater in use for drinking water.

All these factors would not necessarily be present at every location of groundwater, nor be needed in each case to provide some qualities of a barrier. These factors may add value to groundwater and its use.

Unprotective Aspects of the Subsurface

Some groundwater environments may not provide much health protection from contaminants. Factors that contribute to the lack of a subsurface barrier include (but are not limited to)

- Shallow groundwater table – If the depth to groundwater is not great, then contaminants may reach the water table before degradation or die-off.

- Rapid movement – Highly porous gravel and sand or fractured bedrock may provide for rapid movement of groundwater with contaminants through the subsurface to the point of use, allowing little possibility of degradation or die-off.
- No or limited adsorption – Porous gravel and sand or fractured bedrock may not provide for the possibility of soil particles to adsorb contaminants that might be electrically attracted to such particles.
- No confinement – If no confining zone exists to constrain contaminant movement, contaminants may be pulled in toward producing wells.

Furthermore, engineered drains or infiltration basins that do not have sufficient soil zones below them may provide pathways for contaminants to reach groundwater with minimal natural treatment (NRC, 1994; Brumley et al., 2018).

Treatment of Groundwater

A final barrier to protect the health of people and animals from contaminants in groundwater is physical, chemical and biological treatment. Groundwater typically requires less treatment because of the natural features of the subsurface that may provide for its quality improvement, as noted above. In many situations, treatment is a last but necessary resort to ensure that the quality of groundwater is healthful. Water treatment is one of the subjects of the previous chapter.

GROUNDWATER SOURCE QUALITY FOR OTHER LIVING ORGANISMS

Other living organisms, animals and plants, depend on water and, in many cases, groundwater for their survival. Groundwater is used for watering livestock and is used in the irrigation of plants for human consumption. The quality of that groundwater may determine the health and longevity of those organisms. Groundwater with high TDS (also referred to as "salinity" or "electrical conductance") when used for livestock watering can result in low performance, illness and even death (University of Nebraska-Lincoln, 2003) (see Table 1.8.9). Similarly, plants exposed on a continuing basis to higher TDS can have low yields and even die (see Table 1.8.10). Other trace elements may also cause added problems. From least sensitive to most sensitive to salinity, plants resistance to TDS in order are forage crops, field crops, vegetables and fruit crops (Texas Agriculture and Mining University, 1998).

TABLE 1.8.9
Water Quality for Animals

PART 1
A Guide to the Use of Saline Water for Livestock and Poultry

Total Dissolved Solids (Parts per Million)[a]	Comments
Less than 1,000	Water for all classes of livestock.
1,000–2,999	Satisfactory for all classes of livestock. Waters near upper limit may result in watery droppings in poultry, not adverse to bird health or production.
3,000–4,999	Satisfactory for livestock but may take time to adapt. High-sulfate salts may result in temporary diarrhea, but no harm. Poor to unsatisfactory water for poultry resulting in watery feces, and possible increased mortality and decreased growth, especially in turkey poults.
5,000–6,999	Use for livestock except pregnant or lactating cows. Some laxative effects. Unsatisfactory for poultry.
7,000–10,000	Poor for livestock and do not use for poultry or swine; use for older, low-producing grazers not pregnant or lactating with reasonable safety.
Over 10,000	Unsatisfactory for all livestock.

[a] Electrical conductivity expressed in micromhos per centimeter at 25°C can be substituted for total dissolved solids without introducing a great error in interpretation.

PART 2
A Guide to the Use of Waters Containing Nitrate for Livestock and Poultry

Nitrate Content[a] (ppm Nitrate Nitrogen)	Comments
Less than 100[b]	Should not harm livestock or poultry.
100–300	Not harmful to livestock or poultry. Combined with nitrate- containing feed, it will be dangerous. Particularly problematic for cattle and sheep in drought years with waters having elevated nitrate.
Over 300[c]	May cause typical nitrate poisoning in cattle and sheep and is not recommended. Use for swine, horses, or poultry is to be avoided.

Source: University of Nebraska-Lincoln, Cooperative Extension, Institute of Agriculture and Natural Resources, *Livestock Water Quality*, 2003, Web site URL: http://www.ianr.unl.edu/pubs/beef/g467.htm (accessed March 16, 2003).

[a] Includes nitrite nitrogen.

[b] Less than 443 ppm of nitrate or less than 607 ppm of sodium nitrate.

[c] Over 1329 ppm of nitrate or over 1821 ppm of sodium nitrate.

TABLE 1.8.10
Water Quality for Plants

Recommended Limits for Constituents in Reclaimed Water for Irrigation

Constituent Name	Long-Term Use (mg/L)	Short-Term Use (mg/L)
Aluminum (Al)	5.0	20
Arsenic (As)	0.10	2.0
Beryllium (Be)	0.10	0.5
Boron (B)	0.75	2.0
Cadmium (Cd)	0.01	0.05
Chromium (Cr)	0.1	1.0
Cobalt (Co)	0.05	5.0
Copper (Cu)	0.2	5.0
Fluoride (F^-)	1.0	15.0
Iron (Fe)	5.0	20.0
Lead (Pb)	5.0	10.0
Lithium (Li)	2.5	2.5
Manganese (Mg)	0.2	1.0
Molybdenum (Mo)	0.01	0.05
Nickel (Ni)	0.2	2.0
Selenium (Se)	0.02	0.02
Vanadium (V)	0.1	1.0
Zinc (Zn)	2.0	10.0

Source: Texas Agricultural and Mining University, Agricultural Extension Service, *Ions, Trace Elements and Other Problems*, 1998, Web site URL: http://agnews.tamu.edu/drought.drghtpak98/drght59.html (accessed March 16, 2003).

Costs of losses from reduced production of animals and plants in agriculture due to deteriorated or inadequate water quality can be evaluated and compared with costs of treating the water to a quality level that provides a sufficient economic return.

GROUNDWATER SOURCE QUALITY FOR COMMERCIAL APPLICATIONS

Water quality requirements for commercial applications vary widely based on the use (Driscoll, 1986, p. 110). Brackish waters can be used for cooling, but disposal may be a problem if concern exists about the environmental effects from concentrating constituents. Some commercial processes

TABLE 1.8.11
Water Quality Tolerances for Selected Commercial Processes

Industry	Turbidity	Color	Hardness as mg/L of $CaCO_3$	Alkalinity as mg/L of $CaCO_3$	Fe + Mn, mg/L	Total Solids, mg/L	Other
Food Products							
Baked goods	10	10	*	—	0.2	—	a
Beer	10	—	—	75–150	0.1	500–1000	a, b
Canned goods	10	—	25–75	—	0.2	—	a
Confectionery	—	—	—	—	0.2	100	a
Ice	5	5	—	30–50	0.2	300	a, c
Laundering	—	—	50	—	0.2	—	—
Manufactured Products							
Leather	20	10–100	50–135	135	0.4	—	—
Paper	5	5	50	—	0.1	200	d
Paper pulp	15–50	10–20	100–180	—	0.1–1.0	200–300	e
Plastics, clear	2	2	—	—	0.2	200	—
Textiles, dyeing	5	5–20	20	—	0.25	—	f
Textiles, general	2	20	20	—	0.5	—	—

Source: Adapted from Driscoll, F.G., *Groundwater and Wells*, Johnson Screens, St. Paul, MN, 1986, 1089, p. 111. With permission.

Note: Indicated values are general averages only; local variance may be considerable.

* Some hardness is desirable. (a) Must conform to standards for potable water; (b) NaCl no more than 275 mg/L; (c) SiO_2 no more than 10 mg/L, Ca and Mg bicarbonates are troublesome; sulfates and chlorides of Na, Ca and Mg each ≤300 mg/L; (d) no slime formation; (e) noncorrosive; and (f) constant composition; residual alumina ≤0.5 mg/L.

have water quality requirements that exceed drinking water standards because of sanitary protection considerations, such as processing for milk, canned goods, meat and beverages (Driscoll, 1986, p. 110). Because of its low constant temperature, groundwater is useful in some processes such as heat exchange (Driscoll, 1986, p. 784). Natural groundwater hardness may be desirable in some industries, such as distilling, baking and brewing, but not in others, as paper manufacture. Table 1.8.11 indicates water quality tolerances for some commercial processes. Costs to treat groundwater to bring its quality into a tolerable range will depend on the natural quality and the desired use in commerce. For many uses, groundwater that meets potable water standards will be sufficient.

GROUNDWATER PRODUCTION FUNCTION AND HISTORICAL USE

Production Function

Processes to produce groundwater for drinking water can be modeled in an equation format. A typical simple production function (also refer Volume 1, Chapter 5) for this purpose might be:

$$Q = f\left(m_1, m_2\right) \tag{1.8.1}$$

where

Q = quantity produced, in this case the amount of water
m_1 = fixed inputs, or K, capital such as the well, pump and motor
m_2 = variable inputs needed for production, such as people's time or labor (L) to manage the pump and electricity to run the well pump and motor, E

The equation might be re-written then as:

$$Q = f\left(K, E, L\right) \tag{1.8.2}$$

Once the relationship of these factor inputs to the production of water is understood, a more specific relationship might be established among the inputs along with the amount of the input for a unit of output of water, such as a cubic meter or an acre-foot. Other factors may be involved in producing groundwater for community water supply and would need to be included in the model and ultimate production, such as pipe to deliver the water to the consumer. Since the well, pump and motor are an investment the cost of which is fixed, the real production is from the application to the investment of variable inputs, people's management time and electricity. As discussed above, treatment may be needed to respond to water quality improvement requirements, so investment may also be needed in equipment for water processing to remove contaminants. Aspects of treatment may also have variable input needs, such as electricity or chemicals that will be used. New technology can enter the relationship through changes in inputs, such as acquisition of improved equipment and changes to labor to use the new equipment. The variable inputs will be needed in different proportions for different amounts of production. The different proportions will also depend on the prices of the inputs, such as the hourly wage and the price of electricity (or other power source). The important point is to consider the range of inputs in relation to the output needed for producing the water for a community.

Municipal Supply

Fifty-seven percent of municipal water supply is used by residential customers (PI, 2015). The price of groundwater supplied to residential customers from public municipal water systems in the United States averages $1.50 per 3,785 liters (1,000 gallons), about 2 days-worth of water for a family of 4 people, or a household cost of approximately $0.75 per day. This price assumes treatment in a significant portion of cases. North Carolina, which publishes the water rates for public water systems ranging in size in 2019 from a service population of 46 to 954,644, has an average residential water rate for customers using around 1,895 liters (500 gallons) per day of $3.00. At the margin, the small public water systems serving 25 to 500 people in North Carolina, 91% of which have groundwater sources (USEPA, 2019b), supply water to customers using the equivalent amount of a family of four at an average rate of $3.44 per day, 15% more than the state average rate for water supplies serving families of four (UNCEFC, 2019).

Domestic Household Supply

Most domestic household water supplies have groundwater sources draw on a well. The household energy cost to pump an average household water use of 300 gallons/day (1,136 liters/day) is just under $0.02 per day (see Box 1.6.5, Energy Costs To Pump Groundwater). This household cost does not include well and pump installation, pipe costs to the house, treatment of the groundwater or pressurizing the household water system. After an initial investment of $1,200 for a water softener (with an expected life of 15–20 years), the daily cost of operating and maintaining a water softening treatment system is approximately $0.50 per day (WQRF, 2016).

Historical Use and Productivity Change

Combined domestic self-supplied and public water system demand for water was 70 million m^3 (18,410 MGD) in 2015. Data from Table 1.8.12 displayed below show that demand for publicly supplied groundwater declined 3.2% from 2010 to 2015 and during this same time domestically self-supplied groundwater demand declined by 9.3%.

These declining demands continue a trend that started in 2000. Per capita water use fell from 378.5 liters/person/day (100 gallons/person/day) in 2000 to 314.2 liters/person/day (83 gpcd) in 2015 (PI, 2015b; Walton, 2017). This declining demand for residential water was influenced by significant deployment of water-saving appliances and fixtures and by drought, particularly in the

TABLE 1.8.12
Historical Change in Publicly-Supplied Groundwater

Groundwater Use	2010[a]	2015[a]	Change[a]	Percent
(metric units)	(Mm^3/d)	(Mm^3/d)	(Mm^3/d)	(%)
Public supply	59	58	(2)	–3.2%
Domestic	13	12	(1)	–9.3%
	2010	**2015**	**Change**	**Percent**
(U.S. units)	(MGD)	(MGD)	(MGD)	(%)
Public supply	15,700	15,200	(500)	–3.2%
Domestic	3,540	3,210	(330)	–9.3%

[a]Numbers in "metric units" rounded.

Sources: U.S. Geological Survey (USGS). 2014. Estimated use of water in the United States in 2010. USGS Circular 1405. Reston, VA. https://pubs.usgs.gov/circ/1405/pdf/circ1405.pdf U.S. Geological Survey (USGS) 2018. Estimated Use of Water in the United States in 2015. USGS Circular 1441. https://pubs.er.usgs.gov/publication/cir1441

TABLE 1.8.13
Residential Water Use Productivity in the United States

Year	Residential Population Served[a]	Residential Water Use MLD[b]	Residential Water Use MGD[c]	Residential Water Use Productivity Per/MLD[d]	Residential Water Use Productivity Per/MGD[e]	Productivity Change in Residential Water Use (percent)
2000	281,421,906	106,370	28,100	2,646	10,015	---
2015	321,410,820	96,490	25,490	3,331	12,610	25.9% (increase)

[a] Total population of the United States for the indicated years.

[b] MLD = million liters per day

[c] MGD = million gallons per day

[d] Per/MLD = Persons [served] per million liters per day

[e] Per/MGD = Persons [served] per million gallons per day

Assumption: 57% of USGS estimated water use from public water systems (both groundwater and surface water combined) is residential water use and all domestic water use (groundwater and surface water combined) is residential water use.

Sources: U.S. Geological Survey (USGS). 2005. Estimated Use of Water in the United States in 2000. Circular 1268. USGS, Reston, VA; USGS 2018. Estimated Use of Water in the United States in 2015. Circular 1441. USGS, Reston, VA.

U.S. Bureau of the Census. 2000. Census 2000. URL: https://www.census.gov/census2000/states/us.html (Accessed August 2, 2019);

U.S. Bureau of the Census. 2015. Annual Estimates of the Resident Population: April 1, 2010 to July 1, 2015. URL: https://factfinder.census.gov/faces/tableservices/jsf/pages/productview.xhtml?src=bkmk (Accessed August 2, 2019).

western United States. Table 1.8.13 shows the change in residential water use productivity from both groundwater and surface water sources from 2000 to 2015. Residential water use productivity is the number of people served by a given quantity of water. From 2000 to 2015, residential water use productivity increased by 25.9%, as more people's residential water needs were served by a unit of water measured by either megaliters per day or million gallons per day. This percentage increase in residential water use productivity coincides with the declining residential water demand.

SUMMARY

Water is essential to the health of human and other living beings, as well as to commercial processes. Groundwater exists under most land surfaces and is widely used for water supply for people and animals, thereby potentially affecting their health. The resource contains naturally occurring minerals, which in low concentrations are healthful, but high doses may be harmful. Groundwater quality may also be affected by anthropogenic contaminants. For both naturally occurring and anthropogenic contaminants, governments have set health-based standards that define MCLs still considered safe for healthful use. Health-related costs from contaminants in groundwater form the basis of evaluating economic damages to communities and of developing health protection measures. The quality of groundwater may be protected by series of barriers, including wellhead protection to manage risks of contamination, natural features of the subsurface that may facilitate degradation or attenuation of chemicals and die-off of microorganisms and treatment as a final step. Wellhead protection and these natural subsurface features may add value to the groundwater in use, the value of which can be evaluated in terms of the next least-cost health risk-reduction barrier or treatment, or the value of health effects avoided, assuming a fully successful barrier.

Animal and plant health must also be considered in using groundwater in agricultural applications. High salinity levels reduce performance and can cause illness and even death. Costs of losses from inadequate water quality can be determined through comparisons to expected production and with costs of water treatment. Costs to commercial applications to treat groundwater to meet their purposes will vary based on natural quality and use.

A production function for water supply can be constructed considering the various inputs and their cost. Inputs include the well(s), pump(s), treatment equipment, people's management time, pipe and power source, among other factors. Average household water supply of 300 gallons/day (1,136 liters/day) is just under $0.02 per day. Households on wells pay about one-third of the amount small communities pay. Demand for water by households has declined since 2000 with the deployment of household water saving devices and drought, especially in the western United States. Correspondingly, the residential water use productivity increased.

FOR FURTHER CONSIDERATION

1. Obtain a "consumer confidence report" for your water system, or a system nearby. What contaminants were found and reported? Check as to whether the water system provided additional treatment and the costs of that treatment.
2. Considering (1) above results, are the contaminants naturally occurring or of anthropogenic source? Does contaminant source affect type of treatment (e.g., at the point of contaminant source or at point of use)?
3. Has your community or one nearby had a drinking water quality emergency from contamination? If so, check with the water system office or local media on what was done to respond to the emergency and what were the costs of responding to the emergency. Does the response and the costs seem appropriate to you? What alternative response could have been done? Would it have cost more or less than what the water system did?
4. What is the residential water use productivity for your community or state or nation? Pull together the numbers for population and divide them by water used for two different years and calculate the productivity change. Is it increasing or decreasing?

REFERENCES

Alley, W.M. 1993. *Regional Ground Water Quality*. Van Nostrand Reinhold, New York, 634 pp.

Awwa Research Foundation (AwwaRF). 2004. *Demonstrating Benefits of Wellhead Protections Programs*. 90p.

Brumley, J., Marks, C., Chau, A., Lowrance, R., Huang, J., Richardson, C., Acree, S., Ross, R., and Beak, D. 2018. *The Influence of Green Infrastructure Practices on Groundwater Quality: The State of the Science*. U.S. Environmental Protection Agency, Washington, DC, EPA/600/R-18/227.

Bureau of the Census. 2015. *Annual Estimates of the Resident Population*: April 1, 2010 to July 1, 2015. URL: https://factfinder.census.gov/faces/tableservices/jsf/pages/productview.xhtml?src=bkmk (accessed August 2, 2019).

Chapelle, F.H. 2000. *The Hidden Sea: Ground Water, Springs and Wells*. National Ground Water Association, Westerville, OH, 232 p.

Cropper, M.L. and Sussman, F.G. 1990b. Valuing future risks to life. *Journal of Environmental Economics and Management*, 19, 160–174.

Danielopol, D.L., Griebler, C., Gunatilaka, A., and Notenboom, J. 2003. Present state and future prospects for groundwater ecosystems. *Environmental Conservation*, 30 (2), 104–130.

Davis, S.N. and DeWiest, R.J.M. 1966. *Hydrogeology*. John Wiley & Sons, Inc., New York, 463 p.

Davis, S.N. 2007. Early American Medical Doctors Who Studied Ground Water. In *Ground Water (journal)*, Vol. 45, No. 4, pp. 519–522.

Dickie, M. 2002. Workshop at University of Central Florida. March 18–19, 2002. URL: http://www.bus.ucf.edu/mdickie/Health% 20Workshop/Workshop.htm

Driscoll, F.G. 1986. *Groundwater and Wells*. Johnson Screens, St. Paul, MN, 1089 pp.

European Environment Agency (EEA). 1999. *Groundwater Quality and Quantity in Europe*. EEA, Luxembourg, 123 pp.

Fewtrell, L. and Bartram, J. (eds.). 2001. *Water Quality; Guidelines, Standards and Health: Assessment of Risk and Risk Management for Water-Related Infectious Disease*. World Health Organization Water Series. IWA Publishing, London, U.K.

Fisher, A., Chestnut, L.G., and Violette, D.M. 1989. The value of reducing risks of death: A note on new evidence. *Journal of Policy Analysis Management*, 8 (1), 88–100.

Gibert, J., Danielpol, D.L., and Stanford, J.A. (eds.). 1994. *Groundwater Ecology*. Academic Press, San Diego, CA.

Harr, J. 1996. *A Civil Action*. Vintage Books, Random House, New York, 502 pp.

Job, C. 2002. *Regulation of arsenic in drinking water affects ground water systems. Ground Water Monitoring and Review*, Winter, 40–43.

Jorgenson, D., Wireman, M., and Olson, D. 1998. *Assessing the vulnerability of public water supply wells to microbial contamination. Ground Water Monitoring and Review*, Spring, 60–66.

Lenntech, B.V. 2008. Periodic Chart of Elements. URL: http://www.lenntech.com/Periodic-chart.htm (accessed November 30, 2008).

Levin, R. and Harrington, W. 1995b. *The Dilemma of Disinfected Water: Infectious Waterborne Disease and Cancer*, A Quantitative Analysis. Unpublished.

Linn, T. 1893. *The Health Resorts of Europe*. D. Appleton and Co, New York, 330 pp. Cited in Chapelle, F.H. 2000. The Hidden Sea: Ground Water, Springs and Wells. National Ground Water Association, Westerville, OH, 232 pp.

McCabe, W.J., Charles A.J., John J.S., John S.G., and Calvin J.T. 1997. History of the sole source aquifer program: A community-based approach for protecting aquifers used for drinking water supply. *Ground Water Monitoring and Remediation*, 17 (3), 78–86.

National Research Council (NRC) 1994. *Ground Water Recharge Using Waters of Impaired Quality*. National Academy Press, Washington, DC. 283p.

Pacific Institute (PI). 2015b. *Water Use Trends in the United States*. Prepared by Kristina Donnelly and Heather Cooley. URL: https://pacinst.org/wp-content/uploads/2015/04/Water-Use-Trends-Report.pdf (accessed August 1, 2019).

RuralTech Services. 2006. *Organic Facts*. URL: http://www.organicfacts.net/health-benefits/minerals/ (accessed November 30, 2008).

Ryan-Harshman, M. and Aldoori, W. 2005. Health benefits of selected minerals. *Canadian Family Physician*, 51 (5), 673–675. URL http://www.pubmedcentral.nih.gov/articlerender.fcgi?artid=1472933 (accessed November 30, 2008).

Saxena, J. 1996. Personal communication, from scientist with the U.S. Environmental Protection Agency, Office of Ground Water and Drinking Water, December 3, 1996.

Schrecengost, S. 1998. Drinking Eight Glasses of Water a Day (unpublished paper, Vanderbilt University, November 13, 1998). Web site URL: http://www.vanderbilt.edu/AnS/psychology/health_psychology/drinking_8glasses.html (accessed March 1, 2003).

Texas Agricultural and Mining University, Agricultural Extension Service. 1998. *Ions, Trace Elements and Other Problems*. Web site URL: http://agnews.tamu.edu/drought.drghtpak98/drght59.html (accessed March 16, 2003).

University of Nebraska-Lincoln, Cooperative Extension, Institute of Agriculture and Natural Resources. 2003. *Livestock Water Quality*. Web site URL: http://www.ianr.unl.edu/pubs/beef/g467.htm (accessed March 16, 2003).

U.S. Bureau of the Census. 2000. Census 2000. URL: https://www.census.gov/census2000/states/us.html accessed (August 2, 2019); U.S.
U.S. Congress. 1996. *Amendments to the Safe Drinking Water Act.*
U.S. Environmental Protection Agency (USEPA). 1996a. *Ground Water Cleanup at Superfund Sites*. EPA 540-K-96 008. December 1996. URL: http://pubweb.epa.gov/superfund/health/conmedia/gwdocs/brochure.htm (accessed September 21, 2008).
U.S. Environmental Protection Agency (USEPA). 1996b. *Benefits and Costs of Prevention: Case Studies of Community Wellhead Protection-Volume 1*. Washington, DC. EPA 813-B-95-005.
U.S. Environmental Protection Agency (USEPA). 2001a. *Arsenic Web site*. URL: http://www.epa.gov/safewater/arsenic.html (accessed December 2001).
U.S. Environmental Protection Agency (USEPA). 2001b. Cost Analysis for Selected Groundwater Cleanup Projects: Pump and Treat Systems and Permeable Reactive Barriers. EPA 542-R-00-013.
U.S. Environmental Protection Agency (USEPA). 2003. *Drinking Water Standards Web site*. URL: http://www.epa.gov/ogwdw/creg.html (accessed March 1, 2003).
U.S. Environmental Protection Agency (USEPA). 2018. *National Primary Drinking Water Regulations*. URL: https://www.epa.gov/ground-water-and-drinking-water/national-primary-drinking-water-regulations (accessed July 17, 2019).
U.S. Environmental Protection Agency (USEPA) 2019a. *[Drinking Water] Government Performance and Results Act (GPRA) Tool*. URL: https://www.epa.gov/ground-water-and-drinking-water/drinking-water-performance-and-results-report (accessed August 1, 2019).
U.S. Environmental Protection Agency (USEPA). 2019b. *TSCA Chemical Substances Inventory*. URL: https://www.epa.gov/tsca-inventory (Accessed July 17, 2019).
U.S. Geological Survey (USGS). 1998. Estimated Use of Water in the United States in 1995. *Circular 1200*. USGS, Reston, VA, 71 pp.
U.S. Geological Survey. 2002. *Assessing Ground-Water Vulnerability to Contamination: Providing Scientifically Defensible Information for Decision Makers. Circular 1224*. USGS, Reston, VA.
University of North Carolina Environmental Finance Center (UNCEFC) 2019. *Water and Wastewater Rate Structures in North Carolina; Residential Rate Table 2*. URL: https://efc.sog.unc.edu/sites/default/files/2019/NCLM_EFC_AnnualW%26WWRatesTablesExcel_2019.pdf (accessed August 1, 2019).
Viscusi, W. K., and Aldy, J.E.. 2003. *The Value of a Statistical Life: A Critical Review of Market Estimates Throughout the World.* Working Paper 9487, National Bureau of Economic Research, Cambridge, MA. URL: https://www.nber.org/papers/w9487.pdf (accessed July 17, 2019).
Walton, B. 2017. U.S. Household Water Use Continues to Decline. In *Water Management, Water News*, November 8, 2017. URL: https://www.circleofblue.org/2017/world/u-s-household-water-use-continues-decline (accessed August 1, 2019).
Water Quality Research Foundation (WQRF). 2016. *Cost Benefits Of Point-Of-Use Devices In Reduction Of Health Risks From Drinking. Prepared by Marc Verhougstraete, Kelly Reynolds, Akrum Tamimi, Charles Gerba. Environment, Exposure Science, and Risk Assessment Center*, University of Arizona, Tucson, Arizona, USA. 52p.
Wiley Rein, L.L.C. 2019. *TSCA Inventory Reset Confirms Half of Listed Substances are in Commerce*. URL: https://www.wileyrein.com/newsroom-articles-EPA-Publishes-Active-Inactive-Compilation-of-TSCA-Inventory-for-Next-Phase-of-Inventory-Reset-Reporting.html (accessed July 17, 2019).

PART 2: GROUNDWATER AND FOOD PRODUCTION

LARGEST USE OF GROUNDWATER

By far the largest use of groundwater is for irrigation to raise food for an expanding global population (USDA ERS, 2019). World population growth rate from 1950 to 2019 has been between 1 and 2% annually (Pew, 2019) with a projected increase from 7.7 billion in 2019 to 9 billion by 2050 (WPR, 2019). During this same period, while food production is forecast to increase at a rate of 0.8% per year during the same period, a need to increase food production by 70% is projected from 2005 to 2050 (UNFAO, 2009). In 2016, the world had an estimated 815 million undernourished people, 99% of whom were in developing or transitional countries (WHO, 2019). During the time period 2002–2030, irrigated land is projected to increase by 20% (40 million hectares) (UNESCO,

2003a, p. 204). In the United States alone, groundwater is the water source for 15.1 million hectares (37.2 million acres) of irrigated agricultural area or 67% of the U.S. irrigated farmland (Arabiyat, 1999; USDA, 2018). Because water is a basic input for food production, the demand for water is a derived demand. As such it responds to the fundamental laws of economics (Tsur et al., 2004, p. 64).

Irrigating crops allows farmers to manage risk to their incomes. Irrigation reduces the probability of low production because soil moisture is below the necessary level to sustain the crop. Irrigating crops involves significant capital investment. The financial return on such an investment must be beneficial to enable farmers to enter into this form of agriculture (Buchanan and Cross, 2006).

Extent of Irrigation Water Demand

As population expands, food demand increases and, therefore, water demand for irrigation to meet this food demand increases. Irrigation is necessary because rain-fed lands do not appear to be capable of providing sufficient food, especially in countries with high population growth rates and some with low rainfall rates (UNESCO, 2003a, pp. 204–205). Worldwide, irrigated farmland covers approximately 3.7 million km^2 (1.4 million mi^2), about 20% of the world's cropland (Meier et al., 2018). Table 1.8.14 highlights the cultivated land area currently having irrigation infrastructure around the world. The greatest irrigation potential exists in the Near East and India, relative to the lands already being irrigated. The United Nations projects that the portion of food grown using irrigation will increase, with developing countries expanding irrigated lands by 20% by 2030, which will also translate into 60% of lands having "irrigation potential" being brought into cultivation

TABLE 1.8.14
Percentage of Cultivated Areas with Irrigation Infrastructure in Place (By Country, 1998)

Greater than 40%	20%–40%	10%–20%	5%–10%
Albania	Afghanistan	Bulgaria	Algeria
Armenia	Bhutan	Denmark	Argentina
Azerbaijan	China	French Guiana	Bolivia
Bangladesh	Columbia	Indonesia	Cambodia
Chile	Costa Rica	Jordan	France
Djibouti	Ecuador	Laos	Guatemala
Egypt	Greece	Moldova	Guinea
Georgia	Guyana	Morocco	Kazakhstan
Iraq	India	Myanmar	Macedonia
Israel	Iran	Norway	Mauritania
Japan	Italy	Philippines	Mongolia
Korea (North and South)	Libya	Somalia	South Africa
Kyrgyzstan	Madagascar	Spain	Switzerland
Netherlands	Mexico	Sudan	Tunisia
Pakistan	Nepal	Turkey	Ukraine
Saudi Arabia	Peru	United States	**Less than 5%**
Surinam	Portugal	Uruguay	Russia and countries not previously listed
Tajikistan	Romania	Venezuela	
Taiwan	Swaziland		
Turkmenistan	Syria		
Uzbekistan	Thailand		
Vietnam	Yemen		

Source: United Nations Educational, Scientific and Cultural Organization (UNESCO), Water for People, Water for Life, The United Nations Water Development Report prepared by the UN World Water Assessment Programme, 2003a, 207

TABLE 1.8.15
Irrigated Area by Continent Utilizing Remote-Sensing-Based Satellite Data, Estimates of Irrigated Areas Worldwide as of 2012 Are Presented for Each Continent or Region in km²

Continent/Region	Irrigated Area (km²)
Africa	163,783
Asia	2,675,125
Europe	269,190
North America	297,365
Central America & Caribbean	89,339
South America	133,743
Australia & New Zealand	41,821
Oceania	41,844
Other Islands	24
TOTAL	3,712,234

Source: Meier, Jonas; Florian Zabel, and Wolfram Mauser. 2018. A global approach to estimate irrigated areas – a comparison between different data and statistics. In Hydrology and Earth Systems Science (journal), 22, 1119–1133. URL: https://doi.org/10.5194/hess-22-1119-2018 (Accessed July 17, 2019).

during that time (UNESCO, 2003a, p. 204). Note that rain-fed agriculture accounts for 60% of food production in developing countries (UNESCO, 2003a, p. 203). Table 1.8.15 gives a 2012 estimate of land area irrigated by continent based on remote-sensing data.

Irrigated food production requires extensive amounts of water. The amount of water a person consumes in 1 day, ideally, is four liters (about 4 quarts or 1.06 gallons), directly or indirectly through other beverages and food with high water content. Water use by plants that provide food requires 2000 L (528 gallons) of water to produce one person's food for 1 day (Brown, 2004, p. 99), because of plant transpiration leading up to a ripe crop for human consumption. Worldwide, groundwater and surface water used for irrigation equals 3500 km³ (840 cubic miles) of water (approximately equal to covering the combined area of Brazil, Uruguay and Argentina with 0.3 m or 12 inches of water) each year (Sundquist, 2006).

Irrigated agriculture may have one of several bases for being used (Southgate, 2000):

- A shift from lower- to higher-value crops.
- Intensification of production to obtain the results of higher yields or to allow production of successive crops over a longer growing season not limited by rainfall.
- Extensification through bringing new land into production.

Intensification and extensification of irrigation has been a significant factor in land and water use. Globally, from 1950 to the late 1970s, land irrigated for agriculture increased almost 3% per year. From 1970 to 1982, this irrigated land expanded by 2% per year. Over the time from 1982 to 1994, this slowed to 1.3% per year. Overall, however, from 1978 to the present, irrigated land on a per person basis declined 5% (Sundquist, 2006). Factors contributing to this per capita reduction of irrigated farmland include aquifer depletion (Brown, 2004) and waterlogging and salination of agricultural land (UNFAO, 2003a), discussed further later. Adding to pressure to irrigate to increase production, lower soil quality has resulted in declines in productivity on greater than 80% of cultivable land globally (WI, 2005). Furthermore, world grain production area per capita is declining, while per capita fertilizer use is growing to address expanding population needs for food with finite land area (Muir, 2005). From an ecological standpoint, the limiting factors of nutrients and water are being applied with a food production focus.

Notably, as world demand for food grows, the source of water supply to meet the need is changing. Over the period from about 1950 to 1975, an estimated 1000 large dams were completed each year. During the first part of 1990s decade, each year saw approximately 260 large dams coming on line. Additionally, these dams on average fill in with sediment at the rate of about 1% annually. This decline in dam construction represents a major factor in the worldwide increase in reliance on groundwater as a source for irrigating farmland (Sundquist, 2006). Because water from rivers in the world's major agricultural zones has become nearly completely utilized, further expansion of irrigation has drawn on groundwater (Brown, 2004, p. 100).

Table 1.8.16 shows that in the drier climatic areas of the western United States, groundwater was the source for 42% of total western irrigation water use in 2015, a drought year for much of that

TABLE 1.8.16
Change in Western United States Groundwater Use for Irrigation with Comparison to Eastern States

State	Use in 2010		Use in 2015		Percent Change
	Hectare-meter[i] (000s)	Acre-feet[ii] (000s)	Hectare-meter (000s)	Acre-feet (000s)	%
Arizona	234	1,900	273	2,210	16.3
California	1,201	9,740	1,912	15,500	59.1
Colorado	178	1,450	180	1,460	0.7
Idaho	528	4,280	677	5,490	28.3
Kansas	398	3,230	354	2,870	(11.1)
Montana	18	142	8	67	(52.8)
Nebraska	595	4,820	749	6,070	25.9
Nevada	90	732	134	1,090	48.9
New Mexico	171	1,390	146	1,180	(15.1)
North Dakota	11	87	14	115	32.2
Oklahoma	59	481	109	882	83.4
Oregon	264	2,140	168	1,360	(36.4)
South Dakota	27	221	19	156	(29.4)
Texas	704	5,710	619	5,020	(12.1)
Utah	67	544	74	602	10.7
Washington	110	894	100	807	(8.6)
Wyoming	60	490	74	602	22.9
TOTALS Western States groundwater irrigation use	4,715	38,251	5,610	45,481	18.9
Notes on irrigation water use for comparison[iii]:					
Western States surface water irrigation	8,478	68,733	7,627	61,838	(10.0)
Eastern States groundwater irrigation	2,115	17,149	2,297	18,619	8.6
Eastern States surface water irrigation	637	5,167	798	6,462	25.1
U.S.groundwater irrigation	6,833	55,400	7,907	64,100	15.7
U.S. surface water irrigation	9,115	73,900	8,425	68,300	(7.6)
U.S. total irrigation	15,912	129,000	16,282	132,000	2.3

Sources: U.S. Geological Survey. 2014. Estimated Use of Water in the United States in 2010. Circular 1405; 2018. Estimated Use of Water in the United States in 2015. Circular 1441.

[i] One hectare-meter equals 10,000 m^3, one hectare covered with one meter of water, or 8.1071 acre-feet.

[ii] One acre-foot, one acre covered with one foot of water, equals 0.1233 hectare-meter.

[iii] Numbers do not add because of rounding.

region. From 2010 to 2015, the arid western US groundwater irrigation use grew 18.9%, with individual states changes in irrigation water use reflecting varying drought conditions and state policies responding to drought. In the more humid eastern United States, groundwater accounts for 74% of eastern irrigation water (USGS, 2018). Most notably, at the margin, groundwater provided the added irrigation water with surface water use in decline. The amount and distribution of irrigated farmland in the United States is described in Box 1.8.4, totaling approximately 22.3 million hectares. Notably, water intensity of U.S. irrigation by all sources has declined from an average of 1.2 meters (4 feet) per unit area in 1950 to 0.6 meter (2.1 feet) in 2010, while area irrigated has increased from 10.4 million hectares (25.6 million acres) to 25.1 million hectares (62 million acres) during the same time (PI, 2015; USDA, 1950).

Biofuels Demand on Groundwater

In 2018, biofuels production of the ten leading producing countries was 82,367 thousand metric tons. The United States, Brazil, Indonesia, Germany, China and France account for nearly 90% of this production (Statistica, 2019). In the United States from 2007 to 2018, cropland for corn for ethanol production increased from nearly 6 million hectares to 14.6 million hectares (14.8 million acres to 36 million acres) (Hart, 2007; USDA, 2018). Soybean production for biodiesel also has increased and even used for ethanol (Vieira, 2017). Wheat and other grasses contribute to biofuel production.

The production of biofuels is forecast to be 89% of growth of water consumption in the U.S. energy sector. In the United States, most of the biofuels are projected to be grown in the West North Central region and increase by 72 million cubic meters per day, or by 2.5 times, between 2005 and 2030. Sixty percent of this increase is for corn-based ethanol production (ANL, 2008). Typical irrigation for corn production uses groundwater at a rate 25% greater than it is naturally recharged in the United States (Pimentel, 2001). Water consumed for biodiesel production is projected to expand

BOX 1.8.4 IRRIGATED FARMLAND IN THE UNITED STATES, 2013

The 2013 U.S. Farm and Ranch Irrigation Survey reported 229,237 farms with 22.4 million hectares (55.3 million acres) irrigated in the United Sates. The western states had 169,436 irrigated farms or 74% of the total irrigated farms. Irrigated cropland comprised 21 million hectares (52 million acres) and included irrigation for grain and oilseed crops, vegetables, nursery and greenhouse, and hay crops. More area was irrigated with sprinkler systems than with gravity/flood irrigation. In 2013, a total of 10.9 million hectare-meters (88.5 million acre-feet) of water was used for farm irrigation. The average water application was 0.2 hectare-meter (1.6 acre-foot) compared with 0.21 hectare-meter (1.7 acre-foot) in the 2008. Farm expenditures included $2.6 billion for irrigation equipment, buildings, land improvements and computer technology. Average cropland value (irrigated and nonirrigated) was $9,415 per hectare ($3,810 per acre).

Sources: Farms.com. 2014. USDA Reports 55.3 Million Acres Of Irrigated U.S. Farmland. URL: https://www.farms.com/news/usda-reports-55-3-million-acres-of-irrigated-u-s-farmland-84011.aspx (Accessed July 23, 2019)

U.S. Department of Agriculture (USDA). 2013 *Farm and Ranch Irrigation Survey*, Washington, DC, URL: https://www.nass.usda.gov/Publications/AgCensus/2012/Online_Resources/Farm_and_Ranch_Irrigation_Survey/ (Accessed July 23, 2019).

U.S. Department of Agriculture (USDA). Land Values 2017 Summary. URL: https://www.usda.gov/nass/PUBS/TODAYRPT/land0817.pdf (Accessed July 23, 2019.)

from about 6 million cubic meters per day in 2005 to about 28 million cubic meters per day in 2030 mainly in the South Central and Pacific regions (ANL, 2008). Thus, expanding corn, soybean and wheat production through irrigation using groundwater and the associated agricultural chemicals applied to achieve high yields will place both quantity and quality demands on the resource.

Brief History of Groundwater Irrigation

Prior to the 1940s, well-drilling technology consisted largely of using augers to create a hole in the ground from which to extract groundwater, but this technique was primarily for shallow wells with walls that were not supported and were subject to collapse. A second method was to hammer a "drive point" into the ground until it reached the groundwater table. However, the diameter of the hole was typically small, limiting the amount of water that could be produced. In the 1940s, borrowing technology from oil well drillers, the irrigation industry began using drill bits with industrial diamonds in the drill bit which could produce a wider hole. This drilling method forced water down the hole either on the outside or through a pipe on the inside through the hollow drill stem to push the drill cuttings up to the surface. Dense clay was used to make a firm wall for the well. The drills were mounted on the back of trucks, rather than using a stationary derrick, and moved to different sites easily. A screened casing is placed in the bottom of the well to allow water to flow into the well but keep larger solid material out. A solid casing is placed above the screened section to pump the water to the ground surface. Gravel is used to fill in around the casing and stabilize it.

During the same time (1940s), irrigators began installing turbine pumps used in the oil industry to literally push the water up the well. Early shallow wells could be pumped by horsepower. Because water is heavy (1001.2 kg/m^3), powerful pumps are necessary to lift a column of water through deep wells, some of which may be hundreds of meters deep. The pumps were submerged in water at the bottom of the well. Engines at the top of the well that were used to power the pump first used steam and then were developed to use gas, oil, diesel, or electricity, depending on what was available (WLHF, 2006).

Recent Developments in Irrigation Wells

Electric pumps have been made smaller and hand pumps are still used with small diameter wells that can be installed by driving a well point into the ground. The small pumps have been effectively used in developing countries where at present millions of wells have been installed in countries such as China and India. Wells now use widely available polyvinyl chloride (PVC or plastic) pipe. Even bamboo has been adapted to well casing in shallow, low-volume wells. With so many wells in use withdrawing large volumes of groundwater, problems have emerged, such as lowering the groundwater table, producing even more brackish water from deeper subterranean zones, as well as causing land subsidence. Box 1.8.5 covers some of these developments around the world.

Irrigation Wells in Use

The number of irrigation wells in use is large. In the United States, 475,796 irrigation wells were capable of use in 2013 (USDA, 2013). Of the wells in use, 132,700 had meters to check the amount of groundwater used and 310,483 had backflow prevention devices to preclude contamination of the wells and their adjacent groundwater. Table 1.8.17 has more detailed information on irrigation wells in the United States by Water Resource Area (large river basins).

Land Area Irrigated

Land area irrigated by groundwater is significant and increasing, as previously noted. Of agricultural lands irrigated, "India has over 50% of its area irrigated from groundwater, followed by" the United

BOX 1.8.5 IRRIGATION WELL DEVELOPMENT EXAMPLES AROUND THE WORLD

India (2004; 2010; 2012)

- 39 million hectares (96 million acres) irrigated with groundwater in 2010
- 67% of irrigated food production from groundwater in 2010
- 10.5 million dug wells and 6.7 million shallow tube wells in 1994
- Shallow tube wells doubled every 3.7 years from 1951 to 1991
- Groundwater irrigation has doubled the yield of surface water irrigation
- Overexploitation in some states has been significant with water tables declining
- Government subsidies for electricity facilitate greater groundwater pumping

Nigeria (2004)

- About 30,000 ha irrigated with tubewell-pump package
- 30,500 pumps distributed to farmers
- Farmer income rose with reduction in poverty
- Economic rate of return in Fadama project estimated at 40%
- Additional benefits identified:
- Development of simplified well-drilling technology
- Farmers trained to help other farmers construct wells
- Infrastructure for transportation and storage of products
- User association established
- Development of groundwater monitoring and evaluation system

China (2004; 2015)

- Groundwater is the major source of irrigation water in Henan, Shandong, Hebei, Gansu and Xinjiang provinces.
- Rapid increase in irrigated areas resulted in overexploitation of groundwater with serious environmental consequences
- Density of tube wells greater than one per 5 ha with average depth to water level increasing from 3.7 to 7.5 m over 30 years
- One-tenth of the wells go dry in summer
- Farmers' use of plastic tube to convey water has reduced water loss, but basin irrigation still inefficient using twice the standard for North China
- Irrigation is 30% of total farm production costs
- Overexploitation of groundwater resulted in declining profitability because of greater water lift costs and poorer quality water
- Salt content of groundwater increasing soil salinity
- Area facing critical groundwater recharge problem and groundwater use is unsustainable

Sources: Siebert, Stefan; J. Burke; J.M. Faures, K. Frenken; J. Hoogeveen; and P. Doll. 2010b. Groundwater use for irrigation – a global inventory, in Hydrology and Earth Systems Science, 14, pp. 1863–1880.

Badiani, Reena and Katrina Jessoe. 2012. The Impact of Electricity Subsidies on Groundwater

(*Continued*)

BOX 1.8.5 (Continued)

Extraction and Agricultural Production. URL: https://economics.ucdavis.edu/events/papers/Jessoe51.pdf (Accessed July 23, 2019)

World Bank, Investments in Shallow Tubewells for Small-Scale Irrigation, Agricultural Investment Sourcebook, 2004, URL: http://www-esd.worldbank.org/ais/index.cfm?Page=mdisp&m=8&p=04 (accessed March 3, 2006).

Hu, Feng. 2015. Groundwater Under Pressure. In China Water Risk. URL: http://www.chinawaterrisk.org/resources/analysis-reviews/groundwater-under-pressure/(Accessed July 23,2019).

TABLE 1.8.17
Irrigation Wells on Farms in the United States, 2013

United States Water Resource Area	Farms	Wells Used	Wells with Meters	Wells with Backflow Prevention
WRA 01 New England	2,324	3,170	143	1,868
WRA 02 Mid-Atlantic	5,215	9,842	1,476	6,791
WRA 03 South Atlantic-Gulf	12,862	37,228	14,255	26,796
WRA 04 Great Lakes	4,506	10,787	1,284	7,994
WRA 05 Ohio	2,076	4,456	362	2,927
WRA 06 Tennessee	472	714	64	426
WRA 07 Upper Mississippi	5,287	15,746	3,193	12,179
WRA 08 Lower Mississippi	6,869	74,385	3,854	27,261
WRA 09 Souris-Red- Rainy	349	1,459	725	1,191
WRA 10 Missouri	20,893	97,219	41,901	74,136
WRA 11 Arkansas- White-Red	7,528	48,391	22,973	35,349
WRA 12 Texas-Gulf	7,508	65,870	12,602	52,498
WRA 13 Rio Grande	4,620	9,999	5,685	4,796
WRA 14 Upper Colorado	622	1,096	127	269
WRA 15 Lower Colorado	2,204	5,335	1,880	2,679
WRA 16 Great Basin	1,882	4,982	1,417	2,778
WRA 17 Pacific Northwest	12,262	24,953	5,256	14,677
WRA 18 California	23,769	59,851	15,425	35,746
WRA 19 Alaska	114	142	10	60
WRA 20 Hawaii	129	171	68	60
Total	121,491	475,796	132,700	310,483

Source: U.S. Department of Agriculture, Economic Research Service (USDA), Census of Agriculture; Farm and Ranch Irrigation Survey, Table 7, Irrigation Wells Used on Farms 2013, Washington, DC, 2013, URL: https://www.nass.usda.gov/Publications/AgCensus/2012/Online_Resources/Farm_and_Ranch_Irrigation_Survey/fris13_1_007_007.pdf (accessed July 23, 2019).

States (49% [counting lands receiving groundwater solely]), "China (27%), and Pakistan (25%). That percentage can reach as much as 80% in developed countries with mild climate (Germany) and in arid countries (Saudi Arabia, Libya)" (UNFAO, 2003c; USDA, 2013). India irrigates 39 million hectares (96 million acres) with groundwater (Siebert et al., 2010). Of the 22,386,974 hectares (55,319,417 acres) irrigated in the United States, 15,054,162 hectares (37,199,644 acres) receive groundwater for irrigation (USDA, 2013).

IRRIGATION WELL COSTS

Irrigation well costs depend on a range of factors, including the principal factors of size and type of well, depth and geology. A shallow PVC tube well of 45 m in South Asia may cost $240, exclusive of the pump, which adds $580 to the total well cost (2004 US$). By comparison, a 15 m (50 foot) well in India of bamboo construction costs $67 and of cast iron construction costs $150 (2004 US$) (Meghalaya, 2004), while in the United States (Minnesota), a well of the same depth costs $2,210 for PVC casing and screen and $4,020 for steel materials (Job and Gabanski, 1988 (1987 US$ adjusted to 2018)) (see Volume 1, Chapter 6 for other well costs). Shallow wells installed in Africa with truck-mounted drill rigs cost approximately $2250 per well (2006 US$) (LWC, 2006). Figures for irrigation wells elsewhere in the world range from $2,900 to $72,900 per well (adjusted to 2018 US$, based on 1989 estimates from Foster, 1989, p. 59), depending on depth, construction, size and application.

For shallow irrigation wells, the pump may be the largest cost. Navarro indicates that for a typical irrigation well and distribution system in Spain, the proportion of costs is (Navarro, 1989, p. 50):

Well bore and casing: 18%
Pump: 28%
Distribution system: 54%.

The pump costs for shallow irrigation wells may be the largest capital cost as suggested in Box 1.8.6, which describes the costs of irrigation tube wells in India as well as financing those costs for the well. Box 1.8.6 gives example costs for shallow bamboo wells, which are about one-third the cost of wells of iron construction. This box also shows that farmer income from irrigation increases by over 450%, a real business and financial incentive to irrigate. Table 1.8.18 compares an example of bamboo and iron tube well costs. In the case cited, an iron-cased well cost more than two times a bamboo well. Table 1.8.18 gives a component breakdown of a typical tube well.

BOX 1.8.6 EXAMPLE OF IRRIGATION TUBE WELL COSTS AND FINANCING

Unit costs for a shallow tube well (STW) and pumpset in the state of Meghalaya, India, are estimated to be (2004 Indian) Rs. 36200.00 (2004 US$ 788) [June 30, 2004 currency conversion: 1 Indian Rupee = 0.02177 US$]. The cost of a pumphouse is optional.

1. Unit Costs of Shallow Tube wells with Pumpset in Meghalaya State, India, 2004

a. Well

Diameter (mm): 100
Depth (m): 45
Type: PVC
Plain pipe: 33 m
Slotted pipe: 12 m
Spacing (if more than one well): 150 m

		Rate		Amount	
Item	**Quantity (m)**	**Rs.**	**US$**	**Rs.**	**US$**
Drilling charges[3]	50	100/m	2.18/m	5000.00	108.85
Plain pipe (100 mm)	33	100/m	2.18/m	3300.00	71.84
Slotted pipe (100 mm)	12	150/m	3.27/m	1800.00	39.19
Pipe accessories and bail plug				400.00	8.70

(Continued)

BOX 1.8.6 (Continued)

Contingency at the rate of 5%	525.00	11.43
Subtotal – well	11,025.00	240.01
b. Pumpset[b]		
Pit for installation of centrifugal pump: 6.5 m depth (RCC rings 1.25 m dia. × 0.6 m height) at the rate of Rs. 800 per m depth	5,200.00	113.20
Diesel pumping system	18,000.00	391.86
Belt, shaft and pulley arrangement	2,000.00	43.54
Subtotal – pumpset	25,200.00	584.60
Total – tube well with pumpset	36,225.00	788.61
c. Optional Structure		
Pump house (2.5 m × 2.5 m × 2.1 m) @ Rs. 10,000 per unit	10,000.00	217.70

[a] Includes development and gravel packing charges.

[b] Details of pumpset are (1) centrifugal pump, discharge (Q) = 2.5 L/s, total head = 16.5 m, efficiency = 50%; (2) HP of diesel engine = 5; and (3) suction pipe = 100 mm, delivery pipe = 100 mm, reflex valve = 100 mm.

2. Project Financing of 300 Shallow Tube wells in Meghalaya, India

Financial Institution

Reserve Bank of India/National Bank for Agriculture and Rural Development (RBI/ NABARD)

Interest Rate for Ultimate Borrowers

Banks are free to decide the rate of interest within the overall RBI/NABARD guidelines. Project assumption is an interest rate of 12% per annum.

Repayment Period

The repayment period of loan for pumpset may be 9 years, and for shallow tube wells, 11–13 years excluding 11 months grace period. For the purpose of the project, 12 years was used.

Security

Banks may follow the RBI guidelines.

Pre- and Postdevelopment Incomes

The annual incremental income for individual farmers having a 2 ha farm is estimated to be Rs. 12,480/year (2004 US$ 272).

Projected Cash Flows

The discounted cash flows result in the following values:

Discount rate: 1.98%

Internal rate of return: 33%

Benefit-cost ratio: 1.78

Installment Payment

Using the standard amortization equation, we can calculate a farmer's installment payment for a shallow tube well and pumpset. The amortization formula is $C = rB/[1 - (1 + r)^{-n}]$, where

C = annual installment (the amount to be calculated),

r = annual interest rate (12%),

B = initial loan balance (Rs. 36,225) and

n = number of years to repay the loan (12 years).

For the program, C, the annual installment, was calculated to be Rs. 5551 (2004 US$ 121). Please see part 3 below.

Part 4 a and b (below) compares estimated farmer income before and after installing tube wells for irrigation during the growing seasons.

(Continued)

BOX 1.8.6 (Continued)

3. Discounted Cash Flow Analysis of Tube wells in Meghalaya

			End of Year				
Specific Item	**Units**	**Year 0**	**Years 1–8 (Repeats Each Year)**	**Year 9**	**Years 10–12 (Repeats Each Year)**	**Years 13–14 (Repeats Each Year)**	**Year 15**
Investment cost	Rs.	36,200					
Other cost (replacement)	Rs.			20,000			
Total cost	Rs.	36,200	0	20,000	0	0	0
Incremental income	Rs.		12,480	12,480	12,480	12,480	12,480
Other income (salvage)	Rs.			2,000			9,215
Total benefit	Rs.	0	12,480	14,480	12,480	12,480	21,695
Net benefit	Rs.	–36,200	12,480	–5,520	12,480	12,480	21,695
NPV of total costs	Rs.	41,885.25					
NPV of total benefits	Rs.	74,676.18					
Benefit cost ratio		1.78					
NPV of net benefits	Rs.	32,790.93					
Internal rate return	%	33%					
Equal annual repayment	Rs.		5,551.81	5,551.81	5,551.81		
Discount rate			2.25	–0.99	2.25		
Average discount rate		1.98					

4. Pre- and Postdevelopment Cropping Pattern and Income – Analysis for 1 ha Farm Plots

(a) Predevelopment

Season	**Crops**	**Area (in ha)**	**Yield (Qtl per ha)**	**Total Yield (Qtl)**	**Price (in Rs. per Qtl)**	**Gross Income (in Rs.)**	**Cost(per ha)**	**Cost Incurred**	**Net Income (in Rs.)**
Kharif	Paddy	0.50	14.00	7.00	550.00	3,850.00	6,340.00	3,170.00	680.00
(April–June)	Maize	0.50	8.00	4.00	500.00	2,000.00	2,630.00	1,315.00	685.00
								Total	1,365.00
Rabi	Gram	0.75	5.50	4.13	1,200.00	4,950.00	4,350.00	3,262.50	1,687.50
(October–February)	Mustard	0.25	5.20	1.30	1,200.00	1,560.00	4,300.00	1,075.00	485.00
								Total	2,172.5
								Rounded	2,173
Total Predevelopment income									3,538.00

(b) Postdevelopment

Season	**Crops**	**Area (in ha)**	**Yield (Qtl per ha)**	**Total Yield (Qtl)**	**Price (in Rs. per Qtl)**	**Gross Income (in Rs.)**	**Cost (per ha)**	**Cost Incurred**	**Net Income (in Rs.)**
Kharif	Paddy	0.75	20.00	15.00	550.00	8,250.00	6,340.00	4,755.00	3,495.00

(Continued)

BOX 1.8.6 (Continued)

(April–June)	Maize	0.25	11.00	2.75	500.00	1,375.00	2,630.00	657.50	717.50
								Total	4,212.50
								Rounded	4,213
Rabi	Wheat	0.50	21.00	10.50	700.00	7,350.00	8,720.00	4,360.00	2,990.00
(October–	Gram	0.25	7.50	1.88	1,400.00	2,625.00	4,350.00	1,087.50	1,537.50
February)	Potato	0.25	260.00	65.00	150.00	9,750.00	9,890.00	2,472.50	7,277.50
								Total	11,805.00
Total Postdevelopment income									16,018.00

Postdevelopment income of Rs. 3,538 less predevelopment income of Rs. 16,018 equals incremental income of Rs. 12,480 (2004 US$ 272, rounded).

Source: Meghalaya, India (Government of), Shallow Tube wells with Pumpsets in Alluvial Areas, 2004, URL: http://www.megcooperation.gov.in/ (accessed March 4, 2006).

TABLE 1.8.18
Bamboo Tube Well Cost Example

Bamboo hollowed out is used as a wellbore in South Asia. This example gives the capital cost for a 15 m (50 foot) well, with a comparison to an iron-cased well.

Average Cost/Boring (in 2004 Rs.)

	Bamboo Casing	Iron Casing
Iron sheet	258	—
Bamboo	87	—
Coconut coir	308	—
Iron wire	18	—
Blank pipe/iron pipe	1,120	3,020
Bottle tee/socket	120	230
Check valve	115	195
Iron nails	5	5
Socket and nipple	89	—
Sand and gravel	15	—
Gunny bag	10	—
Socket	—	120
Strainer	—	2,000
Labor charge	525	385
Transport of material	55	130
Miscellaneous	50	65
Overhead/supervision	275	620
Total	3,041	6,770
Annual Cost of Bamboo Tube well Investment and Operation (2004)		
	Annual Cost	
Category	Rs./Year	US$/Year
Fixed cost		
Interest	188.35	4.16
Depreciation	1,071.42	23.66
Operating cost		
Fuel	20,108.50	444.00

(Continued)

TABLE 1.8.18 (*Continued*)
Bamboo Tube Well Cost Example

Annual Cost of Bamboo Tube Well Investment and Operation (2004)

	Rs./Year	**US$/Year**
Lubricant	402.17	8.88
Pump maintenance	74.66	1.65
Engine maintenance	75.00	1.66
Operator wage	589.00	13.01
Miscellaneous	186.20	4.11
Total	22,695.50	501.12

Assumes:

Interest on fixed cost at 12%; 8 hp diesel engine uses fuel at rate of 1.25 L/h; pump life of 15 years

Note: In 2004, 1 Indian rupee = 0.02208 US$.

In 2004 in Bihar state, India, 7,946,435 ha were cropped with 61% of that land being irrigated, half with tube wells. The state estimated that 43% of the available groundwater was in use, with the remainder to be planned for further irrigation use. The India Planning Commission calculated a benefit-cost ratio for bamboo tube wells of 2.22, comparing the return from irrigation pumping to the cost of pumping. Farmer income rose 29% after using bamboo tube wells for irrigated agriculture. While bamboo tube wells appear to have cost advantages for poor farmers, some disadvantages include unreliability of electric power in some locations, high cost of metal components and motors, high cost of fuel, potential for vandalism and rotting in 5 years (IPC, 2004).

In terms of land area supplied with irrigation water from shallow tube- wells, a study in India found that a STW could irrigate an area of 4.06.0 ha, but on average served 2.3 ha. The principal factors for this small land size being serviced are that the individual farmers own small farms and the distribution systems are inefficient (Kansakar, 2001).

Source: India (Government of) Planning Commission (IPC), *Economics of Bamboo Boring: A Study of the North-East Region of Bihar*, IPC, New Delhi, India, 2004, 78.

A range of tube well costs extracted from the literature shows expenditures affected by technology and location:

- In India, a 12.2 m tube well with 7.5 submersible pump and pump house cost 2000 Rs. 259,000 (2000 US$ 5,880) (Dhillon, 2000).
- A shallow tube well in Bihar state, India, cost 2006 Rs. 28,864 (2006 US$ 653) (Bihar, 2006).
- Tube well irrigation in South Asia cost 2000 US$ 800–1000/ha (IWMI, 2000).

Well operation requires substantial energy. In some states of India, the electricity used to produce groundwater equals half the power consumed in those states (Brown, 2004, p. 104). However, treadle pump technology (a simple foot pump) provides much lower pumping cost for small plots with a cost of under 2016 US$ 5 in South Asia (Pal, 2016).

IRRIGATION METHODS AND COSTS

Irrigating crops was practiced by the Mesopotamians and Egyptians (WHI, 2005) thousands of years ago using canals to carry water to fields to ensure adequate food supply. Irrigation method refers to the technique of delivering water to the fields being irrigated. The type of irrigation method has a significant effect on the volume of water used. Flood irrigation is still one of the most widely used methods of irrigating crops because it is simple and inexpensive (USGS, 2005). Other principal methods include spray, center pivot, and drip irrigation.

Factors in selecting an irrigation method include (UNFAO, 2006; O'Brien et al., 2011)

- Capital accessibility
- Natural conditions (soil type, slope, climate, water availability, water quality)
- Type of crop
- Type of technology
- Well capacity
- Pumping cost
- Previous experience with irrigation
- Required labor input
- Costs and benefits

Box 1.8.7 reviews the major irrigation methods in use. Some have variations that have evolved as a result of technological advances, scarcity of water and type of crop

BOX 1.8.7 MAJOR IRRIGATION METHODS

Method	Description	Suitability
Flood	*Flood:* The application of irrigation water where the entire or nearly all the surface of the soil is covered by ponded water (USGS, 2005). Several types of flood irrigation methods are used:	
	Furrow: A type of flood irrigation in which partial surface flooding is used with clean-tilled crops where water is applied in furrows or rows made to carry water to irrigate the crop. The crop is usually grown on the ridges between the furrows (USGS, 2005).	*Furrow* irrigation may be applied to many crops, particularly crops planted in rows that would be damaged by water covering the plant's stem or crown. Application is best for uniform flat or gentle slopes and most soil types. Not recommended for coarse sands because of high evaporation rates (UNFAO, 2006).
	Basin: A type of flood irrigation in which water is diverted to basins created by embankments with the basins next to each other (UNFAO, 2006).	*Basin* irrigation may be applied to many field crops. Most suitable for crops which can tolerate roots being submerged in water, such as paddy rice. Flat land surface provides ease in constructing flood basins, which can also be constructed on steeper grades by building a series of terraces. Soil suitability for this method is dependent on the type of crop (UNFAO, 2006).
	Border: A type of flood irrigation with borders of long flat strips of land, separated by shallow embankments to guide the water flow through the field (UNFAO, 2006).	*Border* irrigation is most appropriate to larger farms with long field distances facilitating machine operations. Border distances may be up to 800 m or longer and 3–30 m wide. This method is not as applicable to smaller farms relying on hand labor or cultivation using animals. It is best for closely spaced crops such as pasture or alfalfa (UNFAO, 2006).

Note on a cost of flood irrigation: A significant quantity of water from flood irrigation methods may be lost due to runoff along the sides of the fields. This loss can be minimized by capturing the runoff in ponds and pumping it back up to the front of the field for reuse (USGS, 2005).

(*Continued*)

BOX 1.8.7 (Continued)

Method	Description	Suitability
Sprinkler/ spray	*Sprinkler:* A planned irrigation system in which water is applied by means of perforated pipes or nozzles operated under pressure so as to form a spray pattern (USGS, 2005). A hand-moved portable sprinkler system is moved by uncoupling and picking up the pipes manually, requiring no special tools. A solid set sprinkler system covers the complete field with pipes and sprinklers in such a manner that all of the field can be irrigated without moving any of the system. A side-roll (linear) sprinkler has the supply pipe typically mounted on wheels with the pipe as the axle with the system being moved across the field by rotating the pipeline by engine power.	*Sprinkler* irrigation is applicable to most row, field and tree crops with water sprayed over or under the crop canopy. This method should not be used with delicate crops (e.g., lettuce) since water drops can damage them. The method can be used on farmable slopes, is most appropriate for sandy soils having rapid infiltration, but can be adapted to a range of soil types (UNFAO, 2006).
	Traveling Gun: A variation of a sprinkler irrigation system consisting of a single large nozzle that rotates and is self-propelled. The name refers to the fact that the base is on wheels and can be moved by the irrigator or affixed to a guide wire (USGS, 2005).	
	Center Pivot: An automated rotating sprinkler pipe or boom supported by towers on wheels, supplies water to the sprinkler heads or nozzles, as a radius from the center of the field to be irrigated. Water is applied at a uniform rate by progressive increase of nozzle size from the pivot to the end of the line. The depth of water applied is determined by the rate of travel of the system. Single units are ordinarily about 381–396 m long and irrigate about a 52.6 ha circular area (USGS, 2005).	*Center pivot* irrigation has been proven to be very flexible and can accommodate a variety of crops, soils and topography with minimal modification (USDA, 2001).
Drip	A *drip irrigation* system applies water directly to the root zone of plants by means of applicators (orifices, emitters, porous tubing, perforated pipe, etc.) operated under low pressure with the applicators being placed either on or below the surface of the ground (USGS, 2005).	*Drip irrigation* is best applied to row crops (vegetables, soft fruit), tree and vine crops. Owing to high installation costs, this method is used for high-value crops. The method can be applied to any farmable slope following the land contour and to a range of soil types with application rates based on soil type. Drip irrigation is suitable for most soils with application rates varying on soil type: low rates on clay soils; high rates on sandy soils to provide sufficient lateral wetting (UNFAO, 2006).

Sources: Abstracted from United Nations Food and Agricultural Organization (UNFAO), Irrigation Water Management: Irrigation Methods, 2006, URL: http://www.fao.org/docrep/S8684E/s8684e00.htm (accessed March 11, 2006).

U.S. Department of Agriculture, Economic Research Service (USDA), *Irrigation and Water Use: Glossary; Irrigation Systems and Land Treatment Practices; Onfarm Water Conveyance Systems,* USDA, Washington, DC, 2001, URL: http://www.ers.usda.gov/briefing/WaterUse/Questions/glossary.htm (accessed March 24, 2006).

United States Geological Survey (USGS), *Irrigation Techniques and Some Irrigation Methods, USGS*, Reston, VA, 2005, URL: http://ga.water.usgs.gov/edu/irmethods.html (accessed March 11, 2006).

FIGURE 1.8.4 Center pivot irrigation in Delaware, USA

The differences in capital costs of various irrigation methods reflect the relative reliance on technology to bring water to fields (Figure 1.8.4). Table 1.8.19 indicates that in the United States (Kansas), on a unit basis, flood irrigation has the lowest capital cost ($1,289/ha or $522/ac) and drip irrigation, the highest ($4,435/ha or $1,796/ac) (2010 US$) (O'Brien et al., 2011). In this case and time, drip irrigation cost nearly 2.5 times more to install than a flood irrigation system. As will be seen here, however, drip irrigation systems are considerably more efficient in terms of water use, which translates into lower operating costs for pumping water. In using overhead sprinkler irrigation, often employed with groundwater sources, irrigation efficiency may improve from 65% to 80%

TABLE 1.8.19
Capital Costs of Irrigation Methods in the High Plains of the United States

Area in Hectares/Acres	50.6 ha/125 ac		62.7 ha/155 ac		62.7 ha/155 ac	
Irrigation Method	Center Pivot		Gravity/Flood		Drip System	
	/ha	/ac	/ha	/ac	/ha	/ac
Expense Category						
Well	$632	$256	$509	$206	$509	$206
Pump and Gearhead	543	220	437	177	437	177
Power Unit	247	100	200	81	200	81
Water Meter	27	11	22	9	22	9
Well Connector	49	20	40	16	40	16
Irrigation System	1,420	575	81	33	3,227	1,307
Total	$2,918	$1,182	$1,289	$522	$4,435	$1,796

Assumptions: 300-foot well, 8-inch pump set at 75 meters (245 feet), motor sized to power needed for well depth and pump type and natural gas power. Costs will change with depth and well capacity requirements.

Source: O'Brien et al. 2011. Irrigation Capital Requirements and Energy Costs. Kansas State University Agricultural Experiment Station and Cooperative Extension Service. URL: http://www.waterandenergyprogress.org/library/irrigationcosts.pdf (Accessed July 24, 2019).

FIGURE 1.8.5 Olive grove in the desert watered by drip irrigation in Israel

by converting from a high-energy to a low-energy system (Brown, 2001). Using a low-energy precision sprinkler system may increase the efficiency to 90% or more (Brown, 2001). Investigations of drip irrigation systems in a range of countries indicate potential water reductions of 30%–70% (Brown, 2001).

From Table 1.8.19, economies of scale exist for using irrigation methods on larger areas. This economy of scale can be seen in considering the well, pump and gearhead, power unit, water meter and well connector costs common to all the irrigation systems when comparing the 50.6 hectare (125 acre) costs to the 62.7 hectare (155 acre) costs. These costs decrease on a per hectare (acre) basis by 17% for the larger irrigation area. While drip irrigation systems are more expensive, because of the labor needed for them, they can be easily adapted and profitably applied to the small farm sizes of South and East Asia. Drip systems developed specifically for farms can have a payback period of 1 year (Brown, 2001). Figure 1.8.5 shows an olive grove grown with drip irrigation in Israel in a desert environment.

IRRIGATION EFFICIENCY

The efficiency of irrigation systems is defined as the percent of water used beneficially for irrigation purposes. Clearly, as evapotranspiration is part of the hydrologic cycle, it is also an essential process for crop growth and must be factored into water demand for irrigation. Water may be lost through distribution lines, and this is a cost to the farmer since it represents water pumped that does not reach the field. Box 1.8.8 indicates the efficiency of a range of irrigation methods along with defining the key terms in describing irrigation water use efficiency. Critically, the efficiency of any irrigation method depends on how the delivery system is operated and maintained, and, therefore, each system has a range of efficiency, which may be quite wide. Flood irrigation methods typically use more

BOX 1.8.8 IRRIGATION WATER USE EFFICIENCY

1. Defining Efficiencies

Water Conveyance Efficiency (*Ec*): The percentage of source water that reaches the field:

$$Ec = 100(Wf / Ws)$$

Wf = water delivered to field
Ws = water diverted from source

Water Application Efficiency (*Ea*): The percentage of water delivered to the field is used by the crop:

$$Ea = 100(Wc / Wf)$$

Wc = water available for use by the crop
Wf = water delivered to field

Irrigation Efficiency (*Ei*): The percentage of water delivered to the field that is used beneficially:

$$Ei = 100(Wb / Wf)$$

Wb = water used beneficially
Wf = water delivered to field

Irrigation efficiency is more broadly defined than water application efficiency in that irrigation water may have more uses than simply satisfying crop water requirements. Other beneficial uses could include salt leaching, crop cooling, pesticide, fertilizer applications or frost protection. Most irrigation systems are single purpose, that is to supply water for crop use, which allows water application efficiency and irrigation efficiency to be used interchangeably.

2. Range of Application Efficiencies for Various Irrigation Systems

System Type	Application Efficiency Range[1] (as a Percent, %)
Flood/surface irrigation	
Furrow	50–90
Basin	60–95
Border	60–90
Sprinkler irrigation	
Handmove	65–80
Traveling gun	60–70
Center pivot and linear	70–95
Solid set	70–85
Drip/microirrigation	
Point source emitter	75–95
Line source emitter	70–95

Source: Abstracted from Rogers, D.H. et al., *Efficiencies and Water Losses of Irrigation Systems*, Kansas State University, Research and Extension, 2006, URL: http://www.oznet.ksu.edu/library/ageng2/mf2243.pdf (accessed March 3, 2006).

[1] Efficiencies can be much lower due to poor design or management. These values are intended for general system type comparisons and should not be used for specific systems.

water than spray or drip irrigation. Drip irrigation can reduce water use by farmers from 30% to 60% or more (UNFAO, 2003a, 2003b), as suggested in Box 1.8.8. Such reductions in actual water use translate into substantial operating cost savings to farmers. The land area being irrigated to which farmers have applied more efficient irrigation methods is under 1% in India and China to only as much as 4% in the United States (Brown, 2004, p. 113). Arabiyat et al. (1999) modeled low-energy precision irrigation (drip) systems combined with biotechnology and some crop conversion for the area of Texas (United States) supplied by the Ogallala (High Plains) Aquifer and found significant returns to agriculture and reductions in depletion of the aquifer through more efficient irrigation.

IRRIGATION USE OF GROUNDWATER

Irrigation use of groundwater increased significantly over the latter part of the twentieth century (Brown, 2004, p. 100). In the United States, irrigation use of groundwater has expanded, while that of surface water has declined, even though surface water use is larger. Table 1.8.20 provides an overview of groundwater consumptive use around the world for selected regions and countries. Total consumptive use of groundwater for irrigation in 2010 was 545,359 million cubic meters (442.13 million acre-feet).

IRRIGATION PRODUCTIVITY

Irrigated agriculture can produce more crop than rain-fed farming. "While only 20% of the world's farmland is irrigated, it produces 40% of our food supply. The highest yields obtained from irrigation are more than double the highest yields from rain-fed agriculture - even low-input irrigation is more productive than high-input rain-fed farming" (UNFAO, 2003a). In the example from Meghalaya state, India, in Table 1.8.20, on a per land unit basis using groundwater tube wells, estimated farmer production increased 43% for paddy (rice) and 38% for maize.

A further consideration of the productivity of the range of irrigation techniques is useful. Flood irrigation loses considerable water to evaporation. Current studies have shown that flood irrigation, typically applied to rice production, in certain conditions allowing for lower quantity of application under periodic rather than continuous flooding, may provide the same level of production (Brown, 2001). Investigations of irrigated cotton production comparing flood and drip systems demonstrated

TABLE 1.8.20
Groundwater Consumptively Used for Irrigation in Selected Regions and Countries

Region/Country	Groundwater Consumptively Used For Irrigation (million m^3/year)	Region/Country	Groundwater Consumptively Used For Irrigation (million m^3/year)
Africa	17,863	Eastern Europe	6
Central America and Caribbean	1,252	Russian Federation	856
Mexico	11,386	Northern, Western and Central Europe	2,165
United States and Canada	88,498	Mediterranean Europe	15,179
South America	6,221	Oceania	3,301
Central Asia	4,719	TOTAL	545,359
Middle East	71,261		
South and East Asia	322,651		

Source: Siebert, S.; J. Burke; J. M. Faures; K. Frenken; J. Hoogeveen; P. Doll; and F. T. Portmann. 2010. Groundwater use for irrigation–a global inventory. In Hydrology and Earth Systems Science, 14, 1863–1880. URL: www.hydrol-earth-syst-sci.net/14/1863/2010/ doi:10.5194/hess-14-1863-2010 (Accessed July 24, 2019).

water use reductions from 43% to 79%, while yields improved by 25%–40% (Polak and Sivanappan, 2006). Precision delivery of water to plants through drip irrigation systems has raised yields from 20% to 90% in some applications (Brown, 2001).

Growing more water-efficient crops can also improve irrigation productivity. Wheat, in most situations, provides 50% more grain per unit of water than rice does. At least one country (Egypt) sets controls on rice cultivation because of this factor (Brown, 2001).

Relative to extensification of farmland, irrigation has contributed to expanding food production. From the mid-1960s to the mid-1980s, the expansion of land utilizing irrigation "contributed over 50% of the increase in global food production" (Sundquist, 2006).

Globally, "irrigated cropland is about 3.6 times more productive per unit-area than nonirrigated cropland. The dollar-value of production of a given area of irrigated cropland is about 6.6 times that of nonirrigated cropland and about 36 times that of rangelands. The ratio of the productivity of irrigated land to that of nonirrigated cropland is thus larger in dollar-terms (6.6) than in weight-terms (3.6), reflecting the fact that irrigated croplands must be used to grow higher-value crops to justify the high capital costs involved" (Sundquist, 2006).

FINANCING IRRIGATION SYSTEMS

Irrigation systems, including wells, intakes, pumps and distribution works, are expensive infrastructure and may require financing, including for small systems. Worldwide, lending for irrigation systems declined from a high of nearly US$ 2.3 billion to about US$ 0.8 billion (assumed 2000 US$) (UNESCO, 2003a, p. 194). The United Nations cites a range of typical irrigation system costs from US$ 1,000 to 10,000/ha with extremes of as high as US$ 25,000/ha, depending on location, area and type of system (UNESCO, 2003a, p. 206). In the United States, costs per hectare for groundwater irrigation systems have been reported from $500 to $2,000 depending on irrigation method (O'Brien et al., 2011).

Financing may take different forms, including loans to individuals or groups of individuals in water user associations or subsidies through lower rates or tax credits from governments. Most governments provide assistance to water use in agriculture, but the focus should be on the effectiveness of the investment (INPIM, 2004). Financial institutions may not understand approaches of lending to water user associations, complicating the development of local capacity of small farm irrigators for financially sustainable irrigation projects (INPIM, 2004). For example in a financing program in India, small farmers who could have benefitted from group projects were largely not served by a subsidized rate-financing program, even when subsidies were increased for water user associations (Kansakar, 2001). Often, reporting on investments for irrigation projects does not account for the contribution made directly by the farmer, such as land improvement (e.g., grading or terracing) or on- farm irrigation, which may be as much as 50% of the total investment (UNESCO, 2003a, p. 206).

A range of approaches to financing irrigation of agricultural lands exists. For example, in India, the government-sponsored Agriculture Development Bank (ADB) had provided subsidized rates for well drilling up to certain limits and incentive subsidy rates for water user associations. The ADB has also eliminated these subsidies. In Nepal, farmers installed private shallow tube wells without government assistance to irrigate 21% of groundwater-irrigated area of 11,400ha in 1995, indicating that the private benefit offset the risk (Kansakar, 2001). In the United States, the federal government provides a depletion allowance, which farmers, who irrigate with groundwater, use to reduce their taxable income.

Approaches cited to improve financing to small farm irrigators who often are reliant on groundwater, especially those in water user associations (WUA), include (INPIM, 2004)

- Linking government subsidies to local resource mobilization, through eligibility criteria, matching formulas and other mechanisms.
- Offering grants on a competitive basis, subject to eligibility criteria for local contributions and WUA performance in maintaining irrigation infrastructure.

- Creating an irrigation investment fund.
- Encouraging incremental irrigation system improvements rather than large investments.
- Establishing reserve funds to cover future costs of maintenance of the shared infrastructure.

ECOSYSTEM EFFECTS OF IRRIGATION

Irrigated agriculture, while improving crop yields and production, can have significant effects on the ecosystem and groundwater specifically, if not conducted in a manner compatible with the soils, climate and other water needs of the area. Major impacts of large withdrawals of groundwater for irrigation include

- Waterlogging and salination of farmlands (UNFAO, 1992; UNFAO, 2003a)
- Depletion of aquifers (Brown, 2004, pp. 99–104)
- Land subsidence (Gelt, 1992)
- Accumulation of salts, fertilizer and pesticide residuals in groundwater (UNFAO, 1992; WSLH, 2003; Barbash and Resek, 1996)

These effects may result in damage to the ecosystem, lost agricultural production and costs and reduced income to affected property owners.

Waterlogging and Salination

Waterlogging results from overwatering land during irrigation and lack of adequate drainage (UNESCO, 2003a, p. 219). Salination is the accumulation of salts and minerals in agricultural land because of excess water applied during irrigation. As water on the fields evaporates, dissolved salts from the soil accumulate on the surface, creating a crust. High soil salinity reduces seed germination and limits crop production (UNFAO, 1992) and land value, affecting 75 million hectares of agricultural land worldwide (WAF, 2006). Twenty percent of irrigated lands are affected by salination (about 2% of rain-fed croplands are similarly affected) (WAF, 2006).

Depletion of Aquifers

Five to eight percent of the irrigated area worldwide relies to some extent on groundwater being overdrafted with this proportion increasing. In the United States, this fraction is 20%–25% (Sundquist, 2006). The effects of overdrafting "tend to be permanent and cumulative" (USDA, 1997a, p. 70). In the area of Texas (United States) supplied by the Ogallala (High Plains) Aquifer, an estimated 30%–35% of the predevelopment groundwater resource has already been depleted (Arabiyat et al., 1999). Brown has a descriptive overview of groundwater depletion from irrigation, abstracted in Box 1.8.9. Brown notes that the declining groundwater tables are a relatively recent circumstance historically and that the difference between groundwater use and sustainable levels increases each year translating into ever-deeper water tables. This situation threatens not only water supply but also the world's ability to feed itself (UNESCO, 2003a, p. 13; Brown, 2004, p. 117).

Land Subsidence

Land subsidence from groundwater exploitation is a significant problem in many areas in which the water is used for irrigation (Figure 1.8.6). The problem is so extensive worldwide that the United Nations established a database to track it (UNESCO, 2003b). Subsidence can be a substantial cost to farmers affected by it, resulting in (Gelt, 1992)

BOX 1.8.9 GROUNDWATER DEPLETION FROM IRRIGATION USE IN SELECTED COUNTRIES

Country	Description
Mexico	Aquifers are being depleted in northern and semiarid regions from irrigation.
United States	Overpumping the Ogallala aquifer in the High Plains has caused wells to go dry resulting in a decrease of irrigated area by 24%.
Saudi Arabia	Rapid depletion of large fossil aquifer for irrigation has caused wheat production to drop 61% from 1992 to 2004.
Yemen	With one of the world's fastest growing populations, its water table is falling 2 or more meters per year.
Israel	Coastal and mountain aquifers shared with the Palestinians are being depleted resulting in a ban on irrigated wheat and tightening of water supplies.
India	Most states have falling water tables and thousands of wells go dry each year from extensive irrigation.
China	Water tables are declining across northern China, with dry irrigation wells causing a significant decrease in wheat production.

Source: Brown, L.R., *Outgrowing the Earth: The Food Security Challenge in an Age of Falling Water Tables and Rising Temperatures*, Earth Policy Institute, W.W. Norton & Company, New York, 2004, 101–102.

FIGURE 1.8.6 Fissure has been attributed to land subsidence related to groundwater pumping in the Antelope Valley area, California. Source: U.S. Geological Survey

- Broken irrigation ditches and canals from fissures and crevices
- Alteration of the slope of previously level fields, disrupting the flow of irrigation water and causing fields to be releveled
- Collapse of well casing necessitating expensive repairs and replacement of wells with large irrigation wells costing from $100,000 to $200,000
- Land taken out of production and abandoned because of developing fissures
- Crevices are hazards to people, livestock and wildlife

Subsidence can affect extensive areas as evidenced in Arizona, United States. In the area around Stanfield, Arizona, 1101 km^2 of land have subsided by nearly 3.7 m (Gelt, 1992). This is an area nearly half the size of the Grand Duchy of Luxembourg or the U.S. state of Rhode Island.

Accumulation of Salts, Nutrients and Pesticides Residuals

Salts dissolved from the soil, as well as residual nutrients and pesticides may be carried by irrigation water percolating below the soil zone. Table 1.8.21 shows that world fertilizer use increased over 6 times from 31.2 million metric tons in 1961 to 193.3 million metric tons in 2014, principally representing nitrogen, phosphates and potash products applied in agriculture (Rosser and Ritchie, 2019). Table 1.8.22 indicates that world pesticide use was 5,211 metric tons of active ingredient in 2007 (Fishel, 2013). Pesticide use more than doubled worldwide and in the United States from 2001 to 2007. These salts and residuals can pose health hazards to people as well as wildlife when consumed from the "return flow" groundwater from wells in shallow aquifers and from affected wetlands. This polluted groundwater also can pollute streams to which it discharges (UNFAO, 1992). "Pesticides not taken up by plants, adsorbed by soils or broken down by sunlight, soil organisms, or chemical reactions may ultimately reach groundwater sources of drinking water. This will depend upon the nature of the soil, depth to groundwater, chemical properties of the pesticide, and the amount and timing of precipitation or irrigation in an area. Usually the faster a pesticide moves through the ground, as with sandy soils and heavy rainfall or irrigation, the less filtration or breakdown" (WSLH, 2003). Soil permeability and organic-matter content are the primary factors affecting migration of pesticide residues through soil to groundwater (Barbash and Resek, 1996, p. 274).

TABLE 1.8.21
World Fertilizer Use

	Mil. Metric		Mil. Metric		Mil. Metric		Mil. Metric
Year	**Tons**	**Year**	**Tons**	**Year**	**Tons**	**Year**	**Tons**
Nitrogen		**Phosphate**		**Potash**		**Total**	
1961	11.6	1961	10.9	1961	8.7	1961	31.2
1980	60.5	1980	31.7	1980	24.2	1980	116.4
2000	80.8	2000	32.4	2000	21.7	2000	134.9
2014	108.9	2014	46.7	2014	37.7	2014	193.3

Source: Roser, Max and Hannah Ritchie. 2019. Fertilizer and Pesticides. Published online at OurWorldInData.org. URL: https://ourworldindata.org/fertilizer-and-pesticides (Accessed July 24, 2019)

(Continued)

TABLE 1.8.22
World Pesticide Use

2001	World Market (Million kg of Active Ingredient)	U.S. Market (Million kg of Active Ingredient)
Herbicides	848	251
Insecticides	559	48
Fungicides	215	33
Other	666	214
Total	2288	546
2007		
Herbicides	2096	531
Insecticides	892	93
Fungicides	518	70
Other	1705	439
Total	5211	1133

Sources: Fishel, F.M. 2007. *Pesticide Use Trends in the U.S.: Global Comparison. PI-143,* Pesticide Information Office, Florida Cooperative Extension Service, Institute of Food and Agricultural Sciences, University of Florida, Gainesville, FL. URL: http://edis.ifas.ufl.edu/pdffiles/PI/PI18000.pdf (access March 19, 2009).

Fishel, F.M. 2013. *Pesticide Use Trends in the US: Global Comparison. PI-143*, Pesticide Information Office, Florida Cooperative Extension Service, Institute of Food and Agricultural Sciences, University of Florida, Gainesville, FL. URL: https://edis.ifas.ufl.edu/pi180 (accessed July 24, 2019).

Monitoring studies of nutrients and pesticides in groundwater show occurrence of these chemicals across regions. In a study of over 1400 wells of varying depths across the United States from 1993 to 1995, nearly 50% of the wells less than 30.48 m (100 feet) had nitrate concentrations exceeding the 10mg/L drinking water health standard, especially in areas of high nitrogen input, well-drained soil and less extensive woodland relative to cropland (Nolan et al., 1998). Shallow wells (less than 15 m [50 feet] in depth) screened in unconsolidated aquifers in areas of irrigated crops are nearly twice as likely to be contaminated with pesticide residues than shallow wells in nonirrigated areas (Barbash and Resek, 1996, p. 245). Obtaining representative results in monitoring should be done with caution, which could affect monitoring costs and extrapolation of those results for economic analysis and decision-making, since the soil texture around monitored wells may not be the same as in the recharge zone, from where the contaminated water originated (Barbash and Resek, 1996, p. 245, citing Hallberg, 1995). Field-monitoring studies of pesticides in groundwater in Canada and the United States found their occurrence in the broad ranges of 4.5%–100% in monitoring wells, 18%–43% in irrigation wells, 2.5%–33.6% in domestic supply wells, 6.4%–94% in public water supply wells (Barbash and Resek, 1996, pp. 294–295).

Some pesticides and their breakdown products are known carcinogens and high levels of certain nutrients, such as nitrate, have acute health effects (see Part 1 of Volume 1, Chapter 8; also, USDA, 1987, p. 21). A partial list of pesticides found in groundwater under cropland is listed in Table 1.8.23. In the groundwaters of North America, monitoring studies have detected more than 160 pesticides and transformation products (chemicals resulting from degradation of pesticides) (Barbash and Resek, 1996, pp. 162–171), and many of these have also been detected in groundwaters of Europe (Barbash and Resek, 1996, pp. 101–103). Nitrate from fertilizer and pesticides are not always detected coincidently in groundwater and, therefore, “may impose different costs on society” (USDA, 1987, pp. v, 30–31). In the United States, the Department of Agriculture estimated that over

TABLE 1.8.23
A Partial List of Pesticides Found in Groundwaters under Cropland in the United States with Highest Occurring Pesticides Indicated with Percent of Detection Frequency

Atrazine (28%)	Tebuthiuran	Terbacil	Napropamide
Desethylatrazine (24%)	Alachlor	Carbaryl	Pebulate
Simazine (11%)	Cyanazine	Diazinon	Pendimethalin
Prometon (10%)	Diethylanline	Alpha-HCH	Pronamide
Metachlor (6%)	Dieldrin	DCPA	Propanil
Metribuzin (2%)	Carbofuran	Ethelfluralin	Trifluralin
	EPTC	Linuron	

Source: U.S. Geological Survey (USGS), The Quality of Our Nation's Waters; Pesticides in the Nation's Streams and Ground Water, 1992–2001, USGS Circular 1291, 2000, 172.

19 million people on private wells and 53 million people on public water systems had some risk of exposure to agricultural chemicals in their drinking water (USDA, 1987, pp. 17–20). A partial of list of potential effects of groundwater contaminated by agricultural chemicals, which contribute to costs to individuals and society that are difficult to estimate, include (USDA, 1987, p. 21)

- Livestock poisoning and health problems from nitrate/nitrite
- Decreased crop quality or quantity from salts in fertilizers
- Methemoglobinemia (infant death and illness) from nitrate/nitrite
- Cancer from pesticides
- Miscellaneous health (cardiovascular, blood, kidney, liver, nervous system and other) problems from pesticides and nitrate
- Damage to vegetation, waterfowl and aquatic life in groundwater recharge areas and in surface water contaminated by groundwater discharge.

BEST MANAGEMENT PRACTICES

Implementing best management practices (BMP) for irrigated fields can mitigate some of the problems noted in the previous section on ecosystem effects. Many of the BMPs are similar for nonirrigated agriculture. Applying BMPs have a cost associated with them, but may be less costly than dealing with the other effects. These BMPs include but are not limited to (Seelig and Scherer, 1996; NCAES, 1988):

- Scheduling irrigation appropriately by accounting for the soil moisture and crop water use.
- Timing water applications to avoid water movement beyond the rooting zone.
- Adjusting water application amounts to meet varying crop demands at different growth stages.
- Applying irrigation water uniformly and accurately.
- Using the least amount of water possible to apply chemicals.
- When injecting chemicals into an irrigation system, using chemigation equipment protects the water supply.
- Calibrating the chemigation unit with each use to ensure accurate application of chemicals.
- Using a secondary containment structure where pesticides are stored near the irrigation well when chemigation is practiced.
- Using close-growing crops such as grass or small grains as cover strips to catch runoff and retain pesticides.

BMP focused on conditions with limited or restricted water supply for irrigation can still provide yields that could be attractive to farmers. Recognizing that the amount of water moving through the plant and transpired directly, Schneekloth et al. (2001) show that depending on soil type, climatic conditions, crop prices and production costs, even with reduced amounts of irrigation, yield may be profitable. For example, the following irrigation-yield relationships may result in the High Plains of the United States, assuming a slit loam soil, average rainfall and center pivot sprinkler irrigation (Schneekloth et al., 2001):

- Continuous corn – production increases by 50% with 5.1 cm of irrigation, doubles with 10.2 cm and levels off at 3.5 times dryland production at about 52.4 cm.
- Soybean – production increases more than double with 5.1 cm of irrigation, nearly triples with 10.2 cm and reaches 3.5 times dryland production at 15.2–20.3 cm.
- Winter wheat after soybean – production doubles with 5.1 cm and levels off at about 2.5 times dryland production with 15.2 cm of irrigation water.

In a constrained water situation, and assuming year 2000 prices, Schneekloth et al. (2001) show that winter wheat irrigated at 10.2 cm/unit area and soybean at 15.2 cm/unit area have higher net returns than that of corn at similar levels of irrigation, respectively.

Since the cost of BMPs varies based on the objective (e.g., reduce groundwater use, mitigate fertilizer and pesticide use), the soil type, climate, area of application, technology and other factors, little general cost information is available. However, governments have provided some limited economic incentives to encourage farmers to use BMPs. These include tax credits (limited by amount), loans at low rates and over short time periods (1–10 years) for installation and land rentals and payments for leaving land fallow. See, for example, programs listed under Virginia (United States) Cooperative Extension (2000).

Management practices to mitigate the creation and use of saline groundwater include (UNFAO, 1992)

- Growing suitably tolerant crops
- Managing seedbeds and grading fields to minimize local accumulations of salinity
- Managing soils under saline water irrigation
- Operating delivery systems efficiently
- Irrigating efficiently
- Monitoring soil water and salinity and assessing adequacy of leaching and drainage

While installing drain tiles can reduce waterlogging, their expense relative to increased yield must be considered. Subsurface drainage systems may cost as much as 2001 US$ 988/ha, cited in one report (assuming clay soils and tile alignment on 15.24 m centers) (MAES, 2001). Corn yield in the same study indicated an average increase in yield of 314 kg/ha in Missouri, United States. Tile drainage systems may have a useful life of 50 years (UIUC, 2004). Draining soils in wetlands reduces wildlife habitat and ecological diversity. The United States has converted approximately 48.6 million hectares from wetlands to other uses through drainage and land-filling, mostly through agricultural drainage (USGS, 1997). Twenty-five percent of the farmland in the United States and Canada are drained (UMES, 2001).

Studies of pesticide effects on groundwater from different irrigation methods found that (Barbash and Resek, 1996, pp. 243–245)

- Sprinkler irrigation had the least-downward movement of water and atrazine because of evaporative loss.
- Furrow irrigation produced the largest downward migration of water and atrazine.

- Drip irrigation had very slight downward movement of water and atrazine.
- Increased rates of water application result in pesticide residue concentrations being reduced in the soil and enhancing their downward transport particularly in more permeable soils.
- The studies did not evaluate the occurrence of transformation products of the pesticides.

INSTITUTIONAL FACTORS

Irrigated agriculture is one of the most heavily subsidized sectors by governments internationally (UNFAO, 2003a). Because of the growing demand for food, irrigated agriculture is a key target for government policy. As noted previously, countries such as India have worked to develop lending institutions that will support the small farmer. Tube wells have received government subsidies since the early 1980s in developing countries, such as Bangladesh and Nepal (Hartmann and Boyce, 1981; Kansakar, 2001). However, education of these farmers may also be critical. When groundwater levels decline to a point that small and marginal farmers do not have the economic capability to apply their pumping technology, poorer farmers are penalized and their lands may be removed from production (UNESCO, 2003a, p. 209). This circumstance suggests a real issue with economic assumptions that include perfect knowledge of an intermediate commodity, in this case, an input or factor of production, to use it in a way that maximizes welfare. The use of the groundwater resource in this case is one not of efficiency but of equity and having equal access to participate in the economy. Scale of use and distribution of the resource are both significant considerations, then. Thus, institutions should be developed to ensure that even small farmers understand the effect they have on the water table and how best to manage their water use relative to the larger aquifer use.

Sustainable agriculture (using groundwater or any water source) "requires the farmer to be more knowledgeable about his farm's ecosystem and recognize its place in his life and society's" (Guru and Horne, 2000, p. 28). Irrigated biofuels production is displacing land and energy for food production in some countries and this outcome may not be well understood in those places. More fully informed people at all levels of water use should be able to improve their water practices and in the process provide a better balance in meeting the objectives of those uses (e.g., see UNESCO, 2003a, pp. 347–366). However, while international organizations are focused on consumption of groundwater for irrigation and its positive contribution to feeding the world, nations may desire to consider policies that address the open access to groundwaters that are being depleted and establish institutions, investing in social capital that will foster community organizations to promote access to and share critical information (Markandya et al., 2002, p. 25) that will help manage the resource for the long term and for the greatest benefit of human kind.

GROUNDWATER CONSERVATION AND FOOD PRODUCTION

Agricultural policy in the United States has promoted conserving groundwater through subsidies to farmers for using more efficient irrigation systems, such as sprinkler or drip systems. The US program, the Environmental Quality Incentives Program, initially authorized in 1996, provided financial incentive to farmers to invest in efficient irrigation equipment to save water and reduce depletion of aquifers. The USDA Natural Resources Conservation Service will provide payments up to a specified amount per acre under contract for up to 5 years for water conserving (and other stewardship) practices. Recent studies of have shown that farmers receiving the subsidies have employed more efficient equipment and have used the water savings to irrigate more land and increase crop production (Nixon, 2013). This policy has continued through 2018 law preserving the subsidy as well as introducing incentives for aquifer restoration (USDA, 2019). Economic policies targeting or affecting groundwater use should be carefully evaluated for their unintended consequences.

GROUNDWATER PRODUCTION FUNCTION AND PRODUCTIVITY FOR IRRIGATION WATER

Production Function

Processes to produce groundwater for irrigation can be modeled in an equation format. Irrigation water is a factor input to crop production and its demand is derived from the crop production process. The amount of irrigation water needed for a crop also depends on the amount of rainfall received by the cropland. If the quantity of irrigation water needed as an output is known in advance then a typical simple production function for this purpose might apply:

$$Q_{GW} = f\left(m_1, m_2\right) \tag{1.8.3}$$

where

Q_{GW} = quantity of groundwater produced
m_1 = fixed inputs, or K, capital such as the well, pump and motor
m_2 = variable inputs needed for production, such as labor (L) and electricity, E

However, groundwater might enter into a crop production function as a variable input:

$$Q_{CROP} = f\left(K, F, G, L, D\right) \tag{1.8.4}$$

where

Q_{CROP} = amount of crop produced
K = well, pump, motor and irrigation equipment
F = amount of nutrients applied
G = amount of groundwater applied
L = labor and management
D = area of land irrigated.

The amount of groundwater needed as an input will correspond to soil and crop evapotranspiration, temperature and any rainfall deficit the crop is experiencing. A production function for groundwater to be an input to Q_{CROP} may include a factor for rainfall deficit, and/or temperature or evapotranspiration rate. In 2013, 48.5 million acre-feet of groundwater was pumped and supplied to 37.2 million acres of open farmland. Technological changes may be introduced through modifying inputs, such as drip irrigation replacing flood or sprinkler irrigation as a new capital investment.

Use and Productivity

Demand for irrigation water will be specific to the local farm conditions, weather and type of crop. For example, 0.4 hectare (one acre) of corn will transpire 11,400–15,100 liters (3,000–4,000 gallons) of water each day (USGS, 2019). An analysis of irrigation water productivity of the US High Plains Aquifer indicated that 51% more biomass yield was produced on average, equivalent to $3 billion across the area irrigated by the aquifer (Garcia Suarez et al., 2018). The U.S. Department of Agriculture reports from 2008 to 2013 that nearly 2% more open agricultural was irrigated with groundwater while the amount of total groundwater used was about 1% less, an apparent continuation of improving irrigation water productivity with increased use of more efficient sprinkler and drip irrigation systems (USDA, 2013).

MACROECONOMIC CONSIDERATIONS

The agricultural sectors of developed countries contribute about 3% to their GDP and provide less than 5% of total employment. In Nepal, an example of a less developed country, nearly 40% of its gross domestic product and 80% of the employment are derived from the agricultural sector (Kansakar, 2001). Thus, government policies encouraging irrigation have a significant and positive effect on the economy, and potentially a large number of people experience a direct benefit from improved income and greater food availability. Because of the large effect of irrigated agriculture on national economy, losses of wetlands, polluted groundwaters and depleted aquifers should be accounted as consumed assets in the net national product with the value of the remaining natural capital potentially being increased (Markandya et al., 2002, pp. 66–68). Policies supporting sustainable management of natural resources may address balancing ecosystem, economic and social objectives relative to people's needs (Markandya et al., 2002, p. 15). Macroeconomic policies could even include financial incentives for large users to use water efficiently, reducing current demand, and provide adequately for the needs of the future, ensuring sufficient supply (Markandya et al., 2002).

Macroeconomic policies regarding irrigation affect national income derived from agricultural trade. While water is expensive to transport relative to its value, it can be "traded" as input to food commodities. Water deficient countries can import "virtual water" by trading their products for food grown with other countries' water resources (UNESCO, 2003a, pp. 202–203). In effect, the exporting countries are shipping a portion of the production from their water use in irrigation, as well as a part of their natural endowment of rain-fed agricultural output, to external consumers. Both of these acts – exporting and importing of irrigatively grown food – are sufficiently important to be addressed at the highest levels of governments to ensure that their citizens acknowledge that each group (exporters and importers of virtual water) may be depleting national resources whose possession is comparatively advantageous at present. Depending on how much groundwater is coming out of storage from fossil aquifers receiving little recharge, the exporting countries may be trading away their own future food security, unless they focus policy development on efficient irrigation methods to mitigate derived demand for irrigation water and replenishment of diminished groundwater stocks and other alternative water sources, such as recycled water. By establishing such macroeconomic policies influencing efficient water use and future supply, countries exporting virtual water through food trade provide for their internal needs as well as generating income from global food demand.

SUMMARY

Use of groundwater is expanding for irrigated agriculture in response to increased demand for food. Groundwater provides an important input factor to the much-needed food in countries which have growing populations. This activity is depleting aquifers around the world and potentially contributing to other negative effects on the supporting ecosystem. On a per capita basis, less land is being used and more fertilizer is being applied. Pesticide use has also grown and can be applied through irrigation systems, allowing residuals to reach groundwater. Small farmers have benefitted from low-cost technology to irrigate their crops with groundwater. However, small farmers may not understand appropriate groundwater management actions individually and collectively to ensure their investment in tube wells will provide for their sustenance over time and not just for a limited number of growing seasons. Nations should consider developing their institutions to better inform farmers of the appropriate techniques to manage groundwater for irrigation and thereby maximize its capacity for the near and long terms. Since irrigated agriculture is expanding with positive effect on food supply as well as increasing biofuels production, national governments should consider developing balanced policies that account for ecosystem and economic dynamics. Furthermore, such policies could address national income from virtual water trading through food exports to water deficient countries, while promoting efficient irrigation water use to provide for future water supply in response to direct water need and derived demand from food consumption. Policies subsidizing electricity and using efficient irrigation methods may also promote irrigation of more land resulting in greater food production and aquifer depletion. Policies affecting groundwater resources should be evaluated for unintended outcomes.

While groundwater produced by a well is an output of the irrigation well management process, groundwater is an input to crop production, thus it is demand derived. Factors affecting groundwater as an input include rainfall deficit, evapotranspiration and area irrigated, among other considerations.

FOR FURTHER CONSIDERATION

1. How much groundwater is used in your state or country for irrigation relative to other groundwater uses? Are water tables falling in your state or country? If water tables are falling, check and document from local news sources what the effects and costs on other water users and other activities are.
2. Are there local, state or national policies affecting groundwater for drinking or irrigation use? What are they? Who do they affect? Is there any information on costs and benefits of these policies? If so, what are the costs and benefits that have been documented? Can you identify other effects?
3. Contact a farmer who irrigates and ask him/her what portion of the cost to produce a crop is for delivering the water to the fields.
4. What is the irrigation water productivity in your state or nation? Contact the state or national agricultural university and prepare a graph of productivity trends over time (the last 10 to 20 years). The productivity could be measured by value of agricultural products produced in irrigated areas divided by the amount of irrigation water used. If value of agricultural products is not available, obtain the trends in land area irrigated to see if there are obvious efficiencies in irrigation water use.

REFERENCES

Arabiyat, T.S., Segarra, E., and Willis, D.B. 1999. Sophisticated irrigation technology and biotechnology adoption: Impacts on ground water conservation. *AgBioForum*, 2 (2), 132–136. URL: http://www.agbioforum.org (accessed November 4, 2006).

Argonne National Laboratory (ANL). 2008. *Baseline and Projected Water Demand for Energy and Competing Water Use Sectors*. URL: http://www.ead.anl.gov/pub/doc/ANL-EVS-TM08-8_water_demand_report.pdf (accessed March 20, 2009), 108 pp.

Badiani, R. and Jessoe, K. 2012. *The Impact of Electricity Subsidies on Groundwater Extraction and Agricultural Production*. URL: https://economics.ucdavis.edu/events/papers/Jessoe51.pdf (accessed July 23, 2019)

Barbash, J.E. and Resek, E.A. 1996. *Pesticides in Ground Water: Distribution, Trends, and Governing Factors*. Ann Arbor Press, Inc., Chelsea, MI, 525 pp.

Bihar (Government of) 2006. *New Initiatives for Development of Minor Irrigation Facilities*. URL: http://biharirrigation.nic.in/intro.htm (accessed March 3, 2006).

Brown, L.R. 2001. *Eco-Economy: Building an Economy for the Earth*. W.W. Norton & Co., New York, 322 p.

Brown, L.R. 2004. *Outgrowing the Earth: The Food Security Challenge in an Age of Falling Water Tables and Rising Temperatures*. Earth Policy Institute, W.W. Norton & Co., New York, 239 pp.

Buchanan, J.R. and Cross, T.L. 2006. *Irrigation Cost Analysis Handbook. PB* 1721. 24 pp. http://www.utextension.utk.edu/publications/pbfiles/PB1721.pdf (accessed March 11, 2006).

Cropper, M. and Sussman, F. 1990. Valuing future risks to life. *Journal of Environmental Economics and Management*, 19, 160–174.

Dhillon, G.S.. 2000. Tale of two tubewell schemes. *The Tribune* (Chandigarh, India). July 10, 2000 (Online Edition). URL: http://www.tribuneindia.com/2000/20000710/agro.htm#1 (accessed March 3, 2006).

Dumler, T.J. and Rogers, D.H. 2004. *Irrigation Capital Requirements and Energy Costs*. Kansas State University Agricultural Experiment Station and Cooperative Extension Service MF 836, 4 pp.

Fishel, F.M. 2007. *Pesticide Use Trends in the U.S.: Global Comparison. PI-143*. Pesticide Information Office, Florida Cooperative Extension Service, Institute of Food and Agricultural Sciences, University of Florida, Gainesville, FL. URL: http://edis.ifas.ufl.edu/pdffiles/PI/PI18000.pdf (accessed March 19, 2009).

Fishel, F.M. 2013. *Pesticide Use Trends in the US: Global Comparison. PI-143, Pesticide Information Office*, Florida Cooperative Extension Service, Institute of Food and Agricultural Sciences, University of Florida, Gainesville FL. URL: https://edis.ifas.ufl.edu/pi180 (accessed July 24, 2019).

Foster, S.S.D. 1989. Economic considerations in groundwater resource evaluation. In *Groundwater Economics: Selected Papers from a United Nations Symposium. Barcelona, Spain*, Elsevier Science Publishers B.V., Amsterdam, the Netherlands, pp. 53–65.

Gelt, J. 1992. Land subsidence, earth fissures change Arizona's Landscape. In *Arroyo*. University of Arizona, Water Resources Research Center, Tucson, AZ, Summer 1992, vol. 6, No. 2. URL: http://ag.arizona.edu/azwater/arroyo/062land.html (accessed April 8, 2006).

Guru, M.V. and Horne, J.E. 2000. *The Ogallala Aquifer*, The Kerr Center for Sustainable Agriculture Inc, Poteau OK, 35 pp.

Hallberg, G.R. 1995. *Peer Review of Pesticides in Ground Water Manuscript*. University of Iowa, University Hygenic Laboratory, Iowa City, IA.

Hart, C. 2007. *Bottlenecks, Drought and Oil Price Spikes: Impact on U.S. Ethanol and Agriculture*. Minnesota Economic Association, St. Paul, MN, October 26, 2007. URL: http://www.card.iastate.edu/presentations/hart-mea2007–071026.ppt#1 (accessed March 20, 2009).

Hartmann, B. and Boyce, J. 1981. Checking out the Bank. New Internationalist. 96: February 1981. URL: http://www.newint.org/issue096/bank.htm (accessed March 3, 2006).

Hu, F. 2015. Groundwater under pressure. In *China Water Risk*. URL: http://www.chinawaterrisk.org/resources/analysis-reviews/groundwater-under-pressure/ (accessed July 23,2019).

India (Government of) Planning Commission (IPC). 2004. *Economics of Bamboo Boring: A Study of the North East Region of Bihar*. IPC, New Delhi, India, 78 p.

International Fertilizer Association (IFA). 2006. *Fertilizer Consumption Statistics*. Available as electronic spreadsheet in Excel. URL: http://www.fertilizer.org/ifa/statistics/IFADATA/DATA/world.xls (accessed April 19, 2006).

International Network on Participatory Irrigation Management (INPIM) 2004. *PIM [Participatory Irrigation Management] and Financing*. URL: http://www.inpim.org/leftlinks/Topic%20Page/Topic%20Page/Financing (accessed April 8, 2006).

International Water Management Institute (IWMI) 2000. Pedaling Out of Poverty: Social Impact of a Manual Irrigation Technology in South Asia. Report 45.

Job, C. and Gabanski, G. 1988. Monitoring wells need consistent regulation. *Waste Age*, August, 164–170.

Kansakar, D.R. 2001. Shallow tubewell subsidy. *Kathmandu Post*. April 25, 2001. URL: http://www.cgisp.org.np/articles/kans.htm (accessed March 3, 2006).

Levin, R., and Harrington, W. 1995. Infectious waterborne disease and disinfection by-products in the US: costs of disease. In Reichard, E., Zapponi, G., ed. *Assessing and Managing Health Risks from Drinking Water Contamination: Approaches and Applications. Proceedings of a symposium held in Rome*, September 1994. Wallingford (U.K.): International Association of Hydrological Sciences; publ. no. 233.

Lifewater Canada (LWC). 2006. URL: http://www.lifewater.ca/ (accessed March 11, 2006).

Markandya, A., Harou, P., Bellu, L.G., and Cistulli, V. 2002. *Environmental Economics for Sustainable Growth*. Edward Elgar, Cheltenham, UK, 567 p.

Meghalaya, (Government of) 2004. *Shallow Tubewells with Pumpsets in Alluvial Areas*. URL: http://www.megcooperation.gov.in/ (accessed March 4, 2006).

Meier, J., Zabel, F., and Mauser, W. 2018. A global approach to estimate irrigated areas – a comparison between different data and statistics. Hydrology and Earth Systems. *Science (Journal)*, 22, 1119–1133. doi: 10.5194/hess-22-1119-2018 (accessed July 17, 2019).

Missouri Agricultural Experiment Station (MAES). 2001. Site-Specific Evaluation of a Claypan Soil with Subsurface Drainage-Impact on Corn Grain Yield; Final Report 2001. (Authors: Nelson, K.A., Shannon, D.K., Thompson, A.L., and Anderson, S.H.) URL: http://aes.missouri.edu/greenley/research/sitspec01.stm (accessed April 18, 2006).

Muir, P.S. 2005. *Human Impacts on Ecosystems; Fertilizers; Lectures of Fall 2005*, Oregon State University. URL: http://oregonstate.edu/~muirp/fertlim.htm (accessed April 19, 2006).

Navarro, A. 1989. Economics of groundwater works. In *Groundwater Economics: Selected Papers from a United Nations Symposium, Barcelona, Spain*. Elsevier Science Publishers B.V., Amsterdam, the Netherlands, pp. 25–51.

Nixon, R. 2013. Farm subsidies leading to more water use. *New York Times*, June 6, 2013. URL: https://www.nytimes.com/2013/06/07/us/irrigation-subsidies-leading-to-more-water-use.html (accessedJuly 24, 2019).

Nolan, B.T., Ruddy, B.C., Hitt, K.J., and Helsel, D.R. 1998. A national look at nitrate contamination of ground water. *Water Conditioning and Purification*, 39 (12), 76–79.

North Carolina Agricultural Extension Service (NCAES). 1988. *Fact Sheet: Chemigation Practices to Prevent Groundwater Contamination*. Fact Sheet #1. URL: http://es.epa.gov/techinfo/facts/nc/nc-fs1.html (accessed April 8, 2006).

O'Brien, D. M., Dumler, T. J., and Danny, H. Rogers. 2011. *Irrigation Capital Requirements and Energy Costs*. Kansas State University Agricultural Experiment Station and Cooperative Extension Service. URL: http://www.waterandenergyprogress.org/library/irrigationcosts.pdf (accessed July 24, 2019).

Pacific Institute (PI). 2015. *Water Use Trends in the United States*. Prepared by Kristina Donnelly and Heather Cooley. URL: https://pacinst.org/wp-content/uploads/2015/04/Water-Use-Trends-Report.pdf (accessed August 1, 2019).

Pal, S. 2016. How one man's simple low-cost tools are helping millions of farmers tackle climate change in India. *The Better India*. URL: https://www.thebetterindia.com/79360/amitabha-sadangi-idei-treadle-pump-micro-irrigation-farmers/ (accessed July 23, 2019).

Pimentel, D. 2001. Limits of biomass utilization. In *Encyclopedia of Physical Science and Technology*, 3rd edn., vol. 2., Academic Press, San Diego, CA, pp. 159–171.

Polak, P. and Sivanappan, R.K. 2006 *The Potential Contribution of Low Cost Drip Irrigation to the Improvement of Irrigation Productivity in India*. (Undated) International Development Enterprises, Lakewood, CO. URL: http://www.ideorg.org/Files/The%20Potential%20Contribution%20of%20Low%20Cost%20Drip%20Irrigation.pdf (accessed April 7, 2006).

Pew Research Center. 2019. *World's Population is Projected to Nearly Stop Growing by the End of the Century*. Prepared by Anthony Cilluffo and Neil G. Ruiz. URL: https://www.pewresearch.org/fact-tank/2019/06/17/worlds-population-is-projected-to-nearly-stop-growing-by-the-end-of-the-century/ (accessed July 17, 2019).

Rogers, D.H., Lamm, F.R., Alam, M., Trooien, T.P., Clark, G.A., Barnes, P.L., and Mankin, K. 2006. *Efficiencies and Water Losses of Irrigation Systems*. Kansas State University; Research and Extension. Marhattan, KS, URL: http://www.oznet.ksu.edu/library/ageng2/mf2243.pdf (accessed March 3, 2006).

Schneekloth, J.P., Norton, N.A., Clark, R.T., and Klocke, N.L. 2001. *Irrigating for Maximum Economic Return with Limited Water*. University of Nebraska-Lincoln Extension. G1422. URL: http://elkhorn.unl.edu/epublic/pages/publicationD.jsp?publicationId=194#top (accessed April 17, 2006).

Seelig, B. and Scherer, T. 1996. *Irrigation BMPs for Groundwater Protection from Pesticides. North Dakota State University Extension Service*. AE-1116. URL: http://www.ext.nodak.edu/extpubs/h2oqual/wat-grnd/ae1116w.htm (accessed March 3, 2006).

Siebert, S., Burke, J., Faures, J.M., Frenken, K., Hoogeveen, J., Doll, P., and Portmann, F.T. 2010. Groundwater use for irrigation – a global inventory. In *Hydrology and Earth Systems Science*, 14, 1863–1880. URL: www.hydrol-earth-syst-sci.net/14/1863/2010/ (accessed July24, 2019

Southgate, D. 2000. Best practice methods for valuing irrigation benefits. In *Dams and Development: A New Framework for Decision-Making*. The Report of the World Commission on Dams. Earthscan Publications Ltd, London, U.K., URL: http://www.dams.org/docs/kbase/contrib/eco070.pdf (accessed March 10, 2006).

Statistica. 2019. *Leading countries based on biofuel production in 2018 (in thousand metric tons of oil equivalent)*. URL: https://www.statista.com/statistics/274168/biofuel-production-in-leading-countries-in-oil-equivalent/ (accessed July 23, 2019).

Suárez, G.F., Fulginiti, L., and Perrin, R. 2018. What is the use value of irrigation water from the high plains aquifer? *American Journal of Agricultural Economics* URL: https://irrigationinnovation.org/library/what-is-the-use-value-of-irrigation-water-from-the-high-plains-aquifer/ (accessed August 2, 2019).

Siebert, S., Burke, J., Faures, J.M., Frenken, K., Hoogeveen, J., and Doll, P. 2010b. Groundwater use for irrigation – a global inventory. In *Hydrology and Earth Systems Science*, 14, pp. 1863–1880.

Sundquist, B. 2006. *The Earth's Carrying Capacity—Some Literature Reviews; Irrigation Overview (Chapter 1)*. URL: http://home.alltel.net/bsundquist1/index.html (accessed March 3, 2006).

Tsur, Y., Roe, T., Doukkali, R., and Dinar, A. 2004. *Pricing Irrigation Water*. Resources for the Future, Washington, DC, 319 p.

U.S. Department of Agriculture (USDA). 2013 *Farm and Ranch Irrigation Survey*, Washington, DC, URL: https://www.nass.usda.gov/Publications/AgCensus/2012/Online_Resources/Farm_and_Ranch_Irrigation_Survey/ (accessed July 23, 2019).

U.S. Department of Agriculture (USDA). 2017. *Land Values 2017 Summary*. URL: https://www.usda.gov/nass/PUBS/TODAYRPT/land0817.pdf (accessed July 23, 2019.)

U.S. Geological Survey. 2014. Estimated Use of Water in the United States in 2010. Circular 1405; 2018. Estimated Use of Water in the United States in 2015. Circular 1441.

United Nations Educational, Scientific and Cultural Organization (UNESCO). 2003a. Water for People, Water for Life. The United Nations Water Development Report prepared by the UN World Water Assessment Programme. 575 pp.

United Nations Educational, Scientific and Cultural Organization (UNESCO). 2003b. *Overview of the Activity of the UNESCO-IHP Working Group IV Project M-3.5 (C)ON Land Subsidence*. International Hydrological Programme. (Prepared by Laura Carbognin) Paris, France.

United Nations Food and Agricultural Organization (UNFAO). 1992. *The Use of Saline Waters for Crop Production. FAO Irrigation and Drainage Paper 48*. (Authors: Rhoades, J.D., Kandiah, A., and Mashali, A.M.) Rome, Italy. URL: http://www.fao.org/docrep/T0667E/t0667e00.htm#Contents (accessed April 18, 2006).

United Nations Food and Agricultural Organization (UNFAO). 2003a. Improving irrigation technology. In *Agriculture 21*. URL: http://www.fao.org/AG/magazine/0303sp3.htm (accessed March 3, 2006).

United Nations Food and Agricultural Organization (UNFAO). 2003b. Raising water productivity. In *Agriculture 21*. URL: http://www.fao.org/AG/magazine/0303sp3.htm (accessed March 3, 2006).

United Nations Food and Agricultural Organization (UNFAO). 2003c. Groundwater development has delayed the next food crisis. In *The Irrigation Challenge. Issues Paper 4*. UNFAO, Rome, Italy. URL: http://www.fao.org/documents/show_cdr.asp?url_file=/DOCREP/006/Y4854E/y4854e04.htm (accessed April 8, 2006).

United Nations Food and Agricultural Organization (UNFAO). 2006. *Irrigation Water Management: Irrigation Methods*. URL: http://www.fao.org/docrep/S8684E/s8684e00.htm (accessed March 11, 2006).

United Nations Food and Agricultural Organization (UNFAO). 2009. *Global Agriculture Towards 2050*; High Level Expert Forum. URL: http://www.fao.org/fileadmin/templates/wsfs/docs/Issues_papers/HLEF2050_Global_Agriculture.pdf (accessed July 17, 2019).

U.S. Department of Agriculture. 1950. *1950 Census Publications; Volume 3, Part 1: Irrigation of Agricultural Lands*. URL: http://agcensus.mannlib.cornell.edu/AgCensus/getVolumeTwoPart.do?volnum=3&year=1950&part_id=1074&number=1&title=Irrigation%20of%20Agricultural%20Lands (accessed August 2, 2019).

U.S. Department of Agriculture, Economic Research Service (USDA). 1987. *The Magnitude and Costs of Groundwater Contamination from Agricultural Chemicals; A National Perspective*. Washington, DC (Authors: Elizabeth G.N. and Linda K.L.). Agricultural Economic Report No. 576. Government Printing Office, Washington, DC, 31 pp.

U.S. Department of Agriculture, Economic Research Service (USDA) 1997a. *Agricultural Resources and Environmental Indicators, 1996–97*. Economic Research Service, Natural Resources and Environment Division, Washington, DC, Agricultural Handbook No. 712, 350 pp.

U.S. Department of Agriculture, Economic Research Service (USDA) 1997b. *Data: Western Irrigation*. Washington, DC. URL: http://www.ers.usda.gov/data/westernirrigation/ShowTables.asp?tabList=1#flag (accessed March 24, 2005).

U.S. Department of Agriculture, Economic Research Service (USDA) 2001. *Irrigation and water use: glossary; Irrigation systems and land treatment practices; Onfarm water conveyance systems*. Washington, DC. URL: http://www.ers.usda.gov/briefing/WaterUse/Questions/glossary.htm (accessed March 24, 2006).

U.S. Department of Agriculture, Economic Research Service (USDA) 2002a. *1997 Census of Agriculture*. Washington, DC. URL: http://www.nass.usda.gov/census/index1997.htm (accessed September 20, 2008).

U.S. Department of Agriculture, Economic Research Service (USDA). 2002b. *Irrigation and Water Use: Questions and Answers*. Economic Research Service, Washington, DC. URL: http://www.ers.usda.gov/Briefing/WaterUse/Questions/qa1.htm (accessed April 21, 2006).

U.S. Department of Agriculture, Economic Research Service (USDA). 2013. Census of Agriculture; Farm and Ranch Irrigation Survey, Table 7; Irrigation Wells Used on Farms, 2013. Washington, DC. URL: https://www.nass.usda.gov/Publications/AgCensus/2012/Online_Resources/Farm_and_Ranch_Irrigation_Survey/fris13_1_007_007.pdf (accessed July 23, 2019).

U.S. Department of Agriculture, Economic Research Service (USDA). 2007. The future of biofuels: A global perspective. In *Amber Waves*, November 2007, prepared by William Coyle. URL: http://www.ers.usda.gov/AmberWaves/november07/features/biofuels.htm (accessed October 3, 2007).

U.S. Department of Agriculture, Economic Research Service. 2018. *Feedgrains Sector at a Glance*. URL: https://www.ers.usda.gov/topics/crops/corn-and-other-feedgrains/feedgrains-sector-at-a-glance/ (accessed July 23, 2019)

U.S. Department of Agriculture, Economic Research Service (USDA ERS) 2019. *Irrigation and Water Use*. URL: https://www.ers.usda.gov/topics/farm-practices-management/irrigation-water-use/ (accessed August 2, 2019).

U.S. Department of Agriculture (USDA) 2018. *2013 Farm and Ranch Irrigation Survey; Table 4, Estimated Quantity of Water Applied By Source, 2013 and 2018*. URL: https://www.nass.usda.gov/Publications/AgCensus/2012/Online_Resources/Farm_and_Ranch_Irrigation_Survey/fris13_1_004_004.pdf (accessed July 17, 2019).

U.S. Department of Agriculture (USDA) 2019. *Conservation Stewardship Program.* URL: https://www.nrcs.usda.gov/wps/portal/nrcs/main/national/programs/financial/csp/ (accessed July 24, 2019).

U.S. Geological Survey (USGS). 1997. National Water Summary on Wetland Resources; Technical Aspects of Wetlands; History of Wetlands in the Conterminous United States. (Authors: Dahl, T.E. and Allord, G.J.) Water Supply Paper 2425. URL: http://water.usgs.gov/nwsum/WSP2425/index.html (accessed April 18, 2006).

U.S. Geological Survey (USGS). 2000. The Quality of Our Nation's Waters; Pesticides in the Nation's Streams and Ground Water, 1992–2001, USGS Circular 1291, 2000, 172.

U.S. Geological Survey (USGS). 2004. *Estimated Water Use in the United States in* 2000. USGS Circular 1268. United States Government Printing Office, Washington DC. URL: http://water.usgs.gov/pubs/circ/2004/circ1268/htdocs/text-do.html (accessed March 25, 2005).

U.S. Geological Survey (USGS). 2005. *Irrigation Techniques and Some Irrigation Methods.* Reston, VA. URL: http://ga.water.usgs.gov/edu/irmethods.html (accessed March 11, 2006).

U.S. Geological Survey (USGS). 2006. The Quality of Our Nation's Waters; Pesticides in the Nation's Streams and Ground Water, 1992–2001. *USGS Circular* 1291. 172 pp.

U.S. Geological Survey (USGS). 2007. *The Future of Biofuels: A Global Perspective.* By William Coyle. *Amber Waves.* URL: http://www.ers.usda.gov/AmberWaves/November07/Features/Biofuels.htm (accessed March 20, 2009).

U.S. Geological Survey (USGS). 2014. *Estimated use of water in the United States in 2010.* USGS Circular 1405. Reston, VA. https://pubs.usgs.gov/circ/1405/pdf/circ1405.pdf

U.S. Geological Survey (USGS). 2018. *Estimated Use of Water in the United States in 2015.* USGS Circular 1441, USGS, Reston, VA. https://pubs.er.usgs.gov/publication/cir1441

U.S. Geological Survey (USGS). 2019. *Evapotranspiration and the Water Cycle.* URL: https://www.usgs.gov/special-topic/water-science-school/science/evapotranspiration-and-water-cycle?qt-science_center_objects=0#qt-science_center_objects (accessed August 2, 2019).

University of Illinois at Urbana-Champaign (UIUC) 2004. *Conservation Drainage.* URL: http://www.wq.uiuc.edu/dg/conservation.htm (accessed April 18, 2006).

University of Minnesota Extension Service (UMES). 2001. Planning an Agricultural Subsurface Drainage System. (Authors: Wright, J. and Sands, G.) URL: http://www.extension.umn.edu/distribution/cropsystems/DC7685.html (accessed April 18, 2006).

Vieira, Luis. 2017. Brazilians Are Now Making Ethanol With Soybeans. In *Successful Farming*, November 15, 2017. URL: https://www.agriculture.com/news/crops/brazilians-are-now-making-ethanol-with-soybeans (accessed July 23, 2019).

Virginia Cooperative Extension. 2000. Farm Financial Issues and Options; Conservation and Cost-Share Assistance. (Authors: Shank, R.; Grover, G., and Roberts, M.) Publication Number 446-500W. URL: http://www.ext.vt.edu/pubs/agecon/446–500/conserv.html (accessed April 17, 2006).

Water and Food (WAF). 2006. Crop-Water Productivity Improvement. (Undated) URL: http://www.waterforfood.org/pdf/T_1_Brochure.pdf+salinity+crop+reduce+productivity&hl=en&gl=us&ct=clnk&cd=2 (accessed April 7, 2006).

Wessels Living History Farm (WLHF). 2006. Digging Wells; Irrigation Pumps. URL: http://www.livinghistoryfarm.org/index.html (accessed March 10, 2006).

Wisconsin State Laboratory of Hygiene (WSLH). 2003. *Pesticides in Drinking Water.* Madison, WI. URL:http://www.slh.wisc.edu/ehd/pamphlets/pesticide.php (accessed April 8, 2006).

World Bank. 2004. *Investments in Shallow Tubewells for Small-Scale Irrigation.* Agricultural Investment Sourcebook. URL: http://www-esd.worldbank.org/ais/index.cfm?Page=mdisp&m=8&p=04 (accessed March 3, 2006).

World Health Organization (WHO) 2019. *World hunger again on the rise, driven by conflict and climate change, new UN report says.* URL: https://www.who.int/news-room/detail/15-09-2017-world-hunger-again-on-the-rise-driven-by-conflict-and-climate-change-new-un-report-says (accessed July 17, 2019).

World History International (WHI). 2005. *"Mesopotamia" and "Ancient Egypt."* URL: http://history-world.org (accessed March 11, 2006).

World Population Review (WPR). 2019. *World Population 2019.* URL: www.worldpopulationreview.com/continents/world-population/ (accessed July 17, 2019).

Worldwatch Institute (WI). 2005. *State of the World 2005; Trends and Facts*; Cultivating Food Security. URL: www.worldwatch.org/features/security/tf/10 (accessed April 19, 2006).

Worldwatch Institute (WI). 2002. *State of the World 2002; World Fertilizer Use, 1961–2002.* Washington, DC. Available as electronic spreadsheet in Excel, 2006.

9 Mineral Extraction, Geothermal Exchange and Manufacturing

GROUNDWATER IN INDUSTRIAL ECONOMIC APPLICATIONS

Mineral extraction historically has been an economic activity for millennia. Likewise, people have exploited geothermal water sources over centuries for a range of purposes including heating, cooking and health applications. Early historical groundwater use for manufacturing includes pumping brines for producing salt. In one sense, the earliest water wells were used to extract groundwater, not in a mining approach but as a continuing essential life resource. This chapter will focus on mining for substances underground, drawing on the thermal quality of groundwater and groundwater use in industrial processes.

Mining relies on access to a dependable water source, without which mining could not be done (ICMM, 2012). Relatively, the use of groundwater in mining has been recent. With the advent of steam-driven pumps in the Industrial Revolution, miners solved the problem of getting groundwater out of mine shafts (GKC, 2019). Fresh groundwater use in mining has generally declined while saline groundwater use has increased, mirroring somewhat the trend in total water use in mining (USGS, 2018 and earlier dates). Fresh and/or saline groundwaters are used in mining in all 50 United States and Puerto Rico. While mining varies from place to place and state to state, it includes extraction of solid, liquid and gaseous minerals and comprises quarrying, milling, pumping, injection and other operations (USGS, 2018). The value above all costs and returns of mineral extraction globally is approximately 0.5% of the world's gross domestic product (World Bank, 2019) with mining contributing $494 billion to the gross domestic product of the United States in 2019 (about 2.4% of GDP) (TE, 2019). Mineral extraction also has produced environmental effects in some localities, such as air, soil and water pollution, as well as landscape disturbance and loss of wildlife habitat (AGI, 2019a).

Groundwater thermal exchange (also referred to as 'thermal aquifer potential') transfers heat value from and to the groundwater for the purposes of heating and cooling, relying on the relatively constant temperature of the subsurface environment (Lund et al., 2004; Bottcher et al., 2019). This technology applied to shallower groundwater zones is commonly known as the groundwater heat pump and has been identified as a means to reduce greenhouse gas emissions to mitigate climate change. Use of this shallow geothermal energy requires detailed knowledge of a locale's hydrogeology (Bottcher et al., 2019). Geothermal (or "ground source") heat pumps can be used in all parts of the United States (USDOE, 2019) and are growing in use worldwide (Lund et al., 2004). For a small electrical input to operate the thermal exchange unit, four times the energy input is obtained in output (Lund et al., 2004), with cost savings and reduced emissions resulting. Deep geothermal systems obtain heated groundwater from hot springs or deep wells for a range of direct uses to which the heat is applied. Both ground source heat pumps and direct use geothermal systems offer potential to reduce carbon and greenhouse emissions.

Manufacturing industrial water use may be locally significant. Environmental regulation and constrained water availability in some places have caused industry to consider options to manage reductions in water use. Industrial water reuse offers cost savings in addition to ecosystem benefits of maintaining more water in the environment.

MINERAL EXTRACTION (OTHER THAN OIL AND GAS)

Minerals production supports significant commercial and industrial activity. Total non-fuel mine output in 2015 in the United States was 2.601 billion metric tons (excluding helium and lithium carbonate) with a value of $72.2 billion. Table 1.9.1 provides additional detail. In 2015, the principal metal-mined output was in descending order of amount iron ore, copper, zinc, lead and zircon. The primary industrial minerals in descending quantity were crushed stone, sand and gravel, portland cement, salt and marketable phosphate rock. These five metals and five industrial minerals accounted for 97% of non-fuel U.S. mineral extraction by volume. Twelve minerals comprised 89% of the U.S. production value in descending order of value contributed: crushed stone, portland cement, gold, copper, construction sand and gravel, industrial sand and gravel, iron ore, salt, lime, marketable phosphate rock, soda ash and zinc (USGS, 2019a). Coal production in the United States in 2017 was 703 metric tons (775 short tons) from 24 states. Sixty-five percent of the coal was produced from surface mines which were 64% of coal mines (USEIA, 2018a). The coal production was primarily for fueling 359 electric power plants (USEIA, 2018b).

Coal mining in the United States is an industry in transition. Production in 2017 was down 33% from the peak production of 1,162.7 million tons in 2006. Employment of 50,000 coal miners is down from a peak of 883,000 in 1923. Generation of electricity is the largest user of coal, being used to produce 50% of electric power in 2005 and 27% in 2018. The U.S. is a net exporter of coal. U.S. coal exports, for which Europe is the largest customer, peaked in 2012. In 2015, the U.S. exported 7.0% of its production. (Note that some aspects of oil and gas extraction related to groundwater are addressed in Volume 1, Chapter 6, under "Injection Wells and Hydraulic Fracturing".)

DESCRIPTION

The principal techniques for mineral extraction from the subsurface, exclusive of oil and gas production, are (USGS, 2019b; AGI, 2019b; Deberitz, 2006): (1) underground mining though mine shafts, (2) surface mining using open pits, (3) placer mining, (4) in-situ mining and (5) evaporation ponds. Use of groundwater in oil and gas extraction is covered in Chapter 5. Factors affecting the type of mining include the site and dimensions of the mineral deposit, the rock strength, quality of the ore, costs to install a mine and the mineral market prices, among other considerations. High-quality ores in veins mined through underground shafts and ore in significant layers that are drilled and

TABLE 1.9.1
United States Mine Output in 2015

Material	Mine Output (000s metric tons)	Value (US$000s)
Metals	48,800	$24,400,000
Industrial Minerals	2,552,383	$48,300,000
Total	2,601,183	$72,700,000

Source: U.S. Geological Survey. 2019a. 2015 Minerals Yearbook; Statistical Summary. URL: https://prd-wret.s3-us-west-2.amazonaws.com/assets/palladium/production/atoms/files/myb1-2015-stati.pdf (Accessed September 28, 2019).

excavated below ground are usually more costly to extract. The ore rock is crushed to enable the waste rock to be removed from the profitable metallic ore.

Lower quality ores near the ground surface are mined through excavation from the surface with less costly methods which may include drilling and blasting. Coal is produced both by surface mining and underground mining methods.

Placer mining washes sediments and deposits from current streams and old buried channels, and lake beds and beach sands. The sluicing of the materials with water accumulates the heavier minerals for processing. Placer mining is used to recover valuable minerals from sediments in present-day river channels, beach sands or ancient stream deposits. More than half of the world's titanium comes from placer mining of beach dunes and sands. In placer operations, the mined material is washed and sluiced to concentrate the heavier minerals. Most titanium is recovered by this method.

In-situ mining uses chemical processes to dissolve minerals in the ground which are then pumped to the ground surface for processing. Uranium is mined though this technique as well as through underground and open-pit mining.

Evaporation ponds are used to concentrate minerals where the minerals are in brine solutions that are pumped to the ground surface. Ponds are engineered to hold the solution for evaporation of the water in locations where evaporation rates are higher than precipitation rates, such as in arid places of Nevada USA or Chile. Once concentrated in the pond beds, the minerals are removed from the ponds and are further processed to obtain a high-quality material. Major lithium-containing brines are mined using this method. Lithium is also produced through milling spodumene from open-pit mines (Deberitz, 2006).

Costs in Production

Mining has large capital and operational costs to extract valuable minerals for use in the economy (Ossa-Moreno et al., 2018). In mining investment, water is a significant factor with large volumes withdrawn, used and discharged, resulting in expenditure for water infrastructure and management. A mine often operates as long as 50 years or more and therefore has demands for water over this timeframe (Ossa-Moreno et al., 2018). A major contributing factor is that average revenues per unit of water used is higher than the competing agricultural use (Moran et al., 2008).

In establishing mining operations, a critical first step is securing access to sufficient water. In many locations of arid climate, groundwater is the only reliable water source. Acquiring water rights for the volume of water used in mining is essential. A water right in the western United States that function under prior appropriation doctrine for water provides the holder of the right to withdraw water from the water resources held in trust by the state for a specific use and following conditions set by the state. Rights to use a defined volume of water in a location are separate from land ownership and can be sold a person who is not the landowner. However, owning rights to water does not guarantee that the water be available, as the state may reduce amounts of water for use by water rights holders in times of drought. Water rights for mining may be valued in tens of thousands of dollars per hectare-meter (acre-foot).

Costs in mining for water use may be sorted into direct use, indirect use and potential regional impacts (Ossa-Moreno et al., 2018). Direct use costs include mineral processing including grinding and flotation with chemical additions such as sulfuric acid, movement and shipment of concentrated metals and minerals, placement of tailings in dams and then impoundments, drinking water for miners and dust control (ADMMR, 2010). Water from mineral production processes may be collected and recirculated to earlier points in the process such as to leach heaps with additional infrastructure required. Water may also need to be removed (referred to as dewatering) from deep shaft or open pit mines to enable mineral extraction, which includes costs for removal and then discharge to a stream or other reservoir that requires a permit. Indirect use costs consist of evaporative losses from leach heaps, ponds and organic washing, water piped to neighboring populations and water needed for ecosystem maintenance (Moran et al., 2008; ADMMR, 2010). Figure 1.9.1 gives a generalized mine water flow.

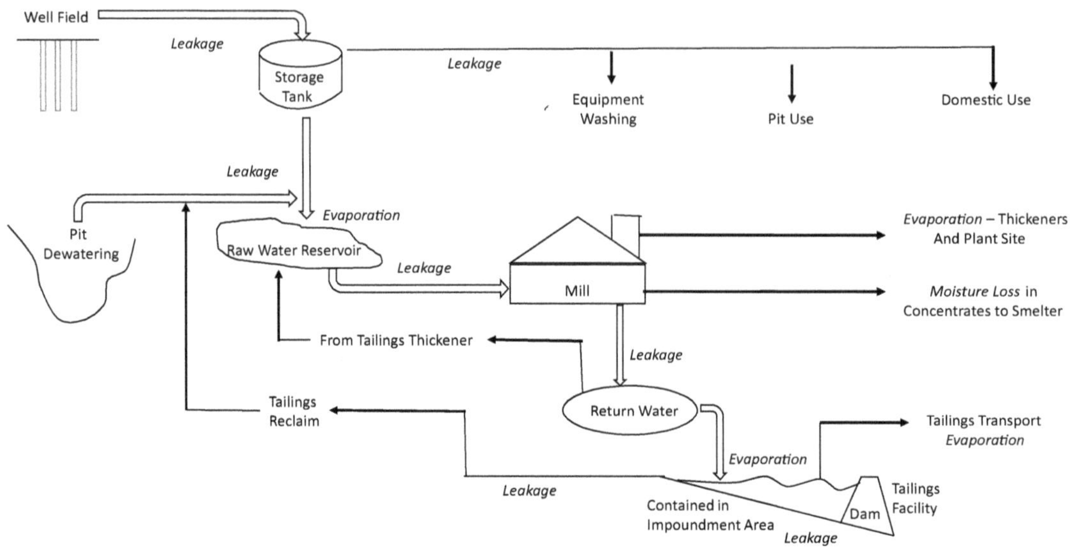

FIGURE 1.9.1 Generalized mine water flow.

Source: Adapted from: Arizona Department of Mines and Mineral Resources (ADMMR). 2010. Water Consumption at Copper Mines in Arizona. Special Report 29. Prepared by Madan M. Singh.

Additional costs are the state mining and federal/state environmental permits.

Regional costs comprise groundwater depletion, surface water quantity and quality impacts and channel creation or modification for water diversion (Moran et al., 2008) as well as losses to opportunities foregone for alternative water uses. Because of high revenue and the cost of investment in mines, the occurrence of water shortage is a significant matter affecting operations. These circumstances may result in mine management having a willingness to make higher payments for water reliability than other potential water users can afford (Young and Loomis, 2014).

Groundwater Source Use

Use of water in mining is mainly for mineral processing, dust control, slurry transmission and water supply for employees (Mining Technology, 2013). Mining water use also includes injection of water for secondary oil recovery and for hydraulic fracturing to support unconventional oil and gas production as well as other operations (USGS, 2018). Groundwater is often a water source, particularly in locations away from streams and lakes. Also, groundwater in underground mines can interfere with mineral extraction and must be removed affecting water tables, reducing stream flow and, if groundwater is discharged to streams, causing pollution of surface waters (Mining Technology, 2013).

In 2015, an estimated 15,141 m^3/day (4,000 MGD) of water were used for mining, about 1% of total water withdrawals in the United States. Groundwater accounted for 72% of total mining withdrawals of water. Sixty-five percent of the groundwater used in mining was saline. Groundwater withdrawals in 2015 were 1% greater than in 2010 while surface-water withdrawals were nearly the same (USGS, 2018). Table 1.9.2 provides an example of water use per mine and average water use per pound for copper production in Arizona (USA). Groundwater use is substantial in the mining and processing of copper relative to the output of metal product.

The mining industry cites opportunities to conserve water. Mining operations can capture and recycle water used in mineral processing which is the largest use of water. Additionally, seepage water from tailings piles can be recycled. Mine dewatering can provide a further water source, rather than discharging this water to surface waters (Mining Technology, 2013).

TABLE 1.9.2

Average Water Use per Mine and per Pound of Copper Mined

Arizona produces more copper than any other of the United States with only Chile and Peru producing more. Arizona has adopted practices to protect its water resources while extracting this valuable metal. Average water use per mine and per pound of copper mined from 2004 to 2008 is given below. Groundwater is the predominant source.

			Average Use per Year[a]		Average Use/Unit[b]	
Mine	**GW Basin**	**Water Source**	**Million m³**	**Acre-Feet**	**Liter/kg**	**Gal/lb**
Bagdad	Big Sandy	Groundwater	18.305	14,840	199.6	23.8
Miami	Salt River	Groundwater	41.790	33,880	519.9	62.3
Mission	Tucson AMA[c]	Groundwater	8.368	6,784	208.6	25.0
	Central Arizona Project		2.151	1,744	(combined above)	
Morenci	Morenci	Groundwater	11.052	8,960	45.1	5.4
Salt River	---	Surface Water	5.131	4,160	n.a.[d]	n.a.
Ray	Lower San Pedro	Groundwater	17.836	14,460	171.9	20.6
Sierrita	Tucson AMA	Groundwater	33.440	27,110	455.7	54.6
Silver Bell	Tucson AMA	Groundwater	1.195	969	55.9	6.7

[a] Averaged over years 2004 to 2008

[b] *ditto*

[c] Aquifer Management Area

[d] n.a. = not available

Source: Adapted from: Arizona Department of Mines and Mineral Resources (ADMMR). 2010. Water Consumption at Copper Mines in Arizona. Special Report 29. Prepared by Madan M. Singh. URL: http://repository.azgs.az.gov/sites/default/files/dlio/files/nid1295/sr29waterconsumptioncoppermines.pdf (Accessed October 3, 2019).

Environmental Factors

While challenging to quantify in all cases, responding to environmental impacts of mining is a cost of this business sector. Ensuring that discharges from mine sites meet standards are considered typical costs of mine operations. Depletion of ground or surface waters may require compensation to other water users. Pollution from acid mine drainage, leach heaps or tailings piles create the need for water quality management, including creation of catchments for this impaired water in order to treat it before discharging water from the mine site. Open water holding ponds for process water can leak contaminating groundwater. One hundred forty mine sites are proposed, currently on or have been remediated and deleted from the U.S. Superfund National Priorities List of hazardous waste sites or are using the Superfund alternative approach for remediation, not including coal mining sites separately addressed by the Department of the Interior's Office of Surface Mining (USEPA, 2018). Soil contamination that can be a source of groundwater contamination and loss of wildlife habitat may also be environmental impacts of mining (WA, 2019). Box 1.9.1 describes a mine site remediation affecting groundwater quality. At the mine site, groundwater remediation met cleanup standards but was not of safe quality for human consumption. As a result, the U.S. Environmental Protection Agency and the Washington (State) Department of Ecology established institutional controls precluding use of groundwater at the site for human consumption because of the presence of cyanide and arsenic (WDEQ, 2019). The state continues to monitor the site, incurring operation and maintenance costs. Box 1.9.2 highlights actions at one mine site to control groundwater quality impacts.

The mining industry is addressing environmental issues through modernization of sites. Upgrading mine sites includes state-of-the-art water reservoir sealing to minimize leakage of contaminated water, advanced water treatment, water reuse and water conservation (Chiaro, 1997). Box 1.9.3 highlights steps to conserve water and protect water quality at mining sites in Arizona,

BOX 1.9.1 MINE SITE REMEDIATION OF GROUNDWATER

Silver Mountain Mine site covers 2 hectares (5 acres) in Horse Springs Coulee near Tonasket, Washington. The mine operated irregularly from 1928 to mid-1960s for silver, copper and gold extraction from soil and ore. During the early 1980s, mine operations at the site used cyanide leach recovery to remove silver and gold from tailings left from past mining when mine owners:

- Created a cyanide heap leach pile
- Applied sodium cyanide in water to the mined tailings
- Collected the cyanide-laden solution in a leachate pond at the base of the heap
- Abandoned heap leach operation in 1981 with no cleanup of cyanide-contaminated waste
- Contaminated groundwater from mine-tailing wastes

The U.S. Environmental Protection Agency:

- Added the mine site to the National Priorities List of abandoned waste sites in 1984
- Developed a remediation plan in 1990
- Completed site remediation in 1992

Steps in the mine site remediation process included:

- Treatment of heap leach pile and an on-site pond to neutralize cyanide
- Consolidated, covered and seeded mine tailings and heap leach piles
- Removed drums of hazardous materials
- Established cleanup standards for the heap pile and mine dump materials and the surrounding soils at 200 milligrams per kilogram (mg/kg) for arsenic and 95 mg/kg for total cyanide
- Monitored groundwater quality on site and to confirm that contaminants found in perched aquifer did not adversely affect regional aquifer
- Blocked mine entrance and diverted runoff
- Fenced mine site for safety and restricting access
- Established institutional control covenant to ensure long-term protectiveness of soil cap and prohibition of groundwater use for human consumption
- Regularly inspected and monitored site through Washington Department of Ecology
- Removed mine site from National Priorities List (NPL) in 1997

Ongoing site status:

- Currently used to graze cattle
- Periodically inspected
- Every five years, site review conducted to ensure remedial measures still protect people and environment.

Sources: U.S. Environmental Protection Agency (USEPA). 2019b. Superfund Site: Silver Mountain Mine, Loomis, WA. URL: https://cumulis.epa.gov/supercpad/SiteProfiles/index.cfm?fuseaction=second.Stayup&id=1000948#Stayup and https://cumulis.epa.gov/supercpad/SiteProfiles/index.cfm?fuseaction=second.cleanup&id=1000948 (Accessed October 5, 2019).

Washington Department of Ecology (WDEQ). 2019. Cleanup Site Details Report; Silver Mountain Mine. URL: https://apps.ecology.wa.gov/gsp/Sitepage.aspx?csid=3033 (Accessed October 5, 2019).

BOX 1.9.2 CONTROLLING GROUNDWATER QUALITY – A MINING EXAMPLE

The Cerro Vanguardia S.A. (CVSA) is the largest gold and silver mine in size and production in the Patagonia region of Argentina. Located in the province of Santa Cruz, the mine includes an extensive zone of quartz veins situated in volcanic rocks.

- Exploration began in 1990 with other mining in Argentina
- Community engagement through information sharing showed acceptance of CVSA in 2009 and 2010 by 54% of the community increasing to 71% in 2011
- Operation of the tailings storage facility (TSF)
    - Receives continuous deposit of processed mine residue as slurry containing water and tailings
    - Separates tailings and water containing traces of cyanide
    - Continually reclaims water for reuse in gold recovery processing
- Recycling and reusing up to 95% of cyanide before water in slurry deposited in TSF
    - Prevents risk of seepage of contaminated water into the soils and groundwater under TSF
- Continuous monitoring of wells around TSF
    - Established baseline groundwater quality at start of mine operation
    - Obtains water samples analyzed for groundwater quality
    - Verifies effects of tailings on groundwater
        - Assesses presence of heavy metals and cyanides
        - Analyses physical and chemical characteristics of groundwater to meet requirement of Argentine law
- Problem identification and mitigation in 2003
    - Monitoring provided early warning of isolated peak in cyanide levels in one well
    - Tailings discharge stopped and investigations initiated
        - Additional monitoring wells installed and geophysical analysis conducted
        - Quartz vein in bedrock served as a conduit for cyanide leak to groundwater
        - Thick two-layer high-density polyethylene (HDPE) liner installed to stop seepage
        - Electronic sensors installed between layers for leak detection
        - Monitoring demonstrating prevention of cyanide release to groundwater
- Continuing environmental assessment and community communication
    - Applying standards more stringent than required by Argentine law
    - Complying with the International Cyanide Management Code
    - Implementing ISO 14001 standard for environmental management systems

Source: International Council on Mining and Metals (ICMM). 2012. Water management in mining: a selection of case studies. Pages 27–28. URL: https://www.icmm.com/website/publications/pdfs/water/water-management-in-mining_case-studies (Accessed on October 4, 2019).

an arid mining environment. Box 1.9.4 provides an example of water conservation steps in practice at a mine through collection and reuse. In the example, the use of water per unit of copper mined is 2.086 liters/kilogram or 0.25 gallons/pound, the equivalent water use for mining in the United States overall for water use for all mineral production in 2015 (exclusive of helium, lithium, coal, oil and gas). A major point of the example is that the water used in copper ore processing is recycled three times on average with 17% of the water required to be made up rather than all the water being lost to release or discharge.

BOX 1.9.3 MINE SITE WATER MANAGEMENT PRACTICE

Measuring water use and quality and factors affecting water are fundamental for water management. Steps for mine site water management over the life of the mine operation may include:

- Monitoring water quality, volumes withdrawn and aquifer level and maintaining electronic records
- Checking level of solution ponds to avoid spills, water infiltration and other water losses
- Testing the quality of underground water sources to ensure no leakage from tailings, leach heaps or other sources
- Modeling aquifer processes that may be useful in mitigating impacts
- Noting the inputs and outputs of treatment plants
- Supervising basin slopes, if applicable
- Checking seismic stability of tailings dams, and other features possibly affected by earthquakes
- Preventive timely maintenance of pipes and controls on evaporation
- Installing instrumentation for timely detection of leaks
- Monitoring and recording flows and quality in distribution lines
- Installing adequate water storage capacity and monitoring stored water levels to avoid emergencies

Water consumption reduction measures may include:

- Reviewing historical water consumption records to identify possible water use reduction options
- Inspecting pipes for water losses and making corrections
- Reviewing current water use procedures for improvement, such as changing practices or equipment
- Checking water recirculation processes and making corrections
- Examining water use for other mine-related purposes, such as road maintenance, dust control, office staff use or landscaped areas.

Practices for water use control at mines generally may include:

- Recovering water wherever possible
- Recirculating cooling water after treatment if necessary
- Installing high density thickeners and pressure filters at concentrators to increase efficiency
- Improving recirculation from tailings impoundments by
  - designing to get better recovery (reducing losses from evaporation, infiltration and retention)
  - decreasing impoundment permeability by using geosynthetic liner or other fine material, installing a basin drain (to reduce filtration) and filtering the tailings
- Improving heap leach water use function by using drip irrigation to reduce evaporation and constructing drains for collection
- Covering process solution ponds to limit evaporation
- Employing spill avoidance procedures to avoid spills, particularly in loading tank trucks or tanks
- Watering at least half of roads in one pass to comply with air quality permits

(*Continued*)

BOX 1.9.3 (Continued)

- Paving or applying chemical stabilizers to roads to reduce water use
- Landscaping with native species
- Conserving water in offices and buildings

Source: Arizona Department of Mines and Mineral Resources (ADMMR) (2010). Water Consumption at Copper Mines in Arizona. Special Report 29. Prepared by Madan M. Singh. URL: http://repository.azgs.az.gov/sites/default/files/dlio/files/nid1295/sr29waterconsumptioncoppermines.pdf (Accessed October 3, 2019).

BOX 1.9.4 WATER CONSERVATION AT A MINE USING GROUNDWATER

The Morenci Mine in Arizona is one of the largest copper mines in the world.
Location and Facilities

- Area covers 24,300 hectares (60,000 acres) with five pits
- Solution extraction/electrowinning (SXEW) facility at the site

Water Availability and Use

- Freeport-McMoRan mining company has a lease agreement with the San Carlos Apache Tribe for up to 17.628 million cubic meters (14,000 acre-feet) per annum and entitlement to 308.37 million cubic meters (250,000 acre-feet) of water from the Black River.
- Water use at Morenci Mine, primarily groundwater, was as follows:
  - 1991-1995
    - Groundwater – 16.899 million m^3 (13,700 acre-feet or AF)
    - Surface Water – 2.991 million m^3 (2,425 acre-feet)
  - 1996-2000
    - Groundwater – 21.956 million m^3 (17,800 AF)
    - Surface Water – 2.596 million m^3 (2,105 AF)
  - 2001-2005
    - Groundwater – 9.991 million m^3 (8,100 AF)
    - Surface Water – 1.407 million m^3 (1,141 AF)

Water Conservation

- Investment in \$9-million water recovery and control system to keep water onsite
- Monitoring wells check downstream water quality
- Mine process uses 2,086 liters per metric ton (500 gallons of water per ton) to treat copper (2.086 liters/kilogram or 0.25 gallons/lb of Cu) in concentrators
  - Of this amount, only 321.8 liters (85 gallons) (0.355 liter/kg or 0.0425 gallons/lb) is fresh make-up water
  - Remaining water is recycled on average three times

Source: Arizona Department of Mines and Mineral Resources (ADMMR). 2010. Water Consumption at Copper Mines in Arizona. Special Report 29. Prepared by Madan M. Singh. URL: http://repository.azgs.az.gov/sites/default/files/dlio/files/nid1295/sr29waterconsumptioncoppermines.pdf (Accessed October 3, 2019).

Benefits

Mining has a high gross value added per unit of water used (the monetary value of the product less the cost of inputs, including water). In Australia in 2014, mining had a gross value added of AUS$200 per gigaliter of water used, higher than other major industrial and commercial sectors (Ossa-Moreno et al., 2018).

GEOTHERMAL EXCHANGE

Geothermal Heat Pumps

Most of the geothermal heat pumps (85%) in the United States utilize a heat exchange process in the ground that circulates the fluid media in a closed-loop system of wells (USDOE, 2017). The pipes are typically made of plastic tubing and are buried horizontally (up to 6 feet deep) or vertically (up to 600 feet deep). The geothermal exchange relies on the temperature difference between the heat in the pipes and the relatively constant temperature of the groundwater. Heat is exchanged in two places: (1) in the ground, heat is either drawn from the aquifer by water in the pipes for heating or transferred to the aquifer for cooling and (2) in the house or building, the heat is given up through a heat exchanger to heat the structure or taken up to cool it. The design of a geothermal exchange system is based on the size of the home or structure, conditions of climate zones, soil, subsurface geology, available land area, access to groundwater (or surface water) and state and local requirements for installation (USDOE, 2017; Bravo, 2019). From 2013 to 2015, new homes constructed with ground source heat pumps averaged 9,000 per year and increased to an average of 12,000 per year from 2016 to 2018 (USCB, 2019).

Description

Typically in the United States, closed-loop systems are required by state code. Closed loop means that water circulating to exchange heat with the aquifer cycles within pipes, only the heat is exchanged with groundwater, not the water, which remains in the pipes. In an open-loop system, water from one well draws groundwater to heat or cool the structure with the indoor heat exchanger and water that has passed through the indoor heat exchanger is pumped into the ground by another separate well. State regulatory concerns with an open-loop system are that minerals may be removed or chemicals may be added that change the chemistry of the water going back into the aquifer that may not be compatible with groundwater or geology and become or cause a contaminating situation for other groundwater uses or users (Yellich, 2019).

The portion of the geothermal exchange closed-loop system in the ground may be of two types: (1) horizontal or (2) vertical. In a horizontal system, a well is connected to horizontal pipes placed across the upper zone of the aquifer. A horizontal installation has to have sufficient land area to accomplish the heat exchange in at least 3-5 trenches that are 40 to 50 meters (130 to 160 feet) long and 4 to 7 meters (12 to 20 feet) apart (Bravo, 2019). The horizontal pipes circulate the water in them, exchanging heat with the groundwater. In a vertical type of system where land area is limited, a number of wells are installed into the aquifer at depths of 45 to 180 meters (150 to 600 feet) (Bravo, 2019; MU, 2012).

The heat exchange unit in the house or building may typically be either a water-to-air heat pump or water-to-water system. Water-to-air systems exchange heat with the air with the heated or cooled are circulated through the structure by a fan blowing air through air ducts. Water-to-water systems exchange heat between the water in the loop system and water of the structure's floor system or baseboard heat system (Bravo, 2019).

Cost

Costs of geothermal systems may range as follows with drilling or excavation expenses being 55 to 75% of the total cost (Bravo, 2019):

Shallow well with horizontal pipe loops

- water-to-water heat exchange: $16,000–$25,000
- water-to-air heat pump: $12,000–$20,000

Vertical looped wells

- water-to-water heat exchange: $20,000–$30,000
- water-to-air heat pump: $16,000–$24,000

Shallow horizontal pipe loops with water-to-air heat pumps are the least expensive, while vertical looped wells with water-to-water heat exchange are the most expensive systems. Larger structures with requirements for larger geothermal exchange systems will cost more. The indoor heat pump life expectancy is 25 years while the underground pipe loops' life is 50 or more years. Even though a geothermal installation may be more expensive than traditional oil or gas heating, geothermal systems may have 65% greater efficiency resulting in a payback period of 10 years or less (USDOE, 2017).

Groundwater Source Use

For the closed-loop system, water is cycled within the sealed loops. The pipe in the loops may be cased in concrete, adding further prevention of water loss (MU, 2012). Groundwater use is minimal. In an open-loop system, water pumped from the ground is returned to the ground, so groundwater use is also negligible.

Benefits

In terms of environmental factors, geothermal exchange systems (or ground source heat pumps) are the only heating and cooling system with a coefficient of performance (COP) greater than one (Brown, 2015). That means that for every unit of energy put into the system, more energy output is produced. At the Miami University (Ohio, USA) geothermal plant that exchanges heat through 690 vertical loop wells, the COP in winter is over 4 while the COP in summer has ranged from 6 to 20 (Hammerle, 2019). The high COP translates to reduced carbon dioxide and other greenhouse gas emissions. For a house of 100 m^2 (1,000 ft^2), five tons of greenhouse gas emissions may be reduced (CCC, 2018). For a large system with an extensive wellfield, the overlying land can be used for park and recreation area (Hammerle, 2019). For closed-loop systems, water is cycled through the system and not consumed.

Direct Use Deep Geothermal Sources

Direct use geothermal sources mine the heat from deeper heated aquifer zones. Wells tap the heated groundwater that averages an increased temperature of slightly more than 0.5 degree Celsius for every 20 meters (1 degree Fahrenheit for every 70 feet) of depth. Heated groundwater sources can be as hot as 93 degrees Celsius (200 degrees F) or more (USEPA, 2016). These waters may rise naturally to the ground surface as hot springs. The hot water may be used directly or may be circulated to a heat exchanger to extract the heat for other purposes. The principal costs would be for wells, distribution pipe and heat exchangers. Low temperature direct use geothermal resources are accessible throughout the United States but have greater availability and higher temperatures in the western United States (NIBS, 2016). One thousand three hundred fifty (1,350) deep geothermal systems are known or identified in the United States (NIBS, 2016). Geothermal production of electricity by utilities in the United States rose from 15.009 million megawatt-hours in 2009 to 16.728 million megawatt-hours in 2018 (USEIA, 2019a).

Description

A direct use system drawing on low-temperature, geothermal energy comprises major parts of (NIBS, 2016):

- Production unit – wells and pumps (either submersible or line-shaft pumps); if the heated water is artesian coming directly to the surface, downhole pumps are not needed
- Water movement and heat exchange and circulation systems – pipes that can tolerate high temperatures 50 to 200 degrees C (180-300 degrees F); pipes to distribute the heated water (dual pipe systems circulate residual heated water to further conserve energy); heat exchangers to convert the heat to useful applications, heat convectors to move heat to the location of use and, if energy for cooling is required, refrigeration units using lithium bromide and ammonia absorption
- Peaking and back-up units - drawn on when additional energy is needed from alternate energy sources
- Used-water recycling process – a reservoir for storing the used water (may be a pond) and an injection well to return the used water to the ground for reheating

Applications of deep geothermal energy include space heating and cooling (including district heating), greenhouse heating crop drying, industrial processing, hot water heating for domestic and commercial purposes including swimming pools and baths or therapeutic use, animal husbandry and aquaculture (NIBS, 2016; USEPA, 2016).

Cost

Deep geothermal energy plants have higher capital expenditure on the front end by lower annual operating cost. In a district heating application, the initial capital investment may include production and injection wells and pumps, pipes and heat exchangers that cost $2–3 million dollars. In comparison, a carbon fuel system may have only a central boiler and pipes to distribute the heated water. While difference in annual O&M costs of the geothermal and oil/gas systems may be small, the carbon-fuel system may have increasing fuel cost but the geothermal heating cost is not volatile. The location of use should be carefully evaluated to support the use of deep geothermal technology. Capital costs can vary by 100% due to the local geology, hydrology, construction requirements and existing infrastructure. Table 1.9.3 provides costs for a range of deep geothermal direct use applications. The unit cost for deep geothermal direct use in the range of applications presented is $0.0223 to $0.0308 in 2005. The average price per kilowatt-hour for residential deliveries of electricity in the United States in June 2005 was $0.0979 and for industrial deliveries was $0.0586. These prices had risen to $0.1334 and $0.0691 per kilowatt-hour, respectively, by June 2019 (USEIA, 2019b).

Groundwater Source Use

Conventional geothermal direct use for electricity production uses from 0.03 to 1.03 liters of groundwater/kilowatt-hour (0.01 to 0.271 gallons/kWh) of lifetime energy output. This water use is vapor

TABLE 1.9.3
Deep Geothermal Direct Use Costs in the United States

Direct Use	Capital Cost $/kW	Annual O&M Cost $/kW	Unit Cost CENTS/kWh
Residential Space Heating*	800	7.1	3.08
Comm./Inst. Space Heating*	500	4.4	2.23
District Heating	650	5.8	2.42

Average costs of direct-use systems in the United States for 2005. Based on kilowatt equivalents.

* Assumes one production and one injection well for a single building

Calculation Assumptions: 30-year life, 8.0% interest and O&M at 10% of capital cost; capital costs include a shallow well (<984 ft or 300 m); no retrofit costs.Source: National Institute of Building Sciences (NIBS). 2016. Geothermal Energy: Direct-Use. URL: http://www.wbdg.org/resources/geothermal-energy-direct-use (Accessed September 24, 2019).

loss in turbines and injection processes. Enhanced geothermal uses more because it injects the used water back to create greater fracture surface for groundwater heating. Enhanced geothermal direct use consumes 1.14 to 2.76 liters/kWh (0.3 to 0.73 gallons/kWh).

Environmental Factors

Use of ground source heat pumps and deep geothermal energy provide substantial environmental benefits (EHO, 2019; Haliburton, 2019):

- Low carbon dioxide emissions
- Fewer on-site chemicals and limited exposure
- Smaller environmental footprint than conventional systems and energy sources
- Closed-loop heat pumps cycle water within system; deep geothermal systems return water to subsurface

Benefits

Overall, conventional geothermal direct use for electricity production uses less water than other electric production technologies: coal, 0.98 to 5.53 liters/kWh (0.26 to 1.46 gal/kWh); coal with carbon capture, 2.16 to 5.79 liters/kWh (0.57 to 1.53 gal/kWh); nuclear, 1.06 to 3.75 liters/kWh (0.28 to 0.99 gal/kWh); and conventional natural gas, 1.44 to 3.71 liters/kWh (0.38 to 0.98 gal/kWh) (ANL, 2010). Deep geothermal direct use for electricity production consumes 4 to 19% of the amount of water that conventional coal technology for electricity consumes. The reduced water use combined with the other environmental factors of less emissions, less chemical use and small environmental footprint can be evaluated on a site basis to demonstrate the economic benefit of this technology.

MANUFACTURING

Description

Manufacturing industry in this discussion includes self-supplied commercial and industrial processes and thermoelectric power plant cooling. Manufacturing industrial water use comprises three categories (1) processing a product including fabricating, washing, diluting, cooling or transporting; (2) physical inclusion of water in a product; and (3) meeting the sanitary requirements of the

TABLE 1.9.4
Groundwater Use in Manufacturing and Thermoelectric Cooling, 2015

	Groundwater-Supplied		Total Water Supplied	
	Fresh	**Saline**	**Fresh**	**Saline**
Manufacturing	10.1 MMSCMD (2,670 MGD)	0.162 MMSCMD (42.9 MGD)	61.7 MMSCMD (14,000 MGD)	3.5 MMSCMD (786 MGD)
Thermoelectric	1.61 MMSCMD (425 MGD)	0.651 MMSCMD (172 MGD)	418.9 MMSCMD (95,100 MGD)	166.5 MMSCMD (37,800 MGD)
TOTALS	11.71 MMSCMD (3,095 MGD)	0.813 MMSCMD (214.9 MGD)	480.6 MMSCMD (109,100 MGD)	170.0 MMSCMD (38,586 MGD)

MMSCMD = Million Metric Standard Cubic Meters Per Day

MGD = Million Gallons Per Day

Source: U.S. Geological Survey (USGS). 2018. Estimated Water Use in the United States in 2015. Circular 1441. Washington, DC.

production plant. The production of food, paper, chemicals, refined petroleum or primary metals requires large volumes of water (USGS, 2019c). Twenty-two percent of worldwide water use is water consumed by industry (UN Water, 2013). Table 1.9.4 gives a perspective on groundwater use in manufacturing and thermoelectric cooling in the United States in 2015.

If wastewater cannot be reused directly, decentralized system processes may be able to reduce contamination of water for particular uses. Decentralized wastewater treatment systems include lined waste stabilization ponds, aerated ponds, constructed wetlands, anaerobic digestion technologies, activated sludge, membrane reactors, oxidation processes and activated carbon (Pain and Spuhler, 2019).

Cost

Manufacturing industrial water treatment cost is affected by processing and site requirements. Water treatment in an industrial setting includes water quality treatment, process water contaminant removal and product mixture separation, and wastewater treatment. Industrial water treatment systems may include the following equipment (STI, 2018):

- Clarifiers to remove suspended solids
- Softeners to reduce total dissolved solids
- Oil/water separators to extract oils
- Particle filtration to separate solids
- Membrane filtration to remove chemical and biological contaminants
- Reverse osmosis or nanofiltration for additional contaminant extraction
- Filter presses to dewater sludge
- Ion exchange columns to remove targeted substances
- Reactors and chemical additives to manage pH and precipitate metals
- Computerized management system
- Pumps, reservoirs, pipe and valves

Thus, water treatment and reuse require investment based on knowledge of water quality of manufacturing processes that may lead to changes in those processes. This investment may lead to cost savings from direct reuse and treat and reuse processes (Pain and Spuhler, 2019).

Groundwater Source Use

Self-supplied manufacturing and thermoelectric plant cooling used 3.8% of total fresh groundwater drawn in 2015 and 9.1% of total saline groundwater drawn, as shown in Table 1.9.4. From 1985 to 2015, total self-supplied manufacturing water withdrawals decreased by 43%. However, fresh groundwater withdrawals declined by 32% and saline groundwater withdrawals increased by 65% over the same 30 years. Similarly, thermoelectric power use of fresh groundwater declined by 30% from 1985 to 2015 while saline groundwater use grew from 0 to 0.651 MMSCMD (172 MGD) (USGS, 2018; 2019c).

Environmental Factors

Decreasing manufacturing industrial water use can be attributed to several conditions in the United States economy. While less manufacturing in the United States occurred with reliance of foreign product sources, more water efficient manufacturing processes in the United States focused on water reuse and recycling because of environmental controls and reduced fresh water available in some regions (USGS, 2019c).

For manufacturing of chemical products derived from petroleum and coal that may involve onsite liquid and sludge chemical waste storage, the principal issues have been "soil and surface water and groundwater contamination from unlined or leaking storage tanks, drums, surface impoundments, and surface water lagoons, and uncontrolled polluted stormwater runoff." (USEPA, 2019b) Additional environmental contamination problems include: abandonment and disposal of contaminated soil and debris; improper storage of hazardous waste; and soil and water contamination from spills and hazardous substance management practices. The U.S. Environmental Protection Agency has determined that modern practices and federal and state regulations have addressed these problems. Comprehensive regulations for hazardous waste management and disposal under the Resource Conservation and Recovery Act specifically target preventing such releases and require past spills to be remediated by facility owners and operators (USEPA, 2019b). Landfills and surface impoundments must install groundwater monitoring, comply with technical requirements, such as double liners and leachate collection, and hold financial assurance (USEPA, 2019a). The cost of groundwater monitoring wells is addressed in Volume 1, Chapter 6, covering groundwater access and supply.

Benefits

Water treatment provides the potential to reuse water rather than to purchase more water and treat it to discharge to a wastewater system beyond the production facility, thereby reducing cost to the manufacturer. Treated water for reuse can be used between manufacturers or other businesses providing on-site and off-site savings (Pain and Spuhler, 2019). Water reuse provides advantages through reduced water consumption, reduced water costs and reduced amount of wastewater produced through the industrial manufacturing process (Pain and Spuhler, 2019). Chemical waste disposal requirements for manufacturers for double liners, leachate collection and groundwater monitoring should protect groundwaters used as a source of water supply.

SUMMARY

Mining uses a small amount of a nation's water use but locally this use can be significant, particularly in arid locations. Without access to water, mining would not exist in most regions. Because mining is significant in arid and remote areas, groundwater is an important and significant source of water for many mining sites. While capital and operation and maintenance costs are high, mining provides substantial revenue from metal and industrial mineral production to enable mining operations to afford water even at higher prices. Because of chemical use in modern mine operations, mining sites can be damaged if chemical use is not controlled, as demonstrated by past abandoned hazardous waste sites that required remediation. Current best practices maintain water and chemicals on-site thereby reducing the cost or need to respond to environmental damage to ground or surface water.

Geothermal exchange for ground source heat pumps and deep direct use geothermal energy offer renewable sources for heating and cooling and energy production. These systems have a high coefficient of performance with low energy inputs providing high energy outputs. Additionally, environmental benefits are significant, including renewability, low carbon emissions, minimal chemical inputs and small environmental footprint minimizing groundwater consumption. While capital and operating costs are higher, these are offset by savings compared to rising costs of carbon-based fuels used in conventional thermal exchange and energy systems.

Manufacturing industrial processes draw on groundwater at particular locations that allow self-supply through wells, rather than a central public community water system. Water reuse has been facilitated by environmental regulation and limitations on water availability in some locations, but requires investment. Savings from water reuse come from lower water demand and less wastewater treatment needed. While water reuse has reduced demand on groundwater sources, saline groundwaters have increased their contribution to manufacturing water supply. Secure disposal sites for chemical wastes that include liners, leachate control and groundwater monitoring should protect water sources.

FOR FURTHER CONSIDERATION

1. What mining operations exist in your area? What type of mining occurs? Have environmental problems been identified from these mining operations? Identify the costs to respond to environmental concerns through an online literature search. Are these costs what you expected?
2. Have ground source heat pumps or direct use geothermal energy projects been installed in your area? If not, why not? If so, contact the company or contractor and inquire about costs to install a system for your community or home. Do the costs seem reasonable? Compare these costs to the price and cost of conventional sources.
3. Obtain a copy of your community's, county's or state's water reuse plan or strategy. What sources of wastewater are identified for reuse? What processes and/or sites are identified for reuse of reclaimed water? Contact a business that reuses water and inquire about the costs of implementing steps to reuse water, included treatment, additional pipe and pumps and other equipment as well as labor operating the equipment. What cost savings can the business identify? What percent of water can be reused?

REFERENCES

American Geosciences Institute (AGI). 2019a. How can metal mining impact the environment? URL: https://www.americangeosciences.org/critical-issues/faq/how-can-metal-mining-impactenvironment (accessed September 5, 2019).

American Geosciences Institute (AGI). 2019b. What are the main methods of mining? URL: https://www.americangeosciences.org/critical-issues/faq/what-are-main-mining-methods (accessed September 7, 2019).

Argonne National Laboratory (ANL). 2010. "*Water Use in the Development and Operation of Geothermal Power Plants.*" URL: http://www1.eere.energy.gov/geothermal/pdfs/geothermal_water_use_draft.pdf. Cited in: "Geothermal electricity: Saving the Earth, It's worth the watt" URL: https://sites.lafayette.edu/egrs352-sp14-geothermal/contexts-and-consequences/water-usage/ (accessed September 25, 2019).

Arizona Department of Mines and Mineral Resources (ADMMR). 2010. Water Consumption at Copper Mines in Arizona. Special Report 29. Prepared by Madan M. Singh. URL: http://repository.azgs.az.gov/sites/default/files/dlio/files/nid1295/sr29waterconsumptioncoppermines.pdf (accessed October 3, 2019).

Bottcher, F., Alessandro C., Gregor G., and Kai Z. 2019. TAP – Thermal Aquifer Potential: A Quantitive Method to Assess the Spatial Potential for the Thermal Use of Groundwater. *Renewable Energy*, vol. 142. November 2019, pp. 85–95. URL: https://www.sciencedirect.com/science/article/pii/S0960148119305701 (accessed September 5, 2019).

Bravo, A. 2019. *Geothermal Heating and Cooling Cost 2019: Pros & Cons, Comparisons*. Inn RemodelingImage.com, January 18, 2019. UsRL: https://www.remodelingimage.com/geothermal-heating-and-cooling-cost/ (accessed September 19, 2019).
Brown, J. 2015. *How to Calculate Coefficient of Performance*. Maritime Geothermal Ltd., New Brunswick. URL: https://www.nordicghp.com/2015/08/how-to-calculate-coefficient-of-performance/ (accessed September 19, 2019).
Chiaro, P.S. 1997. Environmental Strategies in the Mining Industry: One Company's Experience. In *The Industrial Green Game*. National Academy Press, Washington DC. pp. 165–181.
Climate Change Connection (CCC). 2018. *Alternative Heat Energy*; Geothermal. URL: https://www.climatechangeconnection.org/solutions/alternative-heat-energy/geothermal/ (accessed September 19, 2019)
Deberitz, J. 2006. *Lithium: Production and application of a fascinating and versatile element*. Chemetall GmbH, Munich, Germany.
EnergyHome.org (EHO). 2019. *Geoefficiency*. URL: http://www.energyhomes.org/renewable-technology/geoefficiency.html (accessed October 6, 2019).
General Kinematics Corporation (GKC). 2019. *A Brief History of Mining: The Advancement of Mining Techniques and Technology*. URL: https://www.generalkinematics.com/blog/a-brief-history-of-mining-and-the-advancement-of-mining-technology (accessed September 5, 2019).
Halliburton. 2019. *Geothermal Energy Benefits*. URL: https://www.halliburton.com/en-US/ps/solutions/clean-energy/geothermal-energy/geothermal-energy-environmental-benefits.html (accessed October 5, 2019).
Hammerle, D. 2019. *Seminar on Miami University Geothermal Plant*. September 11, 2019. Oxford, Ohio.
International Council on Mining and Metals (ICMM). 2012. *Water management in mining: A selection of case studies*. URL: https://www.icmm.com/website/publications/pdfs/water/water-management-in-mining_case-studies (accessed on October 4, 2019).
Lund, J., B. Sanner, L. Rybach, R. Curtis, and G. Hellstrom. 2004. *Geothermal (Ground-Source) Heat Pumps; A World Overview*. In GHC Bulletin, September 2004. URL: https://www.pdfs.semanticscholar.org/04b7/12d7455d4f12fc18e24c4eeabf64d6db81ddc.pdf (accessed September 7, 2019).
Miami University (MU). 2012. To grow sustainably, Miami digs deep: Western campus geothermal project to supply new housing and dining facilities with efficient heating and cooling. News Release, November 9, 2012. URL: https://www.miamioh.edu/news/article/view/17899 (accessed September 19, 2019).
Mining Technology. 2013. *Managing water consumption in mining*. URL: https://www.mining-technology.com/features/feature-managing-water-consumption-mining-global-shortage/ (accessed September 25, 2019).
Moran, C.J., Evans, R., Silvester, N., and Ringwood, K. 2008. Characterising the values of water in minerals operations. In *Water in Mining, GECAMIN*. pp. 3–18.
National Institute of Building Sciences (NIBS). 2016. *Geothermal Energy: Direct-Use*. URL: http://www.wbdg.org/resources/geothermal-energy-direct-use (accessed September 24, 2019).
Ossa-Moreno, J., McIntyre, N., Ali, S., Smart, J.C.R., Rivera, D., Lall, U., and Keir, G. 2018. *The hydro-economics of mining*. Ecological. Economics., 145, pp. 368–379. URL: https://reader.elsevier.com/reader/sd/pii/S0921800916315701?token=F3B5C979D5B0119838A2F99859BAF5C3A1971FF9D377A79BCB44A994971590011FE2CC17691D00DB56BD39961A8DDA66 (accessed December 21, 2020).
Pain, A. and Spuhler, D.. 2019. *Wastewater Reuse in Industry*. In Sustainable Sanitation.
Samco Technologies, Inc. (STI). 2018. *How Much Does an Industrial Water Treatment System Cost?* URL: https://www.samcotech.com/how-much-does-an-industrial-water-treatment-system-cost/ (accessed November 19, 2019).
Trading Economics (TE). 2019. *United States GDP from Mining*. URL: https://www.tradingeconomics.com/united-states/gdp-from-mining (accessed September 5, 2019).
United Nations Water (UN Water). 2013. UN-Water Annual Report 2012. URL: https://www.unwater.org/publications/un-water-annual-report-2012/ (accessed November 19, 2019).
U.S. Census Bureau (USCB). 2019. *Census of Housing; Characteristics of Housing; Type of Heating System Used in New Single-Family Houses Completed*. URL: https://www.census.gov/construction/chars/historical_data/ (accessed November 21, 2019).
U.S. Department of Energy (USDOE). 2017. *5 Things You Should Know about Geothermal Heat Pumps*. URL: https://www.energy.gov/eere/articles/5-things-you-should-know-about-geothermal-heat-pumps (accessed September 7, 2019).
U.S. Department of Energy (USDOE), National Renewable Energy Laboratory. 2019. *Geothermal Heat Pump Basics*. URL: https://www.nrel.gov/research/re-geo-heat-pumps.html (accessed September 5, 2019).
U.S. Energy Information Agency (USEIA). 2018a. Coal explained; Where our coal comes from. URL: https://www.eia.gov/energyexplained/coal/where-our-coal-comes-from.php (accessed October 2, 2019).

U.S. Energy Information Agency (USEIA). 2018b. Table 4.1 Count of electric Power Industry Power Plants, by Sector, By Predominant Energy Sources within Plant, 2007 through 2017. URL: https://www.eia.gov/electricity/annual/html/epa_04_01.html (accessed October 2, 2019).

U.S. Energy Information Agency (USEIA). 2019a. Table 1.1.A. Net Generation from Renewable Sources: Total (All Sectors), 2009-July 2019. URL: https://www.eia.gov/electricity/data.php (accessed September 25, 2019).

U.S. Energy Information Agency (USEIA). 2019b. Average retail price of electricity, United States, monthly. URL: https://www.eia.gov/electricity/data/browser/#/topic/7?agg=2,0,1&geo=g&freq=M&start=200101&end=201907&ctype=linechart&ltype=pin&rtype=s&maptype=0&rse=0&pin= (accessed September 25, 2019).

U.S. Environmental Protection Agency (USEPA). 2016. *Renewable Heating and Cooling; Geothermal Heating and Cooling Technologies*. URL: https://www.epa.gov/rhc/geothermal-heating-and-cooling-technologies (accessed September 24, 2019).

U.S. Environmental Protection Agency (USEPA). 2018. *Abandoned Mine Lands: Site Information*. URL: https://www.epa.gov/superfund/abandoned-mine-lands-site-information (accessed October 2, 2019).

U.S. Environmental Protection Agency (USEPA). 2019a. Financial Responsibility Requirements Under CERCLA Section 108(b) for Facilities in the Petroleum and Coal Products Manufacturing Industry; Proposed Rule. Federal Register, December 23, 2019, 84 FR 70467.

U.S. Environmental Protection Agency (USEPA). 2019b. *Superfund Site: Silver Mountain Mine*, Loomis, WA. URL: https://cumulis.epa.gov/supercpad/SiteProfiles/index.cfm?fuseaction=second.Stayup&id=1000948#Stayup and https://cumulis.epa.gov/supercpad/SiteProfiles/index.cfm?fuseaction=second.cleanup&id=1000948 (accessed October 5, 2019).

U.S. Geological Survey (USGS). 1988. *Estimated Water Use in the United States in 1985. Circular 1004*. Washington, DC.

U.S. Geological Survey (USGS). 1993. *Estimated Water Use in the United States in 1990. Circular 1081*. Washington, DC.

U.S. Geological Survey (USGS). 1998. *Estimated Water Use in the United States in 1995. Circular 1200*. Washington, DC.

U.S. Geological Survey (USGS). 2004. *Estimated Water Use in the United States in 2000. Circular 1268*. Washington, DC.

U.S. Geological Survey (USGS). 2009. *Estimated Water Use in the United States in 2005. Circular 1344*. Washington, DC.

U.S. Geological Survey (USGS). 2014. *Estimated Water Use in the United States in 2010. Circular 1405*. Washington, DC. URL: pubs.er.usgs.gov/publication/cir1405 (accessed December 21, 2020).

U.S. Geological Survey (USGS). 2018. *Estimated Water Use in the United States in 2015. Circular 1441*. Washington, DC. URL: pubs.er.usgs.gov/publication/cir1441 (accessed December 21, 2020).

U.S. Geological Survey (USGS). 2019a. 2015 Minerals Yearbook; Statistical Summary. URL: https://prd-wret.s3-us-west-2.amazonaws.com/assets/palladium/production/atoms/files/myb1-2015-stati.pdf (accessed September 28, 2019).

U.S. Geological Survey (USGS). 2019b. *How do we extract minerals?* URL: https://www.usgs.gov/faqs/how-do-we-extract-minerals?qt-news_science_products=0#qt-news_science_products (accessed September 7, 2019).

U.S. Geological Survey (USGS). 2019c. *Industrial Water Use*. URL: https://www.usgs.gov/mission-areas/water-resources/science/industrial-water-use?qt-science_center_objects=0#qt-science_center_objects (accessed November 19, 2019).

Washington Department of Ecology (WDEQ). 2019. *Cleanup Site Details Report; Silver Mountain Mine*. URL: https://apps.ecology.wa.gov/gsp/Sitepage.aspx?csid=3033 (accessed October 5, 2019).

WorldAtlas.com. 2019. *What Is The Environmental Impact Of The Mining Industry?* URL: https://www.worldatlas.com/articles/what-is-the-environmental-impact-of-the-mining-industry.html (accessed October 2, 2019).

World Bank. 2019. Mineral rents (%of GDP) (2017). URL: https://www.data.worldbank.org/indicator/NY.GDP.MINR.RT.ZS (accessed September 5, 2019)

Yellich, J. (Michigan Geological Survey). 2019. Personal communication, September 11, 2019.

Young, R.A. and Loomis, J.B.. 2014. *Determining the Economic Value of Water: Concepts and Methods*. RFF Press Routledge/Taylor &Francis Publishers, 358 p.

10 Competing Uses and Value of Groundwater and the Subsurface Environment

KEY GROUNDWATER CONDITIONS

Groundwater use has many different aspects, some complementary and others competitive of the resource's conditions, qualities and capacity. Several points from previous chapters are summarized here because of their significance to production, use and value of groundwater and to the use of the subsurface environment:

- Groundwater is nearly ubiquitous in the subsurface portion of the ecosystem under the ground surface of continents and islands.
- Groundwater exists in a complex subsurface environment, is mobile and travels slowly, but, in some cases, long distances crossing many boundaries.
- Managing groundwater on an aquifer basis in a watershed recognizes the groundwater flow component of the hydrologic cycle and supports a holistic approach to addressing water issues.
- Groundwater represents the largest freshwater source of water supply.
- Many organisms live in and rely on groundwater and the subsurface as their habitat.
- The largest uses of groundwater are for water supply for human beings (in some cases, their only water source) and other life forms (for some, their only habitat) and for irrigation (the largest use) to produce and supply food – in many locations, of necessity, the only affordable source.
- Groundwater is a significant resource input to the economies of communities and nations around the world, supplying at least half the world's population with water.
- Aquifers are under stress from depletion and contamination around the world, with the stress most likely increasing with climate change.
- Groundwater is vulnerable to contamination and is expensive to clean up.
- Not as readily recognized is the use of groundwater and the subsurface as a waste sink.
- Laws provide the basis for use of groundwater relative to who may access and use it, which provides the basis for economic transactions affecting groundwater.
- Laws in some countries and jurisdictions treat groundwater as a separable water source, but other countries and jurisdictions recognize its role in the larger hydrologic cycle, such as in maintaining the baseflow of streams.
- In some locations, groundwater is easily accessed but this is more difficult in other places – this affects the cost to produce it, typically through wells.
- Aquifers around the world are being depleted – used up faster than they are recharged – but have been instrumental through irrigation in ensuring food supply in many countries, especially some densely populated countries.
- The quality of managed recharge of groundwater using excess water at different times and places affects the uses to which groundwater may be applied or needing treatment before use.

- Water reuse is gaining acceptance and offers an alternative water source along with water conservation.
- Because of their many alternative purposes and uses, groundwater and its subsurface environment should be valued as some pecuniary amount significantly greater than zero.

INDICATORS OF COMPETITION FOR GROUNDWATER AND THE SUBSURFACE ENVIRONMENT IN THE UNITED STATES

Below are a listing of use categories and specific uses of groundwater and the subsurface that may serve as indicators of competition for groundwater and the subsurface environment in the United States. Resources are used from the ecosystem, transformed to products and returned as waste. This circumstance is true for groundwater and the subsurface environment as well. This list gives some selected statistics to show a level of competition for the subterranean zones of the earth in the United States only, with an emphasis on groundwater. It is only a partial list to be expanded by others, but may be an indicator of the future challenge and opportunity posed within this portion of the ecosystem. Certainly, many other uses of the subsurface may be identified, and therefore also compete for this portion of the ecosystem.

Supply and Production

Maintenance of stream baseflow: 1.862 billion m^3/day (492 billion gal/day) from groundwater discharge (USEPA, 1996)

Household domestic water: 41,861,000 consumers of groundwater (USGS, 2018) served in 16.1 million households (est.) (ArcGIS Online, 2019)
- 12,151,000 m^3/day (3,210 million gallons/day) (USGS, 2018)

Public water supply: 262,000 wells (est.) (USEPA, 2002)
- 89,790,000 consumers served through 38,250 community groundwater-supplied systems (USEPA, 2019a)
- 57,538,000 m^3/day (15,200 million gal/day) (USGS, 2018)

Irrigation: 462,788 wells in 2018 (USDA, 2018)

Oil and Natural Gas: 1,666,715 active O&G wells (FracTracker, 2015)
- 10,990,000 barrels of oil/day produced in 2018 (USEIA, 2019a)
- 3.3 billion m^3/year (882 billion gallons million/year) of water produced ("produced water") with oil and natural gas each year (Aquatec, 2016)
- 0.728 million m^3 (192.325 million gallons) of water/day used in oil refining (Sun et al., 2018)
- 2.373 billion m^3 (83.8 billion cubic feet) per day of natural gas produced in 2018 (USEIA, 2019b)
- 0.038 million m^3/day of water use projected by 2030 for tight sands and gas shale production (ANL, 2008)
- Natural gas hydraulically fractured wells: over 300,000 wells in 2015 (USEIA, 2019b)
- Hydraulically fractured well natural gas production: 1.5 billion m^3 (53 billion cubic feet)/day in 2015 (USEIA, 2016)

Geothermal power: 141 units (Lund et al., 2000)

Geothermal heat pumps: 450,000 units, increasing at 50,000 units annually (Lund et al., 2000)

Coal: 1,216 active mines (Statistica, 2019)
- 686.0 metric tons (756.2 million short tons) produced in 2018 (USEIA, 2019c)

Transmission

351,460 km (218,387 miles) of oil pipelines (USDOT, 2019)

2,618,337 km (1,626,959 miles) of natural gas pipelines (USDOT, 2019)

1,609,000 km (1,000,000 miles) of hazardous liquid transportation pipelines (est.; USDOT, 2017)
3,359,000 km (2,211,500 miles) of water transmission and distribution lines (est.; USEPA, 2009; 2019a)
1,000,000 km (621,000 miles) of sewer lines (est.) (Vipulanandan, 2008)
2,414 km (1,500 miles) of CO_2 pipeline (Ronca, 2008)

Storage

Underground storage tanks: 552,000 (est., USEPA, 2019b)
Oil (proved reserves): 39.2 billion barrels in 2017 (USEIA, 2018)
Natural gas (proved reserves): 13.148 trillion m^3 (464.3 trillion cubic feet) in 2017 (USEIA, 2018)
Active storage fields: 380 fields with 120 billion m^3 (4,263 Bcf) in 2018 (USEIA, 2018b)
Coal (proved reserves): 431 billion metric tons (475 billion short tons) (USEIA, 2018c)
Managed aquifer recharge sites: 262 sites in 2018 (IGRAC, 2018)

Waste Sink

Underground injection wells regulated: 850,000 (est.) (AGI, 2019)
– Hazardous waste: 140 wells – Class I (approx.) (USEPA, 2016)
– Non-hazardous waste: 660 wells – Class I (approx.) (USEPA, 2016)
Solid waste: 418 land disposal facilities subject to groundwater monitoring (USEPA, 1996)
Heating/air conditioning return flow: 35,000 wells (est., USEPA, circa. 2000)
Oil and gas brine: 167,000 disposal wells (USEPA, 2006)
– 7.571 million m^3 (2 billion gallons) of fluids injected/day (est., USEPA, 2006)
Domestic wastewater septic systems: 21 million (est.) (COB, 2015)
Abandoned waste National Priority List sites: 1,218 sites affecting groundwater (USEPA, 2017)
Releases from leaking underground storage tanks: 278,000 releases (USEPA, 1996)
Power plants and stationary sources annual release of CO_2: 1.763 billion metric tons in 2018 (USEIA, 2019d)
– Carbon dioxide disposal potential underground: 3175 billion metric tons of CO_2 (USDOE, 2007b)
Stormwater infiltrated carrying pesticides, fertilizers, oils, metals, pathogens, salt, sediment, litter and other debris: volume not determined
– 855 medium and large cities and counties (pop. > 100,000) permitted to manage stormwater
– 6,695 smaller communities permitted to manage stormwater (USEPA, 2019c)

Emission Source

Abandoned gas Wells: 600,000 wells emitting 1.4 million metric tons of carbon dioxide equivalents (MMT CO_2e) in 2018 (USEPA, 2020)
Abandoned oil Wells: 2.6 million wells emitting 5.7 MMT CO_2e in 2018 (USEPA, 2020)

USE VALUE IN SUMMARY

An extensive review of freshwater value completed by Frederick et al. (1996) documented the results of 494 water value estimates for both groundwater and surface water uses in nonmarket conditions. A variety of methods were used by the original estimators and both average and marginal values were included, though, in some cases, only average values reported. Where applicable, the values of water as an input (rather than as a final product) were derived from the value of final goods and services. (Methods to derive these values as benefits are covered in Volume 2, Chapter 5). The original values reported vary by region and season of the year but are aggregated here as a national value estimate for water use in the United States. A summary of the water use values in the United States found in this comprehensive review of data reported for 1994 and inflated to 2020$US appears in Table 1.10.1, below.

TABLE 1.10.1

Water Use Values	1994 Average Value in $/1000m^3**	1994 Median Value in $/1000m^3	2020* Average Value in $/1000m^3	2020* Median Value in $/1000m^3
Domestic water supply	157	79	271	136
Irrigation	61	32	105	55
Industrial processing	229	107	396	185
Thermoelectric power	28	24	48	41
Waste disposal	2	1	3	2
Recreation/Habitat	39	4	67	7
Navigation	118	8	204	14
Hydropower	20	17	35	29

Source: Frederick, K. , VandenBerg, T., and Hanson, J.. 1996. Economic Values of Freshwater in the United States. Discussion Paper 97-03. Resources for the Future, Washington, D.C.

* 2020 $/1000/m^3 calculated from 1994 using U.S. Bureau of Labor Statistics CPI calculator at https://data.bls.gov/cgi-bin/cpicalc.pl

** 1000m^3 = 0.811 acre-foot; convert $/1000m^3 to $/acre-foot by multiplying $/1000m^3 by 1.233

Reporting the aggregate value of both groundwater and surface water values is appropriate because they may often be alternatives to each other. Frederick notes that quantity is the dimension of value captured in the estimates while the quality of water may not be captured. Additionally, technology may affect supply and demand and thereby influence its value over time. Significant water conservation technologies were implemented over the 1994–2020 timeframe, so the inflated 2020 estimates may not be as close to values observed in that year.

Other observations about the use values above related to groundwater (Frederick et al., 1996) include (1994$US):

- Domestic water supply – a more valuable use of water but municipal water value is "understated" with respect to other uses considering the data because with negative-sloped water-demand curves, marginal values are less than average values; after subtracting the production costs of treated and delivered domestic water, the marginal value of domestic water may be less than that of some alternative onsite water uses depending on location and other factors; groundwater supply was 39% of total domestic public water supply in 2015; domestic supply had the fourth highest maximum value in its data range of $465/1000m^3.
- Irrigation – groundwater was 48% of total irrigation water use in 2015; irrigation use had the second highest maximum value in its data range of $996/1000m^3.
- Industrial processing – this use relied on fresh groundwater for 19% of its supply in 2015 and had the third highest maximum value of $650/1000m^3.
- Thermoelectric power – less than 1% of U.S. electric power generation is cooled by groundwater; maximum value in its data range was $51/1000m^3.
- Waste disposal – these values do not include underground injection of wastewater; this use had the lowest maximum value in its data range of $10/1000m^3.
- Recreation and fish and wildlife habitat – recreation activities associated with springs and geysers are related to groundwater and wetland habitat may be a result of groundwater discharge to those areas; recreation and habitat use had the highest maximum value in its data range of $2,142/1000m^3.

- Navigation and hydropower – since about 40% of average annual streamflow in the United States is attributable to groundwater discharge to streams, about 40% of these use values may be attributable to groundwater; navigation maximum value was $392/1000m^3$ and hydropower maximum value was $92/1000m^3$.

Additional information on use values appears in Volume 2, Chapter 2.

SEQUEL

The expanding level of competition for groundwater and the subsurface environment need the careful and full consideration of thorough economic analyses. The list above of production, transmission, storage and waste sink is incomplete. From the surface, the underground environment seems a vast, dense mass of resources comprised of soil, rock, water and other substances that may have economic use. The portion nearest the ground surface continues to be mined, excavated, removed, filled in, infiltrated with unwanted residuals, drilled, piped, caverned out and consumed with its qualities modified by land-applied chemicals and wastes. Deeper underground, the rock matrix gives up water, brine, minerals, oil and gas and superheated water, while receiving pressurized fluids to fracture it to give up more oil and gas and injected with pressurized wastewaters. The subsurface environment with its aquifers – the place of groundwater storage and flow – is under economic competition, hopefully for the sustainable betterment of humankind. What is the value of its services? Is the value captured by the use values presented above? Where and what are its vulnerabilities? Can we manage it sustainably and at what cost?

FOR FURTHER CONSIDERATION

1. Which uses of the subsurface environment are in greatest competition with each other in your opinion? Which are in greatest competition with groundwater uses?
2. From your understanding of the subsurface environment, do any uses of the subsurface environment seem to affect the sustainability of groundwater use into the future? Are there any competing uses affecting groundwater sustainability more than others?
3. What vulnerabilities related to quantity or quality impacts for groundwater access and use do you see in the listing of competing uses above? Are there economic consequences to the local or state economy from the vulnerabilities?
4. In your locality or state, are there competing uses of the subsurface environment that have interfered with the use of groundwater and have affected the local or state economy? What are those uses and what effect on the economy occurred? Was there a resolution of the effect and what was the resolution? What were the costs and the benefits of the resolution and who paid and received them?
5. If your residence is supplied by a public water system (or water utility), then your household probably receives a bill for water supply. What is the cost of your water in $ per cubic meter compared to that reported above? Can you find out what it costs to produce, treat and deliver the water to your house? Subtract that cost from the cost of the amount on the bill. What is the difference – the marginal value of your water? If marginal value is higher or lower than the national median value above, why do you think that is? What factors contributed to the value per cubic meter in your location?

REFERENCES

American Geosciences Institute (AGI). 2019. What are underground injection wells used for? URL: https://www.americangeosciences.org/critical-issues/faq/what-underground-injection-wells-used (accessed November 22, 2019).

Aquatec. 2016. About Produced Water (Produced Water 101). URL: http://aqwatec.mines.edu/produced_water/intro/pw/index.htm (accessed November 20, 2019).

ArcGIS Online. 2019. 2019 USA Average Household Size. URL: https://demographics9.argis.com/arcgis (accessed November 20, 2019).

Argonne National Laboratory (ANL). 2008. *Baseline and Projected Water Demand Data for Energy and Competing Water Use Sectors,* ANL/EVS/TM/08-8, December 2008. URL: http://www.ead.anl.gov/pub/doc/ANL-EVS-TM08-8waterdemandreport.pdf (accessed August 28, 2009).

Circle of Blue (COB). 2015. Water News; Infographic: America's Septic Systems. URL: https://www.circleofblue.org/2015e/world/infographic-americas-septic-systems/ accessed November 22, 2019).

FracTracker Alliance. 2015. 1.7 Million Wells in the U.S. – A 2015 Update. URL: https://www.fractracker.org/2015/08/1-7-million-wells/ (accessed November 20, 2019).

Frederick, K. , VandenBerg, T., and Hanson, J.. 1996. Economic Values of Freshwater in the United States. Discussion Paper 97-03. Resources for the Future, Washington, D.C.

International Groundwater Resources Assessment Centre (IGRAC). 2018. Managed Aqujjifer Recharge Portal. URL: https://www.un-igrac.org/special-project/mar-portal (accessed November 22, 2019).

Lund, J.W., Boyd, T.L., Sifford, A., and Bloomquist, R.G. 2000. *Geothermal Energy Utilization in the United States—2000.* Geo-Heat Center, Oregon Institute of Technology, Klamath Falls, OR.

Sun, P., Amgad, E., Michael, W., Jeongwoo, H., and Henderson, R.J.. 2018. Estimation of U.S. refinery water consumption and allocation to refinery products. In Fuel, Volume 221, 1 June 2018, Pages 542–557. URL: https://www.sciencedirect.com/science/article/pii/S0016236117309511 (accessed November 20, 2019).

Ronca, D. 2008. *How Carbon Capture Works.* HowStuffWorks.com, 9 July 2008. URL: https://science.howstuffworks.com/environmental/green-science/carbon-capture.htm (accessed November 22, 2019).

Statistica. 2019. Number of active coal mines in the United States from 2006 to 2017. URL: https://www.statista.com/statistics/949260/number-active-coal-mines-united-states/ (accessed November 21, 2019).

U.S. Department of Agriculture (USDA). 2018. *Census of Agriculture: Table 9, Characteristics of Irrigation Wells Used on Farms: 2018.* URL: https://www.nass.usda.gov/Publications/AgCensus/2017/Online_Resources/Farm_and_Ranch_Irrigation_Survey/fris_1_0009_0009.pdf (accessed November 20, 2019).

U.S. Department of Energy (USDOE). 2007a. *World Proved Reserves of Oil and Natural Gas, Most Recent Estimates.* URL: http://www.eia.doe.gov/emeu/international/reserves.html (accessed February, 19, 2007).

U.S. Department of Energy (USDOE). 2007b. *Carbon Sequestration Sites Found for Over 3,500 Billion Tons.* URL: http://www.eere.energy.gov/news/news_detail.cfm/news_id=10722 (accessed April 23, 2007).

U.S. Department of Transportation (USDOT). 2019. U.S. Oil and Gas Pipeline Mileage. URL: https://www.bts.gov/content/us-oil-and-gas-pipeline-mileage (accessed November 22, 2019).

U.S. Department of Transportation (USDOT), Pipeline and Hazardous Materials Safety Administration. 2017. PHMSA Regulations. URL: https://www.phmsa.dot.gov/phmsa-regulations (accessed November 22, 2019).

U.S. Energy Information Agency (USEIA). 2016. Hydraulically fractured wells provide two-thirds of U.S. natural gas production. URL: https://www.eia.gov/todayinenergy/detail.php?id=26112 (accessed November 20, 2019).

U.S, Energy Information Agency (USEIA). 2018. U.S. Crude Oil and Natural Gas Proved Reserves, Year-end 2017. URL: https://www.eia.gov/naturalgas/crudeoilreserves/ (accessed November 22, 2019).

U.S. Energy Information Agency (USEIA). 2018b. Underground Natural Gas Working Storage Capacity. URL: https://www.eia.gov/naturalgas/storagecapacity/ (accessed November 22, 2019).

U.S. Energy Information Agency (USEIA). 2018c. U.S. Coal Reserves. URL: https://www.eia.gov/coal/reserves (accessed November 22, 2019).

U.S. Energy Information Agency (USEIA). 2019a. Petroleum & Other Liquids. URL: https://www.eia.gov/dnav/pet/hist/LeafHandler.ashx?n=PET&s=MCRFPUS2&f=A (accessed November 20, 2019).

U.S. Energy Information Agency (USEIA). 2019b. U.S. natural gas production, consumption, and exports set new records in 2018. URL: https://www.eia.gov/todayinenergy/detail.php?id=41955 (accessed November 21, 2019).

U.S. Energy Information Agency (USEIA). 2019c. Annual Coal Report. URL: https://www.eia.gov/coal/annual/ (accessed November 21, 2019).

U.S. Energy Information Agency (USEIA). 2019d. How much of U.S. carbon dioxide emissions are associated with electricity generation? URL: https://www.eia.gov/tools/faqs/faq.php?id=77&t=11 (accessed November 22, 2019).

U.S. Environmental Protection Agency (USEPA). 1996. *National Water Quality Inventory: 1994 Report to Congress.* EPA-813-R-96–001, 63 pp.

U.S. Environmental Protection Agency (USEPA) (circa). 2000. *Class V UIC Fact Sheet; Heat Pump/Air Conditioning Return Flow Wells.* URL: http://nsdi.epa.gov/safewater/uic/classv/pdfs/heatpump.pdf (accessed February 19, 2007).

U.S. Environmental Protection Agency (USEPA). 2002. Community Water System Survey 2000. EPA 815-R-02- 005B. December 180 p.

U.S. Environmental Protection Agency (USEPA). 2006. *UIC Program: Classes of Injection Wells.* URL: http://www.epa.gov/safewater/uic/classv/index.html (accessed February 19, 2007).

U.S. Environmental Protection Agency (USEPA). 2009. 2006 Community Water System Survey; Volume II: Detailed Tables and Survey Methodology. URL: https://nepis.epa.gov/Exe/ZyPDF.cgi?Dockey=P1009USA.txt (accessed November 22, 2019).

U.S. Environmental Protection Agency (USEPA). 2016. Underground Injection Control (UIC); Class I Industrial and Municipal Waste Disposal Wells. URL: https://www.epa.gov/uic/class-i-industrial-and-municipal-waste-disposal-wells (accessed November 22, 2019).

U.S. Environmental Protection Agency (USEPA). 2017. Superfund Remedy Report; 15th Edition. EPA-542-R-17-001. URL: https://www.epa.gov/remedytech/superfund-remedy-report (accessed November 22, 2019).

U.S. Environmental Protection Agency (USEPA). 2019a. *Drinking Water Government Performance Results Act Tool.* URL: https://obipublic.epa.gov/analytics/saw.dll?PortalPages (accessed November 20, 2019).

U.S. Environmental Protection Agency (USEPA). 2019b. *Underground Storage Tanks (USTs).* URL: https://www.epa.gov/ust (accessed November 22, 2019).

U.S. Environmental Protection Agency (USEPA). 2019c. *National Pollutant Discharge Elimination System (NPDES)*; Stormwater Discharges from Municipal Sources. URL: https://www.epa.gov/npdes/stormwater-discharges-municipal-sources (accessed December 4, 2019).

U.S. Environmental Protection Agency (USEPA). 2020. Inventory of U.S. Greenhouse Gas Emissions and Sinks: 1990-2018. URL: https://www.epa.gov/sites/production/files/2020-04/documents/us-ghg-inventory-2020-chapter-3-energy.pdf (accessed June 17, 2020).

U.S. Geological Survey (USGS). 2018. *Estimated Water Use in the United states in 2015.* USGS Circular 1441. Washington, DC, URL: https://pubs.usgs.gov/circ/1441/circ1441.pdf. (accessed November 20, 2019).

Vipulanandan, C. 2008. Life cycle cost model for water, wastewater systems. *Underground Construction*, Wednesday, October 1, 2008.

Water Systems Council (WSC). 2006. *Household Wells Data: Number and Types of Wells by State and County or Other Locality*. Washington, DC.

11 Sustainable Development Concepts

ALTERNATIVES FOR FUTURE GROUNDWATER USE

Major questions related to long-term resource availability include: How do macroeconomic considerations affect the natural capital stock of groundwater for the future service of humankind? What is sustainable development? Is sustainable development of groundwater possible? What steps could be considered in obtaining sustainable development? In response to questions like these, a U.S. Council on Sustainable Development prepared a plan, which noted that the rising pace of groundwater depletion in the United States affects a major resource factor as an input to meeting our society's needs (Sitarz, 1998, p. 7). The independent Heinz Center has raised fundamental questions about the ability of society to maintain society's provision of goods and services from the ecosystem sustainably, recognizing that increased human demands are creating stresses on the natural system (Heinz, 2008) and that basic data to manage groundwater is lacking (Heinz, 2006). The European Union has implemented a research plan for sustainable water resources. Thus, sustainable development now has an economic and policy context, with some activities on the international level specifically targeted at groundwater.

The goal of preserving the range of choice of the present for future generations is certainly central to sustainable development, but it would be naively altruistic to think that each person acting on their own would use groundwater in that manner. Protecting the range of options for the future is probably best addressed at the macro, social–political decision level, affected through economy-wide policies such as conservation zoning (Turner et al., 1993, p. 60), limiting the scale of natural capital throughput (Daly, 1996, p. 15) or State growth policy specifically recognizing limited water supply as a factor (WWPRAC, 1998, pp. 5–17). The principal question to be addressed in this section is the following: Which alternatives protect the range of future options for groundwater availability and use while providing the services needed through a market economy?

DEFINING SUSTAINABLE DEVELOPMENT

Sustainable development recognizes the maintenance of necessary and critical resource services that the ecosystem provides while addressing equitable treatment of others in the current generation of people (intragenerational equity) and of future generations (intergenerational equity) (Daly and Farley, 2004; Common and Stagl, 2005). The World Commission on Environment and Development defined sustainable development as "development that meets the needs of the present without compromising the ability of future generations to meet their own needs" (UNWCED, 1987, p. 43). It also underscored that the basic needs of poor people must be a priority. With such a broad definition, a range of approaches responding to sustainable development have evolved. The United Nations defined at the beginning of the 21st century (2003, p. 4)

> Sustainable development [as] development that ensures nondeclining per capita national wealth by replacing or conserving the sources of that wealth; that is, stocks of produced, human, social, and natural capital.

A more specific, if not idealized, definition of sustainability was proposed by Hartwick (1989, p. 96):

> in a special case, accumulation of reproducible capital can offset a steady decline in resource use and allow per capita consumption to be maintained at a constant positive level indefinitely. This case involves society investing all rents from exhaustible resources reproducible capital instant by instant.

Known as the "invest resource rents for sustainability" rule, Hartwick's definition applies to both renewable and nonrenewable resources and requires that society invest an amount equivalent to the value of the remaining stock. As the stock declines with continued resource use through time, its value rises and that change in value is the resource rent (Markandya et al., 2002, p. 20). This required investment approach is a constraint on individual behavior because individuals will not save sufficiently to attain the sustainability envisioned in such an economic model. The point is that "interporal efficiency is not sufficient for sustainability" (Common and Stagl, 2005, p. 352). This result also applies in other related contexts, such as technological advances, which do not ensure some constant consumption level for a resource (Common and Stagl, 2005, p. 352). A further challenge is measuring the resources of concern at some aggregate level so that we might be able to devise an approach or model that could enable achievement of sustainability. Drawing on many of these concepts and more specific to groundwater, its "sustainability" has been defined as (Alley et al., 1999, p. 1)

> development and use of groundwater in a manner that can be maintained for an indefinite time without causing unacceptable environmental, economic and social consequences.

The largest groundwater industry association, the National Ground Water Association, has adopted a similar definition for sustainable groundwater resources (NGWA, 2016).

The measurement of sustainable development is a "sustainable flow of income" (Turner et al., 1993, p. 56). Calculating income "give[s] people an indication of the amount which they can consume without impoverishing themselves" (UN, 2003, p. 4, citing Hicks, 1946, p. 172). A sustainable national income level should not result in depreciation of the national capital, including natural capital, i.e., endowments of the ecosystem. Natural capital and ecosystem inputs in an economics context may be defined as natural resource stocks, land and ecosystems (UN, 2003, p. 5) providing resource, sink and service functions. All these functions are attributable to groundwater, in addition to it being an economic product or commodity. To more properly measure sustainable national income, revisions to the national income accounting system could guide policy for sustainable development, as described by the United Nations handbook on national environmental and economic accounting (2003). These revisions should address (1) depreciation of natural capital (a quantity measure) consumed and depleted by production and (2) degradation of natural capital (a quality measure) calculated through defensive expenditures on undesirable by-products of production and consumption (Turner et al., 1993, p. 56; Daly, 1996, p. 100). Since either gross or net national product treat natural capital depletion or environmental remediation as products and services adding value, rather than as costs to the nation or the ecosystem, society will believe that it has more national product to be consumed than what exists.

A number of authors have provided a range of operational responses or approaches to this definition of sustainable development (Turner et al., 1993; Reid, 1995; Daly, 1996; WRI, 1997; Sitarz, 1998; Markandya et al., 2002; Daly and Farley, 2004; Common and Stagl, 2005). Several key factors affecting the approaches emerge from these perspectives:

1. Scale is important (Daly and Farley, 2004): The scale is the ecosystem and the services it provides for the economy. The economy is a subsystem of the ecosystem, not a separate realm, and the economy draws its raw materials from the ecosystem. Relative to the earth's ecosystem, the world economy has grown over time, consuming an ever increasing amount of ecosystem resources (or natural capital). Significant depletion of aquifers underlying

extensive interbasin, interstate and international areas is a major example of scale effects relative to groundwater use.

2. The ecosystem provides services that are intrinsic and are valued by people (perhaps unknowingly) as providing basic livelihood (Turner et al., 1993, p. 58): "Primary products" such as drinking water; "assimilative capacity" such as the ability to break down or process some pollutants; "life support" such as the hydrologic cycle, which provides rain for crops; and "amenities" such as "satisfactions" from recreation on a river maintained by groundwater or the visual pleasure of seeing water flowing from a spring running down a hillside at the headwaters of a stream (Reid, 1995, p. 90).
3. Natural capital is used up in the production and maintenance of man-made capital. Therefore, natural capital and man-made capital are complements rather than substitutes (Markandya et al., 2002, p. 27; Daly and Farley, 2004, p. 232).
4. Preserving natural capital will provide services and income into the future for a stable population whose demands on the ecosystem are within its carrying capacity (Daly and Farley, 2004).
5. Economic policies and technology must be flexible and adaptable to provide alternatives for responding to ecosystem stresses (UNWCED, 1987, Ch. 2; Daly and Farley, 2004, p. 362).
6. Certain ecosystem functions and services are critical for human survival and cannot be replaced (Turner et al., 1993, p. 56). These functions and services include photosynthesis and the hydrologic cycling of water. More data are needed to evaluate changes in the ecosystem and their effects on people and societies (Heinz, 2008), including data on groundwater (Heinz, 2006).
7. Uncertainty with potential irreversibilities for environmental outcomes are economic costs that must be recognized in decisions involving trade-offs between near-term economic benefits and sustainable flows of income (Turner et al., 1993, p. 57; Daly, 1996, p. 16).

DEFINING ECONOMIC GROWTH

Economic growth is usually defined at the macroeconomic level as the increase in gross or net national product, which measures the value of all goods and services produced and provided. In recent years in the United States, the GNP has grown an average of 2%–4% annually, adjusted for inflation. This measure, used as the yardstick by most economists and financiers, counts all production growth in the same way, regardless of what is produced. By definition of the economists that espouse GNP as the measure of national productivity, more GNP is better, if the goal is economic growth. In the accounting perspective of this process, man-made capital stock, which is produced from consumed natural capital stocks or throughput, provides services recognized economically. As the man-made stock requires maintenance, it consumes more natural capital stock that can no longer provide its ecosystem services. In the economic growth vision, producing and using more natural resources is good, regardless of how well – efficiently – they are used, and regardless of the effects on the future availability of natural capital stocks for the next generations. Based on the present economic policy, then, economic growth must go on indefinitely. This circumstance assumes that technology will address any resource, capital and labor shortages, that natural capital is completely substitutable with man-made capital and, as a result, the ecosystem has no limits or constraints on carrying capacity for economic activities.

AN ALTERNATIVE APPROACH TO MEASURING NATIONAL PRODUCTION

Box 1.11.1 provides one alternative perspective for measuring production for a nation, state or other area. It considers first that natural capital and man-made capital are complementary, with the latter using up the former, in some cases, irretrievably. Its concept is that man-made capital services gained are at the expense of natural capital services lost, and it describes the relationships involved. Importantly, the approach recognizes that natural capital extracted (e.g., produced, as in mining) should not be counted the same as man-made capital in our accounting system of national production.

BOX 1.11.1 NATURAL CAPITAL – MAN-MADE CAPITAL COMPLEMENTARITY

As discussed previously, since man-made capital requires natural capital, the two types of capital are complementary. Because of this complementary relationship, services gained from man-made capital are at the expense of natural capital services lost. Herman Daly describes such fundamental relationships affecting economic growth and sustainable development through the "comprehensive efficiency identity" (Daly, 1996, pp. 84–86; Daly and Farley, 2004, p. 423):

$$\frac{\text{MMK services gained}}{\text{NK services sacrificed}} = \underset{\text{Ratio 1}}{\frac{\text{MMK services gained}}{\text{MMK stock}}} = \underset{\text{Ratio 2}}{\frac{\text{MMK stock}}{\text{thruput}}} = \underset{\text{Ratio 3}}{\frac{\text{thruput}}{\text{NK stock}}} = \underset{\text{Ratio 4}}{\frac{\text{NK stock}}{\text{NK services sacrificed}}}$$

where
MMK is man-made capital
NK is natural capital

Thruput is "the entropic physical flow of matter and energy from nature's sources, through the human economy and back to nature's sinks; it is the flow that is accumulated into [man-made capital] stocks and funds and out of which [man-made capital] stocks and funds are replaced and maintained" (Daly, 1996, p. 111)

Service is "the satisfaction experienced when wants are satisfied" (Daly, 1996, p. 110)

MMK services gained/NK services sacrificed is the "overall ecological economic efficiency" describing the amount of service obtained from each unit of man-made capital per amount of NK services sacrificed for each unit of natural capital used up from being transformed into man-made capital

MMK services gained are the satisfactions received from the application of man-made capital to human wants

MMK stock is the "inventory" of man-made "things" that are used up gradually

NK stock is the "inventory" of natural "things" that are used up over time

Ratio 1 is the "service efficiency of man-made capital stock" derived from technology and resource allocation efficiency as well as distributive efficiency within the population (Daly, 1996, p. 84). For groundwater, this technical ratio might be the number of persons served (for drinking, cleaning, waste disposal and recreation) per unit of water treatment and supply capacity (such as a 5000 m^3/day plant). For distributional equity, a measure of the number of low-income persons served at lifeline rates per 100 customers might be included. Allocative efficiency might indicate the quantity of lower quality water used for nonhuman purposes per unit treatment and supply capacity.

Ratio 2 is the "maintenance efficiency or durability of the man-made capital stock." For example, this could be the treatment and supply capacity per units of water, energy, steel, concrete and chlorine used to provide the ready water supply.

Ratio 3 is "the growth efficiency of natural capital in yielding an increment available for offtake as throughput." This ratio is a measure of economic growth. The more units of water, energy, iron and limestone as concrete that are sent through the economy to maintain water wells and supply systems, along with maintaining the other man-made capital, the larger the economy gets. As throughput increases, NK gets smaller. As groundwater is used for irrigation (throughput) in many parts of the High Plains Aquifer to produce crops in the midwestern United States, the aquifer is depleted (natural capital is smaller).

(*Continued*)

BOX 1.11.1 (Continued)

Ratio 4 "measures the amount of natural capital stock that can be exploited for throughput per unit of other natural services sacrificed." This "ecosystem service efficiency" is defined by the bound or limit of the natural environment as a source of low entropy resources or sink for high entropy wastes. The more natural capital stocks are used, the more natural capital services are sacrificed. In the groundwater environment, the mining of groundwater lowers the water table, which may cause wetlands to dry up, trees to die and streams maintained by groundwater to disappear along with their wildlife, while using more and more energy to pump the water further up the well (e.g., see USGS, 1999b).

Thus, in an economy that values more water production at least cost, the result is gaining MMK services by organizing to produce more water to satisfy more water-related wants while losing groundwater stocks, wildlife and limited hydrocarbon stocks, as well as services provided by the ecosystem that are not easily recognized, quantified or monetized. In depleting groundwater sources, future groundwater services may not occur at all, or occur at a significantly reduced level.

Economic production growth focuses on only one part of the equation: ratio 3, throughput/natural capital stock, ignoring natural capital services losses. Man-made capital services grow only if natural capital services are lost, some in perpetuity, and therefore, irretrievably. Certain aspects of groundwater supply may appear to correspond to irretrievable loss from macrolevel development. For example, zones of salt-water intrusion have expanded around centers of large past groundwater pumpage in the United States and Europe (USEPA, 1999; EU, 1999b). The stock aspect of the groundwater resource as a freshwater supply is gone in these locations. Furthermore, areas of groundwater table decline have expanded in the United States and Europe (USEPA, 1999; EU, 1999b), as well as in other parts of the world. Therefore, nations that are groundwater users cannot continue to focus on only one part of the "ecosystem efficiency" equation, but rather should consider the entire equation and its further refinement. This focus at a macrolevel would reflect the perspective that the economy does not exist independent of the ecosystem, but is in fact highly dependent on the ecosystem as its source of groundwater and other critical resource stocks and their services. This would reflect that the economy is a subsystem of the ecosystem. Nations, states and localities that are groundwater users would need to formulate policies consistent with limits to certain resources, recognizing the fact that technologies, while available, may not be able to be deployed equally everywhere to fully sustain groundwater or any other natural resource stock for future services.

This "comprehensive efficiency identity" suggests, and seemingly the world would benefit, that efficiency be redefined. For sustainability purposes, the redefinition may not be in monetary terms for the final calculations. Current traditional economic methodologies are still needed to perform assessments of production processes and to inform such service and capital flows as in the identity from the underlying economic processes to point to necessary resources adjustments critical to natural capital maintenance.

Source: Daly, H.E., Beyond Growth: The Economics of Sustainable Development, Beacon Press, Boston, MA, 1996, pages 84–87.

Thus, they are not additive. Man-made capital services can be produced at some rate of natural capital services sacrificed. Some natural capital services can be counted but others not currently known cannot. The point of the alternative approach is that we should conceptually get the accounting correct or we will mislead ourselves into believing that we can continue producing more indefinitely.

Another critical point in responding to this alternative approach is that some level of natural capital may be used to replenish the ecosystem services lost where nonrenewable resources are involved.

While this is a "defensive" use of the capital, it prolongs the services of the remaining natural capital. Thus, recycling and reuse are not just "nice to do" but imperative, especially in situations where the natural capital is being depleted to a point of nonexistence or prohibitively expensive extraction. Technological advances are therefore important not solely on economic efficiency grounds, as considered in current economic analyses, but to preserve the natural capital for future support of humankind in the ecosystem.

CAUTIONS ON COST–BENEFIT EVALUATIONS

The use of conventional cost–benefit analysis also assumes that society has a full understanding of the ecosystem relationships, which can be addressed through technology if any of the relationships becomes broken and that correction of market failures resulting in resource misallocation can be accommodated through efficiency pricing; that is, society, through the market, can establish prices for the perfect allocation of capital and labor – a truly unlikely occurrence, since science has not uncovered all the natural balances of micro and macro forces of the earth's ecological system to inform each individual consumer so he or she can make rational choices now or into the future about how much they are worth in the market that will be affected by these environmental balances in the future, or how resources should be valued in the political arena that influences their use. As Sagoff (1990, pp. 92–94) suggests, attempting to price unknowable, or at least currently unknown, natural balances of the ecological system as private transactions suffers from "category mistake." The aggregation of individual decisions in the marketplace is not the same as the establishment of public value and expressly stated objective for such natural balances to be achieved through decisions resulting from discourse among scientists, engineers, theologians, businesspersons, philosophers and preservationists in the public places of congresses, parliaments, legislatures or councils.

Recognizing the limits of cost–benefit evaluation in general and as applied to a particular project or program, decision-making should consider just how efficient the decisions are with positive benefit-to-cost ratios. However, we should continue to pursue understanding the ecosystem balances and requirements as they affect natural capital's goods and services, since they are critical to maintaining the economy overall and to our individual lives. We may obtain information to enable valuing some, albeit, very small, portion of the ecosystem to improve our interaction with it for a sustainable future. In doing so, we should recognize that net benefits of a project or program do not imply sustainability, but rather one possible efficient outcome as defined by our limited but growing understanding of the ecosystem and our economy within it. Notably, ecosystem experts have documented the inadequacy of available groundwater data for decision-making (Heinz, 2006).

NATURAL CAPITAL SHORTAGE

Since in the United States and Europe, there have been few periods in the last 50 years that have been characterized by shortages of natural capital, production growth has appeared to be limitless. However, as shortages have appeared, they may have not received sufficient attention by most people because their magnitude has not been large enough to cause widespread concern. One area of shortage in the United States is groundwater, and its shortage currently often appears somewhat local or regional: the declining water table levels of the High Plains Aquifer in the midwestern United States and of aquifers in California's Central Valley; locally in coastal areas, excessive groundwater production has caused saltwater intrusion, making the aquifers less usable in certain areas for freshwater supply and resulting in higher cost replacement water supply; and declining aquifers across the country because of various local, State and federal policies, such as central wastewater treatment and discharge, with limited regard for the value of this potentially reusable water, which has already received some level of treatment.

PRINCIPLES OF SUSTAINABLE DEVELOPMENT

Principles Applying to All Waters

Based on the considerations outlined here, several principles for sustainable development can be identified (adapted from: Turner et al., 1993, pp. 59–60; see also Sagoff, 1990, pp. 74–98; WWPRAC, 1998):

1. Achieve exigent macrolevel control while maintaining as much microlevel choice and discretion as possible (Daly and Farley, 2004, p. 361) – In situations where groundwater is limited and as a result allocated, people may still choose among the uses they decide for their allocation. For example, Westminster, Colorado, United States, had a legal limit to its groundwater production and implemented a water reuse program with an investment in a parallel water distribution infrastructure that allowed it to achieve local economic objectives: the city supplied more users with the same groundwater allocation through wastewater reclamation for reuse. The city has a limit to the water available to it.

   In a similar vein, California (United States) has developed a policy that provides for conservation and mitigates revenue problems for the water supplier. A law enacted in 2009 provides for (Hildebrand, 2009)
   - Metering water use
   - Establishing "basic" water use appropriate for essential consumer needs through criteria including number of users per dwelling, land use type, irrigated area and meteorological conditions
   - Setting water volume rates for the "basic" allocation based on "proportional cost of water service" considering customer category, basic use, metering and incremental cost of water supplied
   - Implementing a "conservation charge" applied to all water use above the "basic" allocation on a volumetric basis to allow the water supplier to encourage conservation and recoup conservation costs

   This approach provides flexibility to the water supplier and the consumer, allowing for micro- and macrolevel decision-making to arrive at a mutually acceptable result.
2. Correct market failures due to pricing and use rights – Examples might include eliminating water supply subsidies and modifying groundwater use rights that allow withdrawal without consideration of effects on adjacent users (e.g., WWPRAC, 1998, pp. 3–17, 3–20 and 5–10, 6–3[1]; EU, 1999b, p. 113; Daly and Farley, 2004, p. 361). Also, the European Union and many United States require delineation and management of drinking water source protection areas. Localities can choose the level of regulatory activities in these areas, providing rights or constraints to chemical use and water management in these areas within guidelines for groundwater protection.
3. Ensure regeneration and waste assimilation of natural capital – Basically, groundwater use should follow managed yield, and waste disposal should consider specific capabilities of natural degradation prior to the disposal (e.g., WWPRAC, 1998, pp. 3-2, 5–8, 5-13–5-15).
4. Develop technologies that change resource use from nonrenewable to renewable natural capital – For example, in areas of irrigation, embrace managed yield of groundwater through technologies such as drip irrigation gauged on local evapotranspiration rates (e.g., WWPRAC, 1998, pp. 5-4–5-17). Water reclamation and reuse technologies also support this and principle 5 below.
5. Utilize renewable natural capital only at levels allowing replacement by substitutes including recycling – Water use rates above the safe yield that would deplete an aquifer may be offset by artificial recharge of the aquifer by clean water (e.g., WWPRAC, 1998, p. 5–6).
6. Use the carrying capacity of natural capital to bound the scale of economic activity and, where uncertainties of this limit exist, incorporate a margin of safety to ensure maintenance of natural capital and potential irreversible consequences of its loss – For example, when disposing

of sludge on agricultural lands applying less than the agronomic rate may better take into account uncertainties of the subsurface environment to reduce the potential of groundwater contamination in known hydrogeologically sensitive areas, or setting standards for irrigation return flow to protect instream uses (e.g., WWPRAC, 1998, pp. xiv, 5-11–5-15, 6–21; USGS, 1999b; Daly and Farley, 2004, p. 361).

7. Utilize the appropriate political/institutional entity for the area or resource being addressed (Daly and Farley, 2004, p. 363) – For example, early United States policy encouraging development of the western states territory through irrigation of homesteaded land was done with limited knowledge of the groundwater resource capability in the longer term. The United States Congress and state legislatures could debate the level of sustainability of this fundamental resource, considering more recent information in all areas of geological and hydrological sciences and economics, and beliefs stemming from theology, philosophy and culture associated with the resource – all of which affect individual tastes and preferences now and in the future (e.g., Sagoff, 1990, Ch. 5). From a water quality standpoint, the European Union has decided from a social–political perspective that the lowest levels of pesticides and other man-made chemicals possible can only be permitted in drinking water, and the member States have spent billions of dollars to achieve high levels of treatment of public water supplies and for protecting sensitive areas around and upgradient of public groundwater supply wells (e.g., NWRI, 1998, pp. 209–230 and 295–304).
8. Organize around hydrologic systems and their entire water resource (WWPRAC, 1998, p. 6–4) – Applying sustainable principles at the hydrologic system level (aquifers and watersheds) would ensure the entire water resource (groundwater and surface water) (USGS, 1999a, 1999b; EU, 2000; USEPA, 2007a, 2007b) was considered in water use decisions and that effects on other resources are taken into account (WWPRAC, 1998, pp. 1-1, 3-2; USGS, 1999b). This principle reflects a sound science approach to implementing policies of sustainability for groundwater resources. While it may be difficult to address issues that cross political boundaries, governing bodies have developed and implemented such policies (e.g., the operation of the Delaware River Basin Commission provides regulatory control within a river basin that includes water catchment and aquifers in four states (DRBC, 2008)). This principle can embrace the concept of "total water management" (TWM) for all water flows including recycling and reuse of water on the continuum of conservation measures (Jeffcoat et al., 2009). Orange County, California (United States) has implemented this expanded concept through its "groundwater replenishment system" to treat wastewater and use it to create a saltwater intrusion barrier as well as provide water for groundwater recharge to the larger groundwater basin (Markus, 2009). Such an approach may be important to coastal communities' management of groundwater resources in responding to the effects of climate change and rising ocean levels.

Specific Sustainability Principles for Groundwater

Groundwater sustainability was previously defined as "development and use of groundwater in a manner that can be maintained for an indefinite time without causing unacceptable environmental, economic, or social consequences" (Alley et al., 1999, p. 1). Such consequences may embrace a wide range of factors, but using hydrologic principles established on recent years of improved understanding of water in the subsurface, principles of sustainability specific to groundwater may include the following:

1. Understand the status of the resource as the basis for taking actions for its sustainability – The European Union is establishing a groundwater monitoring network to implement the groundwater aspects of the Water Framework Directive for all member states (EU, 2007). From 2007 to 2013, the United States Advisory Committee on Water Information developed and implemented a proposal for a national groundwater monitoring network, recognizing that the

demands of drinking water, food and energy production, and climate change require "information necessary for the planning, management, and development of groundwater supplies to meet current and future water needs, and ecosystem requirements" (ACWI, 2019). In addition to establishing the condition of the subsurface portion of the ecosystem, such monitoring is a necessary and fundamental input to the economic baseline information essential to evaluating efficiency of future projects and programs through cost-benefit analysis.

2. Consider the natural sensitivity and contaminant vulnerability of the overlying unsaturated zone (e.g., USEPA, 1993a, 1993b; USGS, 1999b; EU, 1999b) – "Earth materials vary widely in their ability to transmit and store groundwater" (USGS, 1999b, p. 10). This variability is manifest on the ground surface and in the soil zone, as well as deep in the subsurface. Areas of more rapid infiltration of precipitation in the ground can be in proximity to those that induce more runoff. Contaminants, once in the ground and entering the saturated zone, may present great difficulty in being extracted from aquifers or aquifer zones that are characterized by slow groundwater velocities in some settings and by relatively faster movement in fractures of bedrock in others. Such physical factors heavily affect the cost of removing contamination from groundwater.
3. Consider the status of the well – Whether it is damaged through time, use, abuse or neglect (abandoned), well condition can contribute to contamination of groundwater (e.g., Jorgenson et al., 1998; NGWA, 2015).
4. Consider low groundwater velocities – Because of slow movement of water or contaminant to discharge points, contamination may remain for a long time, slow movement makes it difficult to treat and subsurface conditions may reduce degradation capabilities (e.g., USGS, 1999b, p. 59).
5. Effects of production or recharge occur over long periods of time (USGS, 1999b, p. 3; EU, 2007, p. 15, Annex. 1) – Water balance and quality should be evaluated over time, considering local and extended areas affected, which can be long distances from the point of use.
6. Groundwater may be renewable or nonrenewable, depending on use and recharge rates over specified timeframes – Losses from storage must be considered in the period for which sustainability is to occur (e.g., USGS, 1999b, p. 3; Meyland, 2011).
7. The deeper the well, the more distant the effects on and from other water and associated resources and activities (USGS, 1999b, p. 37) – Deeper wells may provide a larger or cleaner water resource, but deep wells pull water from larger areas and longer distances, potentially affecting more distant streams, wetlands and other surface waters, or are affected by them if contaminated. These situations may be caused by and affect individual water users and the costs of and the benefits derived using these waters. Aquifer modeling is being refined over time to address concerns such as these.
8. Consider all potential recharge areas or zones and contaminant pathways – Recharge zones from streams, wetlands, lakes, coastal interfaces, seepage ponds, adjacent brines, past and current injected wastes and other pathways should be evaluated when projecting future water quantity and quality effects (USEPA, 1990; USGS, 1999b). Box 1.11.2 identifies some of the most significant effects of not considering the entire water resource for planning and management purposes.
9. Consider the degradation potential of any product or waste introduced on land, into the subsurface or discharged to streams recharging groundwater (e.g., USEPA, 1993a, 1993b) – Decisions made ahead of the time of use of products or disposal of wastes relative to whether their capability to degrade in low-oxygen and no-sunlight environmental settings is sufficient or not may affect the long-term use of groundwaters that become sinks for these chemicals. These decisions affect the cost of using these groundwaters and whether future users will derive benefits from their use.
10. Perhaps most importantly and considering the other principles above, establish that groundwater and the subsurface environment have a positive nonzero value – This value can be established through specific resource policy performance objectives and then once human and ecosystem requirements are met, through other economic processes, including market exchange and tradable rights to groundwater and subsurface services.

BOX 1.11.2 SOME POTENTIAL EFFECTS OF POLICIES NOT CONSIDERING THE ENTIRE WATER RESOURCE

Quantity Effects

- Loss of water supply
- Lower groundwater tables from excessive pumping
- Adjacent landowners' property values reduced from inadequate water supply
- Stream flow reduced from nearby or distant groundwater pumping
- Headwaters of streams shortened from lower water tables
- Lake levels reduced from lower groundwater tables
- Vegetation (relying on shallow groundwater) changes or dies from lower groundwater tables
- Loss of habitat for local and endangered species requiring groundwater
- Shallow wells dry up from lower groundwater tables
- Deeper wells needed to reach water when shallow wells dry up
- Wetlands dry up from lower groundwater tables
- Wildlife not supported when streams, lakes and wetlands dry up
- Land subsidence from lower groundwater tables which previously provided structural support Highways, private and public structures, subsurface infrastructure damaged by subsidence

Quality Effects

- Groundwater quality deteriorates from chemical (product/waste) release, microbial infestation or brackish or saline water intrusion
- Groundwater users cannot use water source for drinking water
- Adjacent landowners' property values are reduced by deteriorated groundwater quality Stream quality reduced by deteriorated groundwater discharge
- Lake quality reduced by deteriorated groundwater discharge
- Loss of local and endangered species requiring groundwater within a specified range of quality and/or temperature
- Surface water users cannot use water source for drinking water or swimming
- Riparian landowners' property values are reduced by deteriorated surface water quality Wetland quality reduced by deteriorated groundwater discharge
- Wildlife not supported when streams, lakes and wetlands have deteriorated water quality
- Public use of public lands reduced by unsafe water supply unable to support public use

Sources: Adapted from U.S. Geological Survey (USGS), Sustainability of Ground-Water Resources, U.S. Geological Survey Circular 1186, 1999b.
Adapted from Western Water Policy Review Advisory Commission (WWPRAC), Water in the West: Challenge for the Next Century, 1998.

These general and specific principles may provide a substantial basis for implementing sustainable development and use of groundwater as a component for the ecosystem's natural capital. Each principle has a cost and benefit associated with it. These costs and benefits would need to be evaluated for each project and program to identify the most efficient way to achieve sustainability of the groundwater resource in those situations.

Range of Alternative Approaches

Sustainable development at the macroeconomic level in practice embraces a range of policies from pollution taxes (i.e., raise the price of activities that are not desirable) to regulation and standards (i.e., set a performance objective for the resource and for water-using activities). The basis for these alternative approaches might be characterized at one extreme as very weakly sustainable to very strongly sustainable at the other end of the spectrum (Turner et al., 1993, p. 60). Table 1.11.1

TABLE 1.11.1
Sustainability Approaches and Practices

Characterization (Overlapping Categories)	Features Used in Alternative Management Approaches	Rationale for or Shortcomings of Management Approach	Example Policy Tool Used with Approach	Potential Tools for Groundwater Sustainability
Very weakly sustainable	Cost-benefit analysis applying market correction through efficiency pricing; potential Pareto compensation; unbounded substitution	Technology can solve supply problems; correct pricing can be known and applied; near-term benefits of greatest value; extreme focus on economic methods	Tax on waste release, end subsidies, implement property rights	Taxes based on degradability of contaminants or volume of use; groundwater use rights for all property owners
Weakly sustainable	Cost-benefit analysis modified through use of augmented monetary valuation techniques, specific payment, expected value price adjustments, interdependent considerations, minimal associated requirements	Continued reliance on market and technology for substitutes; relies on discovery of total costs and benefits of system	Tax on waste release, permits; refund for residual recycle; target environmental conditions	Releases to subsurface controlled; set range for ambient conditions for acceptable water table levels or chemical application rates
Strongly sustainable	Set standards relying on precautionary principle, natural capital value; social preference; strengthened minimum requirements; internalize costs of externalities	Relies on standards that are broadly applicable	Environmental regulation of process or results; protective actions; waste release standards; permits; performance bonds	Specific concentration limits for contaminants in subsurface set; bonds required for chemical use and waste disposal on or in the ground
Very strongly Sustainable	Focus on ecosystem function; limited influence of cost-effectiveness information	Relies on knowledge of specific ecosystem relationships to guide decisions	Environmental standards; lifestyle directives	All uses of the subsurface controlled through strict withdrawal and discharge limits and near-real-time water use adjustment requirements

Sources: Modified and expanded from Turner, R.K. et al. 1993. Environmental Economics: An Elementary Introduction, *The Johns* Hopkins University Press, Baltimore, MD. p. 60; Daly, H.E. and J. Farley. 2004. Ecological Economics: Principles and Applications. Island Press. Washington, D.C. pp. 331–333, 363–365, 373–387; Common, M. and S. Stagl. 2005. Ecological Economics: An Introduction. Cambridge University Press. New York, NY. pp. 377–379.

presents this range of approaches and portrays some practical policy outcomes. Notably, approaches that rely on individual preference are characterized as weakly sustainable, whereas approaches that set performance requirements are considered strongly sustainable. In practice, all the approaches may be necessary, depending on the circumstances, emphasizing stronger measures of macrolevel policy where critical natural capital may be irreversibly withdrawn from providing a continued flow of income or where uncertainty about environmental outcomes is high.

Decision rules for inclusion of sustainability in macrolevel actions to maintain natural capital address both renewable and nonrenewable resources (Common and Stagl, 2005, p. 378, citing Costanza and Dely, 1992):

1. For renewable resources – Limit consumption to sustainable yield of the resource (this would be somewhat analogous to managed yield for groundwater) for renewable groundwater, typically shallow aquifers, but this only deals with the quantity supplied. Relative to quality, natural degradation rates would need to be established for different subterranean environments over a range of time periods to satisfy interests for groundwater use in a range of quality conditions.
2. For nonrenewable resources – Reinvest the proceeds from nonrenewable exploitation into renewable natural capital. Nonrenewable groundwater would most likely be deeper aquifers or shallow aquifers being used beyond safety margins for recharge, but could also include brackish and even saline groundwaters to be treated through desalination technology in arid and water resource-limited regions. For example, a portion of profits from mineral mining that deplete groundwater in the production process could be used to treat groundwater and artificially inject it back to the aquifer, if subsurface conditions permitted this.

TARGETING SUSTAINABLE GROUNDWATER POLICIES

Most communities or nations may not be able – politically, institutionally or financially – to address every aspect of sustainability that is of concern in the near term. However, they can be aware of the historical conditions and existing institutions affecting sustainability and begin aligning their policies and activities, after examining the larger picture of resource use and extraction from the ecosystem, to focus on priorities for maintaining their natural capital for *both* current *and* future generations. This has happened in the European Union with the management of its water resources. These policies and activities may also be designed to address resource resilience, the ability of the resource to recover when stressed, particularly to address the needs of critical human and environmental receptors. Several policy alternatives previously discussed and elaborated further below address intermediate steps that localities or nations could take to sustain their groundwater stocks to reduce the possibilities of future ecosystem service losses. These policies should be targeted at the largest categories of groundwater use for maximum results: (1) irrigation, (2) public water supply, (3) industrial use, (4) mining and (5) domestic water use (USEPA, 1996; Solley et al., 1998; EU, 1999b).

Additionally, the United States Council on Sustainable Development identified a range of policy changes that could potentially result in steps toward a sustainable future if action is taken. The policy changes that appear most related to groundwater, although not specifically identified as such in each proposed policy, are addressed in Box 1.11.3.

The European Union has also implemented an "Energy, Environment and Sustainable Development Programme for Research, Technology Development and Demonstration [RD&D] under the Fifth Framework Programme" (EU, 1999a). This RD&D program specifically targeted "Sustainable Management and Quality of Water" as its "Key Action 1" for years 1999–2002, and budgeted 62.5 million Euro specifically for sustainable water projects in 1999, which grew to 175 million Euro in 2002. These projects are summarized in Box 1.11.4 and reflect a value for sustainable water – albeit, unquantified at the natural resource level – within the policy-making European Council.

BOX 1.11.3 SUMMARY OF PROPOSED POLICY CHANGES FOR SUSTAINABLE DEVELOPMENT IN THE UNITED STATES POTENTIALLY APPLICABLE TO GROUNDWATER

- Shift tax burden toward consumption, and goods/services of major environmental risk
- Encourage economic and material-use efficiency
- Offset impact on poor with reductions in payroll taxes
- Determine if subsidies are still needed
- Ensure responsibility for the environmental effects throughout a product's life cycle
- Enhance efficient and clean technologies
- Provide incentives for recycling and finding substitutes for troublesome materials
- Make use of market incentives (e.g., emission reduction sales, charges, fees, deposit return charges and tradable permits)
- Adopt accounting practices that link environmental costs with the products, processes and activities that generate them
- Develop sustainable development indicators
- Develop immigration policy to address negative impacts on sustainable development from national population distribution
- Use the best ecological, social and economic information to manage natural resources
- Users of public resources should pay the full cost associated with the depletion or use of those resources
- Weigh the benefits of public infrastructure projects against the full costs
- Reform natural resource pricing
- Protect prime farmlands production
- Promote education focusing on interrelationships among the environment, economy, social equity and long-term thinking, hands-on projects and life-long learning and schools as sustainable models
- Support community sustainability planning processes and technology information to local governments and communities
- Manage geographical growth of the existing communities and siting of new ones to take into account carrying capacity and protection from natural hazards
- Eliminate governmental incentives that encourage development in vulnerable areas
- Create diversified local economies, which are built on unique local advantages
- Capitalize on economic development opportunities that target environmental technologies, recycling and pollution prevention
- Revitalize brownfields – contaminated, abandoned or underused urban land
- Support research and encourage other nations to participate more in international research on critical issues relevant to health and the environment
- Continue to promote and encourage global trading systems that mutually reinforce environmental protection and other social development goals
- Look for innovative technologies that allow for more cost-effective and efficient pollution protection and resource use
- Pursue efforts to mitigate potential effects of global warming and to adapt to those effects

Source: Abstracted and adapted from Sitarz, D. (ed.), Sustainable America: America's Environment, Economy and *Society in the 21st Century*, EarthPress/Nova Publishing Company, Carbondale, IL, 1998. With permission.

BOX 1.11.4 SUSTAINABLE RESEARCH PROGRAM OF THE EUROPEAN UNION

Key Action 1: Sustainable Management and Quality of Water

1.1 Integrated management and sustainable use of water resources at catchment scale

1.1.1 Strategic planning and integrated management methodologies and tools at catchment scale – To account for complex interactions of natural and man-made environments, for water quality and (quantity in various uses, and for consideration of uncertainties of water availability and quality and effects on aquatic ecosystem functioning

1.1.2 Socio-economic aspects of sustainable use of water – Assessing pressures and barriers hindering sustainable use of water resources; evaluating the dynamic relationships between water management, policy and institutions; understanding people's perception of water as a resource and economic good; analyzing the economic and ecological efficiency of different technological and managerial options; developing guidelines and indicators for sustainable water use

1.1.3 Operational management schemes and decision-support systems – For integrated water management, including validation and demonstration of management schemes, improved awareness of water problems, defining best management practices and effective transfer of knowledge to water users

1.2 Ecological quality of freshwater ecosystems and wetlands

1.2.1 Ecosystem functioning – Through evaluating the differences between natural variability and anthropogenic impacts; developing indices of ecological quality, functionality and biodiversity; improving and validating methods to maintain and restore regulative functions of degraded aquatic ecosystems

1.2.2 Ecological quality targets – In support of water quality surveillance and monitoring schemes, environmental policy and integrated water management, through development of indicators of ecological water quality and ecosystem health

1.3 Treatment and purification technologies

1.3.1 Management of water in the city – Including evaluation and optimum design of water systems and water exchange between supply and use regions; development of advanced treatment, rehabilitation and biohazards removal technologies

1.3.2 Wastewater treatment and reuse – to minimize environmental impacts of waste-water treatment through use of nonpotable water by industry close to source, develop strategies for local, safe, publicly acceptable and economically feasible reuse, novel sludge treatment and impact reduction techniques

1.4 Pollution prevention

1.4.1 Abatement of water pollution from contaminated land, landfills and sediments – Including development of novel in situ and on-site remediation technologies for contaminated sites, groundwaters and sediments and techniques to enhance natural attenuation

1.4.2 Combating diffuse pollution – By developing and validating land practices to prevent or reduce diffuse [nonpoint source] pollution loads

1.5 Surveillance, early warning and communication systems

1.5.1 Pollution surveillance and control – By developing software, measurement systems, sensing devices and early warning systems of hazardous pollution in soils, aquifers and sensitive aquatic ecosystems

1.5.2 Improved flood and drought forecasting – Through integration of processes and use of ground-based and remotely sensed information, in cooperation with other hazard observation techniques

1.6 Regulation of stocks and technologies for arid and semiarid regions and generally water-deficient regions

1.6.1 Water resources use and management – Through assessing environmental impacts of drought, groundwater exploitation, large-scale water transfers and socio-economic factors affecting water supply and demand, and developing criteria for preventing water stresses and water use conflicts

1.6.2 Prevention and mitigation of saline water intrusion – Through improved understanding of subsurface water movement and chemical processes

(*Continued*)

BOX 1.11.4 (Continued)

1.6.3 Technological development and management levels – Through harmonizing water data collection, maintenance and analysis, cost-effective and environment-friendly methods and tools for regulation of groundwater and surface water, improved knowledge base for implementing artificial recharge and alternative water sources, technological and cost-efficient changes to provide alternative supplies and apply water conservation in agriculture and industry and developing *improved techniques to characterize karst aquifers*

Source: Abstracted from European Union, Energy, Environment and Sustainable Development Programme for Research, Technology Development and Demonstration [RD&D] under the Fifth Framework Programme, 1999a.

Since sustainable development has to do more with the long-term existence and viability of goods and services from the ecosystem to maintaining economic functions, the first priority of sustainable groundwater policy has to be for establishing the sustainable scale of the resource, rather than monetary or fiscal policy. This requires information from scientists, engineers, planners and other disciplines to enable political institutions to establish water policies that provide for sustainable water resources and their use. Once the resource has become scarce, ownership and distribution become issues. Just distribution of water is a social, ethical and moral concern to ensure our fellow earth inhabitants have sufficient water to meet their needs, given current technology and institutions for accessing and using it. Policy instruments related to efficient allocation of the resource, especially if scarce, are the priority for sustainable policy implementation once sustainable objectives are set.

Some specific policy approaches for sustainable groundwater use may include a range of policy, science and technology options, but will need further analysis before implementing decisions relative to any particular project or program.

Comprehensive Integrated Water Resource Management

Groundwater should be more completely integrated into water resource and watershed management, planning and implementation with conjunctive management of groundwater and surface water quantity and quality, thus reflecting the full range of economic factors, costs and benefits affecting the use of the water resource (El-Ashry and Gibbons, 1988). As the review of national laws and policies indicated, most nations have a law that integrates the knowledge that groundwater and surface water are on a continuum of the hydrologic cycle, rather than two separate realms, with possible exception of deep groundwaters. Concerns about climate change indicate that meteorologic conditions must also be considered. In arid areas that may suffer further drought in the manifestation of climate change, brackish and saline groundwater (and surface water) should be included in an integrated approach to developing critical resource information and policies. A more comprehensive integrated approach will help ensure that all major alternatives for water supply are considered in establishing water management laws.

Set Clear Aquifer Management Objectives

As part of a comprehensive integrated water resources management policy, aquifers should have clear management objectives, as they are natural capital assets. These objectives are considered "strongly sustainable" in the framework of Table 1.11.1 (Sustainable Approaches and Practices) above. Groundwaters are mobile – flowing under boundaries, supporting baseflow of streams, maintaining the quality of estuaries and wetlands for habitat and influencing local climate as well as flowing to wells for human, plant and animal water supply, as part of the hydrologic cycle. Aquifers are depletable through both consumption and degradation, with consequent loss of the purposes just listed, others identified elsewhere and still some ecosystem functions unknown and yet to be defined by research. Of all these purposes, human water consumption, food production and degradation of aquifers have become the greatest concern relative to the necessary continuance of water supply and irrigation in locations of aquifer depletion. Surface waters are mobile and have clearly defined objectives reflecting public goals in law to be supported through water management based on best science. A comprehensive integrated water policy should likewise consider the public trust associated with the provision of groundwater through physical and legal means, recognizing best science, and incorporate specific and measurable objectives for its management, enforceable measures where appropriate, to ensure groundwater's availability for significant uses and future generations for its range of purposes. Resource resilience may also be an essential management objective to give forethought to the ability of groundwater to recover when stressed to ensure that critical human and ecosystem receptors are served. Certainly, legal entanglements will occur in sorting out the impact of assuming the public trust responsibilities in setting specific resource objectives in relation to individuals affected, but the larger water supply issue should be addressed to support the ecosystem and the multitudinous communities it maintains in a natural-human balance. Such action by the applicable jurisdictions will recognize the value of groundwater to the ecosystem and the societies it supports and incorporate that value in decisions about its use. Once long-term ecosystem objectives are met, the remaining water can be priced through water markets or can have its value substantiated through cost–benefit or cost-effectiveness analyses to provide signals and guides to efficient allocation.

El-Ashry and Gibbons (1988) point out that states having public trust responsibilities for their waters do not necessarily follow management policies for long-term sustainability of the resource. In arid regions, substituting urban use for agricultural use without changing underlying policies affecting the overall resource use may not be responsive to the future, such as a comprehensive water management policy that addresses future population changes in setting specific resource objectives. For example, the Arizona Groundwater Management Act requires achieving safe yields in aquifers by 2025, with the implicit expectation that continued depletion will be stopped and the economic adjustments will have occurred to accomplish this objective.

For the largest groundwater users, investment in efficient water-using technology will be critical for the long-term viability of the resource. However, as El-Ashry and Gibbons (1988) note, no opportunity costs exist for agricultural users, and otherwise no incentives promote change to efficient technology in the agricultural sector. Thus, without clear resource performance objectives, the benefits of more efficient water use may not occur.

Box 1.11.5 provides a perspective on possible ecosystem objectives that could be set for groundwater.

Source Water Protection

This concept is old, at least dating back to an ordinance adopted in the colony of Virginia prohibiting disposal of human waste near town wells. Additionally, the European Union has had policies of delineating protection areas and managing chemical substances around drinking water wells and conducting monitoring in what are now referred to as "drinking water source protection areas" since 1979. Its "Groundwater Directive" (EU, 1979) has been revised by a new directive (EU, 2006) to set standards for groundwater quality. In the United States, these zones are called "wellhead protection areas" under the Safe Drinking Water Act (U.S. Congress, 1986). Monitoring of groundwater in these protection areas is not required under U.S. federal law (however, some states do require it), and federal groundwater quality standards do not exist; however, drinking water standards have been

BOX 1.11.5 SOME POSSIBLE ECOSYSTEM OBJECTIVES FOR GROUNDWATER (LIST NOT EXHAUSTIVE)

Water Source

- Maintain certain minimum water elevation in the aquifer
- Limit boundary for cone of depression
- Maintain hydraulic head of aquifer within specified range (to recognize seasonal variation)
- Permit use at the rate of replenishment or within an accepted margin of variability or safety
- Ensure minimum groundwater discharge to spring, stream, wetland or coastal zone

Water Quality

- Maintain no statistically significant change or a change within a scientifically and publicly accepted range in water quality (or for a contaminant or set of contaminants)
- Establish contaminant concentration limit in groundwater not to be exceeded
- Set limit for contaminant plume advance to control threat to groundwater quality and receptors

Habitat

- Maintain areal extent of endangered species habitat
- Maintain specified biodiversity level
- Maintain wetlands area

applied to groundwater for remedial purposes. The concept from a water quality protection standpoint is to keep wastes and chemical use away from areas that would allow water to carry the wastes and chemicals to the well before they can be bound up in the subsurface or degraded. The challenge is that for some wastes and chemicals, it takes a long time for them to degrade and certain naturally sensitive areas allow such substances to reach groundwater quickly (in hours or days or months), which can then be drawn to the well and consumed without treatment in many cases. Private wells with limited treatment located in more rural areas rely on the capacity of the soil and subsurface environment for natural protection through attenuation and degradation. Zones describing the time it takes for contaminants to reach wells can be developed through hydrogeologic measurement, calculations and models.

From a quantity or supply standpoint, localities can enact zoning measures that control the amount of impervious surfaces per unit area that prevent percolation of precipitation into aquifers and maintain or even increase recharge (Communication, City of Westminister, MD, 1992; MSC, 2007, p. 6). Such a policy can benefit all major groundwater uses and surface waters receiving groundwater discharge supporting baseflow. Setting a standard for implementation of Drinking Water Source Protection is a strongly sustainable policy. Furthermore, use of aquifer storage of water injected into the subsurface for future use increases the need to protect larger water zones in aquifers from contamination.

Beyond protecting underground freshwater resources, because of demands in arid regions or areas experiencing significant groundwater depletion, brackish groundwaters (1,000–10,000 – and higher – parts per million of total dissolved solids) are being used or considered as sources of water supply after treatment (Olsen, 2008), especially as desalination costs are dropping. An integrated water resources policy that includes surface water, groundwater and brackish and saline waters is necessary to provide a comprehensive approach to future water management and protection of current and future water sources. A major drawback to future use of brackish water as a water source is the absence of policies, laws and regulations for its use and management (Al-Hadadi, 1999). Brackish sources are often deeper, which might provide a level of protection. However, with wastes being injected into the ground rather than discharged to streams, brackish groundwaters will need to be protected as future sources of water supply, perhaps even from the wastes that their treatment generate, which may be injected into the subsurface for disposal. While brackish water sources are not sustainable as a water supply and receive no or little recharge, they will have value in arid locations in times of drought or, if extensive, as a supplemental source. Some arid areas (e.g., State of New Mexico, United States) may have more available brackish and saline groundwaters than fresh groundwaters (Livingston, 2004), indicating a need for policies to use and protect these potential water sources.

Restrictions on Chemical Use in Vulnerable Recharge Areas

The practice of restricting chemical use in vulnerable recharge areas is somewhat taken into account in the United States and the European Union by restrictions on certain chemicals' use in areas known to recharge groundwaters used for public water supply (e.g., U.S. Congress, Safe Drinking Water Act provisions for underground injection control; European Union, 1991 Council Directive 91/676/EEC concerning the protection of waters against pollution caused by nitrates from agricultural sources requires Member States to designate vulnerable zones; and Council Directive 80/68/EEC on the protection of groundwater against pollution caused by certain dangerous substances). While setting a standard for this practice might be considered, it ideally would be more thoroughly considered for all groundwaters potentially usable for potable or other economic supplies, particularly to critical human and environmental receptors, including maintenance of flora and fauna. Protecting only those areas that are currently in use and not including the others because of their service for perceived current lower priority uses may be viewed as short-sighted from a sustainable perspective, and potentially costly later when needed, but contaminated. To reinforce the vulnerability of groundwater from a policy perspective, the EU states in the 2006 Groundwater Directive that "Groundwater is the most sensitive and the largest body of freshwater in the European Union … [and] must be protected in such a way that deterioration in the quality of such bodies of water is avoided in order to reduce the level of purification treatment required in the production of drinking water …" (EU, 2006).

Monitoring Groundwater Status

From an economic efficiency standpoint, providing water users with adequate information on which to make rational decisions is a fundamental element of allocating scarce resources. However, these data will be undersupplied if left solely to the private sector. Their existence would be useful as a public good in which the private sector would not invest except for its own purposes, in particular neither motivated to provide monitoring of third party effects nor to give to the central water authority or government as trustee for the public to use in decision-making. Monitoring of aquifers is the equivalent of obtaining essential data on the status of natural capital assets of local, regional and national importance. As noted above, the EU has requirements for monitoring the chemical status of groundwater, especially around drinking water supply wells. The state of Connecticut has a similar requirement in the United States and other states have done selective monitoring in wellhead protection areas. This monitoring creates a baseline with which to measure the physical and economic effects of future actions affecting groundwater levels and quality. Additionally, some states have a network of wells for drought monitoring, such as Pennsylvania (USGS, 2007). Benefits of monitoring include advance warning of pollution or drought conditions to avoid the costs of responding to those conditions and of any health effects to people and animals. Monitoring also helps ensure that contaminant releases are routinely checked for enforcement purposes across the state or nation. This regular checking precludes creation of "pollution zones" that would attract and concentrate releases of harmful substances. Ideally, the value of the costs avoided in the state or nation from the "early warning" information from monitoring would be greater than the cost of monitoring.

Notably, the U.S. Geological Survey and other federal, state and local agencies and nonprofit associations have implemented the National Ground-Water Monitoring Network with a portal for states, tribes and other organizations to share groundwater levels and quality data nationally (ACWI, 2019). The federal-interstate Advisory Committee on Water Information provided a template built on other water data standards. As of 2019, the Network had nearly 30 states providing groundwater data.

Use of Low-Impact Development

Techniques to improve water balances to increase infiltration and to facilitate natural biodegradation are now being installed and used to reduce the impact of development on watershed and aquifer hydrology. Methods include bioretention, cluster building, reduced impervious area, swales, permeable pavement, vegetated landscaping, wetlands and green roofs. Such approaches have been shown to be effective and reduce capital costs from 15% to 80% over conventional approaches to manage stormwater (USEPA, 2007b). Recognition of these "best" practices across the industry is an economy-wide approach to reduce the "scale" of impact on the natural capital of groundwater. A more strongly sustainable approach would be inclusion in a standard for all site design engineering, such as in the state of Maryland, United States (MSC, 2007). The U.S. Environmental Protection Agency has conducted research on the impact of stormwater infiltration on groundwater as a basis for guidance needed by states for application of low impact development technology to protect groundwater quality (Brumley et al. 2018).

Widespread Use of and Incentives for Efficient Water-Using Technology

The technology is available to obtain the benefits of more efficient water use in agriculture irrigation. For example, the irrigation technique known as "drip irrigation" has been used for many years as a documented physically efficient method. With new electronic technology, water can be precisely delivered to plants to meet their water needs at a particular time. Typically, irrigation is done by flooding fields with water, losing valuable irrigation water to infiltration past the soil zone, or evaporated and not used by the plants at all. Flood irrigation has also concentrated pollutants in the soil zone, contaminated groundwater and, through groundwater return flow, surface water (WWPRAC, 1998, pp. 2–30 and 2–31). This technology is very wasteful of water, especially water only valued at the cost of pumping it from the aquifer. The introduction of drip irrigation may be expensive initially, but taxing or use fees can encourage pricing water at a value that makes price in use at a particular location fit its circumstance, rather than wasting it to depletion or contaminating other resources. Clearly, more research is needed to mitigate groundwater losses from irrigation to grow food necessary to maintain the world's population. The U.S. Department of Agriculture Environmental Quality Incentives Program (EQIP) provides financial incentives to farmers irrigating crops to invest in technology to use water more efficiently.

Other improved efficiency technologies can be broadly applied. Water-saving kitchen and bathroom devices have been available for many years. Certainly, these mechanisms could be further refined. Composting toilets have been developed. This equipment may be expensive, but may be worth the cost in situations in which pumping is depleting the water source. In the United States, the Environmental Protection Agency announced in 2006 its "Water Sense" program to reward water conserving devices and practices with recognition, so the larger public was aware of the options available to it (USEPA, 2006).

Sufficient incentives may not be given by the market for water to encourage their application more fully. However, many utilities have distributed at no cost to customers low flow faucets and shower nozzles to conserve water and energy (heating less water flow). The incentive to the utility was to reduce capital in not having to expand water and wastewater treatment facilities and decrease fuel supply needs for hot water. Such actions reduce the ecological footprint of groundwater use by reducing water use, energy consumption and greenhouse gas (carbon dioxide) release.

One macroeconomic approach would be to provide credits to industry and customers for retrofitting their water-using fixtures not only for installing but also for using these water-conserving devices by demonstrating their lower level of water use documented through water bills. These economic incentives would reduce "scale" impacts. Additionally, grants could be given to low-income and poor people to allow them to make such plumbing replacements, or giving larger credits to owners of multidwelling units to expand the effect of water use reduction where people do not own their residences, thereby addressing intragenerational equity. The case study "Balancing Ecosystem, Water Use and Pricing: San Antonio and the Edwards Aquifer" describes one community's efforts to address an ecosystem maintenance requirement through water conservation retrofit devices and continuing reductions in water throughput.

Water Recycling in Manufacturing

Industries manage costs to increase profits. Manufacturing is a large user of water. While recycling technology has been available for many years, in many applications it is less expensive in the short run to continue to use large amounts of water rather than recycle or conserve water. Several approaches are possible to promote more sustainable water use by industry. Governments at all levels can induce recycling by taxing increasing water use or large water uses; however, a more strongly sustainable policy would be for governments to set required water recycling goals for industry, which might ultimately demonstrate technology applicable to and economic for residential applications, based on the objective of groundwater resource maintenance. Then, technologies may evolve to use the water more efficiently, allocated in water markets or substantiated through cost–benefit and cost-effectiveness analyses.

A national industrial water use goal may promote a level playing field for all industries that use water. Otherwise, industries in which water is a significant raw material input will continue to move to locations that offer water at the lowest price. Since most industries are engaged in interstate commerce, national governments can act on this matter within their constitutional authority. The location of industrial water users would also be affected by the policies of other nations willing to deplete their natural capital of groundwater, which may be contrary to the long-term economic viability of those countries. Large water users could also engage in water recycling and aquifer replenishment.

Other options include encouragement of all water using industry to control costs including costs associated with water use. Many companies using ISO 14046 standards report water use and their steps to minimize water use (ISO, 2020), a more strongly sustainable approach. Alternatively, taxing industrial water use to reduce "throughput" of the resource and make water waste more expensive thereby encouraging water conservation would be a weakly sustainable policy.

Water Reuse for Public Supplies

Most communities using public water systems treat their wastewaters and discharge them to a stream, using gravity or pumping stations, which consume large amounts of energy to transmit water to move the wastewater to the treatment plant and then to the stream. Communities relying on groundwater for their public water supplies and having declining water tables may be encouraged to reevaluate their position on use of treated wastewater and pump it back into the ground upgradient of the location of municipal water production. Careful evaluation of the quality of the treated wastewater may be necessary to ensure that other natural flora and fauna will not be adversely affected and/or that injection of treated wastewater will not bring naturally occurring contaminants of the underground environment into solution. Treated wastewater from many treatment plants is of higher quality than the water into which it is being discharged. This water may be ideal for groundwater recharge, as long as federal and state prohibitions against waste injection into an aquifer that is a public water source can be adequately addressed.

As an example, the Orange County (California, USA) Water District (OCWD) has since 1933 treated wastewater and recycled the water for use by communities it serves. Each day, the Orange County Sanitation District treats 757,100 cubic meters (200 million gallons) of wastewater from over 2.5 million customers and pipes 492,100 cubic meters (130 million gallons) to OCWD for further treatment surpassing drinking water quality requirements. Through OCWD's Groundwater Replenishment System, the purified water is then used to recharge the Orange County Groundwater Basin and to create a groundwater barrier to prevent saltwater intrusion from the ocean to the aquifer. The cycle of groundwater withdrawal, water use, wastewater treatment and groundwater recharge is repeated continually to meet the water needs of the Water District's customers (OCWD, 2020). Fountain Valley, California, has a monthly residential water rate of $41.90 per 1.983 cubic meters (500 gallons) that is below the average monthly rate of $42.17 of similar-size California communities supplied by groundwater reporting water rates to the American Water Works Association 2019 Water and Wastewater Rate Survey (AWWA, 2019).

Full Cost Pricing

While classical economists would indicate that, for economic efficiency, price should equal marginal cost (P = MC), only financial marginal costs for water are known in most cases (USEPA, 2007a). If governmental entities are providing the water, they may not be charging their full financial cost of producing the water, thereby subsidizing consumers. In some circumstances, other ecosystem costs of producing the water may be known (but often difficult to quantify), or if not known, could be valued at some level through a user fee or tax, albeit imprecisely established, to provide for resource maintenance. At the very least, the full financial cost of pumping, treating and delivering groundwater should be accounted for along with an informational and educational program on the value of groundwater and the ecosystem externalities of producing it to make consumers aware of these values and costs in their economic decisions. Life-cycle costing, addressed in Volume 2, Chapter 5, can be used to define water processing actions taken and their costs. Full cost financial pricing, while economically theoretically least satisfactory (UN, 2003, p. 301), addresses a portion of the allocative efficient value, but still does not respond to scale or distribution considerations identified for macrolevel policies. However, a groundwater use charge or fee at some level would overcome "free" (no-cost) consumption of this natural capital and its ecosystem services that are not valued in the market, thereby being closer yet to a "fuller" cost price. Revenues from such a charge or fee could then be used in ways to replenish the resource, make it more sustainable and begin to capture a fuller portion of the economic cost of producing it, reflecting groundwater's role in the ecosystem: water supply to communities and wildlife, the baseflow and quality of streams in watersheds and the habitat of wetlands and coastal waters, among others. Unless there would be industry standards for pricing of water, reliance on the utility pricing, which could be done based on political response, is weakly sustainable.

New National Accounting

By establishing a new national accounting practice measuring depletion losses and costs for remediation of groundwater (and other resource) contamination, the government may spur creation of rights in maintenance of natural capital. In the United States, endangered species have been recognized as needing habitat preservation, including species in groundwater (EARDC, 2006; also, see Chapter 2 and Case Studies in Volume 2). The United Nations expressed global concern about loss of ecosystem endowments from economic growth by issuing a handbook detailing a system of national environmental and income accounts that nations can use (UN, 2003) and which some countries have adopted (Markandya et al., 2002, p. 68). The UN system responds to the question: "Is [the] use [of "environmental endowments"] posing a threat to economic development now, either by being used up too quickly with no prospect of replacement or by generating a level of pollution which threatens human health and the existence of species?"

State and national governments may further encourage creation of rights in maintaining groundwater (and other natural) capital. The State of Arizona has established in its Groundwater Management Act the goal of attaining "safe yield" in three major groundwater use areas of the state by 2025 (Smith, 1989, p. 44). While many options may exist for adopting policies for sustainable development, one national approach might be to establish federal boards for water (or other) resource maintenance, similar to the Federal Reserve Board system or establish River Basin Commissions for all large watersheds with actual water authority (such as the Delaware River Basin Commission) in the United States (e.g., see WWPRAC, 1998, pp. 6–4 and 6–28). Different regional boards could specialize in different aspects of water use, costs, pricing and ecosystem sustainability. Such boards could meet on a regular interval to provide guidance on pricing for water based on use, location, environmental and resource effects and weather patterns, with the principal national trust of maintaining natural water capital, elimination or minimization of depletion and degradation, and encouragement of reuse and recycling, where practical. These practices could then be accounted for in a new national accounting system that kept track of natural resources and their values, including losses from depletion and degradation. Such an approach, with State participants on the boards, could then provide a more level policy field for groundwater use and pricing, instead of pricing groundwater at zero or some low value such as the cost of pumping that does not account in some way for third party effects or externalities.

Table 1.11.2 provides one list of potential accounting categories for water and an example accounting table for groundwater assets. Box 1.11.6 explains the possible monetary valuation of environmental accounts, indicating the use of price times physical units is one method of valuing water and other environmental assets. Using these categorizations and valuation techniques (see Volume 2, Chapter 5, "Life-Cycle Cost–Benefit Analysis") and constructing national account tables to address (1) depletion (loss of domestic product from consumption), (2) defensive expenditure (costs to protect the health of humans, fauna and flora and the quality of natural resources on which they depend) and (3) degradation (accounting for the loss of stocks of natural resources from contamination) (UN, 2003) in successive years, environmental accountants will be able to refine approaches to recording the use and maintenance of natural capital and ecosystem services. This accounting should be essential and useful to future generations as they act to maintain their welfare in societies. Several countries have employed an integrated national income and environmental accounting framework as described in the UN handbook to promote more sustainable natural capital. In one case, Mexico found that resource depletion reduced national income by 5.7%. When resource degradation was included, the reduction was 13% for the year analyzed (Markandya et al., 2002, p. 68). Mexico has significant groundwater mining for irrigated agriculture. Adopting such an approach as a national standard of accounting would be strongly sustainable in informing decisions affecting resource use, depletion and degradation.

TABLE 1.11.2
Potential National Account Categories for Water and Water Asset Account Table

Water Account Example Categories

Water Supply	Water Use
Total Withdrawal	Received for use
Onsite use	Available
Leakages during use	Recycled
Available for use	Wastewater - Sewerage
Delivered	- Treated
Distribution leakage	- Untreated
Recycled plus imported	Cooling
Total Supplied	Hydroelectricity
	Irrigation return
	Other returns
	Total Consumption

Asset Account Template for Groundwater—Generalized Example

		Groundwater Million m^3
Opening stocks		150,000
Abstractions		765
Residuals	Returns from	
	Irrigation	50
	Wastewater	268
	Lost water in transport	300
	Others	
Net precipitation (+)		2,275
Inflows (+)		1,100
Net natural transfers (±)		–45
Evaporation from water bodies (–)		
Outflows (–)	To other country	380
	To the sea	1,000
Other volume	Due to natural disaster	
Changes	Discovery (+)	
Closing stocks	Others	149,229

Source: Modified from United Nations (UN). 2003. Handbook of National Accounting: Integrated Environmental and Economic Accounting, Publications Board and Exhibits Committee, New York. Pages 333–335.

BOX 1.11.6 MONETARY VALUATION OF ENVIRONMENTAL ACCOUNTS

Monetary accounts

Monetary valuation

- Flows
- Environmental expenditure
- Resource management accounts
- Issues related to water resources valuation
  - Nature of water
  - Changes to water source
    - Due to policy
    - Due to action

Monetary valuation of flows

- Record of major economic output of industries related to water
- Price as given to the consumer or stated by the supplier, in the case of residuals management
  - Multiply by units of physical stock, flow and quality in an asset account
  - Other methods of valuation possible

Categories of activities for monetary valuation

1. Environmental protection activities – Targeted actions for soil and groundwater for:
   - Reducing or eliminating polluting substances applied to soil and percolated into groundwater
   - Decontaminating soil
   - Monitoring and controlling soil and groundwater pollution
2. Natural resource management and exploitation activities – Expenditures related to:
   - Abstraction and purification of water
   - Current intermediate consumption
   - Compensation of employees, taxes and subsidies related to water, on capital expenditure and on consumption of fixed capital, stock of fixed assets and labor inputs.
3. Environmentally beneficial activities – Expenditures related to:
   - Abstraction and purification of water
   - Saving water, whether for final consumers, industries, services or agricultural users
     - Investment (irrigation systems, facilities and appliances to reduce water consumption, recycle water, etc.)
     - Use of products adapted for lower water consumption
     - Direct abstraction of water by
       - Manufacturing industries for cooling purposes
       - Final consumers for own use

(Continued)

BOX 1.11.6 (Continued)

4. Minimization of natural hazards – Expenditures to minimize flood and drought effects related to:
   - Construction of dams to restrict water flow
   - Management of water retention areas or recharge zones
   - Response needed for effects of alteration of landscapes and water systems or global warming

Source: Modified from United Nations (UN). 2003. Handbook of National Accounting: Integrated Environmental and Economic *Accounting*, Publications Board and Exhibits Committee, New York. Page 340.

MANAGED YIELD AND SUSTAINABLE DEVELOPMENT

The hydrogeologic concept of "managed yield" aligns with "sustainable development." Managed yield is the production capacity of an aquifer with a margin of safety to maintain aquifer levels and quality for critical receptors such as minimum well depths or survival of endangered species (Meyland, 2011). Managed yield is distinct from the safe yield of an aquifer defined as "the amount of water [that can be] withdrawn from [an] aquifer roughly equals the amount of water returning naturally or artificially to the aquifer over an extended period of time" (Smith, 1989, p. 10) and "can be withdrawn from the aquifer without producing an undesired effect" (USGS, 1985, p. 463). The USGS definition of "groundwater sustainability" includes indefinite use without unacceptable environmental, economic or social problems and "must be defined within the context of complete hydrologic system of which groundwater is a part" (USGS, 1999a, p. 2). Thus, managed yield addresses an action (use of an aquifer for water supply or wastewater release) over time that maintains the resource without adverse impacts. Sustainable development, as defined here, concerns maintenance of natural capital to maximize future services. The concepts appear synonymous as applied to groundwater. The question, then, is how to achieve "managed yield"?

Managed yield is imbedded in the water balance equations developed earlier. While this observation may seem obvious, one might ask, why cannot it be achieved if the concept is clear? The concept may be clear, that is, managing water balances to maintain critical or essential functions, but the challenge is with rights of use of the aquifer and the subsurface environment, it is important to have sufficient information on water levels and quality. Unless a state or federal government has adequate data about the groundwater capital and the needs of critical receptors, depletion or degradation of the resource may occur. Depletion occurs when abstraction of groundwater is greater than recharge of the resource. These abstraction and recharge relationships can be established over regions with past and current data. The USGS publishes water balances for areas it has investigated (e.g., see USGS, 1999a, pp. 25 and 28). Climate changes from global warming makes this situation even more challenging. Adequate monitoring must be planned and implemented.

Degradation of groundwater occurs when waste or chemical residue exceeds the assimilative capacity of a subsurface zone and results in contamination of the groundwater resource. Such situations exist when excessive amounts of sludge above agronomic rates are applied to the land surface for disposal, leachate carrying landfill chemicals escapes from a disposal site or pesticides infiltrate beyond the root zone and the control of the crop manager. In these instances, the immediate sink for these waste or chemical releases is the first aquifer, which may be used for public water supply, industry and stream baseflow, in each case potentially causing further unanticipated damage and economic loss. While much work is still needed in understanding "quantity" management for managed yield, this field is more developed than that of "quality" management for sustainable development. Certainly, the subsurface is being characterized relative to its ecological significance and ultimately its economic importance in the larger ecosystem (Gibert et al., 1994). Research has progressed in discovering how previously unknown organisms act to break down wastes in the subsurface environment. The implications for the value of these organisms in the subsurface are enormous, as we have unknowingly relied on them to "process" our wastes placed on or in the ground. Indeed, some chemicals may kill these waste processors. Thus, it is important to find out what the balance for subsurface or groundwater quality is on a sustainable basis that can continue to provide the services the ecosystem requires and the economy needs. Research is necessary to address this balancing of contaminant disposal and processing. Monitoring is needed to provide a baseline and track changes over time for natural capital and its economic management to occur.

Developing plans for managed yield of groundwater as a component of a larger community, state or national sustainable development policy also should incorporate total economic value of both costs and benefits associated with it. However, if a sustainability policy is to be successful, the costs and benefits need to be described as fully as possible to foster understanding within communities and states of groundwater value when sustainably managed, given current science and technology, even if the costs and benefits can only be quantified but not monetized.

MAINTAIN CAPITAL, MAXIMIZE SERVICE, MINIMIZE THROUGHPUT

In summary, sustainable development of groundwater is comparable to managed yield. To implement managed yield, water policy should be modified to recognize the limits of the resource. The limits of the resource can be incorporated in policy by establishing the objective of maintaining the natural capital of groundwater to provide ecosystem and economic goods and services as a cornerstone of future water policy, recognizing the needs of maintaining critical receptors. The water policy should be directed toward maximizing the service of the resource, employing more efficient and conserving technologies in the near and long term. Actualizing the constraint of responding to critical receptor needs by these technologies will minimize throughput, which otherwise transfers groundwater more quickly to the atmosphere and the oceans than would otherwise occur. It will also reduce continued irreversible risks that future generations would have at least essential quantities and qualities of a fundamental resource that will need to provide services to more people.

Is sustainable development a better state than economic production growth? Sustainable development uses currently undefined natural resource limits as targets for supporting the economy. Production growth has targets of increased natural capital throughput as a positive barometer of the economy, assuming no limits on natural capital and unlimited substitution of man-made capital for natural capital. Sustainable development recognizes given resource capacity as a management constraint, which can guide technology to restore, replenish and maintain the resource. Under sustainable development, everyone would ideally have the groundwater they need through qualitative improvements in lifestyle. Under production growth, anyone could take as much groundwater as they can afford or are legally allowed, perhaps leaving some persons without the amount they need. Some path of the economy that takes an intermediate approach in transition may be necessary, as suggested in the policies above. The question of which is "better" may need to be addressed in the context of which addresses the essential needs of the population and environment.

WHO BENEFITS AND WHO PAYS THE COST OF THE DIFFERENT POLICY APPROACHES – AND HOW MUCH?

The production growth approach benefits the individuals and corporations that are allowed to produce crops and manufactured goods that require water while paying little or nothing for the water. The benefit they receive is water of lower or no cost, providing a subsidy to their operations. Whether the water should be used for other social purposes that might have higher social values has not been addressed. However, federal laws providing quality standards and species maintenance are being recognized as expressing and incorporating some of these values (see, e.g., WWPRAC, 1998, p. 5–42). The costs of producing water considered to have low value may be borne by adjacent neighbors at a different time. For example, in the western United States, land subsidence from overdrafting aquifers for many years for irrigation has subsequently caused a range of social and private costs, including damage to an interstate highway and other roads, surface water reservoirs, railroad tracks, use of cropland, wells and stream channels (Smith, 1989, p. 39; USGS, 1999a, pp. 55–58). Pumping resulting in declining water tables can eliminate wetlands dependent on groundwater influx providing habitat for terrestrial and aquatic wildlife (USGS, 1999a, p. 42), and, consequently, to a wide range of outdoor enthusiasts from bird watchers to duck hunters. Management of aquifers, knowing of their depletion, occurs "with little thought of the future consequences and foregone opportunities" (Smith, 1989, p. 15) and is unsustainable in much of the western United States (WWPRAC, 1998, p. 3-2). Other foregone opportunities include the range of advantages of groundwater identified in Chapter 1. All of society – a nation, state and locality – pays the cost of such an outcome in terms of lost production for higher value uses.

Sustainable development, focusing on improvement in the quality of use, provides the benefit to this and future generations of a long-term water source where managed yield is practiced. The managed yield approach ensures that a range of future groundwater uses are preserved, while allowing present uses through a consciously applied, objective-driven decision-making framework of policies designed to supply current needs and preserve the services of natural capital providing for critical receptors. The costs will not only be incident on those whose use of groundwater is subsidized by current water law, but also on others affected in the long term over which groundwater effects take place. These costs serve as potential benefits supporting sustainable development policies. Marginal producers whose existence is predicated on low value of water will most likely go out of business. The United States Western Water Policy Review Advisory Commission recognized that such marginal water uses were part of the challenge the western United States must deal with and that the federal government should not support through subsidies (WWPRAC, 1998, pp. 6–34–6–35). Moreover, the cost to polluters to supply clean water in cases where they are liable for providing alternative safe water will rise.

It is difficult to estimate the value of the costs and benefits of these policy alternatives. This matter should be the focus of considerable research. One study to estimate such values on a macrolevel (Costanza et al., 1997) took the first steps in addressing the web of interrelationships that exist in nature and its economic subsystem to project macroeconomic results.

SUMMARY

Sustainable development is defined as "development that meets the needs of the present without compromising the ability of future generations to meet their own needs." Economic growth usually applies the measure of current national production, gross national product, as a signal of well-being. However, this measure combines national production benefits and costs into one number and assumes that unlimited natural resources may be extracted from the ecosystem for this production. The sustainable development of groundwater is constrained by the lack of recognition of its limited presence even though it is the largest store of freshwater in the world. Groundwater supports the ecosystem in ways we do not understand, having value in these unknown services that manifest certain aspects of balance in nature. Groundwater is undervalued as natural capital and produced and provided in a way that imputes low cost for it. Water is needed for future generations and groundwater will most likely be a critical source at that time. Sustainable development means considering scale and distribution of resources within current and to future generations, recognizing the needs of critical receptors and relying on economic efficiency to deliver the services of groundwater and other critical natural capital in a least-cost way. Policies for providing groundwater include those that are strongly sustainable, such as natural capital maintenance objectives and production and quality standards, and others that are weakly sustainable, such as taxes and subsidies. The benefits of economic growth policy accrue to the present population, leaving fewer resources for the future with greater costs to fend for themselves with what remains. The benefits of sustainable development accrue to the present and the future generations by recognizing that current extraction of services from the ecosystem must be managed to provide for a fair distribution to people and critical environmental receptors in the future. Groundwater can be managed sustainably and policies for states and nations have been devised to do so, including policies to conserve and protect the resource for future use.

FOR FURTHER CONSIDERATION

1. Check with the local or regional planning commission or council as to what sustainable development concepts or principles are used relative to groundwater. What specific sustainable development objectives and actions have been undertaken? What are the costs and benefits identified with these objectives and actions? Can the costs and benefits be monetized?
2. How have stakeholders been involved in establishing the sustainable development concepts and principles? To what extent do the stakeholders have an economic stake in the implementation of the concepts and principles?
3. Do any of the local or regional sustainable concepts or principles involve maintaining groundwater-related natural capital, such as maintaining minimum groundwater or well levels for certain receptors, protecting wetlands or preserving endangered species? What was given up to maintain groundwater natural capital and by whom? Was any value (dollar/monetized amount) specified for the maintenance of the natural capital?

NOTE

1 The WWPRAC (1998, p. 3–17) notes: "The whole system [of western water (management)] encourages inefficient use. Federal water subsidies, hydropower subsidies, crop subsidies, the doctrine of appropriative rights, constraints on water transfers, fixed, or declining block rates—a whole gamut of conservation disincentives has given the American West the most prodigious thirst of any desert civilization on earth" (Reisner and Bates, 1991, p. 7).

REFERENCES

Advisory Committee on Water Information (ACWI). 2019. *Subcommittee on Ground Water*. URL: acwi.gov/sogw/index.html (accessed January 18, 2020).

Al-Hadidi, M.S. 1999. Brackish water management and use in Jordan. *Desalination*, 126, 41–44. URL: http://www.desline.com/articoli/3773.pff (accessed January 18, 2009).

Alley, W.M., Reilly, T.E., and Frenke, O.L. 1999. *Sustainability of Ground-Water Resources*. U.S. Geological Survey Circular 1186, Reston VA, 79 p.

American Water Works Association (AWWA). 2019. *2019 Water and Wastewater Rate Survey*. Denver, Colorado.

Boulding, K. 1949. Income or welfare? *Review of Economic Studies*, 17, 79.

Brumley, J., Marks, C., Chau, A., Lowrance, R., Huang, J., Richardson, C., Acree, S., Ross, R., and Beak, D.. 2018. *The Influence of Green Infrastructure Practices on Groundwater Quality: The State of the Science*. U.S. Environmental Protection Agency, Washington, DC, EPA/600/R-18/227.

City of Chicago. 2003. Chicago's Water Agenda 2003. URL: http://www.egrv.cityofchicago.org/webportal/COCWebPortal/COC_EDITORIAL/wateragenda_1.pdf (accessed November 12, 2007).

Common, M. and Stagl, S.. 2005. *Ecological Economics: An Introduction*. Cambridge University Press. New York, NY. 560 p.

Costanza, R. and Dely, H.E. 1992. Naturel capitel and sustainable development. *Conservation Biology*, 6, 454–464.

Costanza, R., d'Arge, R., de Groot, R., Farber, S., Grasso, M., Hannon, B., Limburg, K., et al. 1997. The velue of the world's ecosystem services and natural capital. *Nature*, 387 (Mey), 253–260.

Daly, H. 1996. *Beyond Growth: The Economics of Sustainable Development*. Beacon Press, Boston, MA, 253 p.

Daly, H.E. and Farley, J. 2004. *Ecological Economics: Principles and Applications.* Island Press, Washington, DC, 454 p.
Delaware River Basin Commission (DRBC). 2008. DRBC Web site; DRBC Overview. URL: http://www.nj.gov/drbc/ (accessed January 11, 2008).
Edwards Aquifer Research and Data Center (EARDC). 2006. *Threatened and Endangered Species of the Edwards Aquifer System.* URL: www.eardc.txstate.edu/endangered.html (accessed April 28, 2007).
El-Ashry, M.T. and Gibbons, D.C. 1988. *Water and Arid Lands of the Western United States: A World Resources Institute Book.* Cambridge University Press, New York, 415 p.
European Union (EU). 1979. Council Directive 80/68/EEC on the protection of groundwater against pollution caused by certain dangerous substances. December 17, 1979.
European Union (EU). 1999a. *European Environment Agency, Ground Water Quantity and Quality in Europe.* Copenhagen, Denmark, 123 p.
European Union (EU). 1999b. *Energy, Environment and Sustainable Development Programme for Research,* Technology Development and Demonstration [R, D and D] under the Fifth Framework Programme.
European Union (EU). 2000. Council Directive 2000/60/EC establishing a framework for water policy. October 23, 2000.
European Union (EU). 2006. Council Directive 2006/118/EC on the protection of groundwater against pollution and deterioration. December 12, 2006.
European Union (EU). 2007. Common Implementation Strategy For the Water Framework Directive (2000/60/EC); Guidance on Groundwater Monitoring; Guidance Document No. 15; Technical Report-002-2007. Office for Official Publications of the European Communities, Luxembourg, 52 p.
Gibert, J., Danielopol, D.L., and Stanford, J.A. 1994. *Groundwater Ecology.* Academic Press, Inc., San Diego, CA.
Hartwick, J.M. 1989. *Non-Renewable Resources Extraction Programs and Markets.* Routledge, New York, 149 p.
Heinz Center for Science, Economics and Environment. 2006. *Filling the Gaps: Priority Data Needs and Key Management Challenges for National Reporting on Ecosystem Condition.* Heinz Center for Science, Economics and Environment, Washington, DC, 104 p. URL: http://www.heinzctr.org/Programs/Reporting/Working%20Groups/Data%20Gaps/Gaps_LongReport_LoRes.pdf (accessed November 22, 2008).
Heinz Center for Science, Economics and Environment. 2008. *The State of the Nation's Ecosystems 2008: Measuring the Land, Waters, and Living Resources of the United States.* Island Press, Washington, DC, 368 p.
Hicks, J.R. 1946. *Value and Capital,* 2nd edn. Oxford University Press, Oxford, UK.
Hildebrand, M. 2009. Water conservation made legal: Water budgets and California law. *American Water Works Association Journal,* 101 (4), 85–89.
International Standards Organization (ISO) 2020. *ISO 14046:2014; Environmental management — Water footprint — Principles, requirements and guidelines.* URL: https://www.iso.org/standard/43263.html (accessed January 18, 2020).
Jeffcoat, S., Baughman, D., and Thomas, P.M. 2009. Total water management: Strategies for utility master planning. *American Water Works Association Journal,* 101 (2), 56–64.
Jorgenson, D., Wireman, M., and Olson, D. 1998. *Assessing the Vulnerability of Public Water Supply Wells to Microbial Contamination: Ground Water Monitoring and Remediation.* National Ground Water Association, Westerville, OH.
Livingston, E.C. 2004. *Considerations for Brackish Groundwater Use as a Water Supply for Small Water Systems and Municipalities. New Mexico Brackish Groundwater Aquifer Assessment Workshop.* Albuquerque, New Mexico, January 15, 2004. URL: http://wrri.nmsu.edu/conf/brackishworkshop/presentations/livingston.pdf (accessed January 18, 2009).
Markandya, A., Patrice, H., Bellu, L.G., and Cistulli, V. 2002. *Environmental Economics for Sustainable Growth.* Edward Elgar Publishing, Inc., Northhampton, Massachusetts, USA. 567p.
Markus, M.R. 2009. The groundwater replenishment system. *American Water Works Association Journal,* 101 (2), 49–51.
Maryland Stormwater Consortium (MSC). 2007. Core Environmental Site Design Principles for the implementation of the Maryland Stormwater Management Act of 2007 (in progress).
Meyland, S.J. 2011. Examining safe yield and sustainable yield for groundwater supplies and moving to managed yield as water resource limits become a reality. In *Water Resources Management,* WIT Press. pp. 813–823. URL: https://www.witpress.com/Secure/elibrary/papers/WRM11/WRM11073FU1.pdf (accessed January 18, 2020).
National Ground Water Association (NGWA). 2015. *Grouting of Water Wells, Position paper; and Water Well Contractor Licensing, Position paper.* URL: https://www.ngwa.org/docs/default-source/default-document-library/advocacy/Grouting-Wells.pdf and https://www.ngwa.org/docs/default-source/default-document-library/advocacy/WW-Contractor-Licensing.pdf (accessed January 18, 2020).

National Ground Water Association (NGWA). 2016. *Sustainable Groundwater Resources*; Position paper. URL: https://www.ngwa.org/docs/default-source/default-document-library/advocacy/GW-Sustainability.pdf (accessed January 18, 2020).

National Water Research Institute (NWRI). 1998. *Source Water Assessment and Protection 1998: A Technical Conference*, Dallas, TX.

Olsen, S. 2008. Brackish water is expensive to clean: Desalination to use more energy, raise water bills. *Albequerque Journal (New Mexico)*, December 29, 2008. URL: http://www.istockanalyst.com/article/viewiStockNews/articleid/2912697 (accessed January 18, 2009).

Orange County (California) Water District (OCWD). 2020. *How water works in Orange County*. URL: https://www.ocwd.com/learning-center/how-water-works-in-oc/ (accessed January 19, 2020).

Reid, D. 1995. *Sustainable Development: An Introductory Guide*. Earthscan Publications Ltd, London, U.K.

Reisner, M. and Bates, S.F. 1991. *Overtapped Oasis: Reform or Revolution for Western Water*. Island Press, Covelo, CA, p. 7.

Sagoff, M. 1990. *The Economy of the Earth*. Cambridge University Press, Cambridge, U.K.

Schumpeter, J. 1954. *History of Economic Analysis*. Oxford University Press, New York.

Sitarz, D. (ed.). 1998. *Sustainable America: America's Environment, Economy and Society in the 21st Century*. EarthPress/Nova Publishing Company, Carbondale, IL.

Smith, Z.A. 1989. *Groundwater in the West*. Academic Press, Inc., New York.

Solley, W.B., Pierce, R.R., and Perlman, H.A. 1998. Estimated use of water in the United States in 1995. U.S. Geological Survey Circular 1200.

Time. 1999. Six billion … and counting. October 19, p. 61.

Turner, R.K., David P., and Ian, B. 1993. *Environmental Economics*, The Johns Hopkins University Press, Baltimore, MD.

United Nations (UN). 2003. *Handbook of National Accounting: Integrated Environmental and Economic Accounting*. Publications Board and Exhibits Committee, New York.

United Nations World Commission on Environment and Development (UNWCED). 1987. Our Common Future, Report of the World Commission on Environment and Development. Annex to General Assembly document A/42/427, Development and International Co-operation: Environment, August 2, 1987. URL: http://www.un-documents.net/wced-ocf.htm (accessed January 5, 2008).

U.S. Congress. 1986. *Safe Drinking Water Act, as amended*. Washington, DC.

U.S. Environmental Protection Agency (USEPA). 1990. *Citizens' Guide to Wellhead Protection*. EPA 440/690-004. Washington, DC.

U.S. Environmental Protection Agency (USEPA). 1993a. A Review of Methods for Assessing Aquifer Sensitivity and Ground Water Vulnerability to Pesticide Contamination, 813-R-93-002, Washington, DC.

U.S. Environmental Protection Agency (USEPA). 1993b. *Ground Water Resource Assessment*, EPA 813-R-93-003, Washington, DC.

U.S. Environmental Protection Agency (USEPA). 1996. Ground Water Chapters. *Water Quality Report to Congress, EPA-813-R-96-001*, Washington, DC.

U.S. Environmental Protection Agency (USEPA). 1999. *Ground Water Chapters. Water Quality Report to Congress*, Washington, DC.

U.S. Environmental Protection Agency (USEPA). 2006. Water efficiency leader awards—Call for applicants. *Federal Register*. 71, 132, Notices. Tuesday, July 11, 2006, p. 39116.

U.S. Environmental Protection Agency (USEPA). 2007a. *Sustainable Infrastructure for Water and Wastewater*. URL: http://www.epa.gov/waterinfrastructure/.

U.S. Environmental Protection Agency (USEPA). 2007b. *Reducing Stormwater Costs through Low Impact Development (LID) Strategies and Practices*. URL: http://www.epa.gov/owow/nps/lid/costs07/ (accessed January 13, 2008).

U.S. Geological Survey (USGS). 1985. National Water Summary 1984, Water Supply Paper 2275.

U.S. Geological Survey (USGS). 1999a. *Ground Water and Surface Water: A Single Resource*. U.S. Geological Survey Circular 1139.

U.S. Geological Survey (USGS). 1999b. *Sustainability of Ground-Water Resources*. U.S. Geological Survey Circular 1186.

U.S. Geological Survey (USGS). 2007. Pennsylvania Drought Condition Monitoring. URL: http://pa.water.usgs.gov/monitor/ (accessed January 11, 2008).

Western Water Policy Review Advisory Commission (WWPRAC). 1998. Water in the West: Challenge for the Next Century.

Westminster (City of), Colorado. 2009. *Reclaimed Water*. URL: http://www.ci.westminster.co.us/353.htm (accessed January 18, 2009).

World Resources Institute (WRI). 1997. *Frontiers of Sustainability*. Island Press, Washington, DC.

12 Sustainable Groundwater Resources Management

SUSTAINABLE MANAGEMENT

Groundwater conditions and use change with personal and business activities, scales of water need, seasons of the year, unexpected and major weather events and long-term climate status (USGS, 2016, 2019; Taylor et al., 2013). Responding to changes in groundwater levels and quality requires resilience enabling recovery to provide for human and ecosystem maintenance through adjustments in decisions and activities affecting use. As major segments of society and its industry rely on groundwater, the water resource should be sustainable and sustainably managed to derive continuing the economic benefits and advantages it can provide. Adaptive management of groundwater, drawing on science to anticipate potential direction and giving local insight to action, will be fundamental to provision of groundwater that meets the intended purposes for communities and nations (Thomas, 2018). Management involves planning, organization, leading and controlling (Barnett, 2019; MSHQ, 2019) that can be applied to groundwater for managed yield (Meyland, 2011). Sustainable yield of groundwater has progressed to focus on groundwater managed to maintain critical human and ecosystem targets or objectives (Alley and Leake, 2004; Meyland, 2011). Box 1.12.1 provides a brief description of management functions that might be applied to groundwater resources.

BOX 1.12.1 MANAGEMENT FUNCTIONS

Principal management functions include planning, organizing, leading and controlling. A brief description of these functions is provided here (adapted slightly to apply to groundwater management).

1. Planning comprises establishing objectives and setting a course of action to accomplish the objectives. Awareness of environmental conditions affecting the resource and forecasts of future conditions guide planning to contribute to good decision-making.
   - Strategic planning involves analyzing applicable opportunities and threats, as well as the strengths and weaknesses of circumstances to position resource management to operate effectively in optimizing resource capability. Long-range thinking applies.
   - Tactical planning is intermediate-range (3–5 year) in effort to develop relatively concrete and specific ways to carry out the strategic plan.
   - Operational planning specifies ways to achieve the objectives in the short-range timeframe by developing specific action steps to support the strategic and tactical plans. Operational planning usually has a short activity horizon of one week to one year.
2. Organizing focuses on developing an organizational structure and allocating human and other needed resources to achieve the objectives.

(*Continued*)

BOX 1.12.1 (Continued)

3. Leading aims to influence others for attaining organizational objectives effectively drawing on motivational and communication techniques and authority.
4. Controlling ensures performance meets standards and targets and includes setting performance standards, comparing results to standards and correcting action as appropriate to adapt to changed circumstances (adaptive management).

Source: Abstracted and modified from Barnett, T. 2019. Management Functions. In Reference for Business. URL: https://www.referenceforbusiness.com/management/Log-Mar/Management-Functions.html (Accessed December 5, 2019).

ADAPTIVE RESOURCE MANAGEMENT

While groundwater may move more slowly than surface waters in most environmental settings other than karstic formations, it is still in motion, interacting with the geo-bio-chemical matter with which it exists and changing the conditions around it. Add to that pumping and injection wells, then flow conditions are affected by human action. Sometimes responses are observed in short periods of time such as days or over longer periods of time as months or decades. Monitoring wells and remote space observation may be applied to observe changes locally, regionally or globally. Even though direct observation is limited by the number of monitoring wells installed, circumstances and impacts from natural and human interactions affecting human and wildlife use of groundwater may need to be managed to support people's needs. This management will need to be responsive to current demands as well as future use of the resource and guide adaptation to changes in groundwater quantity and quality. Adaptive actions will have costs and benefits to groundwater suppliers and receptors – people and wildlife – that require careful analysis.

Adaptive resource management has evolved to support action with limited scientific data and knowledge and recognizing the "complexities and stochastic behavior of large ecosystems" (NAS, 2004). The process of adaptive management is used in undertaking response to unanticipated events and addressing uncertainty of forecasted outcomes by considering a range of possible future results. Adaptive management approaches are flexible, adjusting through an iterative learning process (NAS, 2004; Williams and Brown, 2012). Application of adaptive management addresses not only groundwater resource use under climate change factors (Treidel et al. 2011) but remediation as well (USEPA, 2018).

Adaptive management for water projects can generally be described in six steps that iterate (NAS, 2004; LTA, 2019):

1. A collaborative structure for stakeholder participation and learning
2. Management objectives that are regularly reviewed and revised
3. Assessment and modeling of the system being managed
4. A range of alternatives and forecast consequences informing a plan for decision-making
5. Monitoring and evaluation of outcomes of decisions
6. A mechanism(s) for incorporating learning into future plans, decisions and institutional processes

Box 1.12.2 describes a more detailed process for adaptive management applied to natural resources.

BOX 1.12.2 ADAPTIVE RESOURCE MANAGEMENT ELEMENTS

Adaptive resource management is designed to address complex problems through learning and modifying actions to progress to an expected improved outcome. While many similar approaches have evolved, the steps described below may provide a useful framework to consider applying adaptive management to groundwater projects and issues.

Initiation Components

Stakeholder involvement – Drawing on stakeholders perspectives and knowledge at initiation and throughout a project to develop "common ground" contributes to establishing rationale for decision-making.

Objectives – Establishing clear objectives provide a "benchmark" to examine potential effects of a range of management actions as well as guide strategy development.

Management alternatives – Clear identification of "potential alternatives" for action at each decision point is necessary to recognize differences in the consequences of a range of actions that can guide movement toward the best action.

Predictive models – Models express "understanding of the resource" and environmental interaction, provide output reflecting consequences including effects on critical human and environmental receptors and uncertainties over time and serves as a basis for the costs and benefits of alternative management strategies.

Monitoring protocols – Monitoring gives "direct observation" based on ecological attributes specified that supports learning and evaluation of management effectiveness thereby contributing to decision-making.

Iterative Components

Decision-making – Each decision point over time "incorporates current understanding," future consequences anticipated and comparison of resource status with management objectives.

Follow-up monitoring – "Ongoing monitoring supports estimating resource status, decision making, evaluation and learning" after decisions culminate. Monitoring continues, following protocols established in the initiation phase.

Assessment – The monitoring data are applied to "evaluate management effectiveness," understand resource status and reduce uncertainty about management effects. Monitoring data are compared with model predictions, desired outcomes and objectives to evaluate management effectiveness.

Learning and feedback – Monitoring and assessment facilitates understanding in future management actions. "Iterative decision making, monitoring, and assessment improves understanding" of resource systems supporting adjustments in the management strategy.

Institutional learning – On a regular basis, interrupting the technical cycle to re-evaluate project or program objectives, alternatives and other elements of the initiation phase promotes institutional "learning to equilibrate higher level decision processes."

Source: Adapted from: Williams, B. K., and E. D. Brown. 2012. Adaptive Management: The U.S. Department of the Interior Applications Guide. Adaptive Management Working Group, U.S. Department of the Interior, Washington, DC. URL: https://www.doi.gov/sites/doi.gov/files/migrated/ppa/upload/DOI-Adapative-Management-Applications-Guide.pdf (Accessed December 6, 2019).

ADAPTIVE MANAGEMENT FOR SUSTAINABLE GROUNDWATER RESOURCES

As greater reliance on groundwater continues on all fronts from water supply to irrigation, the need for a management approach that could be applied across jurisdictions, by homeowners and businesses drawing on wells and springs, and over time should be useful. The National Ground Water Association developed adaptive management principles for sustainable groundwater resources to address that perspective (NGWA, 2016). The principles draw on the recognition that "[g]roundwater, a basic life-sustaining and economic resource, must be managed locally through best practices of states and local governments and their residents and businesses recognizing the range of domestic, municipal, industrial, ecological, and agricultural uses it supports for their communities' continued well-being." The specific principles are displayed in Box 1.12.3, Adaptive Management Principles for Sustainable Groundwater Resources.

BOX 1.12.3 ADAPTIVE MANAGEMENT PRINCIPLES FOR SUSTAINABLE GROUNDWATER RESOURCES

In 2016, the National Ground Water Association adopted a position statement on "Sustainable Groundwater Resources" that defined principles for adaptive groundwater management that may provide states, local governments, businesses, individuals and well owners approaches for managing their groundwater resources and associated infrastructure sustainably for the well-being of their communities:

1. Collaborate with adjacent local and regional jurisdictions within and among states and nations.
   a. Recognize groundwater flows under jurisdictional boundaries
   b. Develop a common understanding of the groundwater resource and influential factors
   c. Conduct a local and state groundwater resource audit and develop a water budget
   d. Share water plans, objectives and data for local and state water management
2. Set groundwater resource objectives and forecast water demand.
   a. Establish publicly vetted aquifer objectives for groundwater availability in future uses
   b. Adopt aquifer management/groundwater use rules, codes and laws
   c. Evaluate future water uses and compare with available groundwater and alternatives
3. Incorporate the hydrologic cycle and natural processes in groundwater management.
   a. Store water underground to reduce evaporative loss and protect groundwater quality
   b. Facilitate recharge by use of green infrastructure and native-vegetated landscapes
   c. Maintain groundwater levels to avoid nonrecoverable land subsidence
   d. Preserve groundwater levels, flow and quality for ecosystem biodiversity

(Continued)

BOX 1.12.3 (Continued)

4. Minimize the carbon footprint of groundwater infrastructure and production.
   a. Evaluate water supply alternatives for minimizing life-cycle footprint and costs
   b. Use most efficient pumps, equipment and production processes
   c. Reduce water loss in storing and delivering groundwater
   d. Promote conservation and reuse of water as alternate sources for future use
5. Implement source water protection and remediation for private and public groundwater uses.
   a. Eliminate or reduce sources of contamination
   b. Protect usable groundwater from future contamination through practices and laws
   c. Remediate contaminated groundwater
6. Measure, report and share resource status data with residents and other jurisdictions.
   a. Monitor groundwater levels, quality and use
   b. Report on achievement of groundwater resource objectives
7. Prepare to avoid disruption from extreme events affecting reliable safe groundwater supply.
   a. Consider effects of extreme weather events and human interventions
   b. Identify and address water supply vulnerabilities of infrastructure and operations
   c. Establish processes for timely recovery of water services to communities
8. Reassess groundwater management on regular intervals.
   a. Evaluate availability, source, protection, provision, use and resilience alternatives
   b. Conduct life-cycle assessments of natural and constructed infrastructure
   c. Adjust codes, laws and practices for cost-effective provision of safe sustainable water

Source: National Ground Water Association. 2016. Sustainable Groundwater Resources. URL: https://www.ngwa.org/docs/default-source/default-document-library/advocacy/GW-Sustainability.pdf (Accessed December 9, 2019) (With Permission).

Specific steps that may address adaptive groundwater management start with collaboration and understanding local groundwater conditions and governance, establishing objectives for groundwater purposes, incorporating the science of what is known about groundwater in evaluation of alternatives for its use, measuring key groundwater indicators, addressing groundwater vulnerabilities and reassessing conditions to adjust steps to meet essential benchmarks. The description that follows focuses on steps particular to groundwater in an adaptive management context.

COLLABORATION

Following along the same path as adaptive management of other organizations (NAS, 2004; Williams and Brown, 2012; LTA, 2019), the approach of the NGWA starts with local collaboration to develop a common shared understanding of the resource. This step is one of shared discovery about the local resource and its uses, exchange of perspectives about the resource and its uses and developing forethought together for the benefit of the community(ies) that share the resource flowing beneath them. This understanding can be developed by individuals and wellowners, as well as businesses and governments. Groundwater modeling that provides a 'picture' of a community's aquifer may be a part of this information development to integrate existing data and guide collaborators to cost-effective conclusions about where more data may be useful in developing their extended understanding of their groundwater resource (Anderson et al., 2015). Collaboration may involve meeting with public officials, attending community meetings and convening gatherings to focus on groundwater assessment and identification of critical human and environmental receptors requiring a minimum amount or quality of groundwater. Consulting stakeholders will enable tradeoffs to be identified and addressed, such as balancing demand for safe water with maintenance of a viable aquatic ecosystem (Maimone, 2004).

Shared Objectives and Plans

The adaptive groundwater management approach may then consider what objectives should be set for the resource and whether existing laws and regulations may need to be revised to address those objectives. Evaluating alternatives to achieve the objectives is a significant step in planning for adaptive management of groundwater. The cost of time devoted to defining resource objectives and proposing regulation revisions with properly developed rationale through a plan should be compared to the cost of insufficient or contaminated water supply for an entire community or region. Ensuring that critical human and environmental receptors receive the quantity and quality of water for maintenance and survival is essential. This effort should be future looking to be of greatest use to the community considering the near-term and the long-term (10–30 years or more). The Frederick County, Maryland, water and sewerage plan, with 19 water systems drawing on groundwater exclusively and two systems partly using groundwater sources, considers a 20-year planning horizon (FCMD, 2016). The Texas state water plan evaluates and projects water production and use over 50 years (TWDB, 2019). Elsinore Valley (pop. 125,450 in 2018), California, budgeted \$2.5 million for its 20-year Sustainable Groundwater Plan in 2017 with 9% for plan management, 14% for stakeholder outreach, 15% for groundwater modeling, 58% for plan development and 4% for administrative compliance (EVGSA, 2017). The Orange County Water District in California, covers 906 km^2 (270 square miles) supplying 19 municipalities and special districts drawing on 200 large capacity wells serving 2.5 million people (OCWD, 2015; 2019). The OCWD Groundwater Management Plan for sustainable groundwater addresses: history and governance, outreach, basin hydrologeology, monitoring, recharge management, replenishment, saltwater intrusion management, water quality management, relation to other natural resources, sustainable basin management and financial management. Box 1.12.4 provides more detail on the groundwater management plan.

Alternatives Assessment

Fundamental to an adaptive management approach for sustainable groundwater resources is the assessment of alternative actions, processes and outcomes. For groundwater, this includes but not limited to accounting for groundwater in the flow of hydrologic processes, minimizing groundwater's carbon footprint and source water protection and remediation. The role of science in describing the physical conditions of alternative actions, processes and outcomes is critical to evaluating the advantages, disadvantages, costs and benefits of the range of management paths for and decisions about the resource.

BOX 1.12.4 GROUNDWATER MANAGEMENT PLAN OUTLINE

The Groundwater Management Plan for Orange County Water District has the following outline:

HISTORY AND GOVERNANCE

History of the Orange County Water District
OCWD Governance
Groundwater Producers
Public Education and Events

PREPARATION OF GROUNDWATER MANAGEMENT PLAN

Sustainable Groundwater Management Act
Basin Management Goals and Objectives
Recommendations and Projects Completed 2009–2015
Recommendations for 2015–2020
Planning and Implementation Horizons

BASIN HYDROGEOLOGY

Description of Basin Hydrogeology
Determination of Total Basin Volume
Water Budget
Calculation of Change in Groundwater Storage
Elevation Trends
Land Subsidence
Basin Model

WATER SUPPLY MONITORING

Groundwater Monitoring
Recycled Water Monitoring
Surface Water Monitoring
Water Resources Management System: Database Management
Water Sample Collection and Analysis
Ground and Surface Water Interactions

MANAGEMENT AND OPERATION OF RECHARGE FACILITIES

History of Recharge Operations
Sources of Recharge Water Supplies
Surface Water Recharge Facilities
Maintenance of Recharge Facilities
Recharge Studies and Evaluations
Improvements to Recharge Facilities 2009–2014

(Continued)

BOX 1.12.4 (Continued)

GROUNDWATER REPLENISHMENT SYSTEM

Overview
Advanced Water Treatment Process
Energy Efficient Operations
Plant Optimization and Expansion
Water Quality Monitoring and Reporting
Public Outreach

SEAWATER INTRUSION AND BARRIER MANAGEMENT

Background
Talbert Seawater Intrusion Barrier
Alamitos Seawater Intrusion Barrier
Sunset Gap Investigation
Evaluation of Potential Impacts Due to Climate Change

WATER QUALITY PROTECTION AND MANAGEMENT

OCWD Groundwater Quality Protection Policy
Well Development, Management and Closure
Managing Salinity in Water Supplies
Management of Nitrates in Groundwater
OCWD Prado Wetlands
Amber-Colored Groundwater Management
Regulation and Management of Contaminants
Constituents of Emerging Concern
Groundwater Quality Improvement Projects
BEA Exemption for Improvement Projects

NATURAL RESOURCE AND COLLABORATIVE WATERSHED PROGRAMS

OCWD Natural Resource Programs – Overview
Natural Resource Programs in the Watershed
Collaborative Watershed Programs
Management of Areas Within Basin 8-1 Outside OCWD Boundaries
Orange County Water Resources-Related Plans
Collaboration with Federal and State Agencies
Land Use, Development and Environmental Reviews

SUSTAINABLE BASIN MANAGEMENT

Background
Basin Operating Range
Balancing Production and Recharge
Managing Basin Pumping
Supply Management Strategies

(*Continued*)

BOX 1.12.4 (Continued)

Removing Impediments to Conjunctive Use
Water Demands
Drought Management
Record Keeping

FINANCIAL MANAGEMENT

Background Financial Information
Operating Expenses
Operating Revenues Reserves

Source: Adapted from: Orange County Water District (OCWD). 2015. Orange County Water District Groundwater Management Plan 2015 Update. URL: https://www.ocwd.com/media/3622/groundwatermanagementplan2015update_20150624.pdf (Accessed December 17, 2019).

Accounting for Hydrologic Processes

For groundwater to be appropriately and adequately addressed in adaptive management, groundwater flow and quality must be incorporated in a project or program by evaluating its role in the hydrologic and natural processes affecting and affected by a project or program. Acquisition of data and application of modeling will be important in considering the groundwater aspects of hydrologic relationships to give a total water resource perspective (Anderson et al., 2015). Groundwater–surface water interaction may be accounted for in an adaptive management approach as well as protecting groundwater quality, reducing evaporative loss, aquifer recharge by natural and artificial means and maintenance of groundwater levels to avoid land subsidence and supporting biodiversity. With climate change affecting precipitation patterns, demands for dealing with increased stormwater volumes may affect groundwater storage and quality if polluted stormwaters are transferred to the subsurface environment whether through drains, infiltration ponds or other green infrastructure, rather than allowed to reach streams (Brumley et al., 2018). Reduced precipitation, drought effects and activities that increase impervious surfaces while reducing recharge potential should also be addressed in a hydrologic systems evaluation context affecting groundwater and surface water with adaptive management addressing maintenance of critical ecosystem indicators such as water provision to meet basic human needs and essential groundwater discharge to streams and lakes for aquatic life support (EAA, 2019). Managed aquifer recharge is a groundwater technology to store excess water and reduce evaporative loss (see Volume 1, Chapter 6). Examining the potential hydrologic effects over time of a range of alternatives and strategies to address hydrologic uncertainty is essential in an adaptive robust groundwater management approach (Maimone, 2004).

Minimizing Carbon Footprint

Similar to considering hydrologic process effects, evaluation of alternatives should include minimization of the carbon footprint of projects and programs, including life-cycle costs to address energy use and carbon footprint reduction, efficient equipment and processes, water loss reduction in groundwater storage and delivery, and conservation and reuse of water instead of developing additional water sources. Ground source heat exchange, direct geothermal use and managed aquifer storage of excess water are groundwater technologies that offer reduced carbon footprints (Hayes, 2010).

Protecting and Remediating Source Water

Source water protection for and remediation of contaminated groundwater focus on the quality of the water supply for all uses, including ecosystem requirements. Protecting and remediating the quality of groundwater also maintains its availability for its range of purposes. Source water protection and remediation may be objectives as well as effects of a project or program. To the extent that source water protection and groundwater remediation can be examined in alternatives evaluation and achieved in project completion, they may point to hydrologic process factors and reduction in carbon footprint because high-quality groundwater does not have to be pumped from aquifer zones that are deeper or further away from the intended water purposes. In turn, costs may be avoided.

Resource Status Measurement

Measuring resource status – both levels and quality – provides feedback and the basis for learning about groundwater and its subsurface environment and achievement of resource objectives including groundwater supply and quality servicing critical human and environmental receptors. Measurement of groundwater quantity and quality condition is essential to the adaptive groundwater management approach. This step leads management to consider whether, what, how much and when changes are needed to actions taken based on past decisions. The significant challenge for adaptive management of the groundwater resource is to decide where, how many and how deep monitoring wells are needed to provide the data for feedback and learning to inform future decisions. Since well installation and monitoring may be substantial expenses depending on the project or program objectives and the data needed, modeling the unseen subterranean portion of the ecosystem will be useful if not essential to understand hydrogeologic perturbations over large scales with many users and long timeframes with slow groundwater movement (ACWI, 2013).

Avoiding Disruption of Extreme Events

Extreme events such as flooding, drought, earthquakes and other natural and human-caused harmful incidents can affect both quantity and quality of groundwater. Data derived from these occurrences provide additional feedback and learning for adaptive groundwater management to address vulnerabilities in providing safe, reliable groundwater for its purposes. Fundamentally, though, these extreme events are to be avoided if at all possible due to their injurious nature to people, wildlife and infrastructure. Recovery from such events is often laborious, time-consuming and costly. The National Ground Water Association estimated that the 5 months of rainfall across the midwestern and other areas of the United States during 2019 caused flooding that may have affected 250,000 wells in 835 counties of 28 states (NGWA, 2019). Pathogens tested for in water samples from flooded wells were much more frequently observed in states experiencing flooding and create the possibility of waterborne disease occurrence and thereby health treatment costs for people sickened and disinfection costs for wells impacted. The challenge is to create conditions that reduce groundwater-related impacts and make the aquifers, wells and groundwater use locations more resilient to allow them to withstand negative effects of these extreme events. Managed aquifer recharge is an example of a groundwater technology that may provide cost-effective resilience in local water supply (see Volume 1, Chapter 6). Properly maintaining or relocating wells or elevating wellhead casing are alternatives in a potentially longer list of possible actions to avoid disruption of extreme events and provide options to decision makers who may be business managers or homeowners (Job, 2017).

Reassessment

Reassessment conditions, decisions and consequent results on a regular basis to guide improved management of the groundwater resource is a key step of the adaptive approach. With the information and learning of the previous steps, decision makers can direct evaluation of availability, protection, use and resilience of groundwater. Feedback from data on maintenance of critical human and environmental receptors should be a significant aspect of management learning and adjustment. The data obtained can inform life-cycle assessments of natural and constructed infrastructure as input to policy or project changes, including adjustments to codes, laws and practices for cost-effective provision of safe sustainable groundwater or redirection of project elements to be more responsive to site or regional needs. The reassessment may also include an evaluation of a management entity's objectives to ensure that they are practicable and recognize any changed underlying or less obvious or unanticipated circumstances or behaviors such as land subsidence or disturbance, saltwater intrusion or increased water reuse. Box 1.12.5 briefly describes adaptive management for remediation for a contaminated groundwater site. Adaptive management for sustainability can be applied

BOX 1.12.5 CASE STUDY OF ADAPTIVE GROUNDWATER MANAGEMENT

The Aberdeen Proving Grounds site in Maryland (USA) is a hazardous waste disposal location involving research, engineering and development of munitions. Munitions testing and training has occurred at the site since 1917. The site has contaminated groundwater from munitions development, a missile defense system and land disposal of chemicals and wastes as well as unexploded ordinance, creating significant hazard to removal. The site also contains significant wetlands that support the Chesapeake Bay. Buried waste not able to be removed because of explosive hazards contained metals and chlorinated solvents that were leached to groundwater discharging to a shallow pond.

Collaboration

The site owner, technical staff, regulators and stakeholders engaged in a collaborative process to establish remedial goals and monitor results.

Objectives

Remedial goals for groundwater were set as aquatic life (critical ecosystem receptor) criteria for surface water to be modified based on the results of monitoring.

<u>Alternatives Assessment</u> (for primary alternatives considered over time as learning progressed at the site; costs include discounted operation and maintenance costs to year indicated at 5% in 1996$ and 3% in 2008$ over 30 years)

1. Dredging, solidification and landfill
   - o Significant safety hazards
   - o Year 1997 – Cost $157,000,000
2. Removal of "hot spot" buried waste, solidification and landfill
   - o Significant safety hazards
   - o Year 1997 – Cost $37,000,000
   - o Year 2009 – Cost $36,248,000 - $37,543,000 -
3. Aquatic phytoremediation
   - o Year 1997 – Cost $5,780,000
4. Permeable cover with groundwater bio-enhancement
   - o Partial reduction of toxicity, mobility and volume
   - o Year 2009 – Cost $4,349,000–$4,581,000

(*Continued*)

BOX 1.12.5 (Continued)

5. Capping waste in place with monitored subsurface attenuation barrier or constructed wetland
   - Met regulatory requirements, did not adequately address groundwater and buried waste hazard
   - Year 2009 – Cost $3,671,000–$6,892,000
6. Soil cover with an in-situ bioremediation barrier, constructed wetland buffer and bio-beneficial sediment cover – Final Selected Remedy
   - Permeable cover, bio-beneficial sediment cover, munition in-situ degradation, groundwater bio-remediation with wetland buffer then discharge to surface water
   - Year 2009 – Cost $3,648,000
   - Year 2019 – Contract (7-year) cost $2,038,540
7. Accounting for Hydrologic Processes/Protecting and Remediating Source Water

Degradation of munitions in ground over time reduced hazard. Treatment injections into groundwater bioremediated chlorinated solvents and sequestered metals incorporated groundwater flow. The wetland buffer and sediment cover reduced human exposure to sediments and seeps and completed final treatment of groundwater before discharge to surface waters.

Resource Status Measurement

Monitoring point for performance of the site was the groundwater immediately downgradient of the waste but upgradient of the discharge to surface water.

Reassessment

Learning took place over 20 years as remediation moved from feasibility study, site monitoring and remedial investigation then to remedial construction, monitoring and maintenance. As learning occurred, decisions were modified and updated in the official Records of Decision for the site in 1998, 2009 and 2019.

Site-specific data allowed adaptation of remedial goals through time to transition to sustainable remedial steps modifying goals from removal to dual passive and active remedial functions of in-situ degradation and bioremediation applying technology advances. The resulting project was sustainable and cost-effective.

Sources: Ciarlo, Michael, and Jeffrey Aichroth. 2019. Ten-Year Case Study of Successful Adaptive Management of a Groundwater to Surface Water Discharge Site Regulated under CERCLA. URL: https://www.battelle.org/docs/default-source/conference-proceedings/2019-sediments-conference-proceedings/c2.-adaptive-management-approaches/558.pdf?sfvrsn=d968eb7_2 (Accessed December 12, 2019).

U.S. Environmental Protection Agency (USEPA). 1998. Superfund Record of Decision: Aberdeen Proving Ground (Edgewood Area) Watson Creek Sediment, O-Field OU, Aberdeen Proving Ground, MD (9/23/1997). PB97-963908, EPA/541/R-97/091. URL: https://nepis.epa.gov/Exe/ZyPURL.cgi?Dockey=9100Q8LM.TXT (Accessed January 21, 2020).

U.S Army Garrison, Aberdeen Proving Ground, Maryland (USAGAPG). 2009. Record of Decision for Army environmental Database-Restoration (AEDB-R) Site EAOF04: O-Field Study Area Operable Unit 4: New O-Field Groundwater and Source Area and Other O-Field Areas (OOA); Final July 2009.

(*Continued*)

BOX 1.12.5 (Continued)

U.S Army Garrison, Aberdeen Proving Ground, Maryland (USAGAPG). 2018. O-Field Study Area Remedial Action Completion Report for Army Environmental Database Restoration (AEDB-R) Site EAOF04, O-Field Study Area Operable Unit 4: New O-Field Groundwater and Source Area and Other O-Field Areas; Final September 2018.

to groundwater protection, development, recharge, conservation and habitat for biodiversity, among other resource purposes.

The example of the case study of the Aberdeen Proving Grounds provides some perspective to utilizing adaptive groundwater management over time and being able to reduce costs of the remedy through learning about the site and the potential for application of technology advances with the prospect of achieving a sustainable outcome.

Volume 2 of this series, Groundwater Economics: Cost-Benefit Analysis of Groundwater Policy and Projects, with Case Studies, builds on the material and concepts presented in this volume. Volume 2 focuses on the development, estimation and evaluation of costs and benefits of groundwater projects, programs and policies. The particular perspective it follows is that of examining, where possible, the life-cycle costs and benefits of groundwater actions, considering the stream of factors and results over the duration of groundwater activities. Volume 2 also includes discussion of other considerations that may affect the costs and benefits of a project or program, such as transboundary location, climate change and macroeconomic aspects. Balancing objectives of groundwater decisions in an economics framework and needed groundwater economics research are considered before presenting sixteen case studies covering choices among well types, wellhead protection, agricultural irrigation, managed aquifer recharge, water conservation to sustain ecosystems and other pertinent groundwater economics topics.

SUMMARY

Sustainable groundwater management recognizes the essential features of an adaptive approach with the steps of stakeholder collaboration, resource objectives to be regularly reviewed and revised recognizing needs of critical human and environmental receptors, assessment and system modeling, detailed alternatives with forecast consequences, decision-making, monitoring and evaluation of decision results, and incorporating learning into future actions. A previously developed adaptive approach specific to sustainable groundwater management (NGWA, 2016) focused on eight steps of: (1) collaborating with other stakeholders, (2) sharing objectives and plans to promote understanding as input to assessing alternatives for water supply and demand, (3) incorporating hydrologic processes in assessment and plans, (4) minimizing carbon footprint through efficient processes (e.g., under-ground storage), (5) protecting and remediating source water to maximize resource availability, (6) measuring resource status through monitoring, (7) avoiding disruption of extreme events by proper facility location and maintenance and (8) reassessing purpose and objectives drawing on life-cycle data, technology advances and learning to inform future groundwater decisions. Use of the adaptive management process can foster positive action and may reduce costs through learning as a project or program evolves.

FOR FURTHER CONSIDERATION

1. Contact your local or state water or environmental planning agency and ask whether they have a water plan. What were the steps used in preparing the plan? How were stakeholders involved?

2. Consider a town drawing groundwater for water supply now wanting to evaluate options for its wastewater disposal. What alternatives should the town evaluate?
3. Does your town, county or state monitor groundwater levels and quality? Identify a groundwater monitoring location and study the trends in levels and quality over the last 5 to 10 years. Do you see changes in levels or quality that might be important for decisions about use of the aquifer?

REFERENCES

Advisory Committee on Water Information (ACWI), Subcommittee on Ground Water. 2013. *A National Framework for Ground-Water Monitoring in the United States.* URL: https://acwi.gov/sogw/ngwmn_framework_report_july2013.pdf (accessed December 11, 2019).

Alley, W.M., and Leake, S.A.. 2004. The journey from safe yield to sustainability. *Groundwater (Journal)*, 42, 1, January 2004, 12–16.

Anderson, M.P.; Woessner, W.W., and Hunt, R.J. 2015. *Applied Groundwater Modeling*. Elsevier Press. London, United Kingdom. 564p.

Barnett, T. 2019. *Management Functions*. In Reference for Business. URL: https://www.referenceforbusiness.com/management/Log-Mar/Management-Functions.html (accessed December 5, 2019).

Brumley, J., Marks, C., Chau, A., Lowrance, R., Huang, J., Richardson, C., Acree, S., Ross, R., and Beak, D.. 2018. *The Influence of Green Infrastructure Practices on Groundwater Quality: The State of the Science.* U.S. Environmental Protection Agency, Washington, DC, EPA/600/R-18/227. URL: https://cfpub.epa.gov/si/si_public_record_report.cfm?Lab=NRMRL&dirEntryId=342610 (accessed December 11, 2019).

Ciarlo, M. and Jeffrey, A. 2019. Ten-Year Case Study of Successful Adaptive Management of a Groundwater to Surface Water Discharge Site Regulated under CERCLA. URL: https://www.battelle.org/docs/default-source/conference-proceedings/2019-sediments-conference-proceedings/c2.-adaptive-management-approaches/558.pdf?sfvrsn=d968eb7_2 (accessed December 12, 2019).

Edwards Aquifer Authority (EAA). 2019. *Science and Aquifer Protection*; Habitat Conservation Plan. URL: https://www.edwardsaquifer.org/ (accessed December 11, 2019).

Elsinore Valley (CA) Groundwater Sustainability Agency (EVGSA). 2017. *Elsinore Valley Groundwater Sustainability Agency Sustainable Groundwater Planning Grant Program Proposal.* URL: https://water.ca.gov/LegacyFiles/irwm/grants/sgwp/sgwp_docs/2017_Solicitation/Applications/Elsinore%20Valley%20Municipal%20Water%20District/Att2_2017SGWPC2_EligDoc_1of1.pdf (accessed December 11, 2019).

Frederick County, Maryland (FCMD). 2016. *Frederick County Water and Sewerage Plan.* URL: https://www.frederickcountymd.gov/3742/Frederick-County-Water-Sewerage-Plan (accessed December 10, 2019).

Hayes, P. 2010. Carbon, Energy and Groundwater: A summary from a hydrogeology and physics perspective. UK Groundwater Forum, Groundwater's role in reducing our carbon footprint, 26 May 2010, Birmingham, England, United Kingdom. URL: http://www.groundwateruk.org/downloads/9%20Phil%20Hayes.pdf (accessed December 11, 2019).

Job, C. 2017. *Responding to Flooded Wells; A Sustainability Issue. Water Well Journal*, November 30, 2017. URL: https://waterwelljournal.com/responding-flooded-wells-2/ (Accessed December 11, 2019).

Land Trust Alliance (LTA). 2019. *Conservation in a Changing Climate.* URL: https://climatechange.lta.org/get-started/adapt/adaptive-management/ (accessed December 9, 2019).

Maimone, M. 2004. Dcfining and Managing Sustainable Yield. In *Ground Water*, 42 (6), pp. 809–814.

Management Study HQ (MSHQ). 2019. *Functions of Management.* URL: https://www.managementstudyhq.com/functions-of-management.html (Accessed December 5, 2019)

Meyland, S.J. 2011. Examining safe yield and sustainable yield for groundwater supplies and moving to managed yield as water resource limits become a reality. In *Water Resources Management VI; Transactions on Ecology and the Environment.* WIT Press (URL: https://www.witpress.com/Secure/elibrary/papers/WRM11/WRM11073FU1.pdf (accessed December 5, 2019).

National Academy of Science (NAS). 2004. *Adaptive Management for Water Resources Project Planning.* National Academies Press. URL: https://www.nap.edu/catalog/10972/adaptive-management-for-water-resources-project-planning (accessed December 6, 2019).

National Ground Water Association. 2016. *Sustainable Groundwater Resources.* URL: https://www.ngwa.org/docs/default-source/default-document-library/advocacy/GW-Sustainability.pdf (accessed December 9, 2019).

Orange County Water District (OCWD). 2015. *Orange County Water District Groundwater Management Plan 2015 Update*. URL: https://www.ocwd.com/media/3622/groundwatermanagementplan2015update_20150624.pdf (accessed December 17, 2019).

Orange County Water District (OCWD). 2019. *Service area & board of directors*. URL: https://www.ocwd.com/about/service-area-board-of-directors/ (accessed December 17, 2019).

Taylor, R.G.; Scanlon, B., Döll, P. and Treidel, H. 2013. Ground water and climate change. *Nature Climate Change* 3, 322–329. URL: https://www.nature.com/articles/nclimate1744 (accessed December 5, 2019).

Texas Water Development Board (TWDB). 2019. 2017 State Water Plan. URL: https://www.twdb.texas.gov/waterplanning/swp/2017/index.asp (accessed December 10, 2019).

Thomas, B.F. 2018. Sustainability indices to evaluate groundwater adaptive management: a case study in California (USA) for the Sustainable Groundwater Management Act. *Hydrogeology Journal* September 2018 URL: https://www.researchgate.net/publication/327288587_Sustainability_indices_to_evaluate_groundwater_adaptive_management_a_case_study_in_California_USA_for_the_Sustainable_Groundwater_Management_Act (accessed December 5, 2019).

Treidel, H., Martin-Bordes, J.L., and Gurdak, J.J. 2011. *Climate Change Effects on Groundwater Resources: A Global Synthesis of Findings and Recommendations*. CRC Press. 414 p.

U.S Army Garrison, Aberdeen Proving Ground, Maryland (USAGAPG). 2009. Record of Decision for Army environmental Database-Restoration (AEDB-R) Site EAOF04: O-Field Study Area Operable Unit 4: New O-Field Groundwater and Source Area and Other O-Field Areas (OOA); Final July 2009.

U.S Army Garrison, Aberdeen Proving Ground, Maryland (USAGAPG). 2018. O-Field Study Area Remedial Action Completion Report for Army Environmental Database Restoration (AEDB-R) Site EAOF04, O-Field Study Area Operable Unit 4: New O-Field Groundwater and Source Area and Other O-Field Areas; Final September 2018.

U.S. Environmental Protection Agency (USEPA). 1998. Superfund Record of Decision: Aberdeen Proving Ground (Edgewood Area) Watson Creek Sediment, O-Field OU, Aberdeen Proving Ground, MD (9/23/1997). PB97-963908, EPA/541/R-97/091. URL: https://nepis.epa.gov/Exe/ZyPURL.cgi?Dockey=9100Q8LM.TXT (accessed January 21, 2020).

U.S. Environmental Protection Agency (USEPA). 2018. Superfund Task Force Recommendation #3: Broaden the Use of Adaptive Management; Memorandum of James E. Woolford, July 3, 2018. OLEM 9200.3-120. URL: https://semspub.epa.gov/work/HQ/100001630.pdf (accessed December 6, 2019).

U.S. Geological Survey (USGS). 2016. *Groundwater and Drought*. URL: https://water.usgs.gov/ogw/drought/ (accessed December 5, 2019).

U.S. Geological Survey (USGS). 2019. Groundwater and Aquifers. URL: https://www.usgs.gov/special-topic/water-science-school/science/aquifers-and-groundwater?qt-science_center_objects=0#qt-science_center_objects (accessed December 5, 2019).

Williams, B. K., and Brown, E.D.. 2012. *Adaptive Management: The U.S. Department of the Interior Applications Guide. Adaptive Management Working Group*, U.S. Department of the Interior, Washington, DC. URL: https://www.doi.gov/sites/doi.gov/files/migrated/ppa/upload/DOI-Adapative-Management-Applications-Guide.pdf (accessed December 6, 2019).

Index

central water supply, 240

For Product Safety Concerns and Information please contact our EU
representative GPSR@taylorandfrancis.com
Taylor & Francis Verlag GmbH, Kaufingerstraße 24, 80331 München, Germany

www.ingramcontent.com/pod-product-compliance
Ingram Content Group UK Ltd.
Pitfield, Milton Keynes, MK11 3LW, UK
UKHW050057200926
475702UK00019B/2884
9780367757991